Dairy Cattle Management

Dairy Cattle Management

Editor: Bethany Crawford

STATES
ACADEMIC PRESS
www.statesacademicpress.com

States Academic Press,
109 South 5th Street,
Brooklyn, NY 11249, USA

Visit us on the World Wide Web at:
www.statesacademicpress.com

ISBN: 978-1-63989-144-3 (Hardback)

Cataloging-in-Publication Data

Dairy cattle management / edited by Bethany Crawford.
p. cm.
Includes bibliographical references and index.
ISBN 978-1-63989-144-3
1. Dairy cattle--Management. 2. Dairy farming--Management. I. Crawford, Bethany.
SF208 .D35 2022
636.214 2--dc23

Table of Contents

Preface IX

Chapter 1 **Analysis of Johne's disease ELISA status and associated performance parameters in Irish dairy cows** 1
A. E. Kennedy, N. Byrne, A. B. Garcia, J. O'Mahony and R. G. Sayers

Chapter 2 **Control of *Staphylococcus aureus* in dairy herds in a region with raw milk cheese production: farmers' attitudes, knowledge, behaviour and belief in self-efficacy** 13
Marie-Eve Cousin, Maria Christina Härdi-Landerer, Verena Völk and Michèle Bodmer

Chapter 3 **Comparison of bulk milk antibody and youngstock serology screens for determining herd status for Bovine Viral Diarrhoea Virus** 26
Richard E. Booth and Joe Brownlie

Chapter 4 **Lipopolysaccharide derived from the digestive tract triggers an inflammatory response in the uterus of mid-lactating dairy cows during SARA** 42
Muhammad Shahid Bilal, Juma Ahamed Abaker, Zain ul Aabdin, Tianle Xu, Hongyu Dai, Kai Zhang, Xinxin Liu and Xiangzhen Shen

Chapter 5 **Evaluation of intravenous regional anaesthesia and four-point nerve block efficacy in the distal hind limb of dairy cows** 49
S. Yavari, N. Khraim, G. Szura, A. Starke, E. Engelke, C. Pfarrer, K. Hopster, M. Schmicke, W. Kehler, M. Heppelmann, S. B. R. Kästner and J. Rehage

Chapter 6 **Effects of glucose on lactose synthesis in mammary epithelial cells from dairy cow** 60
Ye Lin, Xiaoxu Sun, Xiaoming Hou, Bo Qu, Xuejun Gao and Qingzhang Li

Chapter 7 **Herd-level seroprevalence of *Fasciola hepatica* and *Ostertagia ostertagi* infection in dairy cattle population** 70
Sławomir J. Kowalczyk, Michał Czopowicz, Corinna N. Weber, Elisabeth Müller, Tomasz Nalbert, Andrzej Bereznowski and Jarosław Kaba

Chapter 8 **Evaluation of arterial digital blood flow using Doppler ultrasonography in healthy dairy cows** 78
H. Müller, M. Heinrich, N. Mielenz, S. Reese, A. Steiner and A. Starke

Chapter 9 **Inflammatory mechanism of Rumenitis in dairy cows with subacute ruminal acidosis** 87
Chenxu Zhao, Guowen Liu, Xiaobing Li, Yuan Guan, Yazhou Wang, Xue Yuan, Guoquan Sun, Zhe Wang and Xinwei Li

Chapter 10 **Evaluation of fecal culture and fecal RT-PCR to detect *Mycobacterium avium* ssp. *paratuberculosis* fecal shedding in dairy goats and dairy sheep using latent class Bayesian modeling**......95
Cathy A. Bauman, Andria Jones-Bitton, Jocelyn Jansen, David Kelton and Paula Menzies

Chapter 11 **The association between serum ß-hydroxybutyrate and milk fatty acid profile with special emphasis on conjugated linoleic acid in postpartum Holstein cows**......104
Pedro Melendez, Pablo Pinedo, José Bastias, Maria Paz Marin, Carolina Rios, Consuelo Bustamante, Natalia Adaro and Mario Duchens

Chapter 12 **Characterization of *Staphylococcus aureus* isolated from milk samples of dairy cows in small holder farms**......109
S. A. Mekonnen, T. J. G. M. Lam, J. Hoekstra, V. P. M. G. Rutten, T. S. Tessema, E. M. Broens, A. E. Riesebos, M. P. Spaninks and G. Koop

Chapter 13 **Estimation of tulathromycin depletion in plasma and milk after subcutaneous injection in lactating goats using a nonlinear mixed-effects pharmacokinetic modeling approach**......117
Zhoumeng Lin, Matthew Cuneo, Joan D. Rowe, Mengjie Li, Lisa A. Tell, Shayna Allison, Jan Carlson, Jim E. Riviere and Ronette Gehring

Chapter 14 **Prevalence and risk factors of bovine tuberculosis in dairy cattle**......127
Michael K. Ghebremariam, V. P. M. G Rutten, J. C. M. Vernooij, K. Uqbazghi, T. Tesfaalem, T. Butsuamlak, A. M. Idris, M. Nielen and A. L. Michel

Chapter 15 **Evaluation of effects of *Mycoplasma* mastitis on milk composition in dairy cattle**......134
Abd Al-Bar Al-Farha, Farhid Hemmatzadeh, Manouchehr Khazandi, Andrew Hoare and Kiro Petrovski

Chapter 16 **Assessment of foot health and animal welfare: clinical findings in 229 dairy Mediterranean Buffaloes (Bubalus bubalis) affected by foot disorders**......142
Jacopo Guccione, Christian Carcasole, Maher Alsaaod, Luigi D'Andrea, Antonio Di Loria, Angela De Rosa, Paolo Ciaramella and Adrian Steiner

Chapter 17 **Serum hepatokines in dairy cows: periparturient variation and changes in energy-related metabolic disorders**......150
Jianguo Wang, Xiaoyan Zhu, Guanghui She, Yezi Kong, Yazhou Guo, Zhe Wang, Guowen Liu and Baoyu Zhao

Chapter 18 **Field application of an indirect gE ELISA on pooled milk samples for the control of IBR in free and marker vaccinated dairy herds**......159
Barbara Colitti, Elvira Muratore, Maria Elena Careddu, Luigi Bertolotti, Bryan Iotti, Mario Giacobini, Margherita Profiti, Chiara Nogarol, Jens Böttcher, Andreino Ponzo, Roberto Facelli and Sergio Rosati

Chapter 19 **Quantification of *Clostridioides* (*Clostridium*) *difficile* in feces of calves of different age and determination of predominant *Clostridioides difficile* ribotype 033 relatedness and transmission between family dairy farms using multilocus variable-number tandem-repeat analysis**......166
Petra Bandelj, Céline Harmanus, Rok Blagus, Marko Cotman, Ed J. Kuijper, Matjaz Ocepek and Modest Vengust

Chapter 20 **Risk factors for the presence of *Fasciola hepatica* antibodies in bulk-milk samples and their association with milk production decreases, in Cuban dairy cattle**......176
Amilcar Arenal, Yipsi García, Lídice Quesada, Dayamis Velázquez, Diamela Sánchez, Mayelin Peña, Asnaldo Suárez, Arnielis Díaz, Yuliet Sánchez, Stijn Casaert, Jan van Dijk, Jozef Vercruysse and Johannes Charlier

Chapter 21 **Chi-miR-3031 regulates beta-casein via the PI3K/AKT-mTOR signaling pathway in goat mammary epithelial cells (GMECs)**......184
Kaiwen Chen, Jinxing Hou, Yuxuan Song, Xiaochuan Zhang, Yuhan Liu, Gonghai Zhang, Kai Wen, Haidong Ma, Guang Li, Binyun Cao and Xiaopeng An

Chapter 22 **Sodium butyrate improves antioxidant stability in sub-acute ruminal acidosis in dairy goats**......194
Nana Ma, Juma Ahamed Abaker, Muhammad Shahid Bilal, Hongyu Dai and Xiangzhen Shen

Chapter 23 **Why do farmers and veterinarians not report all bovine abortions, as requested by the clinical brucellosis surveillance system?**......207
Anne Bronner, Viviane Hénaux, Nicolas Fortané, Pascal Hendrikx and Didier Calavas

Chapter 24 **Non-invasive indicators associated with the milk yield response after anthelmintic treatment at calving in dairy cows**......219
Sien H Verschave, Jozef Vercruysse, Andrew Forbes, Geert Opsomer, Miel Hostens, Luc Duchateau and Johannes Charlier

Permissions

List of Contributors

Index

Preface

Over the recent decade, advancements and applications have progressed exponentially. This has led to the increased interest in this field and projects are being conducted to enhance knowledge. The main objective of this book is to present some of the critical challenges and provide insights into possible solutions. This book will answer the varied questions that arise in the field and also provide an increased scope for furthering studies.

Cattle cows which are bred commercially for the production of milk are known as dairy cattle. Management of such cattle includes hygienic conditions for breeding, dietary regulations, disease management, etc. It can be divided into intensive and extensive management systems. Intensive systems aim to maximize the production per cow in a particular herd. This involves providing the cows with adequate nutrition, housing of the cows, etc. Nutrition plays a crucial role in maintaining the health and strength of the cattle. It also directly impacts the milk production and reproduction performance. In extensive dairy cattle management systems, the cattle is left open on the pasture. They are milked multiple times in a day. Some of the other factors considered in cattle management are infertility and diseases such as mastitis, lameness, among others. From theories to research to practical applications, case studies related to all contemporary topics of relevance to this field have been included in this book. It brings forth some of the most innovative concepts and elucidates the unexplored aspects of dairy cattle management. This book is a complete source of knowledge on the present status of this important field.

I hope that this book, with its visionary approach, will be a valuable addition and will promote interest among readers. Each of the authors has provided their extraordinary competence in their specific fields by providing different perspectives as they come from diverse nations and regions. I thank them for their contributions.

Editor

1

Analysis of Johne's disease ELISA status and associated performance parameters in Irish dairy cows

A. E. Kennedy[1,2], N. Byrne[1], A. B. Garcia[1], J. O'Mahony[2] and R. G. Sayers[1*]

Abstract

Background: Infection with *Mycobacterium avium* subspecies *paratuberculosis* (MAP) has been associated with reductions in milk production in dairy cows and sub optimal fertility. The aim of this study was to highlight the production losses associated with testing MAP ELISA positive in Irish dairy cows. Secondary objectives included investigation of risk factors associated with testing MAP ELISA positive. A survey of management practices on study farms was also conducted, with examination of associations between management practices and herd MAP status. Blood samples were collected from 4188 breeding animals on 22 farms. Samples were ELISA tested using the ID Screen Paratuberculosis Indirect Screening Test. Production parameters examined included milk yield, milk fat, milk protein, somatic cell count, and calving interval. The association between MAP ELISA status and production data was investigated using multi-level mixed models. Logistic regression was used to identify risk factors for testing JD blood ELISA positive at individual cow level and to identify associations between farm management practices and herd MAP status.

Results: Data were available for 3528 cows. The apparent prevalence recorded was 7.4 %. Mixed model analysis revealed no statistically significant association between testing MAP ELISA positive and dairy cow production parameters. Risk factors associated with testing positive included larger sized herds being over twice more likely to test positive than smaller herds (OR 2.4 $P = <0.001$). Friesians were less likely to test positive relative to other breeds. A number of study farmers were engaged in management practices that have previously been identified as high risk for MAP transmission e.g., 73.1 % pooled colostrum and 84.6 % of study farmers used the calving area to house sick animals throughout the year. No significant associations however, were identified between farm management practices and herd MAP status.

Conclusion: No production losses were identified; however an apparent prevalence of 7.4 % was recorded. With the abolition of EU milk quotas herd size in Ireland is expanding, as herds included in this study were larger than the national average, results may be indicative of future JD levels if no JD control programmes are implemented to minimise transmission.

Keywords: Johne's disease, ELISA, Dairy cow, Production

Background

Clinical and sub-clinical manifestations of disease can result in reductions in animal productivity leading to reduced farm profits [1, 2]. Cost-benefit analyses are often conducted to highlight these economic losses in order to promote the use of disease control schemes [3]. Johne's disease (JD) is a chronic granulomatous enteritis of ruminants and is caused by the bacterium *Mycobacterium avium* subspecies *paratuberculosis* (MAP) [4]. Infection with MAP has been associated with production losses at farm level although equivocal results are reported [5]. Due to a prolonged subclinical phase, variable disease progression and immune response [6], diagnosis of MAP infection is challenging. Enzyme linked immunosorbant assay (ELISA), is a popular method of testing for MAP due to its speed and low cost [7]. Despite variable sensitivity and specificity [8, 9], ELISA testing is often the method of choice for epidemiological studies and herd-based diagnosis [10], and forms the basis of a number of international

* Correspondence: riona.sayers@teagasc.ie
[1]Animal & Bioscience Research Department, Animal & Grassland Research and Innovation Centre, Teagasc, Moorepark, Fermoy, Co. Cork, Ireland
Full list of author information is available at the end of the article

control programmes [11]. It is also a common diagnostic tool used in economic studies of Johne's [5].

Economic losses reported due to infection with MAP include decreased slaughter value [12], reductions in milk production in dairy cows [2, 13], sub-optimal fertility [14], and an increase in cow replacement costs [15]. Although losses in clinically affected animals are well defined [15, 16], losses due to subclinical infection appear less well characterised [5]. An association between subclinical MAP infection and decreased milk yield (MY) has been identified in a number of studies [17–19]. In contrast a number of additional studies have identified no such association [20–22]. Similarly, conflicting reports exist regarding an association between subclinical MAP infection and milk fat (MF) or milk protein (MP) content [17, 21, 23]. An increased interval from calving to conception in ELISA positive cows has also been reported [14] in contrast to a different study showing ELISA positive cows to have fewer non-pregnant days [23]. Similarly, JE Lombard, FB Garry, BJ McCluskey and BA Wagner [23] recorded no association between MAP ELISA positivity and somatic cell count (SCC) which again conflicts with other studies [20, 24] that reported an associated increase in SCC in MAP ELISA positive dairy cows.

The variability that exists across diagnostic test methods may, in part, explain the conflicting performance-related data reported across various studies [5, 25]. Geographical location, choice of sample matrix, size of study population, cow breeds, and positive cow classification also differs across studies. Serum samples were used for diagnostic purposes in some studies [22, 23], while others used individual milk samples [19]. MG Gonda, YM Chang, GE Shook, MT Collins and BW Kirkpatrick [18] defined a JD positive cow on the basis of serum ELISA and/or faecal culture results. Study sample sizes ranged from less than 1000 [22] to 35,591 dairy cows [19] with other studies only examining a single cow breed [18, 22]. In this regard, it is important that data continue to be generated on similar cow populations, using similar study designs to improve the degree of confidence that exists in the likely impact on production in MAP positive dairy cows.

The prevalence of JD is believed to be increasing in the Republic of Ireland over the last 10 years [26]. In order to prevent further increases in MAP infection, improvements in control are required on Irish dairy farms. Although a previous Irish study [22] identified no significant effect of MAP sero-status on herd performance in 2004–2005, the increasing prevalence of MAP may now be impacting on Irish dairy production. The dairy landscape in Europe is changing due to the abolition of EU milk quotas in 2015 [27] and Irish farmers have been expanding herds over the past number of years [28]. Given Ireland's increasing herd size, a known risk factor for testing MAP positive [29], this study aimed to investigate the current impact of MAP ELISA sero-positivity on individual cow milk production, SCC and calving interval. Secondary objectives included investigation of risk factors (e.g., breed, parity, calf and calving management) associated with testing MAP ELISA positive in Irish dairy herds and investigating the strength of correlation between milk and serum ELISA results.

Methods

Study population and sampling

Blood samples were collected from all breeding animals aged over 2 years on 22 Irish dairy farms in 2012. This was conducted under licence from the Department of Health and Children, the licencing authority in Ireland at the time of the study. The location of study herds is included in Fig. 1. All but two herds were located in the dairy dense province of Munster, Ireland (south-western region) with an additional herd in each of Ulster and Leinster. A milk sample was also collected from each cow blood sampled on 17 of these farms. All study animals were observed by a veterinary surgeon during sampling visits and none displayed overt clinical signs of JD.

Sample testing

Serum and milk samples were tested by a commercial ISO accredited laboratory (Enfer Labs, Kildare, Ireland) using the ID Screen Paratuberculosis Indirect Screening Test (ID Vet, Montpellier, France). This ELISA has a reported sensitivity (Se) of 41.5 % and specificity (Sp) of 99.42 [30]. The test is an *M. phlei* absorbed ELISA detecting anti-MAP immunoglobulin G (IgG). This ELISA was chosen as it is approved for use in Ireland's national voluntary JD pilot control programme[1] and displayed the highest overall accuracy of four commercial ELISA kits investigated by ROC analysis [30]. Results were reported as sample to positive ratio (S/P ratio) calculated using the formula $S/P\ ratio = ((OD_{Sample} - OD_{Positive\ control}) \div (OD_{Positive\ control} - OD_{Negative\ control}) \times 100)$. Animals were assigned MAP status (positive or negative) according to kit manufacturer interpretation ('kit-interpretation'), with serum results of $S/P \geq 70$ % classified as positive. A more severe test interpretation was also applied to blood ELISA results only in order to achieve increased test sensitivity, similar to the revised ELISA cut-off used by MT Collins, SJ Wells, KR Petrini, JE Collins, RD Schultz and RH Whitlock [31]. In this current study, instead of arbitrarily choosing a more sensitive positive cut-off, the mean S/P ratio of cows classified as negative using kit interpretation plus three times the standard deviation was used. This yielded a new 'severe interpretation' positive cut off of $S/P \geq 51.59$.

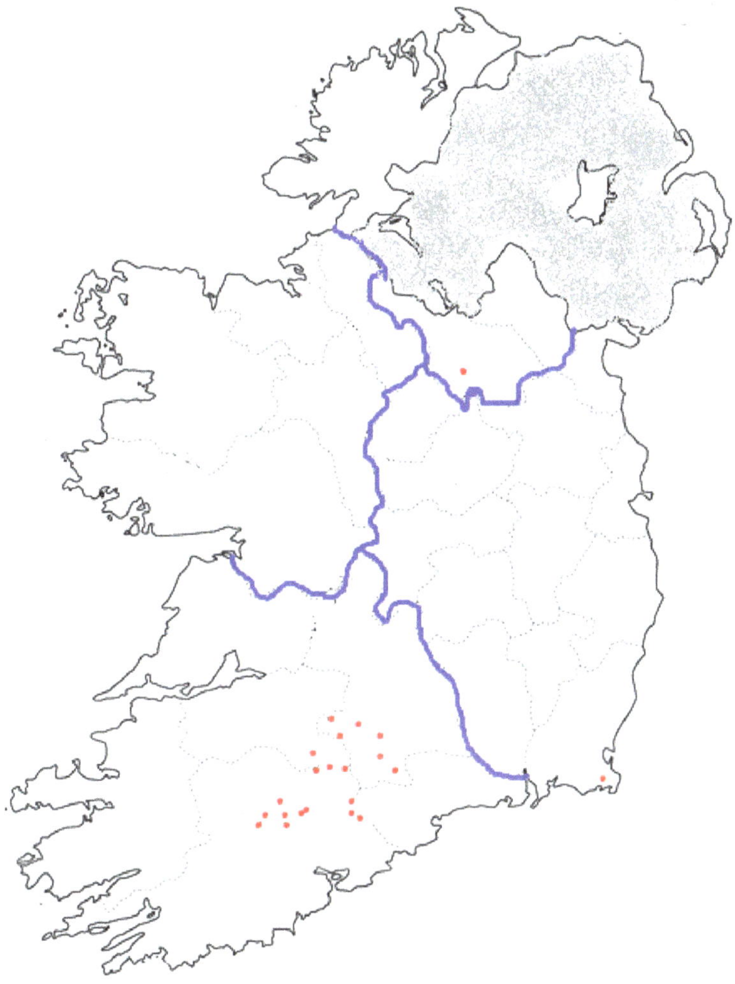

Fig. 1 Map showing location of study farms. The majority of study farms were located in the dairy dense province of Munster, with one farm located in Leinster and Ulster

Individual cow performance data

Production data for each individual cow were downloaded from the Irish Cattle Breeding Federation (ICBF) database. This database holds production data on milk recording herds to support breeding decisions and enhance genetic gain in Irish dairy herds. Production parameters downloaded included 305 day [32] MY, MF, MP, SCC, and calving interval (CI). Only calving intervals of greater than 250 and less than 500 days were retained for analysis to allow comparison with K Hoogendam, E Richardson and JF Mee [22]. Additional cow related data downloaded from the ICBF database included parity, breed, and economic breeding index (EBI; a profit index measured in Euro, aimed at identifying the most profitable cows for breeding dairy herd replacements [33]).

Survey data

Data relating to farm management practices on study farms were collected during scheduled farm visits. Questions focused on management practices which have previously been identified in the literature as being associated with a high risk of MAP transmission.

Dataset construction

The production performance for each cow was matched to her ELISA result to create the dataset of statistical analysis. Both milk and blood ELISA results were included where available and the 'kit-interpretation' used to classify cows as positive and negative. A further dataset was constructed in an effort to identify production losses experienced by cows recording ELISA results near the manufacturers' cut off point. This dataset was constructed using the 'severe-interpretation' ELISA positive cut-off. Both blood data sets were used for the mixed model analysis with the only 'kit-interpretation' dataset used for prevalence investigations and investigating the correlation between blood and milk ELISA results. To investigate the correlation between paired milk and serum ELISA results, Spearman correlation (*rs*) was performed on categorical results i.e., cows classified as positive or negative.

Data analysis

Data manipulation and graphical representations including box plots were completed in Excel (MS 2010). Generalised linear latent and mixed models (gllamm), logistic regression, and Spearman correlations were performed using Stata (Version 12).

Prevalence calculation

The apparent prevalence's (Ap) at animal level and within herd level were calculated as total number of test positive animals out of the total number of animals tested. To estimate the true prevalence (Tp) of positive animals, a Bayesian approach and Gibbs sampling method was applied using an online epidemiological calculator (Epitools) [34]. This calculator requires prior estimates of the true prevalence and test sensitivity and specificity, based on previous data or expert knowledge. These estimates are made as beta probability distributions, with parameters alpha and beta. Alpha and beta can be calculated provided estimates of the mode and 5 or 95 % confidence limits are available from expert opinion. When the mode was between 0.5 and 1, the 95th percentile was entered into the Beta distribution utilities calculator, and the 5th percentile entered if between 0.5 and 1. Prior estimates for true prevalence were based on a national survey which reported a prevalence of 3.3 % from a study population of 15,558 animals aged over 2 years [26]. Beta probability distributions for ELISA Se and Sp were compiled based on estimates from peer-reviewed literature [30, 31], expert opinion from Irish veterinary practitioners, veterinary officers, Teagasc veterinary and agricultural researchers, and a comprehensive longitudinal farm study involving ELISA screening and confirmatory testing using faecal culture and post-mortem examination. The estimates and beta distributions used to determine Tp are outlined in Table 1. Upper and lower confidence limits for Tp calculation were set at 97.5 and 2.5 % respectively. Outputs reported in this study are median values of posterior distributions for prevalence, sensitivity and specificity.

Associations between cow performance and MAP ELISA status

All data were visually assessed for normality using ladder of powers histograms in Stata. The association between MAP ELISA status and production data was investigated using gllamm. Models accounted for random effects of cow nested within herd. Covariates were retained in the final multivariable models on the basis of the highest reduction in the Bayesian information criterion (BIC).

Table 1 Estimates for use in true prevalence calculation

	Estimates of posterior distributions	Beta distribution	
		Alpha	Beta
Animal Prevalence	0.033 (0.072)[a]	5.2021	124.1336
Sensitivity	0.41 (0.587)[a]	9.9689	13.5179
Specificity	0.99 (0.47)[a]	4.0322	1.0306

The online epidemiological calculator (Epitools) used to calculate true prevalence requires prior estimates of the true prevalence and test sensitivity and specificity, based on previous data or expert knowledge. These estimates are made as Beta probability distributions, with parameters alpha and beta. Alpha and beta can be calculated provided estimates of the mode and 5 or 95 % confidence limits are available from expert opinion

Initial values represent the mode, with the value in brackets representing either the 5th or 95th percentile

[a]When the estimated value was between 0 and 0.5 the 95th percentile was chosen, and when the estimate was between 0.5 and 1 the 5th percentile was chosen

Covariates examined in each model were breed (Friesian/Friesian crosses (FRx), Jersey/Jersey crosses (JEx),. Norwegian Red/Norwegian Red crosses (Redx), other), parity (1, 2, 3, 4, 5, 6, 7, 8, 9, ≥10), EBI (categorised into quartiles, category 1 being the highest quartile and category 4 the lowest) and herd size (≤150 cows, >150 cows). This number was chosen as average herd size of study herds was 153 cows. Second-level interactions between covariates were also examined. Values of $P < 0.05$ were considered significant.

Associations between breed, herd size, parity, EBI and MAP ELISA status

Logistic regression was used to identify risk factors for testing MAP blood ELISA positive at individual cow level (dependent variable). Independent variables examined in regression included breed, herd size, parity and EBI. These variables were coded as described for gllamm models. Herd of origin was forced into all models. A manual backwards elimination with a forward step was performed for each model. Interactions between variables were also examined. Variables recording a significance level of $P < 0.05$ were retained in the model and are reported.

Associations between farm management practices and herd MAP status

Logistic regression was used to identify associations between testing MAP blood ELISA positive at a herd level and farm management practices (survey responses). A herd was classified as positive if a minimum of one blood ELISA positive result was identified. Herd size was included in the model as a covariate, as larger herds are more likely to test positive [29]. Again a manual backwards elimination was performed with interactions between variables examined. Variables recording $P < 0.05$ were considered significant.

Results

Descriptive data

Samples were collected from 4188 breeding animals. Production data were available for 3528 dairy cows, the remainder ($n = 660$) being excluded from further analysis. These exclusions consisted of misidentified animals, breeding bulls, and beef cows in herds having mixed beef and dairy enterprises. Of the 22 herds, nine contained >150 cows.

The predominant breed sampled was FRx (82.7 %), the remaining 10.3, 5.5 and 1.5 % being Redx, Jersey, and other, respectively (Fig. 2). The majority of animals tested were parity1 (30.1 %) or parity 2 (26.4 %) (Fig. 3).

Prevalence

The highest within herd Ap recorded on any single farm was 56 %. Box plots showing range of S/P ratios across all herds with at least one positive animal are shown in Fig. 4. The overall study Ap recorded utilising blood MAP ELISA was 7.4 %. Based on the Bayesian analysis, Tp was estimated at 3.8 %, with a median output test Se and Sp of 41.6 and 94 %, respectively. All cows tested in three herds recorded negative blood MAP ELISA results and these herds contained <150 cows. A breakdown of the positives across parity, breed, and herd size are outlined in Table 2. The highest proportion of animals testing ELISA positive were third parity and Redx being proportionally the predominant breed testing positive (Figs. 5 and 6, respectively), although the majority of Redx testing positive were from a single herd. Many of the animals testing positive in this herd were born in the same year.

Milk ELISA results

A total of 131 from 1696 cows available for analysis, tested milk ELISA positive for MAP. Of these, 61 also tested positive on blood. The remaining 70 animals recording positive milk ELISA results tested negative on blood ELISA. It should be noted that samples from 47 'milk positive blood negative' cows were collected in September/October/November. These samples would therefore have been collected during late lactation in the Irish Spring calving dairy system. Further to this a number of the blood results, although classified as negative, were approaching the manufacturer cut off point of 70 S/P, possibly explaining the discrepancy between milk and blood ELISA. Spearman correlation yielded a ρ value of 0.19 indicating poor correlation between milk and blood test results at a categorical level (Fig. 7).

Management practices- survey

The majority (73.1 %) of respondents purchased animals onto their farms. Approximately, three quarters of study farmers fed calves pooled colostrum (73.1 %) and pooled milk (76.9 %). Additionally, milk not fit for sale i.e., milk containing antibiotic residues or milk from sick/mastitic cows was used to feed calves on 65.2 % of study farms. Group calving pens were used by 54 % of study farmers. The majority (84.6 %) of study farmers also used the calving area to house sick animals throughout the year. A high proportion of farmers removed calves from the

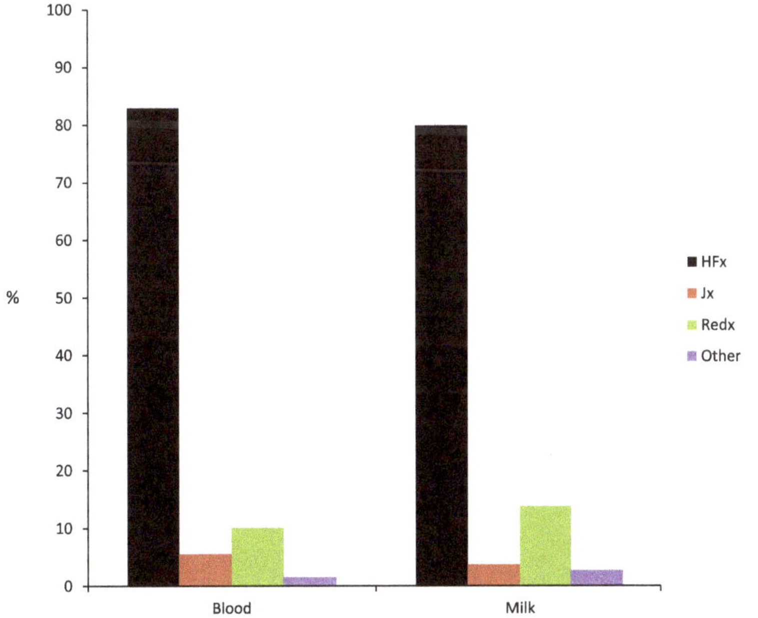

Fig. 2 Proportion of animals tested belonging to each breed. The predominant breed tested using both milk and blood ELISA was Friesian. HFx: Friesian, JX: Jersey, Red: Norwegian red

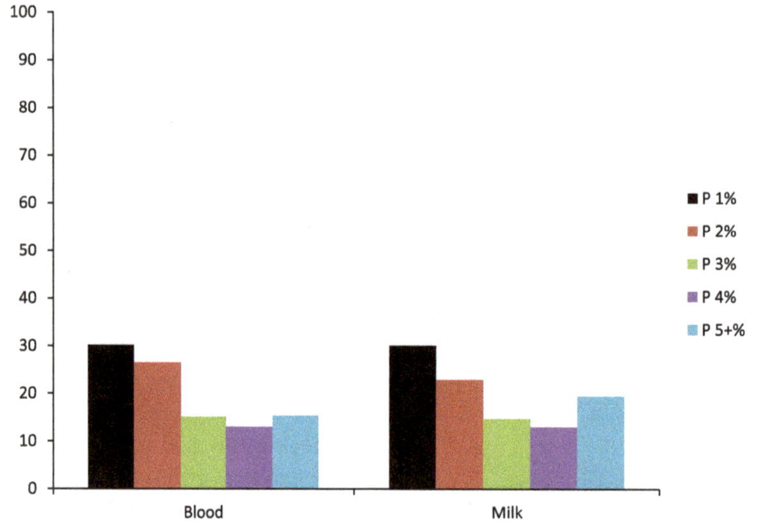

Fig. 3 Proportion of animals tested belonging to each parity. The majority of animals tested were of parity 1 or parity 2. P = parity

calving area within 30 min of birth (65.4 %). Similarly, the majority of study farmers didn't allow calves to suckle the dam (65.4 %) (Fig. 8).

Production data

Mean milk yield was 5494 kgs. Mean milk fat and protein was 240 kgs and 196 kgs respectively. As a percentage of yields this equated to 4.4 % milk fat and 3.6 % protein.

Mixed model analysis of production parameters

Multilevel mixed model analysis revealed no statistically significant association between testing MAP ELISA positive and MY, milk solids, CI, and SCC. Similarly, analysis of the 'severe interpretation' dataset again revealed no statistically significant differences (Table 3). No statistically significant association between testing milk ELISA positive and MY, milk solids, CI, and SCC was identified.

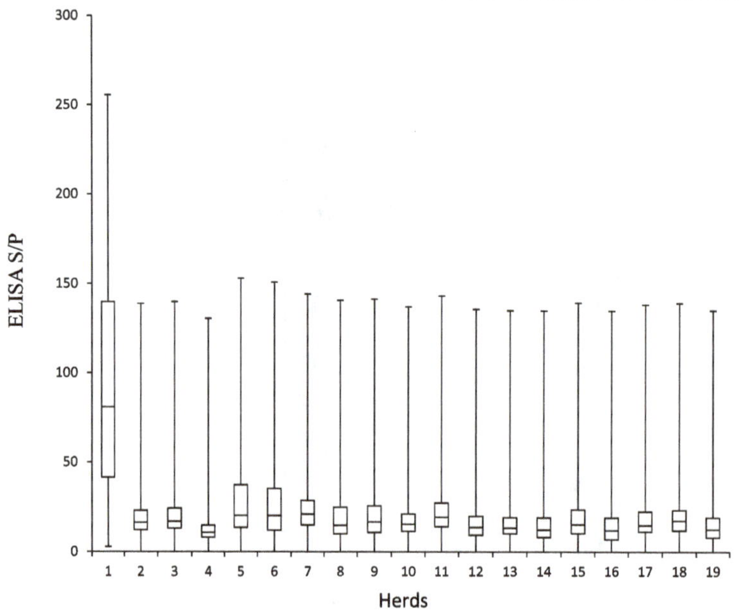

Fig. 4 Box plot showing range of S/P ratios across all herds that recorded at least one positive animal. Over half of the animals in herd 1 tested positive

Table 2 Breakdown of positives across parity and breed

Matrix	Total tested	P 1 %	P 2 %	P 3 %	P 4 %	P 5 %	HFX %	Jx %	Redx %	Other %
Blood	3528	30.1	26.4	14.9	13	15.6	82.7	5.5	10.3	1.5
Milk	1686	30	22.8	14.7	13	19.5	80.5	3.7	14.1	1.7
	Total Positive(n)	P 1 %	P 2 %	P 3 %	P 4 %	P 5 %	HFX%	Jx%	Redx%	Other%
Blood	261	29.5	28.7	14.2	10.0	17.6	53.6	2.4	39.8	4.2
Milk	131	27.5	16.8	13.7	13	29	56.4	3.9	32.8	6.9

P parity

Associations between breed, herd size, parity, EBI and MAP ELISA status

Statistically significant results of logistic regression analysis are included in Table 4. Larger herds were over twice more likely to test positive than smaller herds. Redx and breeds classed as other were more likely to test JD ELISA positive than FRx (OR 6.5, 5.5 respectively). No significant associations were highlighted between parity, EBI and MAP ELISA status.

Associations between farm management practices and herd MAP status

No significant association was identified between calving area, calf feeding management practices and MAP ELISA status.

Discussion

Economic losses are often reported due to JD [35]. Given Ireland's increasing herd size, a known risk factor for testing MAP positive [29], this study aimed to investigate the current impact of MAP ELISA sero-positivity on individual cow performance. Secondary objectives included investigation of risk factors (e.g., breed, parity, calf and calving management) associated with testing MAP ELISA positive in Irish dairy herds and investigating the strength of correlation between milk and serum ELISA results.

Two previous Irish reports have highlighted significant losses in JD clinically affected animals [13, 16], but are limited to individual farm case studies as opposed to across farm studies. An additional Irish study involving 34 herds, however, reported no statistical effect of JD

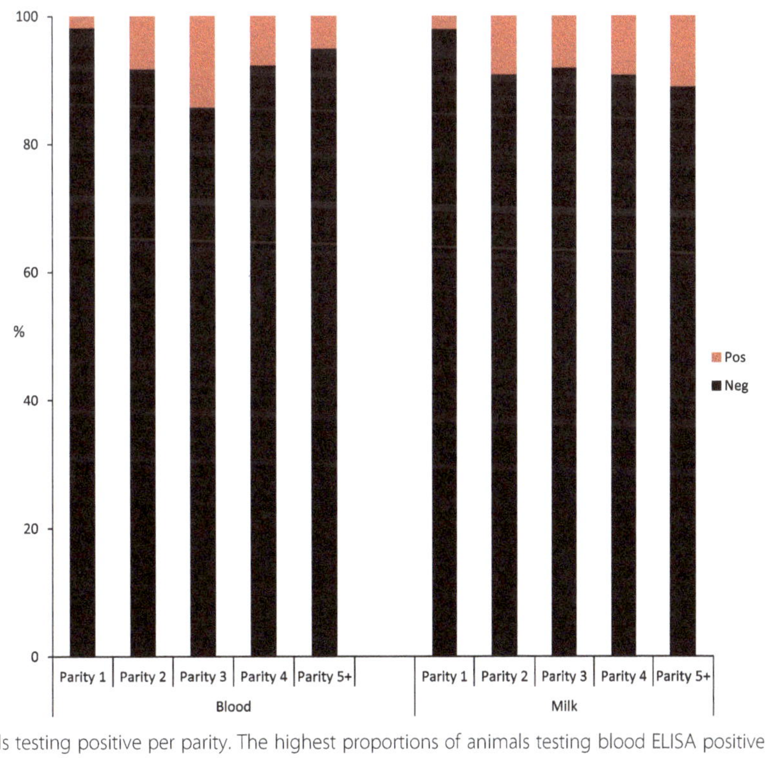

Fig. 5 Proportion of animals testing positive per parity. The highest proportions of animals testing blood ELISA positive were of parity 3

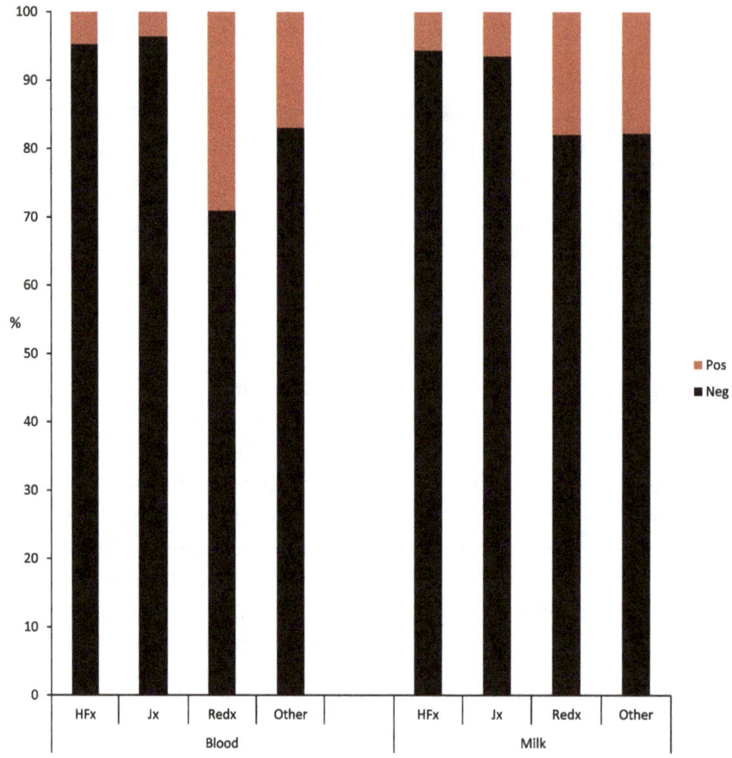

Fig. 6 Proportion of positive results recorded across each breed. Norwegian reds were proportionally the predominant breed testing positive. The majority of this breed testing positive however originated from the same herd. HFx: Friesian, JX: Jersey, Red: Norwegian red

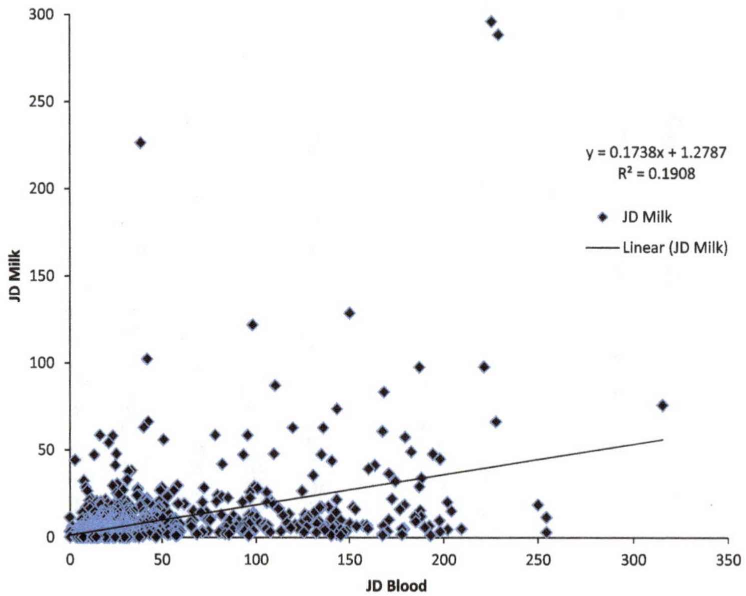

Fig. 7 Scatter plot showing the relationship between matched blood and milk samples. An R^2 value of 0.1908 was obtained

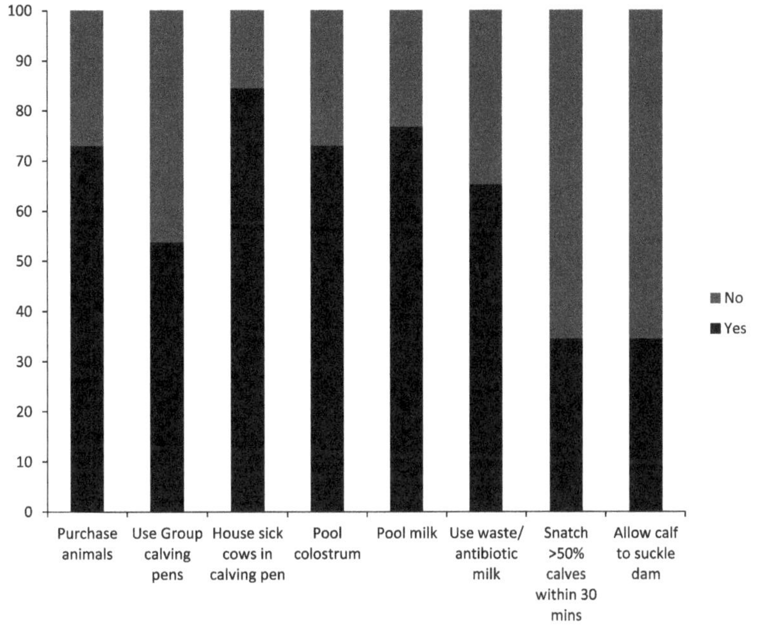

Fig. 8 Responses to survey questions. The questions focus on management practices that have previously been associated with JD transmission. The high risk practices for JD transmission are shown in red

sero-status (sub-clinical cows) on Irish dairy cow milk and fertility performance parameters [22]. As the study conducted by K Hoogendam, E Richardson and JF Mee [22]. involved a sample size of 949 and only 11 serologically positive individuals, it was necessary to conduct a larger study in Irish dairy herds. Both the overall sample size and number of ELISA positive animals detected, were considerably higher in the current study (3528 cows and 261 positives) which would greatly increase the confidence in the findings reported. It was unexpected, therefore, that a continuing lack of statistically significant association between JD sero-positivity and performance in Irish dairy cows was highlighted. This is not particular to Ireland; with additional international studies reporting similar findings [5, 21]. It may be suggested, therefore, that use of poor sensitivity, and possibly specificity, ELISA tests as are currently available is not an ideal study design in order to detect sub-clinical losses due to JD infection. Studies, however, have detected losses in sero-positive individuals [19, 23] and as ELISA testing forms an integral part of many international control programmes [11, 36], studies aimed at highlighting production losses associated with JD ELISA status are important additions to the global JD database. Further such studies may allow an analysis of geographical differences in the impact of sub-clinical JD across various countries, which in turn could assist in identifying protective or stimulatory factors with regard to JD infection.

Table 3 Results from multilevel mixed model analysis

Dataset Name	Coefficient: JD positive vs. JD negative	Stnd Error	*P* Value
Dependent Variable			
Blood			
Milk Kgs	−8.7	47.3	0.854
Protein %	0.14	0.01	0.424
Fat %	−0.03	0.03	0.343
Calving Interval	−3.2	2.0	0.098
SCC	10.3	21.7	0.635
Severe Interpretation			
Milk Kgs	−3.91	52.5	0.9406
Protein %	−0.01	0.01	0.6745
Fat %	−0.02	0.03	0.5257
Calving Interval	−4.52	5.04	0.3695
SCC	10.4	22.4	0.6432

No significant associations between production parameters and sero status were identified utilising the manufacture cut of point of 70 or the severe interpretation cut off point of 51.59
P Value: Significant $P < 0.05$

The impact on production associated with testing MAP ELISA positive will differ depending on choice of diagnostic test. This is because the accuracy of MAP testing differs across both kits and stage of infection in the animal tested [5]. Indeed the poor correlation between the use of milk and blood samples identified in the current study serves to clearly highlight the considerable variability between diagnostic test methods, especially

Table 4 Logistic regression- Significant associations between testing MAP ELISA positive and independent variables

Dependent Variable Independent Variable	Odds Ratio	*P* Value	Conf. Interval (95 %)	Model[a]
Johne's disease ELISA positive				
Herd size				Herd of originBreedHerd sizeParityEBI
Herd size >150 cows vs. herd size < 150 cows	2.4	<0.001	1.7, 3.4	
Breeds				
Red x vs. FRx	6.5	<0.001	4.8, 8.9	
Other vs. FRx	5.5	<0.001	2.5, 12.2	
Red x vs. Jex	12.2	<0.001	5.2, 28.6	
Other vs. Jex	10.3	<0.001	3.3, 32.1	

Larger sized herds were more likely to test positive compared to smaller sized herds. Friesians were less likely to test positive relative to other breeds examined
P Value: Significant *P* <0.05. Only significant results shown [a]Outlines the independent variables included in the logistic regression model

at different stages in lactation as highlighted in this study and others [37, 38]. Indeed as the specificity of MAP ELISA is not 100 % [9], it is possible that a proportion of the animals identified as positive in the current study may be false positives. The Bayesian estimates of Tp would support this in that it indicates a lower level of true infection than the apparent prevalence recorded. Identification of false positives and false negatives due to less than optimal test sensitivity and specificity would lead to miscategorisation of individuals hence reducing the likelihood of detecting association between performance and ELISA status. This again stresses the importance of using large sample size studies for these analyses. The current study strove to correct for poor test sensitivity by using a 'severe-interpretation' yet production differences remained unidentifiable.

A number of environmental mycobacteria have been identified in Ireland [39]. Environmental mycobacteria are known to contribute to false positive ELISA results [40], potentially allowing interference with results in the current study. As faecal culture positive cows have been shown to have consistently larger effects on all production traits compared to MAP ELISA positive cows [18], a limit of this study is the lack of faecal culture. A study conducted in the same region as the current study however, showed moderate agreement between ELISA testing and culture techniques [41], indicating there is a level of agreement between ELISA and culture tests in the region. As, however, results from this study indicate no economic effect of testing ELISA positive, it may indicate that future economic investigations in Ireland should not be conducted utilising ELISA alone. It may also prove worthwhile in future studies to increase the ELISA test kit cut-off as a means of potentially improving test specificity.

It was important to identify whether farmers engaged in good Johne's management in this study to examine the influence of the farmer as a protective influence. Not only were farmers in the current study not engaging in protective practices, a higher level of adoption of certain high risk management practices than previously reported in Ireland was recorded. The current study reported 65 % of study farmers feeding waste/antibiotic milk to calves compared to 59.6 % in the previous national study [42]. Use of the calving area to house sick animals was also reported at a much higher level than previous national [42] and international studies [43]. Results from this study also show higher usage of group calving pens than that reported elsewhere [41, 43] possibly placing study farms at increased risk of the incidence of diarrhoea, including salmonella and JD [44, 45]. Indeed maintenance of a closed herd is a key element of general herd biosecurity [46] however over 70 % of study farmer's reported not operating a closed herd. Given the high level of high risk management practices reported in the current study and a previous study [42] it is important to provide information to farmers about the pathogenesis and transmission of JD to minimise future JD levels. As almost all participants were utilising high risk management practices, it is perhaps unsurprising that no significant differences were identified between management practices adopted on test positive and test negative herds.

Although a high number of MAP ELISA positive animals were identified, no animal was observed to be displaying clinical signs of JD at the time of sampling. As typical results in relation to JD risk factors were identified i.e., larger herds being more likely to test positive, perhaps it indicates an unidentified element exists within Irish dairying systems that limits the production effects and clinical signs on Irish cattle. Potential protective effects may include the widespread use of grass based systems in Ireland or the lower average herd size compared to other countries [47, 48]. It may also however relate to the level of environmental mycobacteria present and the extensive TB testing programme that operates in Ireland. Infection with MAP can lead to false positive reactions on the intradermal skin test for bTB [49]. As a bTB test is administered on at least a single occasion annually to every bovine in Ireland, removal of 'MAP reactors' prior to the development of clinical signs is possible, potentially explaining negligible losses.

Further to this, although the annual culling rate of approximately 21 % is lower than some countries [50], this level of culling may have led to the removal of some MAP infected animals prior to the onset of clinical signs. Currently the dairying sector in Ireland and the EU in general is entering a period of change due to the abolition of EU milk quota restrictions. The Irish dairy sector has set targets to increase dairy output by 50 % by 2020 (Food Harvest 2020). In order to meet this 2020 target, there is a requirement to increase cow numbers by 350,000 in Ireland by 2020, compared to average cow numbers between 2007 and 2009 [51]. With the abolition of EU milk quotas and the intent of many farms to expand in size [27], it may be that animals that would previously have been culled may in future be retained in the herd, potentially leading to an increased risk of MAP transmission. As this study involves herds larger than the national average herd size and results show an increased number of animals testing positive than previous Irish studies, it will be important to monitor the production effects associated with testing JD positive, as national herd size expands.

The greater likelihood of Redx to test positive relative to JEx and HFx was a surprising result from this study as there is a low level of JD in the breed's country of origin and also due to a speculated increased resistance to JD in Norwegian red cattle [52]. Norwegian Reds are commonly utilised in breeding programmes to increase herd genetic merit [53]. In the 1990's a large number of dairy cattle were imported to Ireland from continental Europe [16]. The practice of importing cattle from abroad has been associated with increased risk of testing JD positive [29]. It is possible that farms in the present study that utilised Norwegian Reds to improve herd genetic merit may have previously imported cows from abroad, possibly allowing the establishment of JD within the herd, facilitating current transmission to Norwegian Red cows. Indeed it is possible that herds interested in cross breeding and improving herd genetic merit may be more progressive and potentially larger in size further exacerbating the increased odds of testing positive.

Conclusion

An Ap and Tp of 7.4 and 3.8 %, respectively was recorded in this study and no statistically significant production losses were identified although the majority of study farms engaged in high risk management practices for JD transmission. Although the average dairy herd size in Ireland is relatively small, which may be protective against MAP transmission, the abolition of EU milk quotas is leading to dramatic increases in herd size in Ireland. As this study reiterates the increased risk of JD in larger herds it will be necessary to repeat economic studies to monitor the impact of changing demographics in national herds.

Abbreviations

Ap/Tp: Apparent/True Prevalence; CI: calving interval; JD: Johne's disease; MAP: *Mycobacterium avium* subspecies *paratuberculosis*; MY/MF/MP: milk yield/fat/protein; Se: sensitivity; SCC: somatic cell count; Sp: specificity.

Competing interests

The authors declare that they have no competing interests.

Authors' contributions

RS, NB, JOM participated in the conceptual aspect of the work. NB, AK, RS involved in sample collection. All authors provided consultation and coordination. AK & RS wrote the first draft of the manuscript, with all authors involved in reviewing. All authors read and approved the final manuscript.

Acknowledgements

This Teagasc research was funded by the Irish Dairy Levy. The authors thank the Irish Cattle Breeding Federation, and the farmers who gave their time to participate in this study.

Author details

[1]Animal & Bioscience Research Department, Animal & Grassland Research and Innovation Centre, Teagasc, Moorepark, Fermoy, Co. Cork, Ireland. [2]Department of Biological Sciences, Cork Institute of Technology, Bishopstown, Co. Cork, Ireland.

References

1. Fourichon C, Seegers H, Bareille N, Beaudeau F. Effects of disease on milk production in the dairy cow: a review. Prev Vet Med. 1999;41(1):1–35.
2. Chi J, VanLeeuwen JA, Weersink A, Keefe GP. Direct production losses and treatment costs from bovine viral diarrhoea virus, bovine leukosis virus, Mycobacterium avium subspecies paratuberculosis, and Neospora caninum. Prev Vet Med. 2002;55(2):137–53.
3. Stott A, Jones G, Humphry R, Gunn G. Financial incentive to control paratuberculosis (Johne's disease) on dairy farms in the United Kingdom. Vet Rec. 2005;156(26):825–31.
4. Ayele WY, Machackova M, Pavlik I. The transmission and impact of paratuberculosis infection in domestic and wild ruminants. Vet Med. 2001; 46(7–8):205–24.
5. Hasonova L, Pavlik I. Economic impact of paratuberculosis in dairy cattle herds: a review. Vet Med. 2006;51(5):193–211.
6. Stabel JR. Johne's Disease: A Hidden Threat. J Dairy Sci. 1998;81(1):283–8.
7. Gilardoni LR, Paolicchi FA, Mundo SL. Bovine paratuberculosis: a review of the advantages and disadvantages of different diagnostic tests. Rev Argent Microbiol. 2012;44:201–15.
8. Olsen I, Sigurðardóttir Ó, Djønne B. Paratuberculosis with special reference to cattle A review. Vet Q. 2002;24(1):12–28.
9. Nielsen SS, Toft N. Ante mortem diagnosis of paratuberculosis: a review of accuracies of ELISA, interferon-γ assay and faecal culture techniques. Vet Microbiol. 2008;129(3):217–35.
10. Böttcher J, Gangl A. Mycobacterium avium ssp. paratuberculosis–combined serological testing and classification of individual animals and herds. J Veterinary Med Ser B. 2004;51(10):443–8.
11. Nielsen SS. Danish control programme for bovine paratuberculosis. BCVA. 2007;2007:161–8.
12. Raizman EA, Fetrow JP, Wells SJ. Loss of income from cows shedding Mycobacterium avium subspecies paratuberculosis prior to calving

compared with cows not shedding the organism on two Minnesota dairy farms. J Dairy Sci. 2009;92(10):4929–36.
13. Richardson E, More SJ. Direct and indirect effects of Johne's disease on farm and animal productivity in an Irish dairy herd. Ir Vet J. 2009;62(8):526.
14. Johnson-Ifearulundu YJ, Kaneene JB, Sprecher DJ, Gardiner JC, Lloyd JW. The effect of subclinical Mycobacterium paratuberculosis infection on days open in Michigan, USA, dairy cows. Prev Vet Med. 2000;46(3):171–81.
15. Ott SL, Wells SJ, Wagner BA. Herd-level economic losses associated with Johne's disease on US dairy operations. Prev Vet Med. 1999;40(3):179–92.
16. Barrett DJ, Good M, Hayes M, More SJ. The economic impact of Johne's disease in an Irish dairy herd: A case study. Ir Vet J. 2006;59(5):282.
17. Benedictus G, Dijkhuizen A, Stelwagen J. Economic losses due to paratuberculosis in dairy cattle. Vet Rec. 1987;121(7):142–6.
18. Gonda MG, Chang YM, Shook GE, Collins MT, Kirkpatrick BW. Effect of Mycobacterium paratuberculosis infection on production, reproduction, and health traits in US Holsteins. Prev Vet Med. 2007;80(2–3):103–19.
19. Sorge US, Lissemore K, Godkin A, Hendrick S, Wells S, Kelton D. Associations between paratuberculosis milk ELISA result, milk production, and breed in Canadian dairy cows. J Dairy Sci. 2011;94(2):754–61.
20. McNab WB, Meek AH, Martin SW, Duncan JR. Associations between dairy production indices and lipoarabinomannan enzyme-immunoassay results for paratuberculosis. Can J Vet Res. 1991;55(4):356.
21. Johnson Y, Kaneene J, Gardiner J, Lloyd J, Sprecher D, Coe P. The Effect of Subclinical Mycobacterium paratuberculosis Infection on Milk Production in Michigan Dairy Cows. J Dairy Sci. 2001;84(10):2188–94.
22. Hoogendam K, Richardson E, Mee JF. Paratuberculosis sero-status and milk production, SCC and calving interval in Irish dairy herds. Ir Vet J. 2009;62 Suppl 4:265.
23. Lombard JE, Garry FB, McCluskey BJ, Wagner BA. Risk of removal and effects on milk production associated with paratuberculosis status in dairy cows. J Am Vet Med Assoc. 2005;227(12):1975–81.
24. Dieguez F, Arnaiz I, Sanjuan M, Vilar M, Yus E. Management practices associated with Mycobacterium avium subspecies paratuberculosis infection and the effects of the infection on dairy herds. Vet Rec. 2008;162(19):614–7.
25. Hendrick SH, Duffield TF, Kelton DF, Leslie KE, Lissemore KD, Archambault M. Evaluation of enzyme-linked immunosorbent assays performed on milk and serum samples for detection of paratuberculosis in lactating dairy cows. J Am Vet Med Assoc. 2005;226(3):424–8.
26. Good M, Clegg T, Sheridan H, Yearsely D, O'Brien T, Egan J, et al. Prevalence and distribution of paratuberculosis (Johne's disease) in cattle herds in Ireland. Ir Vet J. 2009;62(9):597.
27. Dillon P. The Irish dairy industry – Planning for 2020. In: Irish national dairy conference: 2011. Ireland: Teagasc; 2011. p. 1–24.
28. Sayers R, Sayers G, Mee JF, Good M, Bermingham ML, Grant J, et al. Implementing biosecurity measures on dairy farms in Ireland. Vet J. 2013;197(2):259–67.
29. Barrett D, Mee J, Mullowney P, Good M, McGrath G, Clegg T, et al. Risk factors associated with Johne's disease test status in dairy herds in Ireland. Vet Rec. 2011;168(15):410–0.
30. Fry MP, Kruze J, Collins MT. Evaluation of four commercial enzyme-linked immunosorbent assays for the diagnosis of bovine paratuberculosis in Chilean dairy herds. J Vet Diagn Investig. 2008;20(3):329–32.
31. Collins MT, Wells SJ, Petrini KR, Collins JE, Schultz RD, Whitlock RH. Evaluation of five antibody detection tests for diagnosis of bovine paratuberculosis. Clin Diagn Lab Immunol. 2005;12(6):685–92.
32. Olori VE, Galesloot PJ. Projection of partial lactation records and calculation of 305-day yields in the Republic of Ireland. Int Bull. 1999;22:149.
33. Berry DP, Shalloo L, Cromie AR, Veerkamp RF, Dillon P, Amer PR, Kearney JF, Evans RD, Wickham B. The economic breeding index: a generation on Technical Report to the Irish Cattle Breeding Federation. 2007, 29–34.
34. Sergeant E. Epitools epidemiological calculators. AusVet Animal Health Services and Australian Biosecurity Cooperative Research Centre for Emerging Infectious Disease. In., vol. 2015. http://epitools.ausvet.com.au. 2015.
35. Garcia A, Shalloo L. Invited review: The economic impact and control of paratuberculosis in cattle. J Dairy Sci. 2015;(98):5019–5039.
36. Benedictus G, Verhoeff J, Schukken Y, Hesselink J. Dutch paratuberculosis programme history, principles and development. Vet Microbiol. 2000;77(3):399–413.
37. Kennedy AE, Da Silva AT, Byrne N, Mac Sharry J, Govender R, O'Mahony J, Sayers RG. The single intradermal cervical comparative test interferes with Johne's disease ELISA diagnostics. Front Immunol. 2014;5:564.
38. Lombard JE, Byrem TM, Wagner BA, McCluskey BJ. Comparison of milk and serum enzyme–linked immunosorbent assays for diagnosis of Mycobacterium avium subspecies paratuberculosis infection in dairy cattle. J Vet Diagn Investig. 2006;18(5):448–58.
39. Cooney R, Kazda J, Quinn J, Cook B, Muller K, Monaghan M. Environmental mycobacteria in Ireland as a source of non-specific sensitisation to tuberculins. Ir Vet J. 1997;50(6):370–3.
40. Osterstock JB, Fosgate GT, Norby B, Manning EJ, Collins MT, Roussel AJ. Contribution of environmental mycobacteria to false-positive serum ELISA results for paratuberculosis. J Am Vet Med Assoc. 2007;230(6):896–901.
41. Cashman W, Buckley J, Quigley T, Fanning S, More S, Egan J, Berry D, Grant I, O'Farrell K. Risk factors for the introduction and within-herd transmission of Mycobacterium avium subspecies paratuberculosis (MAP) infection on 59 Irish dairy herds. Ir Vet J. 2008;61(7):464.
42. Kennedy AE, O'Doherty EF, Byrne N, O'Mahony J, Kennedy E, Sayers RG. A survey of management practices on Irish dairy farms with emphasis on risk factors for Johne's disease transmission. Ir Vet J. 2014;67(1):1–11.
43. Vasseur E, Borderas F, Cue R, Lefebvre D, Pellerin D, Rushen J, Wade K, De Passille A. A survey of dairy calf management practices in Canada that affect animal welfare. J Dairy Sci. 2010;93(3):1307–16.
44. Losinger W, Wells S, Garber L, Hurd H, Thomas L. Management factors related to Salmonella shedding by dairy heifers. J Dairy Sci. 1995;78(11):2464–72.
45. Pithua P, Espejo LA, Godden SM, Wells SJ. Is an individual calving pen better than a group calving pen for preventing transmission of Mycobacterium avium subsp paratuberculosis in calves? Results from a field trial. Res Vet Sci. 2013;95(2):398–404.
46. Van Winden S, Stevens K, Guitian J, Mcgowan M. Preliminary findings of a systematic review and expert opinion workshop on biosecurity on cattle farms in the UK. Cattle Pract: J Br Cattle Vet Assoc. 2005;13(2):135–40.
47. C.S.O. Census of Agriculture 2010- Final Results. In. Edited by Office CS. http://www.cso.ie/en/media/csoie/releasespublications/documents/agriculture/2010/full2010.pdf; 2012.
48. New Zealand Dairy Statistics 2011–2012; http://www.dairynz.co.nz/media/434165/new_zealand_dairy_statistics_2011-12.pdf. Accessed 10 Jan 2016.
49. de la Rua-Domenech R, Goodchild AT, Vordermeier HM, Hewinson RG, Christiansen KH, Clifton-Hadley RS. Ante mortem diagnosis of tuberculosis in cattle: A review of the tuberculin tests, γ-interferon assay and other ancillary diagnostic techniques. Res Vet Sci. 2006;81(2):190–210.
50. Maher P, Good M, More S. Trends in cow numbers and culling rate in the Irish cattle population, 2003 to 2006. Ir Vet J. 2008;61(7):455.
51. Shalloo L, Ryan A, French P: Getting ready for Expansion! Lessons from the Greenfield project. In: Proceedings. Teagasc National Dairy Conference. Is Ireland ready for more milk. November 2012, Ireland.
52. Holstad G, Sigurdaroottir O, Storset A, Tharaldsen J, Nyberg O, Schonheit J, Djonne B. Description of the infection status in a Norwegian cattle herd naturally infected by Mycobacterium avium subsp. paratuberculosis. Acta Vet Scand. 2005;46(1–2):45–56.
53. Buckley F, Begley N, Prendiville R, Evans R, Cromie A. Crossbreeding the dairy herd–a real alternative. Ir Grassland Assoc J. 2008;42:5–17.

2

Control of *Staphylococcus aureus* in dairy herds in a region with raw milk cheese production: farmers' attitudes, knowledge, behaviour and belief in self-efficacy

Marie-Eve Cousin[1], Maria Christina Härdi-Landerer[2], Verena Völk[3] and Michèle Bodmer[4*]

Abstract

Background: Contagious mastitis is an important disease in dairy cattle, and the causative agent *S. aureus* can also impair raw milk cheese quality. In a confined region in eastern Switzerland attitude, knowledge and behaviour towards *S. aureus* und *S. aureus* control was assessed in 90 dairy farmers with communal alpine pasturing including raw milk cheese production with the aid of a questionnaire.

Results: Forty-three out of 90 questionnaires were returned (48% return rate). Farmers perceived reproductive problems as most important in their dairy herds followed by respiratory disease and diarrhoea in young stock. Most frequently stated as important motivating factors to participate in *S. aureus* control were "avoiding negative news about cheese quality in the press" followed by "I want to be proud of my somatic cell counts again". Most frequently chosen and identified as important constraining factors were "I fear that the authorities dictate and the farmers are not heard" followed by "costs to control *S. aureus* are too high because of premature culling" and "I am afraid to be forced to cull genetically valuable cows". Farmers with an experience of a *S. aureus* problem in their dairy herds had a significantly better knowledge about contagiosity and clinical manifestation of different *S. aureus* genotypes than farmers with no self-reported experience of a *S. aureus* problem. Veterinarians were indicated as the most important experts, farmers seek advice in case of mastitis and most farmers suggested subsidising bacteriological milk analysis as an incentive to motivate farmers towards *S. aureus* control.

Conclusion: According to the results an improved knowledge transfer on *S. aureus* to dairy producers and an integrative approach to a *S. aureus* control program with subsidising milk analysis will be most promising to improve the *S. aureus* situation in this confinded region of eastern Switzerland. Veterinarians should cover a key role in consulting farmers during the control program.

Keywords: *S. aureus*, Control-program, Attitude, Behaviour, Raw milk cheese quality

Background

Mastitis is one of the most important diseases in dairy herds [1]. This is especially true for herds selling milk for raw milk cheese production. For this particular niche production, milk needs to fulfill strict quality criteria to guarantee food safety in Switzerland [2].

* Correspondence: michele.bodmer@vetsuisse.unibe.ch
[4]Clinic for Ruminants, Vetsuisse-Faculty, University of Bern, Bremgartenstrasse 109a, 3001 Bern, Switzerland
Full list of author information is available at the end of the article

In alpine parts of the country, communal alpine pasturing takes place during summer (May/June to September). The milk of lactating dairy cows is traditionally used for raw milk cheese production.

In the past 5 years, several cases of contamination of raw milk cheese with *S. aureus* enterotoxins were recorded in the eastern part of Switzerland and can be a threat for human health [3]. Staphylococcal enterotoxins are described as a cause of food poisoning leading to vomiting and diarrhoea in humans and cases of food poisoning by *S. aureus* enterotoxins deriving from dairy products were reviewed [4]. A recent report describes an outbreak of *S. aureus* enterotoxin poisioning due to raw milk cheese consumption in a boarding school [5]. In the year 2016, 4 confirmed cases were reportet to the Federal office for Public Health Switzerland [6].

S. aureus genotype B (GTB) [7] was one of the main isolates to be identified as cheese contaminants [3]. There are a variety of factors triggering enterotoxin production of *S. aureus* very much depending on the type of the enterotoxin and on other circumstances. However, the density of *S. aureus* bacteria within the milk (> 10^5 cfu/g) seems to be crucial to activate accessory gene regulators [8]. Therefore, if *S. aureus* enterotoxins are detected in cheese, we may assume that the raw milk used for cheese production was already heavily contaminated with *S. aureus,* in particular with GTB [3]. In a recent study [9], it was shown that communal alpine farming during summer is a risk for the rapid spread of a particular *S. aureus* strain within the herds leading to high counts of *S. aureus* GTB in the raw milk destined for cheese production.

Until now, no targeted intervention for *S. aureus* control during alpine pasturing was implemented in the affected region, although effective control programs including testing, following a strict milking order according to *S. aureus*-status and culling of non-curable animals were suggested earlier [10]. The need for control programs on herd level is utterly important, since cure rates during lactation are disappointing [11]. They depend on different factors of the affected cow [11], on treatment duration [12] and on distinct characteristics of *S. aureus,* among others internalization into mammary gland cells [13], and forming of biofilms [14].

Although farmers think that their knowledge on current prevention practices is sufficient [15], they very often lack compliance for implementation of control measures [16]. The extent, to which mastitis is perceived as a problem is significantly different between dairy producers [17, 18], and it is likely, that a group of farmers cannot agree on a specific threshold and the measures to be implemented if the threshold is overcome as long as the official regulations are met. The most important factors for motivation to improve mastitis management were shown to be: job satisfaction, overall situation of the farm, economic loss, animal health and welfare awareness, ease in meeting regulatory requirements, extra financial incentives by bonus payments based on bulk tank somatic cell counts (BTSCC), dairy product quality, image and recognition for "job well done" [19]. The relatively low importance of economical aspects for motivation might as well explain the not fully rational reaction of farmers towards economical consequences of controlling somatic cell counts [20]. A study conducted in the United Kingdom showed, that extrinsic factors (i.e. financial barriers) were not relevant to farmers lacking intent to control zoonotic disease, as long as their believe in self-efficacy and their normative beliefs of responsibility towards consumers were absent [21]. Dutch dairy farmers rated the effectiveness of penalty strategies higher than bonus strategies regarding improvement of the mastitis situation in their herds [19]. Most of the above mentioned studies were conducted in the framework of a dairy production regulated by quota. Swiss dairy production is characterized by a milk market without quota, different local niches of raw milk cheese production with high demands of quality and an elaborate governmental direct payment system to support ecological production. It was shown before, that particularly self-reported attitude and knowledge of farmers contributed to a decrease clinical mastitis [22, 23]. Therefore, the aim of the present study was to investigate potential causes for insufficient compliance with *S. aureus* control measures by 1) assessing attitude towards *S. aureus* control considering communal alpine pasturing during summer, 2) assessing knowledge of farmers on contagious mastitis caused by *S. aureus* and in particular GTB, 3) assessing behaviour in the context of mastitis management and *S. aureus* control and 4) assessing their believe in self-efficacy in a confined region with a niche production of raw milk cheese and high amount of governmental subsidies.

Methods

Participants

Ninety farmers in a confined alpine region in eastern Switzerland who sent their dairy cows to 9 distinct communal alpine farms during 90 days in summer were included in the survey. Because of the low number of eligible respondents no random selection of participants was performed.

Questionnaire

In order to assess attitude, knowledge, behaviour and self-efficacy of the farmers a structured questionnaire was created.

The questionnaire was based on the insight of 4 semi-structured interviews conducted with volunteers among the farmers of the respective region and contained 1) items about demographic farm-data, 2) items on general aspects of bovine health including fertility, claw health and calf health 3) items on knowledge about mastitis in general and *S. aureus* mastitis in particular, 4) items on behaviour concerning management of mastitis cases and mastitis prevention, 5) items on attitude towards mastitis prevention and *S. aureus* control, 6) items on self-efficacy and 7) items about different aspects of desired support during a potential *S. aureus* control program. Questions about infectious disease were not included since Switzerland is officially free of IBR, TBC, EBL and BVD. Additionally, no questions about endoparasitic disease were included since there are regionally agreed regulations on deworming of all heifers for communal alpine pasturing.

The interviewees agreed on a written consent that the provided information is exclusively used for research and only published in an anonymised fashion. The repsondents to the written questionnaire were also informed, that data they provided was only published in an anonymised fashion.

Demographic information concerning the farms was collected as continuous data (milk yield, herd size) or as categorical data (e.g. housing system, surface). Variables were categorised before analysis; details about categories are given in Table 1. The full questionnaire translated into English is added as Additional file 1.

The questionnaire was pretested by four farmers not included in the study in order to assess the comprehensiveness of the questions.

Knowledge was assessed with specific statements to which the participants could answer with "true", "false", and "don't know". The attitude, behaviour and self-efficacy items were measured by a 6-point Likert's scale [24] on how much they agreed or disagreed with the statement given (1 = completely agree, 6 = completely disagree). The preference for potential support during eradication was measured by multiple choice questions with the possibility to choose more than one of the suggested answers.

The questionnaire was mailed and a reminder was sent by mail 6 weeks later.

Data analysis

Data analysis was performed with SPSS (SPSS Inc. Chicago IL, USA). A descriptive data analysis was performed, and results are given as means including standard deviations (SD) for normally distributed continuous data and frequencies were estimated for categorical data. The participants were divided into 2 groups based on their experience with a *S. aureus* herd problem (i.e. ≥ 1 positive cow in the herd within the past year). Group 1 were farmers who indicated that they never experienced a herd problem with *S. aureus* before and group 2 consisted of farmers, who indicated that they already had experienced a herd problem in the recent past.

The groups were compared using non-parametric tests i.e. χ^2-test and Fishers exact test for frequencies ≤ 5. The siginficance level was set at $P \leq 0.05$. The small sample size did not allow for logistic regression models.

Results

A total of 90 questionnaires were sent out by mail and 43 were returned, resulting in a return rate of 48%. A total of 38 (42%) questionnaires were completed and were subjected to futher analysis.

Description of sample

Among the respondents there was only one female (2.6%). In the following paragraph, the crude numbers and percentages of answers out of the total answers given to demographic items are listed. The majority of the respondents (16/31, 51.6%) operated farms of > 25 ha of surface, 32.2% (10/31) of 16-25 ha and 16.1% (5/31) of ≤ 15 ha. Twenty-one percent (8/37) of the respondents owned less than 10 cows, 37.8% (14/37) had 10-15 cows, 32.4% (12/37) had 16-25 cows and only 8.1% (3/37) owned > 25 cows. The main source of income was farming in 89.1% (33/37), and 54.1% (20/37) indicated that dairy production was the most important production area on their farm, whereas 24.1% (9/37) stated that veal calf production (using the whole milk of their dairy cows) was their main production area. The remaining respondents (8/37, 21.6%) chose different options such as veal and beef production, dairy and veal production on equal terms and egg production. Only 11% of the farmers produced according to the guidelines of BioSuisse (organic production). During winter, 47.2% (17/36) farmers used the milk of their dairy cows for the fattening of veal calves, 44.4% (16/36) sold it to a dairy factory and 8.3% (3/36) sold it to a raw milk cheese producer. Thirteen (36.1%) of 36 answers stated that the man power on their farms consisted of 2 persons with operating levels of 100% and 50%, respectively. Sixteen out of 36 farmers reported not having experienced a *S. aureus* problem (i.e. ≥ 1 positive cow in the herd, group 1) up to now, whereas 12 respondents indicated that they had dealt with a *S. aureus* problem during communal alpine pasturing in summer, 4 indicated that they had experienced a problem within the last 2 years, 1 respondent indicated that he currently dealt with a *S. aureus* problem and 2 respondents did not know if they ever have had a *S. aureus* problem in their herds.

Attitude

General aspects of animal health

The self-reported importance of different cattle health aspects with the mean and SD are displayed in Fig. 1. Mastitis seemed to be less important than reproductive disorders and calf diseases. Metabolic disorders such as milk-fever and ketosis were of much lower importance.

Attitude towards different udder-health aspects

The distribution of the answers asking participants about self-reported importance of different udder health aspects with means and SD are shown in Fig. 2. The majority of the respondents recorded, that a bulk tank milk somatic cell count (BTMSCC) < 150,000 cells/ml, a low percentage of cows with elevated individual somatic cell

Table 1 List of variables used for assessing demographic farm data

Variable	Unit categories
Surface of farm	< 10 ha
	10-15 ha
	16-25 ha
	> 25 ha
Number of dairy cows on farm	< 10; 10-15; 16-25; > 25
Number of calves on farm	continouus
Number of heifers on farm	continouus
Expenditure of time for farming by the farm owner	100%
	< 50%
	> 50%
Total equivalents of manpower on farm	Only 1 person......%
	Additional person 1 ......%
	Additional person 2 .....%
	Additional person 3......%
Main income	Dairy
	Veal calves
	Beef
	Egg production
	Broilers
Special production label	None
	BIO Suisse (organic)
	IP Suisse (integrated production)
	Terra Suisse
	Naturaplan
Heifer rearing	Only own heifers on own farm
	Only own heifers at external rearing farm
	Own heifers and purchases at external rearing farm
	Own heifers and purchases at own farm
	Only purchases
Number of purchases previous 12 months	None
	Heifers 1-3
	Heifers > 3
	Cows 1-3
	Cows > 3
	Heifers 1-3 and cows 1-3
Housing dairy cows	Loose housing
	Tie stall
Housing heifers	Loose housing
	Tie stall
Number of cows going to communal alpine operation	1-5
	6-10
	11-15
	> 15

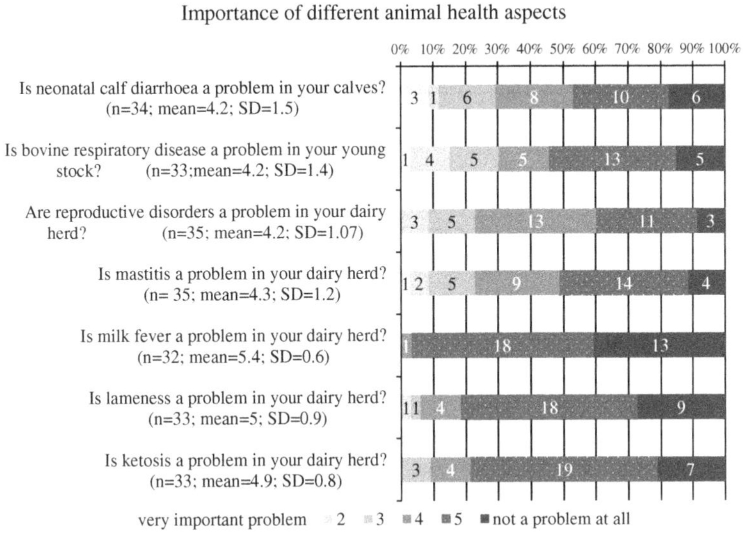

Fig. 1 Distribution of self-reported importance of different cattle diseases in the study herds

counts, a low number of clinical mastitis needing treatment and a low number of teat lacerations were very important to them. Asked about a personal threshold of bulk tank somatic cell count considered as satisfactory, the majority claimed that they aimed at being < 100,000 cells/ml. The distribution of the answers is given in Fig. 3.

Twenty-nine participants out of 34 (85%) answering the questions stated that they would like to decrease their BTMSCC, only 5 out of 34 (15%) stated they dindn't want to decrease the BTMSCC. Thirty-four out of 35 answers indicated, that the first person to seek advice in case of a mastitis problem in their dairy herd was their veterinarian. Only one respondent indicated to seek advice at the consultant for milking and milking equipment. The options "Swiss bovine health service", "nutritionist", "salesman for farm products" or "a colleague" were not chosen.

Attitude towards S. aureus control

The respondents were asked if they were willing to contribute to the control of *S. aureus* in the future. Twenty-three out of 34 (68%) respondents were willing to control *S. aureus*, 7 were not willing to participate in a control program and 4 chose the option "don't know".

Figure 4 shows the responses to questions about respondent's attitude towards the effectiveness of different *S. aureus* control measures with means and SD. The measure "culling of infected animals with failure of treatment" was perceived as the most important measure followed by "sticking to a milking order according to the

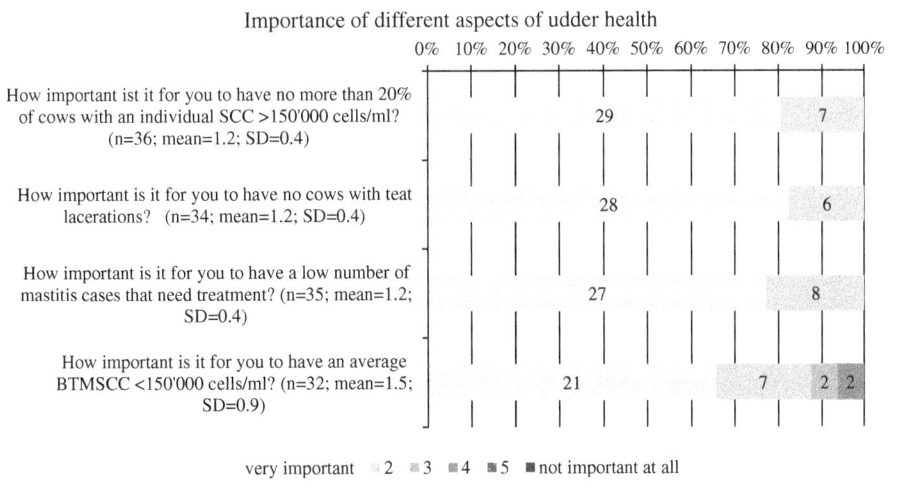

Fig. 2 Distribution of self-reported importance of different aspects of udder health

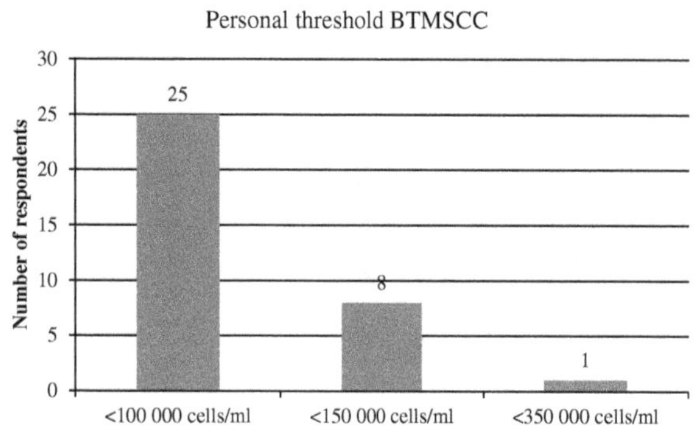

Fig. 3 Frequencies of personal BTMSCC-threshold for satisfactory udder health

udder health status of the individual cows" and "milking according to a correct routine".

Motivation

Motivation was assessed with 7 items. "Avoiding negative news on raw milk cheese quality in the press" was one of the main motivators for respondents to control *S. aureus,* and they also saw some potential in the marketing of raw milk cheese free of *S. aureus*. Two slightly more emotional motivating factors were: "I want to be proud of my BTMSCC" (21/ 34) and "bad milk quality depresses me" (20/34). The distribution of the answers including the mean and SD is displayed in Fig. 5. The frequency of preferred financial support strategies i.e. external motivation factors is shown in Fig. 6, only one answer was possible. Seventeen out of 38 repondents preferred a discount on laboratory costs and 13 a premium for culling infected cows. A bonus for *S. aureus* free milk was preferred by 6 respondents, and only 3 thought a penalty system for *S. aureus* containing milk could serve as a motivating factor.

The strongest constraint for participation in a *S. aureus* control program was the fear of an authority-dictated, top-down eradication program without taking into account the producers' perspective. Distributions with means and SD are displayed in Fig. 7.

Preferred support in future control program

The most frequent answer to the question asking for preferred support in a potential future *S. aureus* control program was "regular support by the vet" (18 out of 35 answers), 8 of 34 respondents stated that no support was necessary, and 7 thought that a special education of personnel seasonally employed by communal alpine operations would be beneficial.

Knowledge on *S. aureus*

In Fig. 8, answers to specific questions about *S. aureus* and *S. aureus* GTB are displayed. The participants had to indicate, if they thought the statement was correct, not correct or if they did not know the answer.

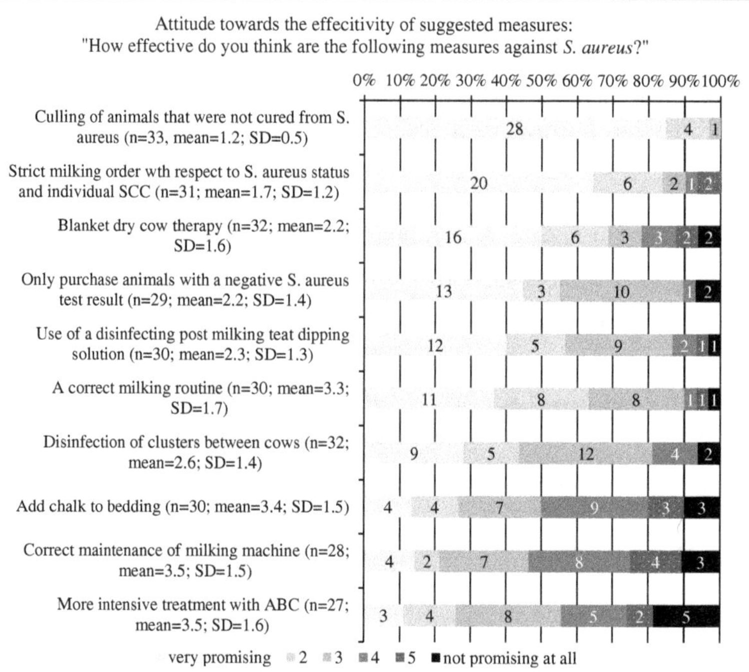

Fig. 4 Distribution of answers to items on attitude towards effectiveness of generally suggested *S. aureus* control measures

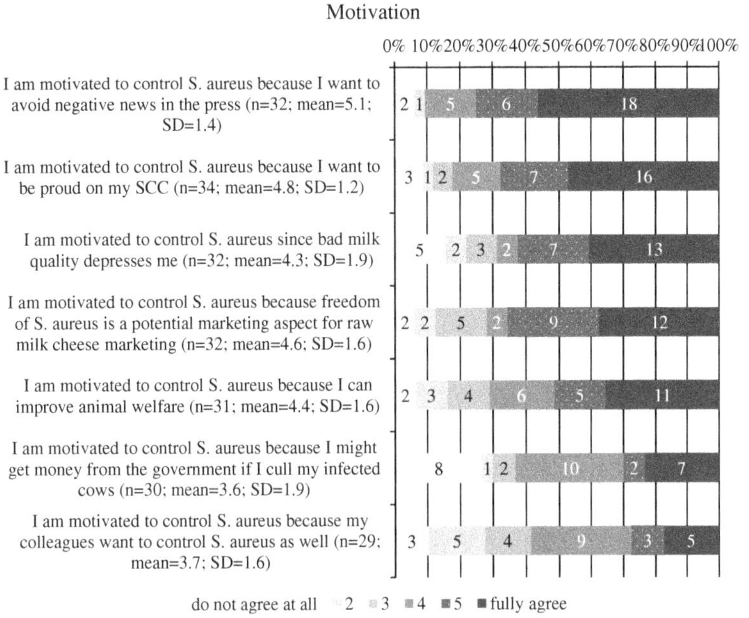

Fig. 5 Distribution of factors motivating producers to participate in a *S. aureus* control program

Behaviour

Mastitis management on cow level

The producers were asked which actions they take in case of an acute clinical mastitis. To avoid misinterpretation of terminology an example of a cow with a high body temperature, anorexia, swollen and painful quarter and changed milk secretion was given: "One of your cows is off feed, she has a body temperature of 41°C and her front right quarter is swollen and painful. What do you do first?" Frequencies of the different actions (options included: call vet, perform CMT, strip off contaminated milk, immediate treatment with antibiotics, homoeopathic treatment, wait and see, collection of milk sample) are shown in Fig. 9. Multiple answers were allowed.

First actions taken in case of subclinical mastitis are displayed in Fig. 10 (options included: CMT immediately, CMT in the next few days, collection of aseptic milk sample, immediate treatment with antibiotics, homoeopathic treatment, milk last, wait for next test day result, no intervention). Again, to avoid misunderstanding of terminology an example of a typical case was given: cow without any systemic signs, elevated individual somatic cell count, positive CMT, no changes of quarter or milk secretion.

Monitoring udder health on herd level

In Fig. 11 the frequencies of tools and combinations of different tools for monitoring of udder health given by

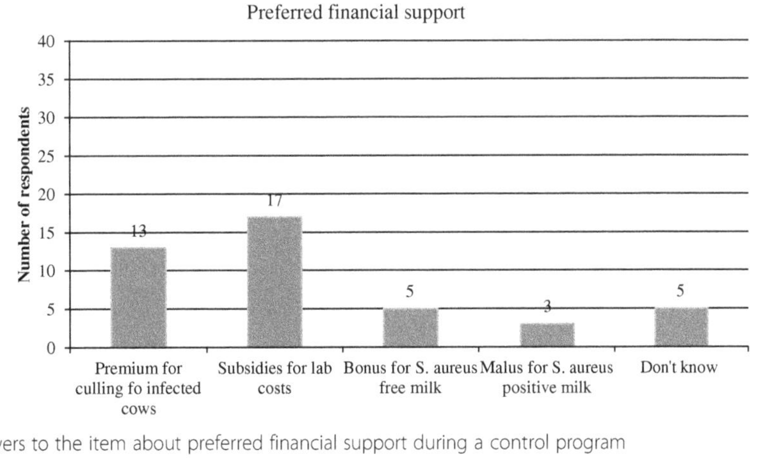

Fig. 6 Frequencies of answers to the item about preferred financial support during a control program

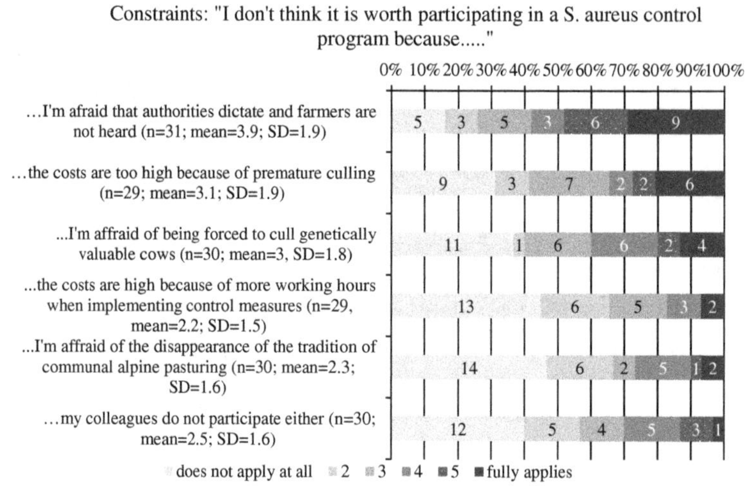

Fig. 7 Distribution of constraining factors for participation in a *S. aureus* control program

37 respondents are displayed. The farmers were asked "Which tools do you use to monitor udder health of your herd?". The options included milk-recording data (individual somatic cell count, milk production), results of official BTMSCC measurements including weighed herd somatic cell count from milk recording and the BTM samples collected twice monthly by the milk company, CMT, bacterial culture (BC) and identification of mastitis pathogens by polymerase chain reaction (PCR). Multiple answers were allowed.

Belief in self-efficacy

The answers to items asking participants about their trust in *S. aureus* control measures (means and SD) are displayed in Fig. 12. Although 21 of 32 respondents believed that consequent implementation of control measures will eradicate *S. aureus* from their herd, 18 of 30 respondents believed that they will never get rid of *S. aureus* despite implementation of control measures, and 11 of 31 respondents indicated that they do not believe in the existence of effective control measures at all. Out of 30 respondents, 13 were convinced that the majority of *S. aureus* infections occur during communal alpine pasturing in summer.

Comparison of groups of respondents

Using χ^2 tests to compare groups of farmers revealed that there were significant differences in terms of

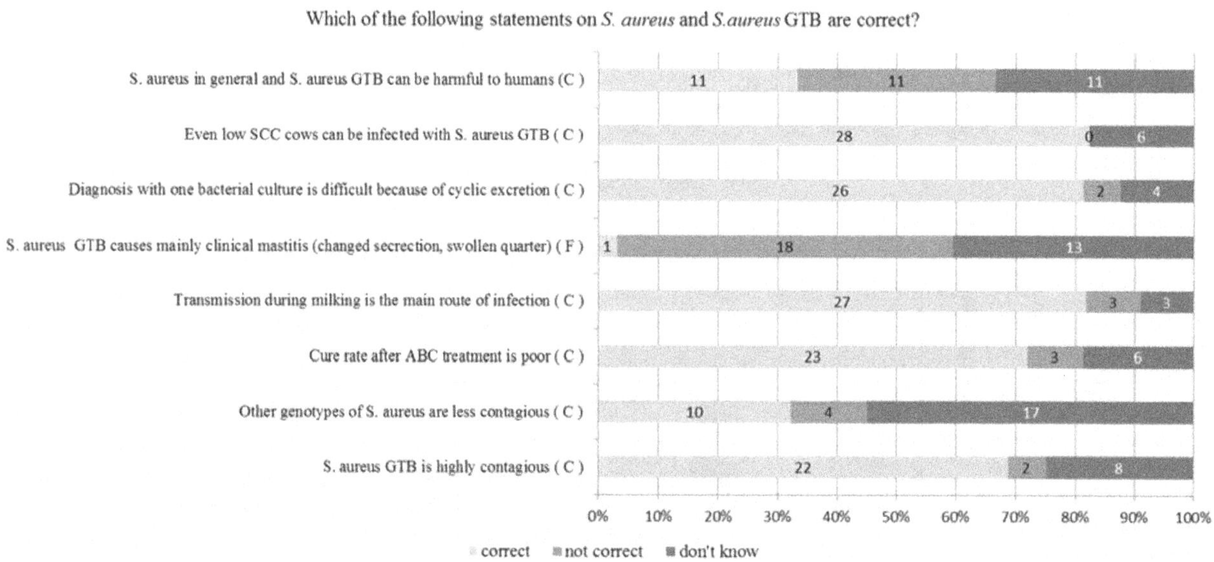

Fig. 8 Frequencies of answers given by the farmers to items on knowledge (*S. aureus* and *S. aureus* GTB). The answers are marked either by C = scientifically correct or by F = scientifically incorrect, false

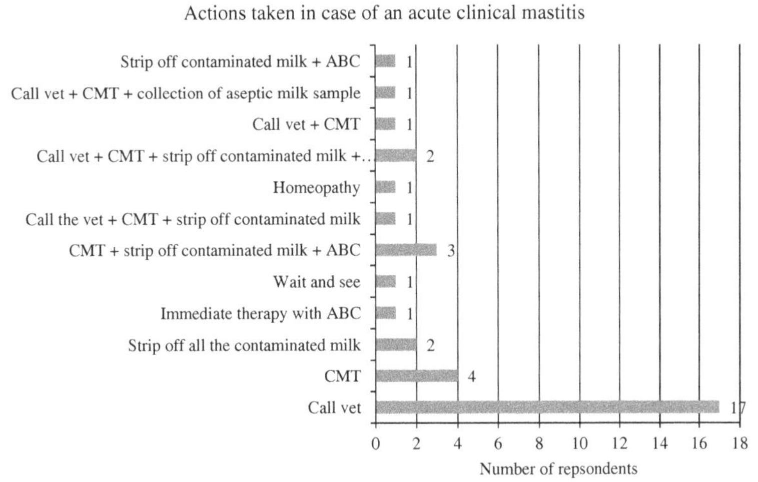

Fig. 9 Frequencies of actions taken in case of acute clinical mastitis

knowledge between farmers with a self reported experience of *S. aureus* within the previous year and farmers without such an experience. The knowledge items were recoded using categories according to the scientific correctness of an answer allocating 1 for scientifically correct answers and 0 for scientifically incorrect and "don't know" answers. The number of scientifically correct answers to the statement "other *S. aureus* genotypes are less contagious than genotype B" was significantly higher in the group with an experience of a *S. aureus* problem (8 out of 13, 61.5%) than the number of correct answers given by farmers without an experience of a *S. aureus* problem (2 out of 18, 11.1%; $P = 0.006$). Farmers without a *S. aureus* experience gave significantly less scientifically correct answers (7 out of 18 38.9%) to the statement "*S. aureus* mainly causes clinical mastitis" and to the statement "Also low cell count cows can be *S. aureus* positive" (14 out of 20, 70%) than farmers with a *S. aureus* experience (11/14, 78.6%, $P = 0.04$ and 14 out of 14, 100%, $P = 0.03$, respectively).

The farmers who had experienced a *S. aureus* problem in the past year, had a tendency towards less rational constraints than farmers that had not been confronted with a *S. aureus* problem in the past year (Fisher's exact $P = 0.055$).

Discussion

This was the first study to assess attitude, knowledge and behaviour towards *S. aureus* control among raw milk cheese producing dairy farmers without a quota system.

Looking at the different attitude items, the respondents stated that mastitis was not the most important

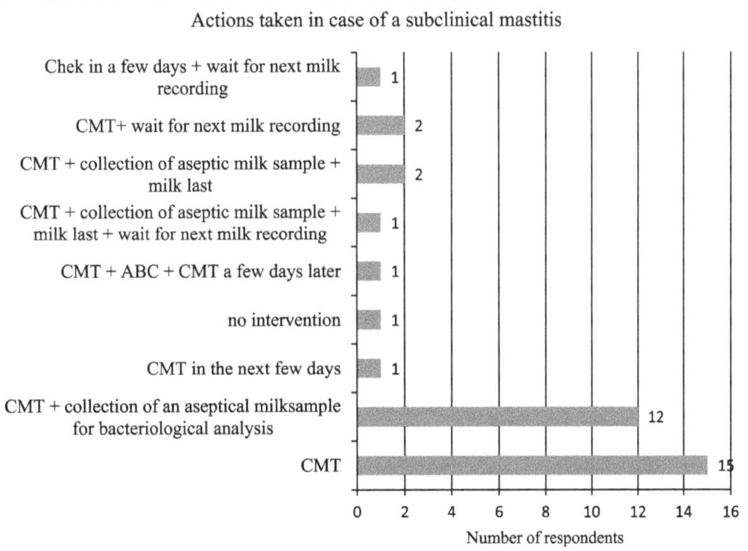

Fig. 10 Frequencies of actions taken in case of subclinical mastitis

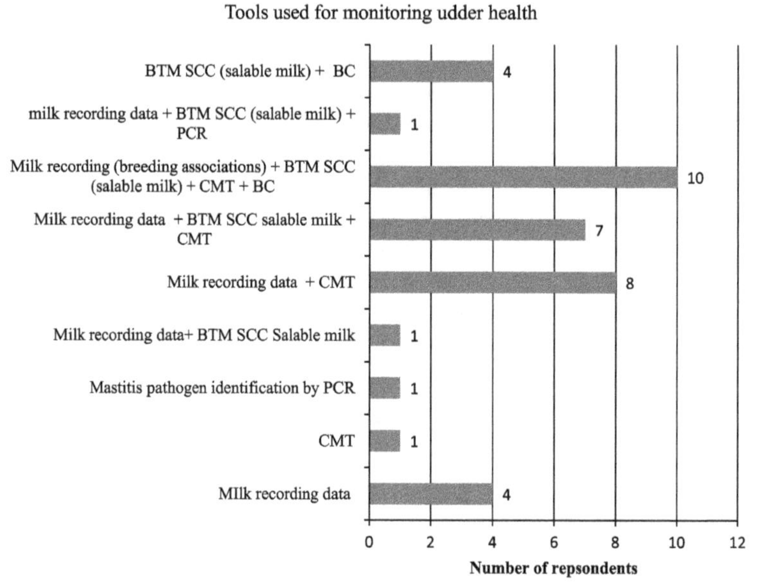

Fig. 11 Frequencies of tools used to monitor udder health on herd level by the respondents

problem in their herds; reproductive failure and calf health seemed to be more important. This confirms results from a study conducted in the Netherlands stating that for most dairy farmers, mastitis is not first priority among herd problems they have to deal with [18].

Most participants indicated that a BTMSCC below 150,000 cells/ml was very important to them, this might be because some milk processers have established a bonus system for milk with a cell count ≤ 100,000 cells/ml [25]. The two respondents who indicated that a low BTMSCC was not important to them, were farmers, that used the milk of their cows exclusively for veal calf production during winter and, therefore, were not eligible to penalty rules of the milk companies. Although percentage of high SCC cows and BTMSCC are strongly linked, none of the respondents chose an option below 4 on the Likert scale when asked about the importance of a low percentage of high SCC cows in their herd. We can only speculate on the reasons, but almost all dairy farmers of the study including the veal producing farms were affiliated with a breeding organisation, and low SCC on the cows performance records are favourable for selling the animals.

Asked about the personal threshold of BTMSCC, the vast majority indicated a limit of 100,000 cells/ml, which was probably again influenced by the bonus system established in some milk companies. Additionally, it cannot be excluded that the answers given were at least partially socially desired. However, there was variation in

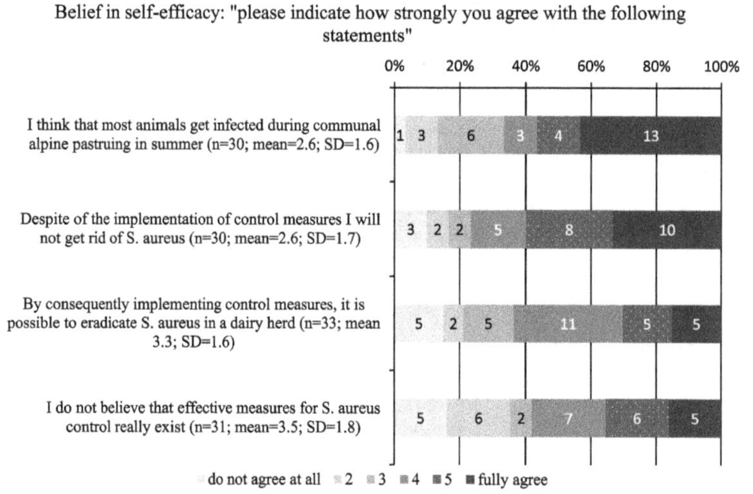

Fig. 12 Distribution of answers concerning belief in self-efficacy

the level of the personal threshold, confirming results of Kuiper et al. (2005) [15].

The majority of respondents (66%) were willing to improve the *S. aureus* situation, but 21% indicated that they saw no necessity in setting up a control program and 12% did not know. The strongest motivation factors to participate in a potential future *S. aureus* control program were: "avoiding negative press concerning cheese quality", "*S. aureus* freedom of the milk might be a strong marketing argument for cheese sales" , "I want to be proud of my SCC" and "bad milk quality depresses me". The strongest factors are not purely economically motivated, and confirm work that was done on lameness control in cattle where the strongest motivator was "I want to be proud of my herd" [26]. Valeeva et al., [19] reported as a major motivating factors to improve udder health "job satisfaction" and "overall situation of the farm" which is in line with the results of the present study. When asked about preferred financial support strategies, which can be interpreted as external motivation factors, a high number of participants chose "discount on laboratory costs" indicating that they were willing to invest in diagnosis of the causative agent of an udder health problem.

The major constraints for participation in *S. aureus* control programs reported by the respondents included: "I'm affraid that we will be forced into a top–down control program by the authorities without asking producers", "economic loss because of premature culling of infected cows" and "lack of time to implement all the measures". The factor "lack of time" was also found by others, asking about barriers to conrol lameness in dairy cattle [26]. The most frequent constraining factor identified in the current study should be taken into account when developing future disease control programs. It indicates that an integrative approach might be beneficial for the compliance with a control program, since the attitude of the farmers seems to be a key element in changing behaviour as described by others [18, 23]. "Culling of infected cows", "no purchase of infected cows" and "milking infected cows last" were indicated as the three most useful measures to control *S. aureus*, but a large percentage did not believe in their ability to reach the desired outcome i.e. to control *S. aureus* by implementing control measures ("I believe that I will not get rid of *S. aureus* despite implementation of measures" and "I do not believe that effective measures to control *S. aureus* exist"). This might be associated with the low cure rate of *S. aureus* mastitis during lactation [11] and with the context of communal alpine pasturing, where comingling of animals from different farms of origin acts as a recurrent risk factor for infection with *S. aureus* every summer [27]. In a study investigating perceptions, circumstances and motivators to control zoonotic disease in cattle [21], a lack of belief in self-efficacy was associated with farmers without intention to control disease. Since we did not ask if specific herd-level control measures were implemented in order to assess intention to control, we cannot confirm the results of Ellis-Iversen et al. (2010) [21].

The knowledge concerning some key elements (such as contagiousness of different genotypes and clinical manisfestation) of *S. aureus* control differed significantly between the group of farmers with an experience of *S. aureus* herd problem compared to the farmers that reported no experience with *S. aureus* herd problem, although they all had attended an information event prior to the first mailing of the questionnaire. This difference might be explained by the fact that farmers, who dealt with a mastitis problem due to *S. aureus* in the past were actively looking for information. We can only speculate on that matter, because we did not include any questions on information sources. However, a recent Dutch study found, that increased knowledge on milking procedures and milking equipment was associated with decreased incidence of clinical mastitis [23]. It also indicated, that owners of affected herds were increasing their knowledge by consulting information sources or seeking advice by their veterinarians [23]. Ellis-Iversen et al. (2010) [21] found in their study that one of the most important extrinsic barriers for lack of disease control was lack of accessible knowledge. Unfortunately, we did not assess this in our study, but in the interviews conducted prior to the mailing of the questionnaire (data not shown), 2 out 4 farmers did not know what pathogens were involved in the mastitis cases occuring in their herds, because veterinarians only communicated how to treat the individual case but neither the result of bacteriological analysis nor how to proceed on herd level. This offers opportunities for udder health consulting by veterinarians as reviewed by Jansen & Lam (2012) [28]. Veterinarians were also reported to be the first persons, farmers would seek advice in case of an udder health problem confirming results of former studies [29] and emphasizing the important role of well educated practicioners in consulting and translation of knowledge.

Looking at the items about self-reported behaviour of mastitis management, the majority of respondents indicated to apply the recommended procedures in the management of mastitis cases, but no information on implemented herd-level measures were collected.

The present study is based on self-reporting and, therefore, socially desired answers cannot be excluded. Additionally, the data set of the present study was limited to very few respondents allowing only descriptive analysis and non-parametric tests for comparison of group 1 (experience with *S. aureus* problem) and group 2 (no experience with a *S. aureus* problem). Therefore, more sophisticated statistical analyses such as logistic regression models were not applicable and the present results should be confirmed on a larger scale. Nevertheless, our results on attitude and knowledge correspond well with other studies. The perception of farmers producing raw milk cheese under subsidised conditions without a quota system towards improving udder health by controlling *S. aureus* seem to be similar to perceptions of those producing under more industrial conditions.

Conlcusions

The results of this study indicate that although 66% of farmers stated that they were willing to participate in *S. aureus* control, they were mainly constrained by the fear of an authority-dictated program without taking into account their individual situation. Therefore, it is crucial to integrate producers in the development of a regional control program. There is a need for harmonized knowledge transfer by veterinarians in order to fill knowledge gaps and to improve intention to control *S. aureus* mastitis.

Acknowledgements

We would like to thank all the farmers, who took the time to being interviewed and to complete the questionnaire. Additionally we would like to thank Prof. Michael Siegrist from the Institute for Consumer Behaviour of the ETH in Zürich and Prof. Adrian Steiner, Clinic for Ruminants, University of Bern for their comments on the manuscript. We would also like to thank the Veterinary Services of the Canton of Graubünden for their support.

Funding

No special funding was needed for the current study.

Authors' contributions

MEC: Design of questionnaire, support data analysis. W: Data management and preliminary analysis. MCH: Organising interviews and input to questionnaire. MB: Planning of the study, performing interviews, creation of the final questionnaire, statistical analysis and preparation of manuscript. All authors read and approved the final manuscript.

Ethics approval and consent to participate

The Swiss Federal act on Research involving Human Beings –Articles 3, 4, 7, 11 and 16 regulates the obligation for clearance by a regional ethics committee [30]. According to the above mentioned legislation, surveys which do not ask for personal data including personal health data and are analysed anonymously do not require clearance by the ethics committee. However, interviewees agreed to participate in the study signing a written form stating that all given information is confidential and will only be published in an anonymised fashion.

Attached to the questionnaire all participants obtained a signed form stating that participation in the questionnaire study is voluntary and all given information is confidential and will only be published in an anonymised fashion.

Consent for publication

With returning the completed questionnaire the participants agreed with anonymised publication of the given information.

Competing interests

The authors declare that they do not have any competing interests.

Author details

[1]Epidemiology, Biostatistics and Prevention Institute, University of Zürich, Hirschengraben 84, 8001 Zürich, Switzerland. [2]Institute of agricultural science, Animal Physiology, Swiss Federal Institute of Technology (ETH), Universitätsstrasse 2, 8092 Zürich, Switzerland. [3]Tierarztpraxis Dr. med. vet. Markus Nydegger, Scherlihalde 5, 3145 Niederscherli, Switzerland. [4]Clinic for Ruminants, Vetsuisse-Faculty, University of Bern, Bremgartenstrasse 109a, 3001 Bern, Switzerland.

References

1. Halasa T, Nielen M, RBM H, Hogeveen H. Stochastic bio-economic model of bovine intramammary infection. Livest Sci. 2009;124:295–305.
2. Swiss ordinance on the hygiene in dairy production. https://www.admin.ch/opc/de/classified-compilation/20051436/index.html. Accessed 15 March 2017.
3. Hummerjohann J, Naskova J, Baumgartner A, Graber HU. Enterotoxin-producing Staphylococcus aureus genotype B as a major contaminant in Swiss raw milk cheese. J Dairy Sci. 2014;97:1305–12.
4. De Buyser ML, Dufour B, Maire MLV. Implication of milk and milk products in food-borne diseases in France and in different industrialised countries. Int J Food Microbiol. 2001;67:1–17.
5. Johler S, Weder D, Bridy C, Huguenin M-C, Robert L, Hummerjohann J, Baumgartner A, Stephan R. Outbreak of staphylococcal food poisoning among children and staff at a Swiss boarding school due to soft cheese made from raw milk. J Dairy Sci. 2015;98:2944–8.
6. Health SFO of P. Monitoring report of zoonotic disease and food related diases outbreaks. 2016. https://www.bag.admin.ch/dam/bag/de/dokumente/mt/infektionskrankheiten/zoonosen/Zoonosenbericht2016BAGBVL.pdf.download.pdf/Zoonosenbericht_2016_DE.pdf. Accessed 1 Nov 2017.
7. Fournier C, Kuhnert P, Frey J, Miserez R, Kirchhofer M, Kaufmann T, Steiner A, Graber HU. Bovine Staphylococcus Aureus: association of virulence genes, genotypes and clinical outcome. Res Vet Sci. 2008;85:439–48.
8. Derzelle S, Dilasser F, Duquenne M, Deperrois V. Differential temporal expression of the staphylococcal enterotoxins genes during cell growth. Food Microbiol. 2009;26:896–904.
9. Voelk V, Graber HU, van den Borne BHP, Sartori C, Steiner A, Bodmer M, Hädi-Landerer MC. A longitudinal study investigating the prevalence of Staphylococcus aureus genotype B in seasonally communal dairy herds. J Dairy Sci. 2014;97:4184–92.
10. Kirchhofer M, Kaufmann T, Guélat-Brechbühl M, Michel A, Syring C, Bodmer M. Sanierung von Milchviehbeständen mit Staphylococcus aureus-Mastitiden. Schweiz Arch Tierheilkd. 2011;153:361–8.
11. Sol J, Sampimon OC, Snoep JJ, Schukken YH. Factors associated with bacteriological cure during lactation after therapy for subclinical mastitis caused by Staphylococcus aureus. J Dairy Sci. 1997;80:2803–308.
12. Sol J, Sampimon OC, Barkema HW, Schukken YH. Factors associated with cure after therapy of clinical mastitis caused by Staphylococcus aureus. J Dairy Sci. 2000;83:278–84.
13. Kerro-Dego O, Prysliak T, Perez-Casal J, Potter AA. Role of GapC in the pathogenesis of Staphylococcus aureus. Vet Microbiol. 2012;156:443–7.
14. Veh K, Klein RC, Ster C, Keefe G, Lacasse P, Scholl D, Roy J-P, Haine D, Dufour S, Talbot BG, Ribon AOB, Malouin F. Genotypic and phenotypic characterization of Staphylococcus aureus causing persistent and nonpersistent subclinical bovine intramammary infections during lactation or the dry period. J Dairy Sci. 2015;98:155–68.
15. Kuiper D, Jansen J, Renes RJ, Leeuwis C, van der Zwaag HG. Social factors related to mastitis control practices: The role of dairy farmers' knowledge, attitude, values, behaviour and networks. Mastit. Mastitis in Dairy Prodduction - Current Knowledge and Future Solutions. Maastricht: Proceedings of the 4th IDF International Mastitis Conference; 2005. p. 576–82.
16. Olde Riekerink RGM, Barkema HW, Scholl DT, Poole DE, Kelton DF. Management practices associated with the bulk-milk prevalence of Staphylococcus aureus in Canadian dairy farms. Prev Vet Med. 2010;97:20–8.
17. Huijps K, Lam TJ, Hogeveen H. Costs of mastitis: facts and perception. J Dairy Res. 2008;75:113–20.

18. Jansen J, van den Borne BHP, Renes RJ, van Schaik G, Lam TJGM, Leeuwis C. Explaining mastitis incidence in Dutch dairy farming: the influence of farmers' attitudes and behaviour. Prev Vet Med. 2009;92:210–23.
19. Valeeva NI, Lam TJGM, Hogeveen H. Motivation of dairy farmers to improve mastitis management. J Dairy Sci. 2007;90:4466–77.
20. van Asseldonk M, Renes R, Lam T, Hogeveen H. Awareness and perceived value of economic information in controlling somatic cell count. Vet Rec. 2010;166:263–7.
21. Ellis-Iversen J, Cook AJC, Watson E, Nielen M, Larkin L, Wooldridge M, Hogeveen H. Perceptions, circumstances and motivators that influence implementation of zoonotic control programs on cattle farms. Prev Vet Med. 2010;93:276–85.
22. Jansen J, van Schaik G, Renes RJ, Lam TJGM. The effect of a national mastitis control program on the attitudes, knowledge, and behavior of farmers in the Netherlands. J Dairy Sci. 2010;93:5737–47.
23. van den Borne BHP, Jansen J, Lam TJGM, van Schaik G. Associations between the decrease in bovine clinical mastitis and changes in dairy farmers' attitude, knowledge, and behavior in the Netherlands. Res Vet Sci. 2014;97:226–9.
24. Likert R. A technique for the measurement of attitudes. Arch Psychol. 1932;22:50.
25. https://www.swissmilk.ch/de/produzenten/services-fuer-milchproduzenten/fachportal-milchproduktion/milch-verkaufen/richtlinien-hilfsmittel/-dl-/fileadmin/filemount/p/milchverkauf-qualitaetsbezahlungssystem-der-milchbranche-richtlinie-2011-12-20-de.pdf. Accessed 22 Jan 2018.
26. Leach KA, Whay HR, Maggs CM, Barker ZE, Paul ES, Bell AK, Main DCJ. Working towards a reduction in cattle lameness: 1. Understanding barriers to lameness control on dairy farms. Res. Vet. Sci. 2010;89:311–7.
27. Berchtold B, Bodmer M, van den Borne BHP, Reist M, Graber HU, Steiner A, Boss R, Wohlfender F. Genotype-specific risk factors for Staphylococcus aureus in Swiss dairy herds with an elevated yield-corrected herd somatic cell count. J Dairy Sci. 2014;97:4886–96.
28. Jansen J, Lam TJGM. The role of communication in improving udder health. Vet. Clin. North am. Food Anim. Pract. 2012;28:363–79.
29. Jansen J, Steuten C, Renes R, Aarts N, Lam TJGM. Debunking the myth of the hard-to-reach farmer: effective communication on udder health. J Dairy Sci. 2010;93:1296–306.
30. Federal act on research involving human-beings. Available from: https://www.admin.ch/opc/en/classified-compilation/20061313/index.html. Accessed 11 Jan 2018.

3

Comparison of bulk milk antibody and youngstock serology screens for determining herd status for Bovine Viral Diarrhoea Virus

Richard E. Booth* and Joe Brownlie

Abstract

Background: This paper examines the use of Bulk Milk antibody (BM Ab), Youngstock (YS) serology (Check Tests) and Bulk Milk PCR (BM PCR) for determining the presence or absence of animals persistently infected (PI) with Bovine Viral Diarrhoea Virus (BVDV) within a herd. Data is presented from 26 herds where average herd sizes were 343 and 98 animals for dairy and beef units respectively. Seventeen herds had sufficient data to analyse using Receiver Operating Characteristic (ROC) and probability curves enabling calculation of the sensitivity and specificity of BM Ab and YS Check tests for determining the presence of PI animals within herds in this dataset.

Results: Using BM Ab to screen a herd for the presence of PI animals, achieved a herd level sensitivity and specificity of 80.00 % (44.39–97.48 %) and 85.71 % (42.13–99.64 %) respectively (95 % confidence intervals quoted). Sensitivity and specificity of YS Check Tests at a cut off of 3/10 Ab positive YS were 81.82 % (48.22–97.72 %) and 66.67 % (22.28–95.67 %) respectively (95 % confidence interval). These results were achieved by comparing the screening tests to whole herd PI searches that took place 1–19 months after the initial screen with a mean interval of 8 months. Removal of this delay by taking BM samples on the day of a whole herd test and simulating a YS Check Test from the herd test data produced improvements in the reliability of the Check Tests. BM Ab sensitivity and specificity remained unchanged. However, the Check Test sensitivity and specificity improved to 90.9 % (58.72–99.77 %) and 100 % (54.07–100 %) respectively (95 % confidence interval) at a cut of off 2.5/10 Ab positive animals. Our limited BM PCR results identified 5/23 dairy farms with a positive BM PCR result; two contained milking PIs, two had non-milking PIs and another had no PIs identified.

Conclusions: Delaying a PI search following an initial herd screen decreased the diagnostic accuracy and relevance of our results. With careful interpretation, longitudinal surveillance using a combination of the techniques discussed can successfully determine farm status and therefore allow changes in BVDV status to be detected early, thus enabling prompt action in the event of a BVDV incursion.

Keywords: Bovine Viral Diarrhoea Virus, Herd screen, Herd status, Bulk milk antibody, Youngstock check test

Abbreviations: Ab, Antibody; AHPA, Animal Health and Plant Agency; BM, Bulk milk; BVDV, Bovine Viral Diarrhoea Virus; CHeCS, Cattle Health Certification Standards; PCR, Polymerase chain reaction; PI, Persistently infected; ROC, Receiver operating characteristic; WHT, Whole herd test; YS, Youngstock

* Correspondence: rbooth@rvc.ac.uk
Royal Veterinary College, Hawkshead Lane, North Mymms, Hertfordshire AL9 7TA, UK

Background

Bovine Viral Diarrhoea Virus (BVDV) is an economically important pestivirus recognised for causing infertility, immunosuppression and, as a consequence, high levels of secondary disease in cattle herds worldwide [1–7]. The losses associated with BVDV infection are well documented and the disease is commonly quoted to cost the UK farming industry £40 million per year primarily as a consequence of sub-optimal fertility and immunosuppression [8]. At the individual animal level, the most recent published estimates of losses due to BVDV infection are €32 and €63 per cow per year in beef and dairy systems respectively within the Irish cattle sector [9]. Similar figures of £37 per cow per year exist for beef suckler herds in the UK [10], but less detail is reported at the individual cow level within the UK dairy sector. Significantly, a number of European countries have recognised the losses caused by BVDV and undertaken national eradication; whilst Norway, Sweden, Finland and Denmark have eradicated it [11, 12], other countries e.g. Austria, Switzerland, Germany, Belgium, Ireland and Scotland are in the process of eradication [11, 13–15]. The Scottish and Irish programmes include control measures with regulations to prevent the sale of persistently infected (PI) carrier animals [13, 14]. This will impact on further eradication efforts throughout England and Wales since clearly it would be beneficial for these programmes to be compatible with those underway in Scotland and Ireland in order to facilitate trade.

The persistently infected (PI) animal is infected as a foetus in the first trimester of pregnancy and born immunotolerant to the infecting strain of BVDV; thereafter becoming a lifelong shedder of the virus [16, 17] excreting large quantities of BVDV in all body secretions [18, 19]. Control of PI animals is critical to the successful control of BVDV transmission both within and between herds. Lindberg et al. 1999 [20] drew conclusions from epidemiological studies of BVDV and stated that "in practice, a herd is not infected until one or more PIs have been established" and, in addition to this, "BVDV cannot persist in a herd where contacts between PI animals and susceptible animals in early pregnancy do not occur". This highlights the pivotal role of PI animals in the epidemiology of BVDV and the need to both identify and cull them as part of any successful BVDV eradication programme. The more rapid the culling of PIs from infected herds, the greater the health and productivity advantage.

Two approaches to BVDV eradication have been utilised by those European countries that have programmes in place. In some, often where seroprevalence was deemed to be high, the decision was made at the outset to test directly for PI animals at the national level without establishing status at the herd level first i.e. Switzerland and Ireland [14, 21]. In both programmes, specialised ear tags were used to collect ear notch tissue samples for testing for BVDV antigen or RNA. Within the Swiss programme, the aim was to test the whole cattle population [21] whereas, in Ireland it has been compulsory to test newborn animals as part of the official tagging process from January 1st 2013 onwards [14]. In contrast, the Scandinavian and Scottish programmes have elected firstly to establish individual herd status and then proceed with eradication in infected herds [13, 20]. At the outset of these eradication programmes, effective herd level screening was essential in order to distinguish between BVDV-infected (contains PI animals) and BVDV-free herds (no PI animals present). This allows resources to be focussed on either identification and culling of PI animals or on surveillance and protection. It is always essential that effective biosecurity is implemented to prevent the re-introduction of infection [20, 22, 23]. Depending upon local circumstances and, in particular cattle density, enhanced biosecurity may be combined with vaccination to protect against herd reinfection [24]; this is especially important where biosecurity does not meet the stringent standards outlined in the Cattle Health Certification Standards (CHeCS) technical document [25].

Within England and Wales, many farmers will not know their current BVDV status and it is hoped that this manuscript will provide practitioners with further advice on the use of appropriate herd level diagnostics to determine whether PI animals are likely to be present. Current recommendations for UK herd BVDV accreditation are documented in the CHeCS technical document [25]. Two options are permissible for herd BVD accreditation. The first involves screening successive calf crops for BVDV antigen and accreditation is achieved following negative test results over a 2 year period. The second method involves antibody (Ab) screening of bulk milk samples and youngstock (YS) cohorts (commonly referred to as 'Check Tests' or 'Spot Tests') in order to accurately assess herd BVDV status. BM Ab has played a major role in the control of cattle diseases since the mid-1980s [26] and was widely used within the Scandinavian BVDV control programmes to establish dairy herd status. The CHeCS technical document recommends the use of quarterly BM Ab screening tests at the point of establishing and monitoring herd status although it recognises a positive result may occur in a herd that has been historically infected; at this point the screening of first lactation animals may provide a more up to date status of the dairy herd [25]. The recommendations within CHeCS for YS screening in the UK are largely based on the work conducted in Denmark and the USA by Houe 1992, Houe 1994 and Houe et al. 1995 [27–29]. The first study determined that herds ranging in size from 96 to 324 animals (with a mean of 135) could be grouped into BVDV infected and BVDV-free herds by testing only three or five YS 6–18 months of age [27]. In a further study, forty-two herds were investigated ranging in size from 40 to 157 animals and in a detailed survey of seven of these herds, low levels of YS seroprevalence were found to correlate with the absence of PI animals (0–1/10 YS Ab positive at the spot test) and high levels of YS seroprevalence correlated with the presence of PIs (8–10/10 YS Ab positive at the spot test) [28]. An additional test available for milk samples is bulk milk PCR (BM PCR) which enables the screening of bulk tank milk for the presence of PI animals. This allows the user to establish the status of the milking animals that contribute to the sample, but not necessarily the entire herd. Initially BM PCR was reported to be sensitive enough to detect one PI animal in 160 milking animals [30], but more recently this has increased to an upper limit of 300 recommended by many diagnostic laboratories [31] with further laboratories reporting a higher diagnostic sensitivity allowing this upper limit to increase further to 1000 animals (Karen Bond, National Milk Laboratories, Personal Communication).

This paper examines and discusses the use and practical implications of BM Ab, YS Check Tests and BM PCR for determining the presence or absence of PI animals using data collected from 26 working UK farms involved in a pilot BVDV eradication programme [32] where average herd sizes were 343 (interquartile range: 224 to 431) and 98 (interquartile range: 58 to 121) animals for the dairy and beef units respectively.

Methods

Farm recruitment began in April 2006 and was largely complete by April 2007. The farms involved and further details of their recruitment are described by Booth & Brownlie 2012 [32]. Farm ID numbers used throughout this manuscript are consistent with those used by Booth & Brownlie 2012 [32] and Booth et al. 2013 [33] enabling cross referencing between publications. All samples were collected by veterinary surgeons as part of the routine infectious disease surveillance on the farms involved and the herd owners gave signed consent for the data collected to be reported anonymously.

Upon recruitment to the pilot BVDV eradication programme, each farm was screened to determine herd status (BVDV-infected/BVDV-free) using a combination of BM Ab, BM PCR and YS Check Tests. Throughout this study, Check Tests consisted of 10 animals of approximately 9 months of age (range 6–18 months, with a preference of 9–12 months) from each separate management group on the farm. If fewer than ten animals were available in the age range required for the Check Test then all available animals were sampled. All samples were submitted to and tested by the Animal Health and Plant Agency (AHPA) via AHPA Starcross. BM samples were submitted in universal containers with Bronopol preservative (AHPA, Weybridge) and blood samples were submitted in heparinised vacutainers. AHPA test codes for the BVDV BM Ab ELISA and PCR were TC0123 and TC0709 respectively. All BM Ab results were reported as Optical Density (OD) ratios and interpreted such that OD ratios <0.1 = negative, 0.1–0.35 = low positive, 0.35–0.7 = mid positive and >0.7 = high positive. BM PCR results were reported as either positive or negative. All blood samples tested for Ab (AHPA test code TC0390) were reported as OD ratios and interpreted such that <0.2 = negative and > 0.2 = positive. Further details of the laboratory tests used in this paper including test sensitivity, specificity and further details regarding cut off values have been described previously by Booth and Brownlie 2012 [32]. Since the data presented in this paper was collected the TC0390 and TC0123 ELISA tests have been superseded and it should be noted that the stated cut-off values are not applicable to the ELISA tests currently offered by the AHPA.

Following each initial screen, the results were assessed in order determine herd exposure and whether BVDV was likely to be active on the farm and thus whether PI animals were likely to be present. For the purposes of this project, active infection based upon the initial screens was defined cautiously as any herd with >1/10 Ab positive YS and/or a positive BM PCR. Where evidence of exposure to BVDV indicated that the infection was likely to be active, a whole herd test (WHT) was performed where the entire herd was blood sampled in order to identify any PI animals present. Due to the longevity of the immune response following field infection with BVDV [7, 34] high BM Ab levels alone were not deemed sufficient to advise WHTs in the absence of a positive BM PCR or Ab positive YS and farms with this combination of initial results were subjected to further regular surveillance (described below). In beef herds, initial herd screens consisted solely of Check Tests. The laboratory tests and testing regimes used for WHTs have been described previously by Booth and Brownlie 2012 [32]. If identified, PIs were either culled or retained on the farm of origin; this was the farmer's choice. Where PI animals were identified, the recommendation was to cull them - sale of PI animals (unless direct to slaughter) was not permitted as a condition of membership of this study.

In herds where initial screening did not indicate an active BVDV infection, the farmer was given the choice of either blood sampling the whole herd to confirm the accuracy of the screen, or implementing regular herd surveillance. If a WHT was selected, once it was confirmed that there were no PI animals within the herd, the farm implemented regular herd surveillance from that point onwards. Within this study, regular herd surveillance consisted of BM Ab testing (at least quarterly), BM PCR testing (at least yearly) and Check Tests performed at least once a year. In herds undergoing regular surveillance, if active infection (as defined above) was suspected at any point, surveillance frequency was initially increased meaning that a further Check and BM Ab test was performed within 4–8 weeks of the preceding one. In the event that active infection was still suspected, a WHT was performed to search for PIs. Again, in beef herds, surveillance consisted solely of Check Tests.

All farms with data reported in this manuscript were monitored for at least 3 years (the majority joined the study in 2006/2007 and remained members until it ended in 2014), thus the data collected accurately reflects the number of PI animals present on each of the study farms over the period presented. On farms that did not undergo WHTs because ongoing herd surveillance indicated that there was no source of BVDV exposure, the assumption was made that no PI animals were present at any time.

For herds that underwent WHTs and had full BM Ab and Check Test results from their initial screen, Receiver Operating Characteristic (ROC) curves were generated in SPSS (SPSS Inc., Chicago) to compare the ability of BM Ab and Check Tests to determine the presence of PI animals in a herd. The optimal cut-off value, sensitivity and specificity of each test was then determined. Using logistic regression, predicted probability curves were produced in 'R' (The R Foundation for Statistical Computing, 2008) using "xyplot <lattice>" for BM Ab and YS Check Tests to illustrate how the probability of identifying a herd containing a PI(s) alters according to the results of the initial screens.

For some herds there were considerable delays between the initial herd screen and the WHT. This consisted of time needed to persuade the farmer to undertake a WHT and agreement on a suitable date (at times, harvesting or silaging meant a considerable delay was incurred). In order to investigate the effect of this delay on the accuracy and relevance of the initial herd screens, BM Ab and BM PCR results from the day of the WHT were collected. In order to simulate a YS Check Test as if performed on the day of the WHT, ten Ab results from YS aged as close to 9–12 months as possible were randomly selected from those sampled during the WHT. The YS chosen were selected using a random number generator in Microsoft Excel (2007) which was used to select ten animals from a list of those present on the day of the WHT arranged in age order starting with the animal closest to 9 months of age on the day of the WHT. Where insufficient animals were present within the 9–12 month range, the range was expanded to 7–13 months. The ROC and probability curve calculations described above were then repeated using data from the day of the WHT in effect removing any delay between conducting a herd screen and acting on the results.

Results

Twenty-six farms (Table 1) from the original 41 study members described by Booth and Brownlie 2012 [32] had results for both an initial herd screen to determine BVDV status and either whole herd tests (WHT) to identify PI animals or sufficient surveillance (described above) to determine that PI animals were unlikely to be present during the study period. Individual details of the 26 farms analysed in this manuscript are shown in Table 1. The group consisted of three beef herds and 23 dairy herds. Average herd sizes were 343 (interquartile range: 224 to 431) and 98 (interquartile range: 58 to 121) animals for the dairy and beef units respectively. Of the farms involved, the majority, 20/26 (77 %), were vaccinating against BVDV using either Pregsure BVD (Pfizer Animal Health, UK) or Bovilis BVD (MSD Animal Health, Milton Keynes, Buckinghamshire) (Table 1). All vaccines were administered according to the datasheet recommendations from the manufacturer with the primary course completed at an appropriate time prior to first service; a minimum of 14 days and 4 weeks for Pregsure and Bovilis BVD respectively.

With the exception of Farms 2, 5 and 15, all farms reported in this manuscript reared homebred youngstock on the farm where they were born.Farms 2, 5 and 15 removed animals at, or shortly after, weaning to separate units where they were reared to return to the home farm for first calving. Farms 5 and 15 only reared their own animals at these separate sites. Farm 2 reared bull calves and non-replacement heifers on the main farm unit as beef animals, however they also utilised a heifer rearer for their replacement dairy heifers. These animals were moved to the heifer rearer after weaning where they were kept with animals from another farm with no attempt to separate them. Once at service age, a bull was introduced and the heifers returned to the main farm to calve.

The results of the initial herd screens for all 26 farms are presented in Fig. 1 alongside the number of PIs found in each herd at the WHT. The total number of PIs identified on each farm and the initial number found at the WHT are shown in Table 1. With the exception of Farm 2, in all herds where PIs were identified, at least one of those PIs was present from the outset of the study. For Farm 2, the initial herd screens identified 40 % seroconversion at the YS Check Test (conducted at the heifer rearing unit) and a mid-positive BM Ab titre.

Table 1 Study Farm Details

Farm ID	Dairy/ Beef	Vaccinating over study period?	Suspicion of active infection following initial screen?^	Was whole herd testing performed?	PIs present at WHT	Total number of PIs identified	Time between initial herd screen and WHT (Months)	Age range of random sample (Months)	Herd size at recruitment
1	Dairy	Y*	Y	Y	2	3	11	9–12	349
2	Dairy	N	Y	Y	0	12	7	9–12	542
3	Dairy	Y*	N	Y	0	0	10	9–12	450
5	Dairy	Y*	Y	Y	2	2	7	9–12	527
6	Beef	Y*	N	N	-	0	-	-	61
7	Dairy	Y*	Y	Y	0	0	9	9–12	381
8	Dairy	Y*	Y	Y	0	0	3	11.5–13	381
11	Dairy	Y*	N	N	-	0	-	-	217
13	Dairy	N	N	N	-	0	-	-	201
15	Dairy	Y*	N	Y	1	1	9	7.7–12	633
17	Beef	Y*	N	N	-	0	-	-	180
18	Dairy	Y*	Y	Y	4	6	9	10–12.6	307
19	Dairy	N	Y	Y	1	1	5	9–12	214
20	Dairy	Y*	Y	Y	3	3	4	9–12	309
21	Beef	N	N	N	-	0	-	-	54
24	Dairy	N	N	N	-	0	-	-	286
25	Dairy	Y*	N	Y	0	0	9	9.3–13.5	300
26	Dairy	Y*	Y	Y	2	3	5	8.5–12.3	170
27	Dairy	Y*	N	Y	1	1	11	9–12	192
28	Dairy	Y+	Y	Y	0	0	12	9–12	335
29	Dairy	Y+	Y	Y	0	0	12	8–11	412
30	Dairy	Y+	N	N	-	0	-	-	360
37	Dairy	Y+	Y	Y	3	5	2	10.1–12.3	402
38	Dairy	Y+	Y	Y	2	2	3	8.3–10.9	542
39	Dairy	N	N	Y	0	0	19	9–12	143
40	Dairy	Y*	Y	Y	5	9	1	9–12	230
				Mean			8		315
				Max			19		633
				Min			1		54

Details of the farms analysed in this manuscript: Farm type, vaccination status (* vaccinating using Pregsure BVD and + vaccinating using Bovilis BVD), Suspicion of active infection upon initial screening (^ according to the cautious initial interpretation), Total number of PIs identified during the study, Whether a whole herd test was performed, Total number of PIs identified at the WHT (where performed), The delay between initial screening and WHT if conducted, Age range of animals selected for the random screening at the WHT and herd size at recruitment. Farm numbers can be cross referenced with Booth and Brownlie 2012 [32] and Booth et al. 2013 [33]

The seropositive youngstock initiated a WHT on this farm at both the main farm where the milking animals were kept and the heifer rearing unit. Permission was not given to test the animals at the heifer rearing unit that were mixed with those from Farm 2. No PIs were identified at the Farm 2 WHT in year 1, however, due to the seroconversion noted in the youngstock, newborn animals were tested as they were born onto the farm and the first PI was identified in year 2 of the study as a 'trojan' PI born to an non-PI heifer returning from a heifer rearing unit where mixing with animals from multiple sources had occurred [32]. Indeed the subsequent two PIs detected on Farm 2 were also born to non-PI heifers returning from the heifer rearer. For these reasons, within the ROC and probability curve analysis described below, Farm 2 was treated as positive when analysing Check Test results, but negative in the BM analyses. Following the birth of the third PI the adult herd had been infected long enough to produce a fourth PI from a 3rd lactation animal. In total, twelve PIs were confirmed on this farm (with two antigen tests four weeks apart) with a further 9 that either died or were culled before they could be confirmed with a second antigen test [32].

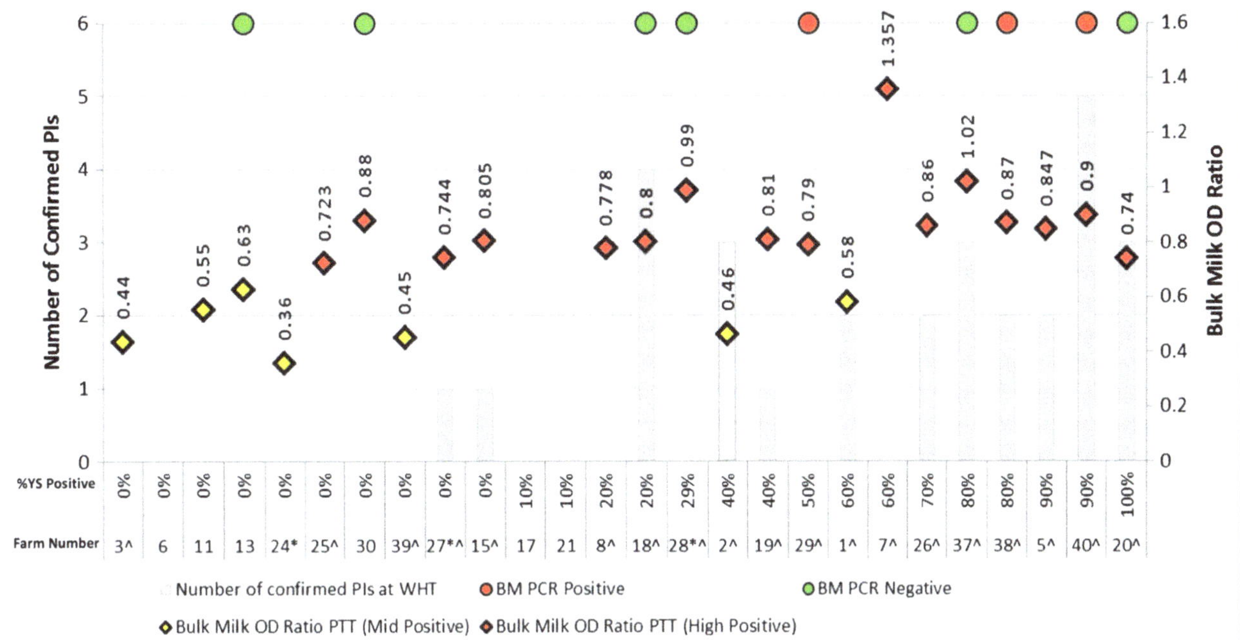

Fig. 1 The results of the initial herd screens of the 26 study farms. The figure illustrates the results of the initial herd screens for the 26 farms presented in this manuscript. ^ indicates the 19 herds where WHTs were performed. The left y-axis displays the number of PIs identified at the WHT (where performed) depicted for each farm by the solid bars - the Farm 2 bar is hashed to represent in-utero PI animals. For the 7 farms that did not undertake WHTs because their regular surveillance did not indicate exposure, it was assumed that no PIs were present. The right y-axis displays the BM Ab OD ratio. Along the x -axis, each column represents an individual farm stating both the percentage of antibody positive youngstock (YS) at the Check Test (* all farms except Farms 24, 27 & 28 submitted ten animals who presented 9, 8 & 7 youngstock respectively) and the Farm number. (Graph Key attached to the bottom of Fig. 1)

Seven of the 26 study herds (the three beef units; Farms 6, 17 & 21 and four of the dairy units; Farms 11, 13, 24 &30) returned regular surveillance results which indicated that they were unlikely to have been exposed to a source of BVDV infection. These herds were excluded from the statistical analysis because they did not undertake WHTs. The remaining 19 herds undertook WHTs however two of these, Farms 27 and 28, were excluded from the statistical analysis since only 7 & 8 YS respectively were available in the appropriate age range for their initial screens. Therefore, of the 26 study herds, a total of 17 were suitable for ROC and probability curve analysis since they had both 10 YS and a BM Ab sampled at initial recruitment and also conducted a WHT. Furthermore, it is only these 17 farms that contribute to the sensitivity and specificity results stated below.

Bulk milk antibody analysis

BM Ab titres on the 23 dairy farms observed in Fig. 1 range from mid to high positive. Interestingly, high positive BM Ab results occur on six farms where no PIs were identified (Farms 7, 8, 25, 28, 29 & 30) and mid positive results occur on two farms where PIs were identified (Farms 1 and 2). On Farm 2 however, it should be noted that the dairy herd would not have had contact with PI animals by this point. Of the dairy farms observed in Fig. 1, none began the study with low positive or negative BM Ab titres, yet 11/23 (48 %) had no PI animals identified. The BM Ab ROC curve in Fig. 2 and its coordinates (Table 2) indicate that the optimal cut-off point for distinguishing whether or not a herd contained a PI animal using the BM Ab titre is achieved at an OD ratio of 0.7950 units with a sensitivity and specificity of 80.00 % (44.39–97.48 %) and 85.71 % (42.13–99.64 %) respectively (95 % confidence intervals quoted). Of the 17 farms analysed in Fig. 2, using a BM OD ratio cut off of 0.7950 there were two false negative farms (i.e. would be declared free of PIs when they were present) (Farms 1 & 20) and one false positive (i.e. would be thought to contain PI animals when there were none present) (Farm 7). For Farm 2, the BM Ab OD ratio of 0.46 units and interpretation that there were not PI animals present can be considered correct for the adult dairy herd given that it had not been exposed to PI animals at this point. Fig. 3a displays the predicted probability curve for BM Ab OD ratio calculated for the 17 farms that were used for the ROC curve analysis. The curve shows a gradual but limited increase in probability of detecting a PI as BM Ab OD ratio increases.

Check test analysis

In Fig. 1, antibody positive YS identified at the initial Check Test are presented as a percentage of the total

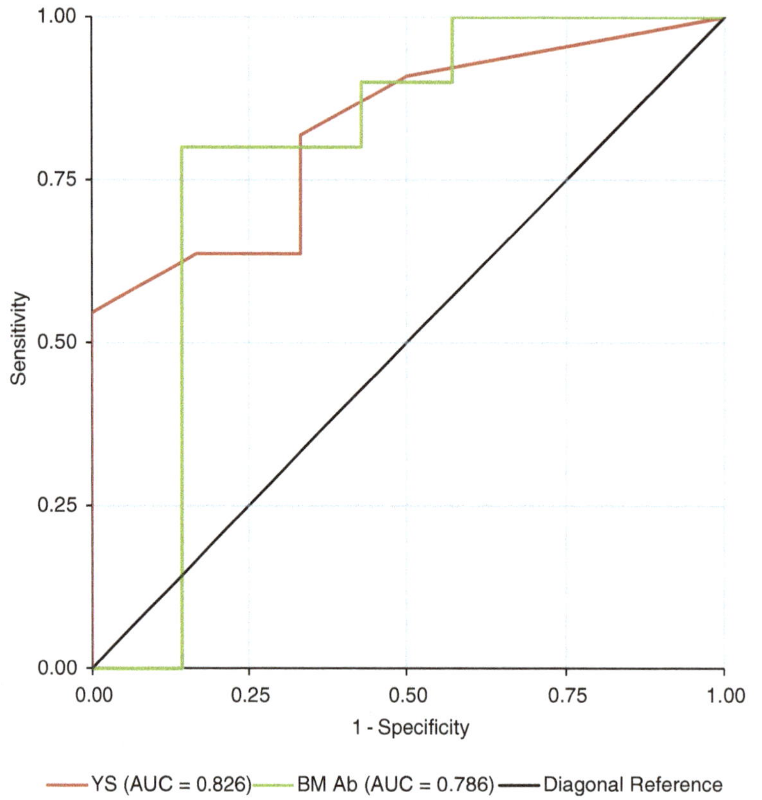

Fig. 2 ROC curves demonstrating the performance of BM Ab and YS Check Tests as predictors of the presence of a PI animal(s) where delays occurred between the initial herd screen and the WHT. The ROC curves were generated from the seventeen dairy farms in Fig. 1 with full initial herd screens and WHT results. YS = Youngstock check testing and BM Ab = Bulk milk antibody testing. Farm 2 is considered negative for BM Ab analysis and positive for YS analysis. AUC = Area under curve. Coordinates of the curves are given in Table 2

tested since Farms 24, 27 and 28 only presented 9, 8 and 7 YS respectively in the required age range. All other farms had ten YS available for the Check Test. When observing the number of antibody positive YS identified on initial herd screening in Fig. 1, the number of antibody positive YS increases with the likelihood of identifying a herd containing a PI animal. Using the Check Test results alone and the original cut-off (defined above) of >1/10 Ab positive animals signifying a BVDV infected farm results in two false negative farms (Farms 15 & 27) and four false positives (Farms 7, 8, 28 & 29). Analysing the Check Test results from the seventeen dairy herds that undertook WHTs, with Farm 2 included as infected since three heifers amongst the group sampled were carrying in-utero PIs at this point, the coordinates (Table 2) of the ROC curve in Fig. 2, indicate two potential cut off points. The first uses a cut off of 5.5/10 YS BVDV Ab positive (thus interpreting results with either ≤5 YS or ≥6 YS as herds that are negative and positive respectively) and returns a sensitivity and specificity of 63.64 % (30.79–89.07 %) and 83.33 % (35.88–99.58 %) respectively (95 % confidence intervals quoted). The second cut off is at 3/10 YS BVDV Ab positive and returns a sensitivity and specificity of 81.82 % (48.22–97.72 %) and 66.67 % (22.28–95.67 %) respectively (95 % confidence intervals quoted). The second cut off of 3/10 YS BVDV Ab positive may be preferred marginally due to the slightly higher sensitivity and thus smaller likelihood of mis-diagnosing a farm that contains a PI as negative. However, neither cut-off can be considered ideal. The predicted probability curve in Fig. 3b is generated for the same seventeen herds that underwent ROC curve analysis and shows how the probability of detecting a herd containing a PI increases as the number of Ab positive YS out of ten tested increases.

Bulk milk PCR analysis

Not all dairy farms had BM PCR results at this stage of the study - this was simply due to the fact that the test was not offered by the AHPA at the start of farm recruitment. With the exception of Farm 29, all farms returning a positive BM PCR result in Fig. 1 had PI animals identified; Farms 38 & 40, where Farm 38 had a milking PI animal yet Farm 40 did not. There were also six farms which were BM PCR negative and of these, three (Farms 18, 37 & 20) contained PIs none of which

Table 2 Coordinates of the ROC curves in Fig. 2

Test result variables	Positive if greater than or equal to a or b[#]	Sensitivity	1 - Specificity
[a]Number of Youngstock antibody positive out of ten tested with delays between the YS screen and whole herd test	−1.00	1.000	1.000
	1.00	.909	.500
	3.00	.818	.333
	4.50	.636	.333
	5.50	.636	.167
	6.50	.545	0.000
	7.50	.455	0.000
	8.50	.273	0.000
	9.50	.091	0.000
	11.00	0.000	0.000
[b]Bulk milk antibody OD ratio with delays between the BM Ab screen and whole herd test	−.5630	1.000	1.000
	.4440	1.000	.857
	.4575	1.000	.714
	.5225	1.000	.571
	.6520	.900	.571
	.7315	.900	.429
	.7590	.800	.429
	.7840	.800	.286
	.7950	.800	.143
	.8025	.700	.143
	.8075	.600	.143
	.8285	.500	.143
	.8535	.400	.143
	.8650	.300	.143
	.8850	.200	.143
	.9600	.100	.143
	1.1885	0.000	.143
	2.3570	0.000	0.000

[#] The smallest cutoff value is the minimum observed test value minus 1, and the largest cutoff value is the maximum observed test value plus 1. All the other cutoff values are the averages of two consecutive ordered observed test values

were of milking age. The ages of the PI animals identified on these farms (and throughout this study) are discussed in Booth and Brownlie 2012 [32].

At this point, Farm 29 is worth highlighting since it returned a positive BM PCR result, a high positive BM Ab result and 5/10 animals Ab positive at the Check Test yet no PIs were identified on the premises either at the WHT or in the testing following this. The delay between the initial herd screens and WHTs to identify PI animals is shown in Table 1 and ranges from 1–19 months a mean interval between initial screens and WHTs of 8 months. The delay for Farm 29 was 12 months between the initial screens presented in Fig. 1 and the WHT to identify PI animals. The impact of these delays between initial screens and what was actually found at the WHT could be considerable and therefore the initial screens presented in Fig. 1 may not accurately reflect herd status at the time the WHT was conducted. For this reason, further data was collected from the farms presented in Fig. 1 so that for those farms that underwent a WHT, the equivalent of an initial screen was generated for that day meaning that the delay between the initial screen and the WHT was effectively 0 days. The age groups of the randomly selected YS are detailed in Table 1 and for most farms, it was possible to select animals from within the 9–12 month range. The YS selected for Farms 18 and 37 were just outside of the 9–12 month range whilst for Farms 8, 15, 25, 26 and 38 it was necessary to extend the range to 7–13 months. These data are presented in Fig. 4. Farms 6, 11, 13, 17, 21, 24, & 30 did not undergo WHT since the results of their regular surveillance (BM Ab, BM PCR and Check Tests) did not justify testing all stock. These farms are included in Fig. 4 and the data presented for each are the year 2 Check Test results which coincided with a quarterly BM Ab and yearly PCR test (where available).

Bulk milk antibody analysis without delay

In the 23 dairy farms represented in Fig. 4 high positive BM Ab results occur on nine infected farms and two that are BVDV-free (Farms 29 & 30), mid positive BM Ab results occur on six BVDV-free farms, Farm 2 and two infected farms (Farms 15 & 19). Finally, three BVDV-free dairy farms in Fig. 4 now return low positive BM Ab results. On Farm 2, the dairy herd was still unexposed at this point hence should be considered a negative herd when assessing BM Ab. Analysis of BM Ab results from the 17 dairy herds that undertook WHTs produced the ROC curve in Fig. 5, the coordinates of which (Table 3) demonstrate that the optimal BM Ab OD ratio cut-off for distinguishing whether a herd contained a PI animal or not occurs at 0.7 OD ratio units. This also coincides with the AHPA cut off for a high positive result using this test. With this dataset, a sensitivity and specificity of 80.00 % (44.39–97.48 %) and 85.71 % (42.13–99.64 %) respectively (95 % confidence intervals quoted) was achieved. The predicted probability curve in Fig. 3c shows better distinction using BM Ab OD ratio at this cut off compared to Fig. 3a.

Check test analysis without delay

Analysing the farms in Fig. 4 using the cautious cut off (defined earlier) of >1/10 Ab positive youngstock results in the false positive misclassification of Farms 13 & 29 and the false negative misclassification of Farm 15. Analysing the Check Test results from the 17 farms produces the ROC curve in Fig. 5, the coordinates of which

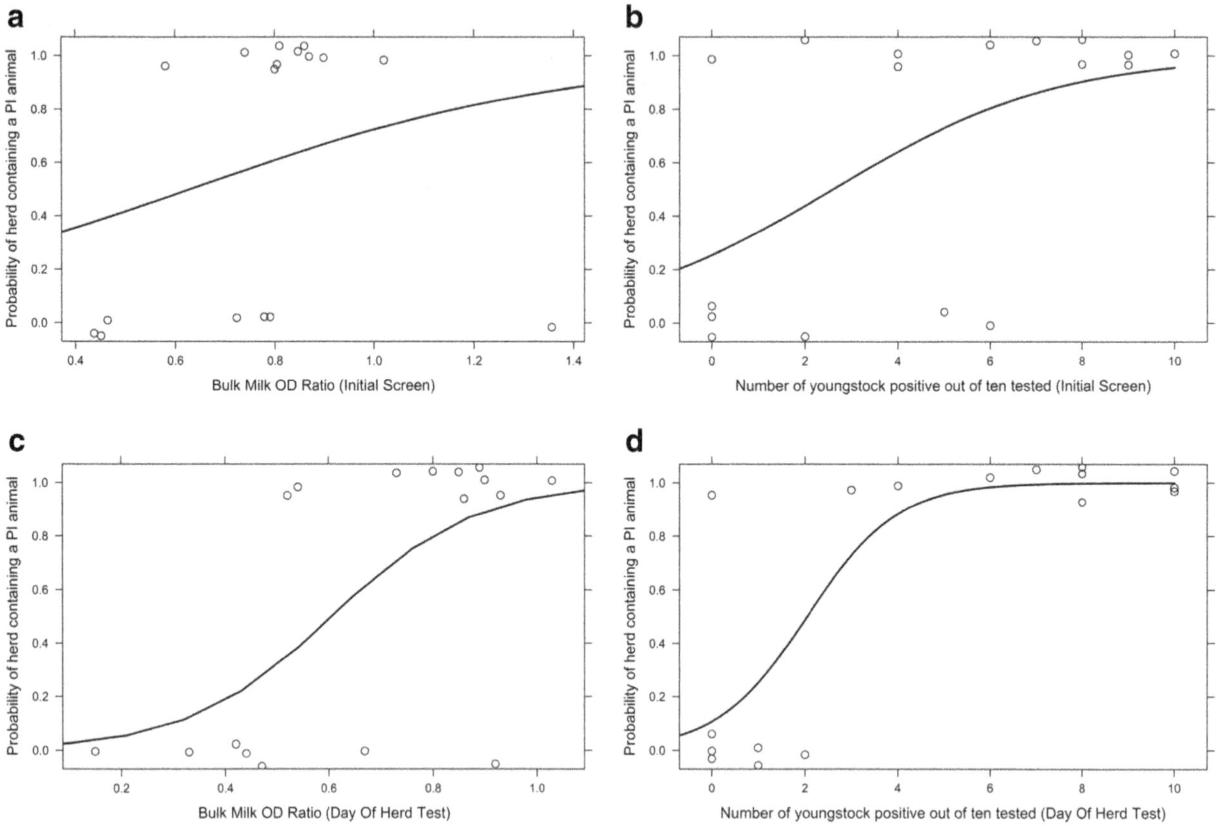

Fig. 3 Predicted Probability Curves indicating the probability of a study herd containing a PI animal with the stated BM Ab and YS Check Test results. Data points demonstrating PI presence are 'jittered' around 0 or 1 on the y axis to show the number of farms that contribute to the curve at each point, **a** BM Ab OD Ratio and the probability of identifying a herd containing a PI animal with delays between performing the BM test and the WHT; **b** Number of YS positive out of the ten tested and the probability of identifying a herd containing a PI animal with delays between performing the Check Test and the WHT; **c** BM Ab OD Ratio and the probability of identifying a herd containing a PI animal with no delay between performing the BM test and the WHT; **d** Number of YS positive out of the ten tested and the probability of identifying a herd containing a PI animal with no delays between performing the Check Test and the WHT. Farm 2 is considered negative for BM Ab analysis and positive for YS Check Test analysis

(Table 3) indicate that an optimal cut-off point occurs at 2.5/10 animals with a sensitivity and specificity of 90.9 % (58.72–99.77 %) and 100 % (54.07–100 %) respectively (95 % confidence intervals quoted). Utilising this cut-off would result in the false negative misclassification of one farm (Farm 15) from the 17 analysed in Fig. 5. The predicted probability curve in Fig. 3d illustrates an improved distinction compared to the tests explored in Fig. 3a-c.

Bulk milk PCR analysis without delay

At this point all farms that returned a positive BM PCR in Fig. 4 had PIs identified (Farms 1, 19 & 38) where Farms 1 & 38 contained milking PIs, but Farm 19 did not. Five out of the twelve farms that returned BM PCR negative results also contained PI animals at the time the BM was sampled (Farms 20, 26, 27, 37 & 40) however, none of the PI animals on these farms were of milking age.

These results indicate that when herd screening is performed with no delay between that and the WHT to identify PI animals, the sensitivity and specificity of both the BM antibody and the YS cohort screens for determining PI presence in a herd are improved. Furthermore, the predictive probabilities of both tests improve when the delay between screening and WHTs is removed.

Discussion

This manuscript presents BM Ab, BM PCR and YS Ab (Check Test) data from twenty six working farms that were recruited to a pilot BVDV eradication programme. Although there is general agreement in the current literature that BM Ab levels and youngstock Check Tests are appropriate ways to assess BVDV status at the herd level [20, 25–27, 35, 36], there remains some confusion about their reliability and the impact of historic infection and vaccination. We discuss these issues in the context of our study herds below.

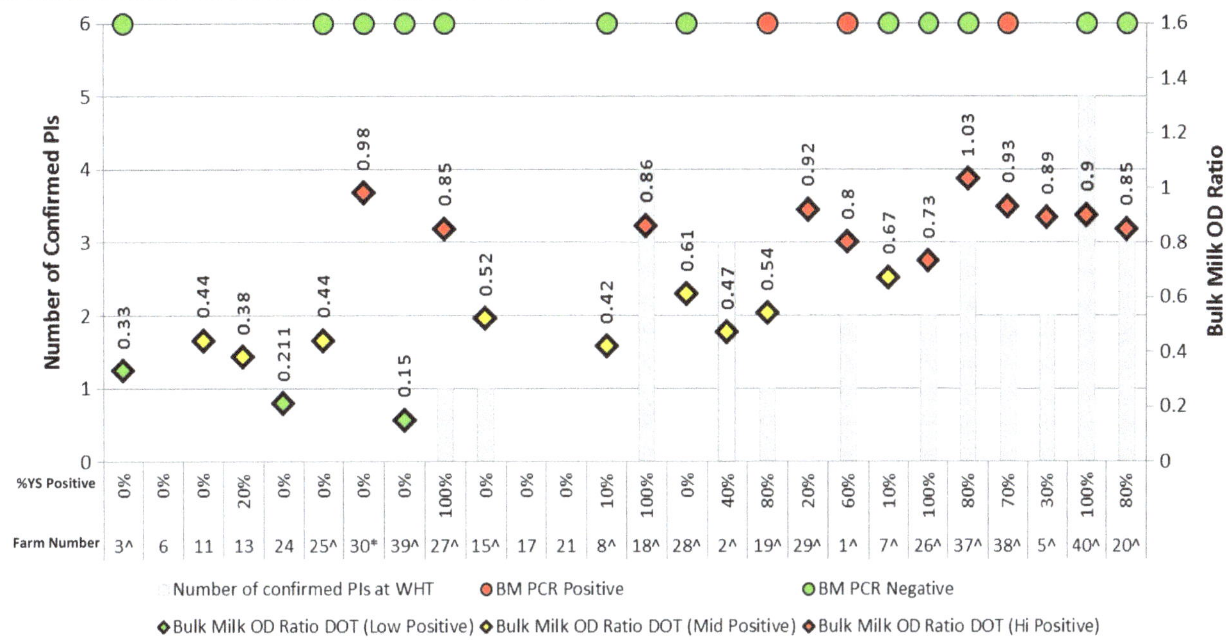

Fig. 4 The results of simulated initial herd screens of the 19 herds that undertook WHTs and the Year 2 screens of the remaining 7 herds. The figure illustrates the results of the simulated initial herd screens for the 19 herds that undertook WHTs and the Year 2 screens for the 7 herds that did not. ^indicates that the farm underwent whole herd testing. The left y-axis displays the number of PIs identified at the WHT (where performed) depicted for each farm by the solid bars – the Farm 2 bar is hashed to represent in-utero PI animals. For the 7 farms that did not undertake WHTs because their regular surveillance did not indicate exposure, it was assumed that no PIs were present. The right y-axis displays the BM Ab OD ratio. Along the x -axis, each column represents an individual farm stating both the percentage of antibody positive youngstock (YS) at the Check Test (* all farms submitted ten animals except Farm 30 where only 9 animals of appropriate age were available) and the Farm number. (Graph Key attached to the bottom of Fig. 4)

The recruited herds were predominantly located in Somerset, UK and the large proportion of dairy herds in the study population reflects the region from which herds were sampled. The average study herd size appears large when compared to the average herd size of 120 animals quoted by Defra for the Taunton area at the time of recruitment [37]. However, the Defra figure may be falsely low as this will include small holdings and in the authors' experience, the herd sizes represented here are typical of the range of farms in this region.

Within the analyses conducted, the wide 95 % confidence intervals quoted throughout highlight that there is a degree of uncertainty surrounding the results and thus the figures should be used with caution. This is largely due to the small sample size of 17 herds which underwent statistical analysis. Despite these potential limitations, the improvements demonstrated in sensitivity and specificity and also in the ROC and probability curves by removing the delay between screening a herd and acting on those results are substantial and worth discussion. Of the 26 herds that were recruited, only 17 were suitable for statistical analysis since BM and YS samples were collected in addition to individual animal samples to determine the BVDV status of all animals on this subset of farms. Two farms that had conducted a WHT were deemed unsuitable for inclusion in the statistical analysis as ten animals were not available for the initial YS screens and are included in Figs. 1 and 4. The seven remaining farms did not undergo WHTs and were therefore excluded from the statistical analysis, but are presented for completeness in Figs. 1 and 4. Whilst we believe that the regular surveillance would have detected infection on these farms over, and beyond, the course of the study period analysed here, it remained an assumption that they were BVDV free and so it would have been inappropriate to influence the ROC, probability, sensitivity and specificity analyses on this basis. This point is especially pertinent when we consider that Farm 15 returned a mid positive BM Ab titre at the WHT and also negative YS results in both the initial Check Test and the simulated Check Test from the WHT yet a PI was present throughout the period. This was primarily due to herd dynamics and is discussed in relation to this herd's results below.

The Use of BM Ab Tests for determining Herd BVDV Status

BM Ab tests provide an initial, rapid and cost effective assessment of the BVDV status in dairy cattle at the herd level. It is largely accepted that high BM Ab levels correlate with a high probability of the presence of a PI.

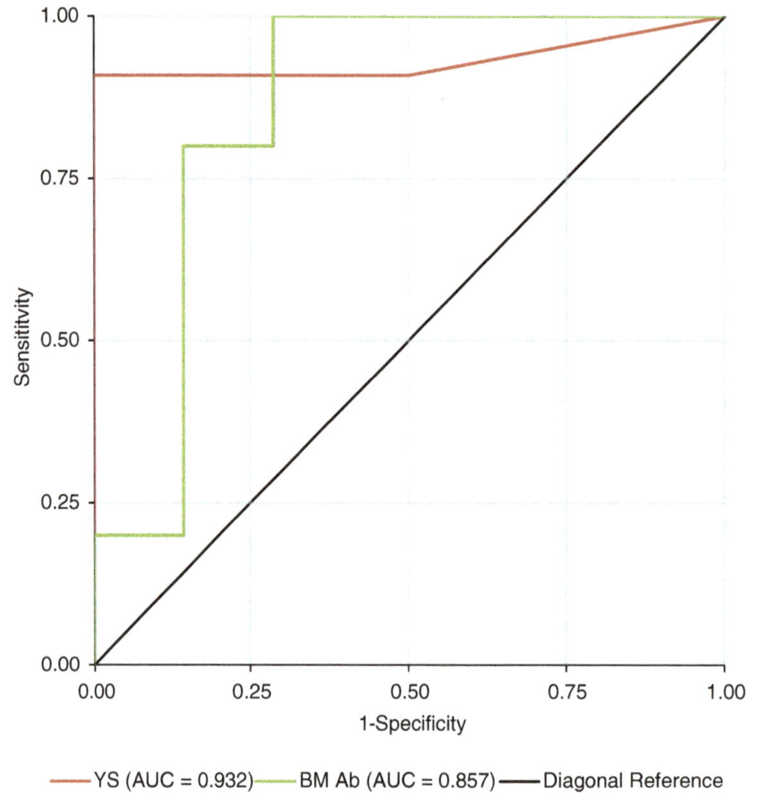

Fig. 5 ROC curves demonstrating the performance of BM Ab and YS Check Tests as predictors of the presence of a PI animal(s) where there is no delay between the initial herd screen and the WHT. The ROC curves were generated from the seventeen dairy farms in Fig. 4 with full initial herd screens and WHT results. YS = Youngstock check testing and BM Ab = Bulk milk antibody testing. Farm 2 is considered negative for BM Ab analysis and positive for YS analysis. AUC = Area under curve. Coordinates of the curves are given in Table 3

However, we found several instances where herd BVDV status would have been incorrectly classified if based on an interpretation of the BM Ab result alone. The sensitivity and specificity of the test throughout this study was 80.00 and 85.71 % respectively however, the probability curves indicate that it is difficult to accurately define a cut-off point where we can say with reasonable certainty that a herd does or does not contain a PI animal. The removal of the delay between the BM Ab screening and further investigations does improve the confidence that one can have in the interpretation of the test but as a sole screening method, the usefulness of BM Ab remains limited in many circumstances. We know that both historic BVDV infections and herd vaccination can have a significant effect on correct BM Ab interpretation [33, 38] and both of these factors are likely to have played a significant role in the test results and interpretation here.

BVDV antibodies following a natural infection may persist for 3 or more years in an individual animal [7, 34]. When one considers that BM Ab assesses all milking animals including the oldest in the herd, it explains why it can take in excess of 1000 days for BM Ab to show any real decrease following the removal of PI animals [38]. Booth et al. 2013 [33] demonstrated that Farms 13 & 24, which were BVDV-free farms that neither vaccinated nor experienced active BVDV infection during the study period experienced a slow yet consistent decrease in the BM Ab titre; in this type of herd BM Ab can be a sensitive surveillance test for monitoring for disease incursion into the milking group since an increase in Ab titre can be considered of significance. Vaccination can also confound interpretation of BM Ab results further. With the exception of Farms 13, 24 & 39, all BVDV-free dairy herds in this study vaccinated during the period data was collected (Table 1) and all began the study with mid to high BM Ab titres (Fig. 1). In the longitudinal analysis of BM Ab trends by Booth et al. 2013 [33], BVDV-free vaccinating herds did not demonstrate a clear decrease in BM Ab. The fact that the majority of the study herds were utilising BVDV vaccines, in combination with the likelihood that many of them may have experienced historic infection and thus contain older seropositive animals may explain the higher level of misclassification of false positive farms observed with the BM Ab tests compared to YS Check Tests. In

Table 3 Coordinates of the ROC curves in Fig. 5

Test result variables	Positive if greater than or equal to a or b[#]	Sensitivity	1 - Specificity
[a]Number of youngstock antibody positive out of ten tested without delay	−1.00	1.000	1.000
	.50	.909	.500
	1.50	.909	.167
	2.50	.909	0.000
	3.50	.818	0.000
	5.00	.727	0.000
	6.50	.636	0.000
	7.50	.545	0.000
	9.00	.273	0.000
	11.00	0.000	0.000
[b]Bulk milk antibody OD ratio without delay	−.8500	1.000	1.000
	.2400	1.000	.857
	.3750	1.000	.714
	.4300	1.000	.571
	.4550	1.000	.429
	.4950	1.000	.286
	.5300	.900	.286
	.6050	.800	.286
	.7000	.800	.143
	.7650	.700	.143
	.8250	.600	.143
	.8550	.500	.143
	.8750	.400	.143
	.8950	.300	.143
	.9100	.200	.143
	.9250	.200	0.000
	.9800	.100	0.000
	2.0300	0.000	0.000

[#] The smallest cutoff value is the minimum observed test value minus 1, and the largest cutoff value is the maximum observed test value plus 1. All the other cutoff values are the averages of two consecutive ordered observed test values

practice, careful thought needs to be given to the application and timing of BM Ab testing in herds utilising BVDV vaccines and the use of this test may be deemed inappropriate in some systems. This is discussed in further detail by Booth and et al. 2013 [33].

Whilst vaccination and historic infection would explain those farms that returned high positive BM Ab results in the absence of PIs, it is more difficult to explain those herds (Farms 15 and 19) that returned only mid-positive BM Ab results in Fig. 4 when PI animals were present within the herd. For these farms, the reasons are likely due to the location of the PI animal(s) within their herds:

- For Farm 15, the lack of a high positive BM Ab result is harder to explain since the PI was of milking age. She was however a poor animal and often separated from the milking herd in a hospital pen due to illness.
- Farm 19 contained a PI animal of approximately 1 year of age and yet returned a mid positive BM Ab result. Upon initial interpretation, it may simply be possible that the PI did not have contact with the milking herd, however, analysis of the BM PCR results bring this interpretation into question since the herd was positive on BM PCR testing. It is difficult to provide an explanation for this combination of results especially given that at the initial screen for Farm 19 (conducted 5 months prior to the WHT), the herd returned a high positive BM Ab result (Fig. 1).

Farm 2 highlights the need for the practitioner to consider the epidemiological units that make up the herd structure in detail both before testing and before coming to a decision on the status of the herd based on the results of the tests. At both the initial herd screen and at the point of the WHT, the milking herd on Farm 2 had not been recently exposed to BVDV, hence the mid positive result. If a single assessment had been performed on a bulk milk sample, and decision reached that the herd was uninfected, then infection in the heifers would have gone unnoticed until subsequent BM Ab titres had increased. Our Check Test results indicated that there was likely exposure in the heifer group at the calf rearer and that this was almost certainly due to the biosecurity breakdown involving mixing with animals from another farm. The return of three heifers carrying PI foetuses to the main farm in combination with the farmers reluctance to cull PI animals once identified and the failure to engage the owner of the other animals at the heifer rearer in the scheme eventually resulted in infection of the main herd and an increase in the BM Ab titre to a high positive (data not shown). This farm is discussed in more detail by Booth and Brownlie 2012 [32].

The use of young stock (YS) check tests for determining herd BVDV status

The real value of YS antibody tests is that young animals will be seronegative if unexposed and maternally derived Ab (MDA) have waned; if they do have antibodies demonstrating an active immune response (in contrast to MDA), they will have been infected only within their short lifetime (e.g. 9 months). That would indicate a recent, or present, infection on the farm. Assessing animals at, or after, 9 months of age is a valuable time to initiate a YS screen as MDA will have has waned and it is typically pre-vaccination.

YS below six months of age should not be selected for Check Testing since MDA are highly likely to be present. Chamorro et al. 2014 [39] demonstrated that in extreme cases, BVDV MDA can take 11 months to wane and that MDA are commonly present up to 6 months of age; hence have the potential to confound interpretation of the test. In the authors' experience, testing animals at least 9 months of age, especially in beef suckler herds where colostral transfer of MDA is more reliable and higher quality than in dairy systems, provides a safer margin for ensuring that MDA is not present.

Whilst the majority of the herds in this study did vaccinate, they only initiated the primary course of vaccine prior to first service. The result was that we were able to assess the YS cohorts in these herds without the influence of vaccine. There is the possibility on Farms 8 and 25, that some of the older YS selected from the WHT in order to simulate a Check Test on that day could have been vaccinated as they were at the older end of the 7–13 month range, but in all the other herds, this was unlikely to have influenced the results. In this study, we achieved both a high sensitivity and specificity when assessing the ability of the Check Test to determine whether PI animals were present in the seventeen study herds analysed when the Check Test was simulated from the WHT results. Furthermore, the probability curves displayed an increasing ability to predict the presence of a PI animal within the study herds at the suggested cut off of 2.5/10 animals Ab positive once the delay between screening and the WHT had been reduced. Whilst we have sampled randomly within the animals tested at the WHT in order to generate the 'without delay' Check Test, it is important to mention that in the field, careful thought should be given to the contact of the cohort(s) tested with the rest of the herd and this is discussed in the context of Farm 15 below. Previous epidemiological studies have shown that when two or fewer YS of the ten tested are Ab positive, there are unlikely to be PIs present on a farm [27–29] and the optimal cut off from this manuscript of 2.5/10 YS Ab positive supports this finding, even in larger UK herds. The authors would urge readers to use some caution when applying this cut-off value in those herds that return 1/10 or 2/10 positive animals since this may be the start of seroconversion in a group, detection of a declining Ab response to vaccine or MDA. The most effective way to distinguish this is to re-test the same animals at least 28 days later to determine whether seroconversion is underway.

Farm 15 was consistently classified as negative when interpreting YS Check Tests when, in fact, it contained one adult PI animal. The explanation is that the PI animal was bought in as a heifer and, after taking considerable time to get in calf, finally produced a dead calf that was quickly removed. All calves born on Farm 15 were routinely raised on a different unit to adult stock and thus the adult PI animal never had contact with young animals and did not produce a live PI calf that was mixed with the YS. This again highlights the importance of considering the mixing of groups within a farm unit. If the cohorts selected for Check Tests have had reasonable exposure to the rest of the herd, the results may be extrapolated to provide an indication of herd status. However, cohorts of animals reared as isolated groups (or even in some cases as separate herds until first calving) are not suitable for selection for Check Tests to determine whole herd status but only group status. Furthermore, Farm 15 highlights the importance of testing purchased stock however, this animal was purchased before Farm 15 was recruited and was therefore only detected once the study commenced.

In a herd with a high positive BM Ab titre due to historic infection and/or vaccination, sampling cohorts of YS may be the only way to accurately obtain an up to date assessment of herd status and the predicted probability and ROC curves in Figs. 3d and 5 respectively support that this should provide a relatively accurate analysis. In beef herds, monitoring antibody levels in YS cohorts is the only method of assessing the herd hence it is re-assuring that this test performs well. In traditional cow:calf suckler units, Check Tests are potentially much more reliable than in dairy units since there is a greater degree of mixing of animals thus separate epidemiological groups are less likely to occur.

The use of bulk milk PCR for determining herd BVDV status

BM PCR provided a highly sensitive and cost-effective test for adult PI animals contributing to the milk tank. Unfortunately, our PCR data is incomplete as the test was not widely available at the time of recruitment of many farms.

Over the course of the study, our limited BM PCR results did identify 5 farms with a positive test; on two of these farms (Farms 1 & 38), PIs were found within the milking herd. On Farm 40, no PIs of milking age were identified and only two older animals had left the milking herd in the time between screening and the WHT. Whilst it is possible that either/both of these culled animals had been PI, another potential hypothesis for the positive BM PCR could be that a number of pregnant animals were carrying PIs and thus the level of virus circulating in the milking herd was sufficient to create detectable levels of viral RNA in the bulk milk. On Farm 19 a positive BM PCR was also returned at the whole herd test, but no milking PI animals were identified, nor were further PIs beyond the initial one identified in the youngstock so it would seem unlikely that this could be due to pregnant animals carrying PIs in this instance and may simply be due to contact between the milking animals and YS group containing the PI on this unit. This concept would be supported if the matching milk sample had returned a high positive Ab titre, but as described above, it is difficult to reconcile why this was not the case and a mid positive titre was returned. In contrast, Farm 29 returned a positive BM PCR result yet no PIs were identified at the WHT. On this farm it is highly likely that the delay in conducting a WHT (12 months) resulted in the death or removal of any PIs prior to identification. The fact that 5/10 YS on Farm 29 tested Ab positive at the initial screen and that the herd had a high positive BM Ab titre support the conclusion that PIs had been present on the farm.

It is important to note that BM PCR can only test those animals contributing to the bulk tank on the day the sample was taken. Our results show that some of the herds that did contain PIs did not return a BM PCR positive test result. PIs later identified on these farms had not contributed to the bulk tank. For this reason, provided the upper limit of milking contributors to a sample, as recommended by the testing laboratory, is not exceeded, a negative BM PCR is a reliable indication that any animals contributing to the BM tank, are not PI. A negative result is not, however, a reliable indication that there is not a PI(s) animal elsewhere in the herd.

The effect of delays between herd screens and whole herd testing

Apart from the 19 month delay between screening and WHTs on Farm 39, most herds were investigated within 8 months of their initial screen. Farm 39 consistently appeared BVDV-free in the regular surveillance conducted and so there was no urgency to perform a WHT in this herd so perhaps it skews the recorded delays unnecessarily. The mean interval is not ideal but in part reflects the reality in practice where results are awaited, reported to the farmer and then a period of discussion and organisation, often avoiding harvesting or silaging, has to occur in order to arrange whole herd testing. If the delay between screening the herd and taking action based on those results is prolonged, the results of the initial screens may not accurately reflect the BVDV status of the herd in question. The data presented here reflects this and in all cases shows an increase in the diagnostic accuracy and relevance of the screens once these delays are removed.

Farms 7, 8, and 29 had delays between initial screens and WHT of 9, 3 and 12 months respectively. Based upon the original YS criteria of >1/10 YS Ab positive indicating an infected herd all were classified as infected. Despite this, no PIs were identified on these three farms during the course of the study (Fig. 1). Farm 8 had a relatively low number of YS seropositive at the initial screen (2/10) which declined further to 1/10 YS Ab positive at the WHT three months later and so is unlikely to have been an infected herd. Whilst no PIs were found, it is highly likely that both farms 29 & 7 were infected due to the number of seropositive YS detected at the initial Check Test; 6/10 and 5/10 YS Ab positive for each farm respectively (Fig. 1). An infected status is further supported for Farm 29 since it returned a positive BM PCR result at the initial screen (Fig. 1). Due to the delays in conducting WHTs on these two farms, it is possible that any PIs had unknowingly been removed from these units prior to undertaking WHTs since both farms show less evidence of BVDV exposure in Fig. 4. Figure 4 demonstrates that, based on Check Tests, herd screens performed closer to the WHT would have indicated that herds 7 & 29 did not contain PI animals at the point of the WHT since their results were 1/10 and 2/10 YS Ab positive respectively (Fig. 4) – this is below the cut off of 2.5/10 animals indicated in the ROC curve in Fig. 5. The result for Farm 29 is further supported by the fact that it was also BM PCR negative at this stage. These results indicate that significant alterations in herd infection status may have occurred on Farms 7 and 29 in the time between the initial screen and WHT.

On Farms 15 & 27 the delay between the initial screens and whole herd tests was 8 & 9 months respectively. Both farms 15 & 27 bought PI animals shortly before the screens presented in Fig. 1 were performed (data not shown). It is evident that for Farm 27, the initial Check Test was likely performed before enough time had elapsed for seroconversion to have occurred within the group screened since continued surveillance on Farm 27 would have detected the presence of the PI animal, since the YS seroprevalence increased from 0/10 YS Ab positive in Fig. 1 to 10/10 YS Ab positive (Fig. 4). This further highlights the need to develop an ongoing surveillance plan. Farm 15 would however remain undetected through YS surveillance and the reasons for this have been discussed above.

Conclusions

The data presented indicate that the approach to establishing herd status endorsed by CHeCS [25] is robust. Check Tests, BM Ab and BM PCR, if used and interpreted appropriately are useful tools for establishing a herds BVD status. Minimising the delay between screening a herd and acting quickly on the results obtained will increase their diagnostic accuracy and relevance. Whilst a good starting point, a high positive BM Ab result alone is not enough to classify a herd as infected but it should trigger further investigation. In the same way a mid positive BM Ab result does not definitively mark a herd as BVDV-free. BM Ab can prove a useful surveillance tool in established BVDV free herds with low or negative BM Ab titres however; such results were rarely encountered in this study. For this reason, if the study population is representative of herds in cattle dense areas where there is no systematic control of BVDV and biosecurity is difficult to implement, then low or negative BM Ab samples could be considered rare. In this paper, Check Tests performed with the greatest accuracy when diagnosing herd BVD status; only one farm was misclassified when using YS screens (Fig. 4). BM PCR may be used at the same point of the initial screen of YS and BM antibody although, on cost grounds, it would be reasonable to postpone this test until Check Test and BM Ab results indicate the likely presence of a PI.

Finally, it is paramount that longitudinal surveillance using a combination of the techniques discussed in this paper is performed on farms wishing to monitor their BVDV status since this allows for changes in status to be detected early thus enabling prompt action in the event of a BVDV incursion.

Acknowledgements
This study was funded by Defra under project grant SE0777. The authors would like to thank Mick Cranwell (AHPA, Starcross) for his help and support with the laboratory testing, Paddy Gordon for his help during the study design and setup, and the practitioners and staff at Shepton Veterinary Group and Synergy Farm Health for their extensive involvement in this work.

Funding
This study was funded by Defra under project grant SE0777.

Authors' contributions
RB and JB conducted the sample collection from all herds involved. RB performed the analysis of the datasets. RB and JB contributed jointly to the drafting of the manuscript. All authors read and approved the final manuscript.

Competing interests
The authors declare that they have no competing interests.

Consent to publish
Not applicable.

References

1. Potgieter LN. Immunology of bovine viral diarrhea virus. Vet Clin North Am Food Anim Pract. 1995;11:501–20.
2. Virakul P, Fahning ML, Joo HS, Zemjanis R. Fertility of cows challenged with a cytopathic strain of Bovine Viral Diarrhea virus during an outbreak of spontaneous infection with a noncytopathic strain. Theriogenology. 1988;29: 441–9.
3. Edwards S, Wood L, Hewitt-Taylor C, Drew TW. Evidence for an immunocompromising effect of bovine pestivirus on bovid herpesvirus 1 vaccination. Vet Res Commun. 1986;10:297–302.
4. Wray C, Roeder PL. Effect of bovine virus diarrhoea-mucosal disease virus infection on salmonella infection in calves. Res Vet Sci. 1987;42:213–8.
5. Ellis JA, Davis WC, Belden EL, Pratt DL. Flow cytofluorimetric analysis of lymphocyte subset alterations in cattle infected with bovine viral diarrhea virus. Vet Pathol. 1988;25:231–6.
6. McGowan MR, Kirkland PD, Richards SG, Littlejohns IR. Increased reproductive losses in cattle infected with bovine pestivirus around the time of insemination. Vet Rec. 1993;133:39–43.
7. Moerman A, Straver PJ, de Jong MC, Quak J, Baanvinger T, van Oirschot JT. A long term epidemiological study of bovine viral diarrhoea infections in a large herd of dairy cattle. Vet Rec. 1993;132:622–6.
8. Bennett R, Ijpelaar J. Economic Assessment of Livestock Diseases in Great Britain. Final Report to Defra. In: Book Economic Assessment of Livestock Diseases in Great Britain. Final Report to Defra. 2003.
9. Stott AW, Humphry RW, Gunn GJ, Higgins I, Hennessy T, O'Flaherty J, Graham DA. Predicted costs and benefits of eradicating BVDV from Ireland. Irish Vet J. 2012;65:12.
10. Gunn GJ, Stott AW, Humphry RW. Modelling and costing BVD outbreaks in beef herds. Vet J. 2004;167:143–9.
11. Lindberg A, Brownlie J, Gunn GJ, Houe H, Moennig V, Saatkamp HW, Sandvik T, Valle PS. The control of bovine viral diarrhoea virus in Europe: today and in the future. Rev Sci Tech. 2006;25:961–79.
12. Uttenthal A, Stadejek T, Nylin B. Genetic diversity of bovine viral diarrhoea viruses (BVDV) in Denmark during a 10-year eradication period. APMIS. 2005; 113:536–41.
13. The Scottish Government. The Scottish BVD Eradication Scheme [http://www.gov.scot/Topics/farmingrural/Agriculture/animal-welfare/Diseases/disease/bvd/eradication]. Accessed 8 Dec 2016.
14. Animal Health Ireland. BVD - National Eradication Programme [http://animalhealthireland.ie/?page_id=220]. Accessed 8 Dec 2016.
15. Sarrazin S. Epidemiological reflections on bovine viral diarrhoea virus control in Belgian cattle herds based on experimental infections and observational studies. Ghent: Ghent University, Department of Reroduction, Obstetrics and Herd Health, Faculty of Veterinary Medicine; 2015.
16. Brownlie J, Clarke MC, Howard CJ. Experimental infection of cattle in early pregnancy with a cytopathic strain of bovine virus diarrhoea virus. Res Vet Sci. 1989;46:307–11.
17. Peterhans E, Jungi TW, Schweizer M. BVDV and innate immunity. Biologicals. 2003;31:107–12.
18. Mars MH, Bruschke CJ, van Oirschot JT. Airborne transmission of BHV1, BRSV, and BVDV among cattle is possible under experimental conditions. Vet Microbiol. 1999;66:197–207.
19. Barlow RM, Nettleton PF, Gardiner AC, Greig A, Campbell JR, Bonn JM. Persistent bovine virus diarrhoea virus infection in a bull. Vet Rec. 1986;118:321–4.
20. Lindberg AL, Alenius S. Principles for eradication of bovine viral diarrhoea virus (BVDV) infections in cattle populations. Vet Microbiol. 1999;64:197–222.
21. Presi P, Struchen R, Knight-Jones T, Scholl S, Heim D. Bovine viral diarrhea (BVD) eradication in Switzerland–experiences of the first two years. Prev Vet Med. 2011;99:112–21.
22. Moennig V, Houe H, Lindberg A. BVD control in Europe: current status and perspectives. Anim Health Res Rev. 2005;6:63–74.
23. Niskanen R, Lindberg A. Transmission of bovine viral diarrhoea virus by unhygienic vaccination procedures, ambient air, and from contaminated pens. Vet J. 2003;165:125–30.
24. Houe H, Lindberg A, Moennig V. Test strategies in bovine viral diarrhea virus control and eradication campaigns in Europe. J Vet Diagn Invest. 2006;18: 427–36.
25. CHeCS. Cattle Health Certification Standards Technical Document [http://www.checs.co.uk/wp-content/uploads/2015/08/CHeCS-Technical-Document-final-2015.pdf]. Accessed 8 Dec 2016.
26. Pritchard G. Milk antibody testing in cattle. In: Practice. 2001.
27. Houe H. Serological analysis of a small herd sample to predict presence or absence of animals persistently infected with bovine viral diarrhoea virus (BVDV) in dairy herds. Res Vet Sci. 1992;53:320–3.
28. Houe H. Bovine virus diarrhoea virus: detection of Danish dairy herds with persistently infected animals by means of a screening test of ten young stock. Prev Vet Med. 1994;19:241–8.
29. Houe H, Baker JC, Maes RK, Ruegg PL, Lloyd JW. Application of antibody titers against bovine viral diarrhea virus (BVDV) as a measure to detect herds with cattle persistently infected with BVDV. J Vet Diagn Invest. 1995;7:327–32.
30. Drew TW, Yapp F, Paton DJ. The detection of bovine viral diarrhoea virus in bulk milk samples by the use of a single-tube RT-PCR. Vet Microbiol. 1999; 64:145–54.

31. Animal Health and Plant Agency: Disease Surveillance Tests - Bovine viral diarrhoea virus in bulk milk samples (up to 300 contributors) [http://science.vla.gov.uk/Tests/SearchResultsDetailed.aspx?SiteName=DST&Search=&TestCode=TC0709&PriceListCategoryId=16]. Accessed 8 Dec 2016.
32. Booth RE, Brownlie J. Establishing a pilot bovine viral diarrhoea virus eradication scheme in Somerset. Vet Rec. 2012;170:73.
33. Booth RE, Cranwell MP, Brownlie J. Monitoring the bulk milk antibody response to BVDV: the effects of vaccination and herd infection status. Vet Rec. 2013;172:449.
34. Fredriksen B, Sandvik T, Loken T, Odegaard SA. Level and duration of serum antibodies in cattle infected experimentally and naturally with bovine virus diarrhoea virus. Vet Rec. 1999;144:111–4.
35. Niskanen R. Relationship between the levels of antibodies to bovine viral diarrhoea virus in bulk tank milk and the prevalence of cows exposed to the virus. Vet Rec. 1993;133:341–4.
36. Valle PS, Wayne Martin S, Skjerve E. A Bayesian approach to estimating the performance of a bovine virus diarrhoea virus (BVDV) antibody ELISA bulk-tank milk test. Prev Vet Med. 2001;50:71–87.
37. The Cattle Book 2008. [https://www.gov.uk/government/uploads/system/uploads/attachment_data/file/69220/pb13572-cattlebook-2008-090804.pdf]. Accessed 8 Dec 2016.
38. Houe H. Epidemiological features and economical importance of bovine virus diarrhoea virus (BVDV) infections. Vet Microbiol. 1999;64:89–107.
39. Chamorro MF, Walz PH, Haines DM, Passler T, Earleywine T, Palomares RA, Riddell KP, Galik P, Zhang Y, Givens MD. Comparison of levels and duration of detection of antibodies to bovine viral diarrhea virus 1, bovine viral diarrhea virus 2, bovine respiratory syncytial virus, bovine herpesvirus 1, and bovine parainfluenza virus 3 in calves fed maternal colostrum or a colostrum-replacement product. Can J Vet Res. 2014;78:81–8.

4

Lipopolysaccharide derived from the digestive tract triggers an inflammatory response in the uterus of mid-lactating dairy cows during SARA

Muhammad Shahid Bilal†, Juma Ahamed Abaker†, Zain ul Aabdin, Tianle Xu, Hongyu Dai, Kai Zhang, Xinxin Liu and Xiangzhen Shen*

Abstract

Background: The aims of the current study were to evaluate the inflammatory response in cow uterus and to explore the molecular mechanism triggered by high concentrate-induced subacute ruminal acidosis (SARA) in mid-lactating dairy cows. Twelve mid-lactating Holstein cows with an average weight of 455 kg were allocated into two groups subjected to two diets for 18-weeks either a low-concentrate (LC) group containing 4:6 (NDF: NFC) and a high-concentrate (HC) group containing 6:4 (non-forage carbohydrates, NFC): (neutral detergent fiber, NDF) ratio based on dry matter.

Results: The HC group showed lower ruminal pH and higher lipopolysaccharide (LPS) concentrations in both the rumen and peripheral plasma compared to the LC group. The LPS concentrations in the rumen fluid and the peripheral plasma were found significantly increased in the HC group compared to the LC group. The concentrations of IL-1β, TNF-α and IL-6 were significantly higher in the HC group compared to the LC group. The uterus of SARA cows revealed elevated mRNA concentrations of nuclear transcription factors and pro-inflammatory cytokines, which confirmed the presence of inflammation. The occurrence of uterine inflammation was further validated by the increased protein expression of NF-κB-p65 and its active phosphorylated variant in the uterus of SARA cows. Similarly, the inflammatory genes TLR4, LBP, MyD88, TRAF-6, NF-κB, IL-6, IL-8, TNF-α and IL-1β were significantly upregulated in the uterus of the HC versus the LC group.

Conclusion: Therefore, the results indicated that LPS derived from the rumen triggered the genes associated with inflammation in the uterus of mid-lactating dairy cows fed a high-concentrate diet, causing endometritis.

Keywords: Lipopolysaccharide, Sub-acute ruminal acidosis, Cow, LBP, Immune gene expression

Background

Feeding a high-concentrate diet for a longer period of time can cause subacute ruminal acidosis (SARA). It is a digestive disorder with a pH less than 5.6 lasted for more than 3 h per day [1, 2]. General clinical signs of SARA comprise of reduced dry matter intake (DMI), decreased ruminal pH and diarrhea [3, 4]. Another crucial outcome of SARA is the decline of milk quantity and quality [2, 3], and earlier studies also revealed that SARA decreased milk protein yield [4, 5]. The persistent consumption of a HC diet by dairy animals enhances the production of organic acids and significantly results in a decline in pH values of the rumen and intestine [5–7]. This decrease in pH due to excessive high concentrate diet [8] may affect the alterations of the type of fermentation [9] and the structure of the microbes in the rumen [10, 11]. Moreover, the decline in pH also perturbs the balance of microbial population in the rumen causing substantial release of free endotoxin (lipopolysaccharide, LPS) from gram-negative bacteria [12]. Due to possible alterations in permeability and disruption of the gastrointestinal barrier, LPS can be translocated into

* Correspondence: xzshen@njau.edu.cn
†Equal contributors
College of Veterinary Medicine, Nanjing Agricultural University, Nanjing 210095, People's Republic of China

the blood stream [2, 13, 14]. There are also other reports revealed that elevated circulating LPS cause a systemic inflammatory response [15, 16]. Possibly, being an element of an immune response to LPS, SARA has been reported to increase the concentration of acute phase proteins in the peripheral blood, such as serum amyloid A (SAA) and haptogloblin (Hp) [2, 12, 14, 17]. The SAA has many functions but mainly alters innate immune responses, particularly in the migration of neutrophils and monocytes, while Hp has anti-inflammatory effects; LPS-binding protein (LBP) is activated by microbial infections and facilitates in the neutralization of LPS and in the triggering the release of pro-inflammatory cytokines, Such as TNF-α, IL-6 and IL-1β [6, 18–20]. The innate immune system is the first line of defence against invading pathogens; it involves various types of transmembrane and secreted molecules, recognized as pattern recognition receptors (PRR). Toll like receptors, such as TLR4, that join the LPS-LBP complex, are located on the outer surfaces of a variety of cells and contribute to the sensing of microbial pathogens in the intracellular endosomes and lysosomes [21, 22]. TLR4 recognizes LPS with the help of the LBP and cluster of differentiation antigen14 (CD14) [23, 24]. LPS activates myeloid differentiating factor 88 (MyD88) after joining with TLR4 on the host cell surface [25]. MyD88 activates TNF receptor-associated factor 6 (TRAF6), which stimulates the IκB kinase (IκK) complex, and phosphorylated NF-κB is consequently translocated from the cytoplasm to the nucleus [26, 27]. The NF-κB-dependent cascade pathway regulates the production of pro-inflammatory cytokines, including interleukin (IL)-1β, TNF-α, IL-6, chemokines (IL-8) and other inflammatory mediators [28]. The principal objectives of this study were to evaluate the inflammatory response in cow uterus and to explore the molecular mechanism triggered by high concentrate-induced SARA in mid-lactating dairy cows.

Methods

Animals, diets and experimental design

Twelve healthy multiparous Holstein cows, which were purchased from the Experimental Farm of Nanjing Agricultural University, with an average body weight of 455 kg were selected for the current study. The experimental cows were randomly divided into two groups, each group containing six cows: the high-concentrate (HC) group and the low-concentrate (LC) group. The HC group was provided a high-concentrate diet containing 6:4 (non-forage carbohydrates, NFC): (neutral detergent fiber, NDF) ratio, while LC group was fed low concentrate diet containing 4:6 (NDF: NFC) for the 18-week experimental period. The details of the nutritional ingredients and their compositions are shown in Table 1. The cows were surgically fixed with a ruminal fistula two weeks prior to the experiment. The cows were kept in individual tie stalls and feed was offered three times per day at 0400, 1200 and 2000 h. DMI was ~21.7 kg/d/head, and fresh drinking water was available to them during the entire experimental period. The experiment was started after the animals made a full recovery from the ruminal fistula fixation, and no cows showed any clinical signs of infection during the experiment.

Table 1 Ingredient and nutrient composition of diets

Ingredients, % of DM	LC[a]	HC[a]
Corn silage	30.0	20.0
Alfalfa	30.0	20.0
Maize	22.7	33.6
Wheat bran	5.1	15.0
Soybean meal	9.8	9.0
Calcium phosphate dibasic	0.9	0.5
Limestone	0.0	0.5
Salt	0.3	0.3
Premix[b]	1.0	1.0
Total	100.0	100.0
Nutritional Composition[c]		
NE MJ/kg	6.3	6.7
CP %	16.0	16.2
EE %	3.9	4.1
NDF %	37.7	31.9
ADF %	22.7	17.5
NFC %	33.4	40.3
Starch, %	25.3	32.2
Ca %	0.9	0.8
P %	0.4	0.4

[a]Treatment LC, 40% grain on DM basis; HC, 60% grain on DM basis
NE Net energy mega joules, *CP* crude protein, *EE* ether extract, *NDF* neutral Detergent fiber, *NFC* non-forage carbohydrates, *ADF* Acid detergent fiber
[b]The premix contained VA,1.900,000 IU/kg; VD, 250,000 IU/kg; VE, 3484.32 IU/Kg; Niacin, 4,000 mg/Kg; Cu, 1,200 mg/Kg; Fe, 525 mg/Kg; Zn, 13,000 mg/Kg; Mn, 5,500 mg/kg; I, 170 mg/Kg; Co, 50 mg/Kg; Se, 27 mg/Mg
[c]The calculated nutritional composition values

Collection of samples and analysis

The rumen fluid was collected at one-hour intervals for 12 h daily for 3 continuous days during the 18th week. Blood samples were collected 4 h after feeding on the sampling days (5, 6 and 7) of the 18th week. 10 mL of rumen fluid was immediately centrifuged at 10,000 × g for 45 min, filtered through a disposable 0.22-μm filter and stored in Pyrogen free glass at −20 °C for LPS analysis. The blood samples were collected from the jugular vein into 5-mL vacuum tubes containing sodium heparin as an anticoagulant. The plasma was separated from the blood samples through centrifugation at 3000 × g at 4 °C for 15 min and stored at –20 °C to determine the concentrations of the pro-inflammatory cytokines IL-1β, IL-6, and TNFα in the peripheral blood. At the end of the experiment, the cows were slaughtered after overnight fasting. The slaughtering was performed according to the law of Jiangsu Provincial People's Government, China and the study was approved by the Animal Ethics Committee (AEC) of Nanjing Agricultural University, China as described in the end of the manuscript. After slaughtering, the uterine tissues were removed from all the cows aseptically, as close to the cervix as possible for the analysis. All collected tissues were immediately frozen in liquid nitrogen and stored at –70 °C within 30 min of slaughtering [29].

Determination of ruminal pH

Rumen fluids were taken through a rumen fistula at one-hour intervals up to 12 h on days 5, 6 and 7 of the 18th week of the trial. The pH of rumen was measured each hour immediately following the sample collection by pH-meter (Sartorius, Basic pH Meter PB-10, PB-21, Goettingen, Germany).

Determination of LPS

The concentration of LPS in the rumen fluid and plasma of the jugular vein was determined by Chromogenic Endpoint Limulus Amebocyte Lysate Assay Kits, CE64406 and CE80545 (Chinese Horseshoe Crab Reagent Manufactory Co., Ltd., Xiamen, China). Briefly, the pre-treated rumen fluid and plasma samples were diluted until their LPS concentrations were in the range of 0.1 to 1 endotoxin units (EU)/mL relative to the reference endotoxin and assayed as described by Gozho et al. [2, 6].

Radioimmunoassay

Radioimmunoassay was used to calculate the concentrations of IL-6, IL-1β, and TNF-α in peripheral blood using radioimmunoassay (RIA) kits (IL-1β, C09DJB; IL-6, C12DJB; TNF-α, C06PJB; Beijing North Institute of Biological Technology, Beijing, China).

RNA extraction and real time PCR (RT-q PCR)

Total RNA was prepared from 100 mg uterus tissue using TRIzol (Takara Co., Otsu, Japan) via homogenization on ice as described by the manufacturer's instructions. The quality of mRNA was assessed by both agarose gel (1%) electrophoresis and spectrometry (A260/A280) using the Eppendorf Bio photometer Plus (Eppendorf AG, Hamburg, Germany). Only samples with a ratio between 1.8 and 2.1 were used in subsequent experiments. The first-strand cDNA was synthesized using 250 ng/μl of the total RNA template using Prime Script RT Master Mix Perfect Real Time (Takara Co., Otsu, Japan) according to the manufacturer's instructions.

Quantitative real-time PCR (qRT-PCR) amplification was performed to evaluate the expression of the selected genes. The primers for genes IL-1β, TLR4, IL-6, IL-8 [30], and LBP [31] were used according to previously published work, while MyD88, TNF-α, TRAF6, and NF-κB were designed using the Premier primer 5.0 software. The primers were assessed for their amplification efficiencies. The reaction conditions included 2 μl of cDNA and 0.4 μM primers in a total volume of 20 μl of super mix. Thermal cycling parameters consisted of initial denaturation at 95 °C for 15 s, followed by 40 cycles of annealing at 95 °C for 5 s and primer extension at 60 °C for 31 s. All the reactions were run in triplicate. Each cDNA sample was amplified using SYBR Green (Takara Co., Otsu, Japan) in the ABI 7300 Fast Real-Time PCR System (Applied Bio systems, USA).

The data were normalized to the mean of housekeeping gene β-actin to control the unpredictability in expression levels and were analysed using $R = 2^{-\Delta\Delta Ct}$ method (Ct $_{target\ gene}$ - Ct $_{\beta\text{-actin}}$) treatment - (Ct $_{target\ gene}$ - Ct $_{\beta\text{-actin}}$) control), as previously depicted [32]. The primer sequences used in this study are presented in Table 2.

Western blotting analysis

The protein was extracted from 100 mg uterus tissue with radio immunoprecipitation assay (RIPA) buffer using a Dounce Homogenizer (Polytron PT 1200 E, Switzerland). After 30 min of incubation on ice, cell extracts were subjected to centrifugation (12,000 × g) at 4 °C for 15 min to obtain cell protein. The supernatant was assayed for protein concentration using the bicinchoninic acid (BCA) protein assay kit (Pierce, Rockford, IL, USA) and was adjusted to 4 μg/μl. Equal amounts of proteins were isolated using 10% SDS-polyacrylamide gel electrophoresis (PAGE) and shifted to nitrocellulose membrane (NC) (Bio sharp, China) at 4 °C. Then, 10% non-fat dry milk was used to block the nitrocellulose membrane. The membrane was incubated at 4 °C overnight with monoclonal goat antibodies against NF-κB and GAPDH (Santa Cruz Biotechnology, Santa Cruz, CA, USA, diluted by 1:200), (Bio world, CA, USA, diluted by 1:10000), respectively. After incubation, the membrane was washed and incubated with horseradish peroxidase-conjugated goat anti-rabbit secondary antibodies. Immunoreactive proteins were detected by chemiluminescence using an ECL reagent (Super Signal West Pico Trial Kit, Pierce, USA) and subsequently autoradiographed. The results were quantified using Bio-Rad Gel Doc 2000™ Systems (Bio-Rad, USA) and analysed with Bio-Rad TDS Quantity One software (Bio-Rad). The relative quantities of proteins were evaluated by a densitometer and expressed as absorbance units (AU).

Table 2 The list of primer sequences used for amplification of qRT-PCR

Gene	Forward primers	Reverse primers	Accession number	Product size (bp)
TLR-4	GGACCCTTGCGTACAGGTTG	GGAAGCTGGAGAAGTTATGGC	NM_174198.6	244
TRAF-6	GCGGCCTTCAAGTTAGGAGA	TCATCAACTGCTCGTTCGGG	NM_001034661.2	141
NF-κB	ACGATCGTCACCGGATTGAG	GGTGCTGAGAGATGGCGTAA	XM_005699996.1	194
β-actin	CTCTTCCAGCCTTCCTTCCT	GGGCAGTGATCTCTTTCTGC	AY141970	178
IL-6	GGAGGAAAAGGACGGATGCT	GGTCAGTGTTTGTGGCTGGA	EU276071.1	226
IL-8	CCTCTTGTTCAATATGACTTCCA	GGCCCACTCTCAATAACTCTC	NM_173925.2	170
TNF-α	GCTCTTACCGGAACACTTCG	GGACACCTTGACCTCCTGAA	NM_173966	238
IL-1β	AACCGAGAAGTGGTGTTCTGC	TTGGGGTAGACTTTGGGGTCT	NM_174093	167
MyD88	AAAGCCCGAGTGTTTTGATG	TCACATTCCTTGCTTTGCAG	NM_001014382	234
LBP	GCAAGATCACTGGATTCTTGGA	AAAACAGGAAGTCCTTGTGGATC	NM_001038674.2	228

Statistical analyses

The mixed procedure of SAS was used to analyze pH with a repeated measures design performed with SAS software (SAS version 9.2, SAS Institute Inc.). The effects of cows were considered random and the effects of diet and time were considered as fixed factors. The time within treatments and cows were considered as repeated measurements and the residuals for each variable were used to assess the normality. The days were averaged within a cow before the analysis for all the data including the pH, radioimmunoassay, ruminal LPS, plasma LPS and the gene expression data were assessed by independent *t*-test and expressed as mean ± SEM. Differences were concluded to be significant at $p < 0.05$.

Results

Rumen pH and LPS content of rumen fluid and blood plasma

The results of ruminal pH fed a high concentrate diet were published previously by Xu et al. [33] and he confirmed the presence of SARA by measuring the ruminal pH at different time intervals. In this study, consuming a high-concentrate diet (HC group) for 18 weeks resulted in a significant decline in ruminal pH (<5.6) than that in the LC group ($p < 0.05$) and lasted for more than three hours [2]. The effect of diet × hours on pH was significant (0.05); however no effect was observed on hours alone.

The LPS concentrations in the rumen fluid and the jugular vein plasma were significantly higher in the cows fed a high-concentrate diet compared to cows fed a low-concentrate diet. The concentration of LPS in the rumen fluid of the HC group was 79040 EU/mL, compared to 47170 EU/mL in the LC group ($p < 0.01$); similarly, in the peripheral plasma, the LPS level in the HC group was 860 EU/mL, whereas it was 470 EU/mL in LC group ($p < 0.001$), which indicated that the LPS level was significantly increased in the HC group relative to the LC group (Table 3).

Levels of pro-inflammatory cytokines in the peripheral blood

The results of the measurement of pro-inflammatory cytokines, together with IL-1β, IL-6 and TNF-α are shown in Fig. 1. The concentration of IL-1β was significantly higher in the HC group compared to the LC group ($p < 0.007$) (Fig. 1a), the concentration of IL-6 was significantly increased in the HC group versus LC group ($p < 0.004$) (Fig. 1b) and the concentration of TNF-α was elevated in the HC group relative to the LC group ($p < 0.002$) (Fig. 1c). Generally, the concentration of pro-inflammatory cytokines in the peripheral blood was significantly higher in the HC group compared to the LC group.

Expressions of genes involved in the inflammation of the uterus

SARA significantly increased the expression level of several genes involved in the inflammation of the uterus as shown in Fig. 2. The gene expression of TLR4 ($p < 0.006$) and MyD88 ($p < 0.01$) in the HC group was significantly upregulated compared to the LC group. Similarly, TRAF-6 ($p < 0.04$) and NF-κB ($p < 0.005$) were highly expressed in the HC group relative to the LC group. The gene expression of TNF-α ($p < 0.005$) and IL-8 ($p < 0.005$) was significantly increased. The gene expression of IL-1β ($p < 0.01$), IL-6 ($p < 0.01$) and LBP ($p < 0.03$) was also significantly higher in the HC group compared with the LC group.

Table 3 LPS concentration in rumen and plasma of dairy cows fed low concentrate (LC) and high concentrate (HC)

LPS concentration (EU/mL)	Treatment[a]			
	LC	HC	SEM[b]	*p*-Value
Rumen LPS	47170	79040	7966.25	<0.01
Jugular vein Plasma LPS	470	860	81.26	<0.001

[a]*HC* high concentrate diet, *LC* low concentrate diet, *EU* endotoxin unit
[b]*SEM* Standard error of the mean between the two groups
The LPS data were compared using Student's *t*-test between HC and LC groups
$P \leq 0.05$ was considered significant

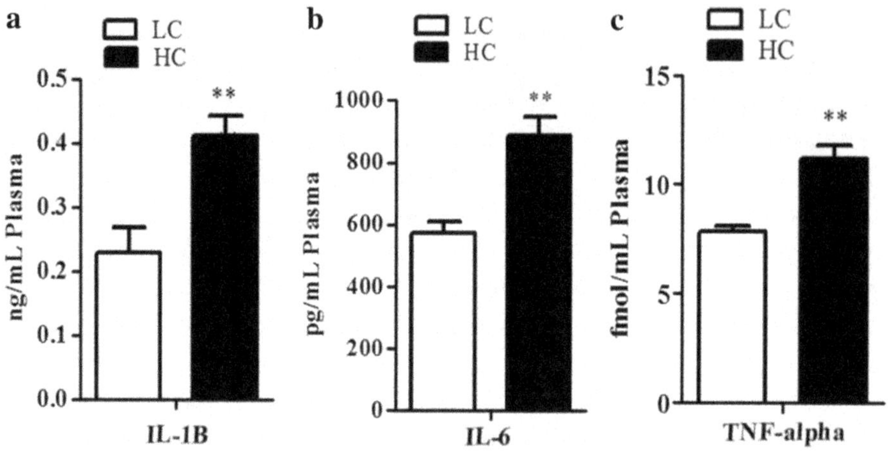

Fig. 1 a-c The concentration levels of TNF-α, IL-1β, and IL-6 in the peripheral blood. The data were expressed as the mean ± SEM.; asterisks indicate the differences between the low-concentrate (LC) group and high-concentrate (HC) group. (*$p < 0.05$, **$p < 0.01$, $n = 12$)

Expression of NF-κB protein in the cow uterus of the HC group and LC group

Western blot analysis demonstrated that high concentrate diet-induced inflammation through the increaseduterine concentrations of the NF-κB (p65) proteins in the HC group compared with the LC group ($p < 0.03$; Fig. 3a). Importantly, the amount of phosphorylated NF-κB (p-p65) was higher in the uterus of the HC group than in the LC group ($p < 0.04$; Fig. 3b). Therefore, it was confirmed that NF-κB was activated in HC group in the TLR4 signalling pathway.

Discussion

The present study assessed the uterine inflammation at the molecular level in dairy cows. The inflammatory response in the uterus due to circulating LPS derived from rumen triggered the expression of pro-inflammatory cytokines through TLR4 signaling. To evaluate the inflammatory response in cow uterus and to explore the molecular mechanism triggered by high concentrate diet, SARA was confirmed by measuring the pH of the rumen. Xu et al. [33] also confirmed the presence of SARA by measuring the ruminal pH at different time intervals. In our study, the low rumen pH (>5.6) lasted for

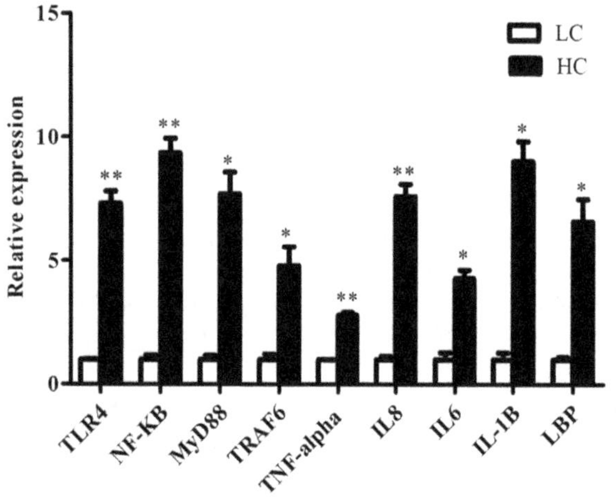

Fig. 2 The uterine gene expression profile analysed by qRT-PCR. The genes involved in transcriptional regulation were measured in the uterine tissues. The error bars indicate the standard error of the mean and ** indicates significance at $p < 0.01$

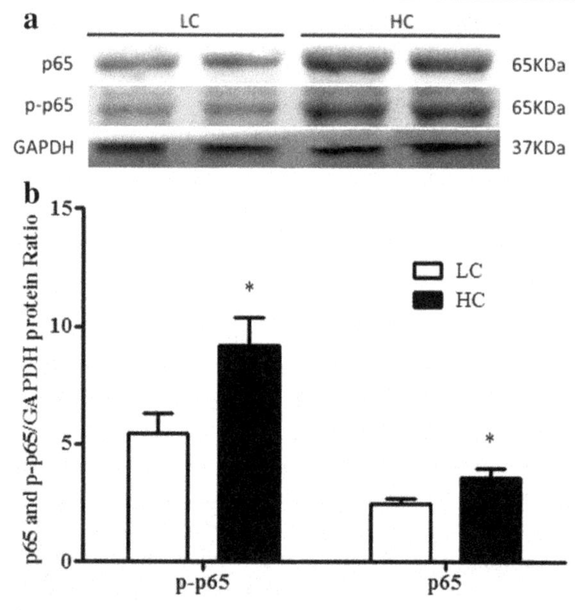

Fig. 3 a The western blotting assay of NF-κB (p65) and phosphorylated-p65 protein in the uterus. The NF-κB contents in uterus were evaluated through western blot of the low-concentrate (LC) group and high-concentrate (HC) group. **b** The quantities of proteins are measured as arbitrary units relative to GAPDH; fold alterations in NF-κB (p65) and phosphorylated NF-κB (p-p65). The data were expressed as the mean ± SEM., asterisks indicate the differences between the high-concentrate (HC) group compared to the low-concentrate (LC) group (** $p < 0.01$)

223 min and high concentrations of LPS were observed in the rumen fluid and peripheral blood. The recent reports also indicated the successful induction of SARA, if theduration of low rumen pH was more than 180 min/day [22, 31]. However, there is little literature revealing that SARA not only depends on pH, but also on the alteration in the microbial population that develops secondarily to the feed type [34]. As a matter of fact, low ruminal pH values led to the lysis of gram-negative bacteria which released LPS into the blood circulation [31, 35]. Consistent with previous studies in dairy cows and feedlot, our results presented higher concentrations of LPS in the peripheral plasma (860 EU/mL) and ruminal fluid (79040 EU/mL) of cows fed a high concentrate die trelative to the peripheral plasma of the LC group [14, 36]. Moreover, we also found that high concentrate diet feeding increased the plasma concentrations of pro-inflammatory cytokines (IL-1β, TNF-α and IL-6). Similar results have been reported by other researchers [6, 19, 37]. The increment in pro-inflammatory cytokines concentration in the plasma may provide confirmation for LPS translocation into the blood circulation and eliciting of inflammatory responses. This implies that inflammation in the HC group was due to the high level of LPS in the bloodstream, which in turn triggers the pro-inflammatory gene to cope with stress conditions [37].

The RT-qPCR-based findings revealed the presence of inflammation-associated markers in the uterus of SARA-induced cows and confirmed that the dairy cows suffered from uterine inflammation at the molecular level. According to our results, the increased expression of TLR4 indicated that TLR4 recognized the LPS and triggered the inflammatory pathway [9, 38]. Identification of inflammatory-stimuli by the innate immune system is organized by PPR that identify external stimuli for example pathogen associated molecular patterns (PAMPs) [39]. Furthermore, our research demonstrated the increased expression of LBP, MyD88, and TRAF6 in the uterus; which provides more clues that the TLR4 signaling pathway was stimulated by LPS [25, 40]. Previous studies demonstrated that as a main transcription factor, NF-κB may bind to the promoter regions of the many inflammatory genes to regulate their transcription activities [37, 41]. Similarly, in the current study, NF-κB expression was increased in the HC group, which also confirmed the activation of TLR4 signalling pathway. Our results of increased expression of pro-inflammatory cytokines (IL-1β, TNF-α and IL-6) and chemokine (IL-8) in the HC group due to SARA-induced LPS were consistent with the previous studies [25, 42–46].

The western blot results further confirmed that the expression levels of NF-κB (p65) and phosphorylated p65 proteins were distinctly increased in the HC group compared to the LC group. Taken together, the results suggested that LPS triggered the activation of TLR4 signalling, which resulted in an inflammatory response in the uterus of cows suffering from SARA. Conclusively, LPS activates TLR4-NF-κB signalling. Together with the existing study and the previous reports [33, 37], the protein expressions of NF-κB (p65) and p-p65 were magnified in the HC group.

Conclusions

According to our results, long-term feeding of high concentrate diet causes uterine inflammation in mid-lactating dairy cows. The inflammatory response in the uterus induced by increased level of circulating LPS triggers the expression of pro-inflammatory cytokines in the uterus through the TLR4-NF-κB signalling pathway.

Abbreviations

AMP: Antimicrobial peptides; APP: Acute phase proteins; CD14: Cluster of differentiation antigen14; EECs: Endometrial epithelial cells; GIT: Gastrointestinal tract; HC: High concentrate; IL-1 α: −1 β, −6, Interleukin-1 α, 1 β, −6; IκK: IκB kinase; LBP: LPS-binding protein; LC: Low concentrate; LPS: Lipopolysaccharide; MyD88: Myeloid differentiating factor 88; NF-κB: Nuclear factor kappa-light-chain-enhancer of activated B cells; PAGE: Polyacrylamide gel electrophoresis; PAMPs: pathogen-associated molecular patterns; PRR: Pattern recognition receptors; qRT-PCR: Quantitative real-time PCR; SARA: Subacute ruminal acidosis; TLRs: Toll-like receptors; TRAF6: TNF receptor-associated factor 6

Acknowledgements

The authors are grateful to Dr. Usman Waheed (Department of Pathobiology, University of Veterinary and Animal Sciences, Sub-campus Jhang, Pakistan) for helping in the revision of the manuscript.

Funding

This study was supported by the National Natural Science Foundation of China (31172371) and the Priority Academic Programme Development of Jiangsu Higher Education Institutions (PAPD).

Authors' contributions

XS and MSB conceived the idea and proposed the experiment. MSB, JAA, ZA, HD, KZ and XL conducted the majority of measurements for the parameters in this work and drafted the manuscript. TX and HD contributed to the statistical analyses. All authors read and approved the final manuscript.

Competing interests

The authors declare that they have no competing interests that would have predisposed the content of the paper or intervened with their objective consideration of the manuscript.

Consent to publish

Not applicable.

References

1. Steele MA, Croom J, Kahler M, AlZahal O, Hook SE, Plaizier K, et al. Bovine rumen epithelium undergoes rapid structural adaptations during grain-induced subacute ruminal acidosis. Am J Physiol Regul Integ Comp Physiol. 2011;300(6):R1515–23.
2. Gozho G, Plaizier J, Krause D, Kennedy A, Wittenberg K. Subacute ruminal acidosis induces ruminal lipopolysaccharide endotoxin release and triggers an inflammatory response. J Dairy Sci. 2005;88(4):1399–403.
3. Kleen J, Hooijer G, Rehage J, Noordhuizen J. Subacute ruminal acidosis (SARA): a review. J Vet Med Ser A. 2003;50(8):406–14.
4. Colman E, Fokkink W, Craninx M, Newbold J, De Baets B, Fievez V. Effect of induction of subacute ruminal acidosis on milk fat profile and rumen parameters. J Dairy Sci. 2010;93(10):4759–73.
5. Li S, Khafipour E, Krause D, Kroeker A, Rodriguez-Lecompte J, Gozho G, et al. Effects of subacute ruminal acidosis challenges on fermentation and endotoxins in the rumen and hindgut of dairy cows. J Dairy Sci. 2012;95(1):294–303.
6. Dong H, Wang S, Jia Y, Ni Y, Zhang Y, Zhuang S, et al. Long-term effects of subacute ruminal acidosis (SARA) on milk quality and hepatic gene expression in lactating goats fed a high-concentrate diet. PLoS One. 2013;8(12), e82850.
7. Garrett E, Pereira M, Nordlund K, Armentano L, Goodger W, Oetzel G, et al. Diagnostic methods for the detection of subacute ruminal acidosis in dairy cows. J Dairy Sci. 1999;82(6):1170–8.
8. De Nardi R, Marchesini G, Plaizier JC, Li S, Khafipour E, Ricci R, et al. Use of dicarboxylic acids and polyphenols to attenuate reticular pH drop and acute phase response in dairy heifers fed a high grain diet. BMC Vet Res. 2014;10(277):014–0277.
9. Chang G, Zhuang S, Seyfert HM, Zhang K, Xu T, Jin D, et al. Hepatic TLR4 signaling is activated by LPS from digestive tract during SARA, and epigenetic mechanisms contribute to enforced TLR4 expression. Oncotarget. 2015;6(36):38578–90.
10. Emmanuel D, Madsen K, Churchill T, Dunn S, Ametaj B. Acidosis and lipopolysaccharide from *Escherichia coli* B: 055 cause hyperpermeability of rumen and colon tissues. J Dairy Sci. 2007;90(12):5552–7.
11. Mao S, Zhang R, Wang D, Zhu W. The diversity of the fecal bacterial community and its relationship with the concentration of volatile fatty acids in the feces during subacute rumen acidosis in dairy cows. BMC Vet Res. 2012;8(237):1746–6148.
12. Emmanuel D, Dunn S, Ametaj B. Feeding high proportions of barley grain stimulates an inflammatory response in dairy cows. J Dairy Sci. 2008;91(2):606–14.
13. Jiang J, Bahrami S, Leichtfried G, Redl H, Ohlinger W, Schlag G, et al. Kinetics of endotoxin and tumor necrosis factor appearance in portal and systemic circulation after hemorrhagic shock in rats. Ann Surg. 1995;221(1):100–6.
14. Khafipour E, Krause D, Plaizier J. A grain-based subacute ruminal acidosis challenge causes translocation of lipopolysaccharide and triggers inflammation. J Dairy Sci. 2009;92(3):1060–70.
15. Gozho G, Krause D, Plaizier J. Ruminal lipopolysaccharide concentration and inflammatory response during grain-induced subacute ruminal acidosis in dairy cows. J Dairy Sci. 2007;90(2):856–66.
16. Plaizier J, Khafipour E, Li S, Gozho G, Krause D. Subacute ruminal acidosis (SARA), endotoxins and health consequences. Anim Feed Sci Technol. 2012;172(1):9–21.
17. Gozho GN, Krause DO, Plaizier JC. Rumen lipopolysaccharide and inflammation during grain adaptation and subacute ruminal acidosis in steers. J Dairy Sci. 2006;89(11):4404–13.
18. Jia Y, Wang S, Ni Y, Zhang Y, Zhuang S, Shen X, et al. High concentrate-induced subacute ruminal acidosis (SARA) increases plasma acute phase proteins (APPs) and cortisol in goats. Animal. 2014;8(09):1433–8.
19. Zhou J, Dong G, Ao C, Zhang S, Qiu M, Wang X, et al. Feeding a high-concentrate corn straw diet increased the release of endotoxin in the rumen and pro-inflammatory cytokines in the mammary gland of dairy cows. BMC Vet Res. 2014;10(1):172.
20. De Nardi R, Marchesini G, Plaizier JC, Li S, Khafipour E, Ricci R, et al. Use of dicarboxylic acids and polyphenols to attenuate reticular pH drop and acute phase response in dairy heifers fed a high grain diet. BMC Vet Res. 2014;10(1):1.
21. Beutler B, Rietschel ET. Innate immune sensing and its roots: the story of endotoxin. Nat Rev Immunol. 2003;3(2):169–76.
22. Li S, Yoon I, Scott M, Khafipour E, Plaizier J. Impact of Saccharomyces cerevisiae fermentation product and subacute ruminal acidosis on production, inflammation, and fermentation in the rumen and hindgut of dairy cows. Anim Feed Sci Technol. 2016;211:50–60.
23. Sohn MJ, Hur GM, Byun HS, Kim WG. Cyclo (dehydrohistidyl-l-tryptophyl) inhibits nitric oxide production by preventing the dimerization of inducible nitric oxide synthase. Biochem Pharmacol. 2008;75(4):923–30.
24. Bannerman DD, Paape M, Hare W, Hope JC. Characterization of the bovine innate immune response to intramammary infection with Klebsiella pneumoniae. J Dairy Sci. 2004;87(8):2420–32.
25. Ju J, Li L, Xie J, Wu Y, Wu X, Li W, et al. Toll-like receptor-4 pathway is required for the pathogenesis of human chronic endometritis. Exp Ther Med. 2014;8(6):1896–900.
26. Turner M, Healey G, Sheldon I. Immunity and inflammation in the uterus. Reprod Domest Anim. 2012;47(s4):402–9.
27. Cohen J. The immunopathogenesis of sepsis. Nature. 2002;420(6917):885–91.
28. Cronin JG, Turner ML, Goetze L, Bryant CE, Sheldon IM. Toll-like receptor 4 and MYD88-dependent signaling mechanisms of the innate immune system are essential for the response to lipopolysaccharide by epithelial and stromal cells of the bovine endometrium. Biol Reprod. 2012;86(2):51.
29. D'Agata M, Russo C, Preziuso G. Effect of Islamic ritual slaughter on beef quality. Ital J Anim Sci. 2009;8(2):489–91.
30. Chang G, Zhang K, Xu T, Jin D, Guo J, Zhuang S, et al. Epigenetic mechanisms contribute to the expression of immune related genes in the livers of dairy cows fed a high concentrate diet. PLoS One. 2015;10(4), e0123942.
31. Zhang K, Chang G, Xu T, Xu L, Guo J, Jin D, et al. Lipopolysaccharide derived from the digestive tract activates inflammatory gene expression and inhibits casein synthesis in the mammary glands of lactating dairy cows. Oncotarget. 2016;7(9):9652–65.
32. Livak KJ, Schmittgen TD. Analysis of relative gene expression data using real-time quantitative pcr and the 2 – δδct method. Methods. 2001;25(4):402–8.
33. Xu T, Tao H, Chang G, Zhang K, Xu L, Shen X, et al. Lipopolysaccharide derived from the rumen down-regulates stearoyl-CoA desaturase 1 expression and alters fatty acid composition in the liver of dairy cows fed a high-concentrate diet. BMCVet Res. 2015;11(1):52.
34. Calsamiglia S, Blanch M, Ferret A, Moya D. Is subacute ruminal acidosis a pH related problem? Causes and tools for its control. Anim Feed Sci Technol. 2012;172(1):42–50.
35. Gott P, Hogan JS, Weiss WP. Effects of various starch feeding regimens on responses of dairy cows to intramammary lipopolysaccharide infusion. J Dairy Sci. 2015;98(3):1786–96.
36. Dong G, Qiu M, Ao C, Zhou J, Wang X, Zhang Z, et al. Feeding a high-concentrate corn straw diet induced epigenetic alterations in the mammary tissue of dairy cows. PLoS One. 2014;9(9), e107659.
37. Chang G, Zhang K, Xu T, Jin D, Seyfert HM, Shen X, et al. Feeding a high-grain diet reduces the percentage of LPS clearance and enhances immune gene expression in goat liver. BMC Vet Res. 2015;11(67):015–0376.
38. Akira S, Uematsu S, Takeuchi O. Pathogen recognition and innate immunity. Cell. 2006;124(4):783–801.
39. Bieghs V, Trautwein C. The innate immune response during liver inflammation and metabolic disease. Trends Immunol. 2013;34(9):446–52.
40. Sheldon IM, Cronin J, Goetze L, Donofrio G, Schuberth HJ. Defining postpartum uterine disease and the mechanisms of infection and immunity in the female reproductive tract in cattle. Biol Reprod. 2009;81(6):1025–32.
41. Kawai T, Akira S. Signaling to NF-κB by Toll-like receptors. Trends Mol Med. 2007;13(11):460–9.
42. Fu Y, Liu B, Feng X, Liu Z, Liang D, Li F, et al. Lipopolysaccharide increases Toll-like receptor 4 and downstream Toll-like receptor signaling molecules expression in bovine endometrial epithelial cells. Vet Immunol Immunopathol. 2013;151(1):20–7.
43. Wang G, Sun B, Gao Y, Meng QH, Jiang HC. The effect of emodin-assisted early enteral nutrition on severe acute pancreatitis and secondary hepatic injury. Mediat Inflamm. 2007;2007:29638.
44. Zhang Y, Guo F, Ni Y, Zhao R. LPS-induced inflammation in the chicken is associated with CCAAT/enhancer binding protein beta-mediated fat mass and obesity associated gene down-regulation in the liver but not hypothalamus. BMC Vet Res. 2013;9(257):1746–6148.
45. Krikun G, Trezza J, Shaw J, Rahman M, Guller S, Abrahams VM, et al. Lipopolysaccharide appears to activate human endometrial endothelial cells through TLR-4-dependent and TLR-4-independent mechanisms. Am J Reprod Immunol. 2012;68(3):233–7.
46. Galvão K, Santos N, Galvão J, Gilbert R. Association between endometritis and endometrial cytokine expression in postpartum Holstein cows. Theriogenology. 2011;76(2):290–9.

5

Evaluation of intravenous regional anaesthesia and four-point nerve block efficacy in the distal hind limb of dairy cows

S. Yavari[1], N. Khraim[2], G. Szura[1], A. Starke[3], E. Engelke[4], C. Pfarrer[4], K. Hopster[5], M. Schmicke[1], W. Kehler[1], M. Heppelmann[1], S. B. R. Kästner[6] and J. Rehage[1*]

Abstract

Background: Intravenous regional anaesthesia (IVRA) and hindfoot four-point nerve block anaesthesia (NBA) are recommended for local anaesthesia (LA) in the distal limb of dairy cows. Two studies were conducted to compare the efficacy, time until onset and stress responses to IVRA and NBA in dairy cows. In the first cross-over designed study, eight healthy unsedated German Holstein cows, restrained in lateral recumbency (LR) on a surgical tipping table, were treated with IVRA and NBA using procaine 2% as a local anaesthetic. Distal limb desensitization was tested by electrical (e-), mechanical (m-) and thermal (t-) nociceptive stimulation 10 min before and 15 and 30 min after LA. Hormonal-metabolic (blood concentrations of cortisol, lactate, non-esterified fatty acids, and glucose) and cardio-respiratory (heart and respiratory rate, mean arterial blood pressure) stress responses to treatment were assessed at predetermined intervals. In the second study, six healthy, unsedated German Holstein cows in LR were treated (crossover design) with IVRA and NBA. Short-interval e-stimulation was measured by the time until complete distal limb desensitization.

Results: In the first study, four of eight cows responded to e-stimulation 15 min after IVRA, while none of the cows treated with NBA responded until the safety cut-off level was reached. E-stimulation revealed complete desensitization of the distal limb 30 min after LA in all cows. Half of the cows did not respond to m- and t-stimulation before LA, so no further evaluation was performed. Stress reactions to IVRA and NBA treatment were similar, but differences may have been masked by stress response to LR restraint. In the second study, complete desensitization was achieved 12.5 min after NBA, while one of the six cows still responded to e-stimulation 20 min after IVRA.

Conclusion: Hindfoot nerve block anaesthesia and intravenous regional anaesthesia induced complete desensitization of the distal hind limb in dairy cows. However, the anaesthesia onset after NBA was significantly faster than that of IVRA, which may be clinically relevant in the field, particularly when distal limb anaesthesia is required for major claw surgeries under time constraints.

Keywords: IVRA, NBA, Cattle, Hind limb, Nociception

* Correspondence: juergen.rehage@tiho-hannover.de
[1]Clinic for Cattle, University of Veterinary Medicine Hannover, Foundation, Hannover, Germany
Full list of author information is available at the end of the article

Background

Claw diseases are frequent health disorders in dairy cattle that are mainly caused by claw horn lesions, such as sole ulcers and white line disease, or by inflammatory alterations of the adjacent soft tissue [1–5]. Claw horn lesions are commonly treated in their early stages by therapeutic claw trimming [6–8]. Major surgical interventions, such as claw amputation or resection of the distal interphalangeal joint, are often necessary in advanced cases when inflammatory purulent alterations penetrate the pododerma and reach the inner structures of the horn shoe, resulting in purulent arthritis of the distal interphalangeal joint, osteomyelitis or necrosis of tendons and ligaments [9, 10]. Claw surgeries are painful for affected cows and require adequate pain management including local anaesthesia (LA) and analgesic application to control postoperative pain [11–16].

For LA at the distal limb, intravenous regional anaesthesia (Bier block [17, 18]; IVRA) or hindfoot nerve block anaesthesia (NBA) are recommended [9, 13, 19–21]. For IVRA of the hindfoot, a tourniquet is placed around the metatarsus to interrupt blood flow before local anaesthetics are injected into a vein to desensitize the limb distally to the tourniquet [18, 19]. Hindfoot nerve blocks require perineural injection of local anaesthetics to the appropriate branches or continuations of the fibular and tibial nerves [20, 21]. Although lameness is a significant health problem in cattle, treatment recommendations for claw lesions are mostly based on "experience" and "anecdotal" reports in journals or in textbooks. Few controlled studies have been reported in peer-reviewed journals [22]. To the best of the authors' knowledge, this also includes LA techniques to the distal limb in cattle.

An advantage of the IVRA tourniquet is reduced intraoperative bleeding, which improves the surgical field visibility. However, as in humans, the tourniquet can cause traumatic nerve injury and ischaemia, which can lead to tissue damage and increased risk of infectious surgical complications after prolonged application times [23, 24]. In humans and horses, both placement and removal of a tourniquet induce moderate pain [23–25].

Due to the lack of literature on this subject, this study compared IVRA and NBA in dairy cows in terms of their efficacy, time until onset and hormonal-metabolic and cardio-respiratory stress responses.

Methods

After ethical review, the Office of Consumer Protection and Food Safety of the Federal State of Lower-Saxony, Germany (Ethics Committee for Animal Experiments; file number: 33.19–42,502–04-1,511,970) approved the study. All experimental cows used in the study were owned by the Clinic for Cattle of the University of Veterinary Medicine Hannover.

Study 1

Experimental animals and study design

In a crossover designed study, eight non-lactating, non-pregnant, healthy and unsedated German Holstein Friesian dairy cows (aged 3–6 years) were used to compare the efficacy of anaesthesia, as well as the hormonal-metabolic and cardio-respiratory stress responses, after IVRA and NBA (8 cows × 2 treatments). The washout period was two weeks. Four cows received IVRA first and two weeks later NBA and in four cows the order of treatments was vice versa. Experimental animals were housed together in a straw yard and fed a diet of grassland hay and 1 kg of concentrate for maintenance with ad libitum access to fresh water. To apply IVRA and NBA, test for anaesthesia efficacy, and evaluate the stress responses to anaesthesia, cows were positioned in left lateral recumbency (LR) on a hydraulic surgical tipping table padded with a rubber-foam mattress (Werner GmbH, Höhenkirchen, Germany).

Local anaesthesia at the distal hind limb

Intravenous regional anaesthesia (IVRA) In all cows, IVRA was administered in the right distal hind limb using the common dorsal digital vein for aseptic intravenous administration of local anaesthetics [9, 18, 20, 26, 27]. The injection site was washed, shaved, and degreased with medical alcohol and disinfected with iodine. A trained surgeon (N.K.) placed a rubber tourniquet (Esmarchbinde, 6 cm width, 1.1 mm thickness, Wirtschaftsgenossenschaft deutscher Tierärzte, Garbsen, Germany) circumferentially in the middle of the metatarsus and punctured the vein distally from the tourniquet with a hypodermic needle (1.0 × 40 mm, Henry Schein, Gillingham, UK). Under gentle flexion of the fetlock joint, blood was drained through the needle until the pressure dropped as indicated by slow dripping rather than blood running out of the hub. Twenty millilitres of procaine hydrochloride 2% (Procasel 2%, Selectavet, Weyarn-Holzolling, Germany) were injected slowly. The injection site was compressed with an iodine soaked cotton swab for 30 s after removing the needle to avoid unintended drainage of the local anaesthetic from the punctured vein or formation of a haematoma.

Hindfoot nerve block anaesthesia (NBA) For the slightly modified technique of hindfoot four-point nerve block anaesthesia according to Raker [28] in the right hind limb injection sites were surgically prepared as described above. Procaine (Procasel 2%, Selectavet) was injected perineurally (hypodermic needle, 0.8*40 mm, Henry Schein) next to the medial and the lateral plantar nerves (each 10–15 ml), both continuations of the tibial nerve (at approximately the middle of the metatarsus, between the suspensory ligament (interosseous m.) and the flexor tendons on the medial and lateral aspect of the limb). Anaesthesia of the superficial and deep fibular (peroneal) nerves was little more proximal than described by Raker [28]. Perineural injections of about 15 ml procaine anaesthetized the superficial fibular nerve (subcutaneously, latero-dorsally, 2 cm proximally to the tarsocrural joint) and of about 10 ml the deep fibular nerve (dorso-medially in the proximal third of the metatarsus, between the tendons of the long digital extensor and the metatarsal bone, in the longitudinal sulcus of the metatarsus).

Testing of distal limb desensitization Desensitization was tested at room temperature (21 °C) in LR by means of electrical, thermal and mechanical stimulation. Stimulation was stopped and the actual electrical, thermal, and mechanical nociceptive threshold was recorded as soon as the cows expressed aversion to the stimulation as indicated by attempts to withdraw the leg, claw movements, muscle twitching or when the cut-off value was reached.

Electrical nociceptive stimulation For constant current (CC) electrical nociceptive stimulation [29] a commercial nerve stimulator (Grass S48, CCU1; Grass Medical Instruments, MA, USA) was used as previously described [30]. At the lateral coronary band of the lateral claw of the right hind limb, two surface electrodes (Neuroline® 70,005-J/12, Ambu GmbH, Bad Nauheim, Germany) were attached to the shaved, cleaned, dried and degreased skin with an inter-electrode distance of two cm. Skin resistance measured before stimulation was always below 3 kΩ (digital multimeter, VC260, Voltcraft, Conrad Electronic SE, Germany). Noxious stimuli consisted of a 25 ms train of five 1 ms CC square-wave pulses. The trains of five were delivered at a frequency of 5 Hz. The voltage was continuously increased (5 V/s) until a maximum voltage of 150 V. For statistical evaluation, mean values of duplicate tests (2 min intervals) were used.

Thermal nociceptive stimulation Thermal nociceptive stimulation was performed by means of the Thermal Threshold Testing System (TT1) (Topcat Metrology Ltd., Ely, Cambridgeshire, UK) with a safety cut-off set to 55 °C and a heating rate of 0.6 °C/s as previously described for horses [31, 32]. A 5 g thermal probe was attached to the shaved, cleaned, dried, and degreased skin at the lateral coronary band of the lateral right claw and was fixed with an elastic bandage (Vetrap, width 10 cm; Wirtschaftsgemeinschaft deutscher Tierärzte) to guarantee constant skin contact with the thermal probe. The thermal probe was equilibrated with the skin temperature for 5 min.

Mechanical nociceptive stimulation For mechanical nociceptive stimulation, a handheld Medio-Line 40,025 (Schumann GmbH, Sillerup, Germany) with a blunt tip of 1.7-mm diameter and a safety cut-off set at 20 N was used. The force was increased with a rate of 1 N/s until the animal had an aversive reaction [33]. Stimulation was applied laterally and medially to the flexor tendons, approximately 5 cm above the dew claws, at the soft skin immediately proximal to the lateral and medial heel bulb of the right lateral claw, and medio- and latero-dorsally of the first phalanx. Nociceptive testing included two consecutive mechanical stimulations with 2 min interstimulation intervals.

Macroscopic examination of the skin at nociceptive stimulation sites Before the cows were returned to standing from LR, the adhesive electrodes and thermal probe were removed. The skin was examined for alterations, visually and by palpation (reddening, swelling, lesions), where the electrodes and thermal probe were located and at the site of mechanical stimulation.

Evaluation of cardio-respiratory, hormonal, and metabolic stress responses

Instrumentation One hour before the start of the experiment, cows received indwelling catheters under aseptic conditions into their right jugular vein (Stericlin®, 2.4 × 200 mm, Walter Veterinär-Instrumente e.k., Baruth/Mark, Germany) and their right auricular artery (Vygon, 20G–L, Leader-Flex, Ecouen, France) as previously described [34]. For placement, cows were fixed with a head halter and two abdominal belts to the surgical table in standing position.

Cardio-respiratory stress responses A patient monitor (IntelliVue MP50; Philips Medizin Systeme, Germany) continuously recorded mean arterial blood pressure (MAP; IntelliVue MP50; Philips Medizin Systeme, Germany) after connection of the auricular artery catheter to a pre-calibrated fluid-filled pressure transducer (Smith pvb, REF ST-37, Critical Care GmbH, Germany) [35]. The level of the shoulder joint was considered as the zero pressure point in standing cows, and the centreline of the sternum was the reference point during LR. The auricular artery catheter was continuously flushed (3 ml/h) with heparinized saline solution [34]. The patient monitor also recorded the heart rate (HR). The respiratory rate (RR) was determined by counting thoracic expansions for a period of 1 min.

Hormonal and metabolic stress responses Venous blood samples were collected in predetermined intervals from the jugular vein catheter and centrifuged at 1500 x *g*. Serum from tubes without anticoagulant (Sarstedt, Nümbrecht, Germany) and plasma from Li-Fluorid tubes (Sarstedt) were stored at −80 °C until analysis. Plasma concentrations of glucose and lactate, and serum concentrations of non-esterified fatty acids (NEFA), were measured using commercial test kits on an automated analysing system (ABX Pentra 400, Moriba ABX Diagnostics, Montpellier, France), and serum cortisol concentrations were measured by a commercial chemiluminescent enzyme immunoassay (Siemens Diagnostics, Germany).

Experimental protocol of study 1

To apply LA, test distal limb desensitization by nociceptive stimulation, and record hormonal-metabolic and cardio-respiratory stress responses to LA, cows were restrained for approximately one hour in lateral recumbency on a surgical tipping table without sedation [34]. During LR front legs remained loose and hind limbs were fixed with a shackle (5 cm width) at the metatarsus to a foam cushion (7 cm height, 10 cm width and length) padded leg halter. The IVRA tourniquet was removed approximately 8 min after the cows were returned to standing position.

Cows were stimulated (electrical, thermal, and mechanical nociceptive stimulation) 10 min before and 15 min and 30 min after LA.

Hormonal-metabolic and cardio-respiratory stress responses to local anaesthesia (5 min before, and 2, 14, 29, and 40 min after LA) and to restraint in LR (20 min and 5 min before and 7 min after restraining) were studied in predetermined intervals. After cows were returned to standing stress responses to removal of the tourniquet were recorded in IVRA cows (5 min before and 10 and 20 min after removal of the tourniquet) and in NBA cows (no tourniquet application) at corresponding time points.

Study 2

Experimental animals and study design

Six healthy, non-lactating, non-pregnant and unsedated German Holstein cows (aged 4–6 years) were used in a cross-over design (6 cows × 2 treatments) to evaluate the time until onset of anaesthesia after IVRA and NBA by electrical nociceptive stimulation. Cows were fixed in left LR on a surgical tipping table for LA and nociceptive stimulation at the right hind limb.

Electrical nociceptive stimulation electrodes were placed on the dorso-lateral coronary band and on the soft skin immediately proximal to the lateral heel bulb of the right lateral hind claw. Electrical stimulations were performed as described above. Cows were stimulated five minutes before, and 5 min, 7.5 min, 10 min, 12.5 min, 15 min, and 20 min after LA.

Statistical evaluation

Statistical evaluation was performed using the Statistical Analysis System package (SAS version 9.3 for Windows, SAS institute Inc., Cary, NC, USA). Results of the nociceptive electrical stimulation and stress response to anaesthesia, change in body position, and tourniquet removal were evaluated by 2f–ANOVA in a general linear model (Proc GLM) with repeated statements (factors: group [1: IVRA, 2; NBA], time, group*time). The paired *t*-test was used to evaluate differences between mean values at specific time points within groups compared to baseline values. Group differences in the frequencies of non-responses to maximal electrical nociceptive stimulation were determined by Fisher's EXACT Test (Proc freq) for significance. The significance level was set at $p < 0.05$.

Results

Study 1

Nociceptive stimulation

Electrical nociceptive stimulation Before LA in both groups, the mean threshold was 36.4 ± 8.8 V (range: 25 V – 60 V; Fig. 1) until aversive responses to electrical nociceptive stimulation were noticed. All eight cows in the NBA group, and four of the eight cows in the IVRA group expressed no aversive reaction to the cut-off level of 150 V fifteen min after placement of the local

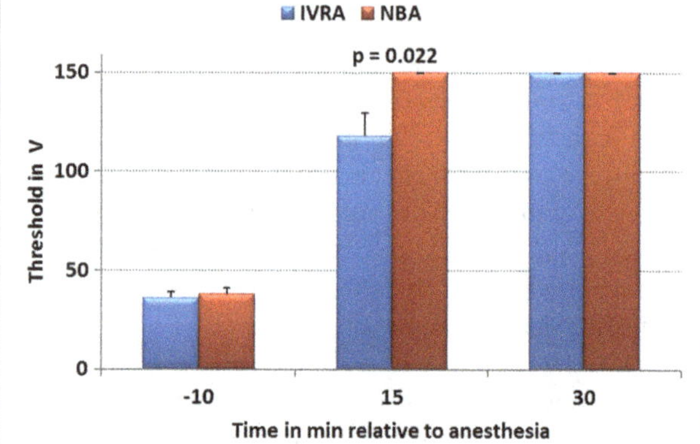

Fig. 1 Mean thresholds (mean ± SEM) of electrical nociceptive stimulation at the lateral coronary band of the lateral claw of the right hind limb until an aversion response was noticed in dairy cows ten minutes before and 15 and 30 min after intravenous regional anaesthesia (IVRA; *N* = 8) and hindfoot nerve block anaesthesia (NBA; N = 8; study 1). Cut-off of electrical stimulation was 150 V (ANOVA effects: group: $p = 0.033$, time: $p < 0.001$, group x time: $p = 0.019$; significant group differences in mean values at specific time points are indicated by *p*-values within graph)

anaesthetic (four cows responded at 75 V, 80 V, 90 V, and 100 V, respectively) (LA) (Fisher's EXACT Test: 8 NBA cows vs. 4 IVRA cows with no response: $p = 0.077$). No cows responded to electrical stimulation at the 150 V level 30 min after anaesthesia. The ANOVA revealed significant effects for group ($p = 0.033$) and time ($p < 0.001$; group x time: $p = 0.019$).

After removing the adhesive electrodes from the coronary band macroscopically, no skin alterations were detected except for an adelomorphic hyperaemia of the skin in all cows, which completely disappeared within 24 h after stimulation.

Thermal nociceptive stimulation Nine of the 16 cows showed no aversive response at the safety cut-off level of 55 °C upon thermal stimulation before LA. Four of the remaining cows responded to thermal stimulation at 47 °C, 49 °C, 50 °C, 51 °C, and three cows responded at 52 °C. No cows responded at the safety level of 55 °C 15 min and 30 min after LA. Before anaesthesia, over 50% of the tested cows showed no response to maximal thermal stimulation; therefore, these results were not used for further statistical evaluation.

After removing the thermal probe from the coronary band, the skin area was reddened but without obvious swelling or exsudation. The next day, the skin reddening disappeared almost completely, and swelling and exudation were not obvious, but a very thin, dry, transparent, corrugated tissue layer separated from the skin was observed. The alteration was macroscopically diagnosed as a desquamated superficial corneal layer of the epidermis.

Mechanical nociceptive stimulation Before LA application 7, 10, 13, 14, 13, and 14 cows of the 16 cows had no response to mechanical nociceptive stimulation at the skin of the lateral and medial heel bulb, the skin lateral and medial to the flexor tendons and dorsal at the lateral and medial first phalanx, respectively, at the cut-off level of 20 N. Cows responding to mechanical stimulation reacted at the different sites between 14 and 19 N. No cow responded at the safety cut-off level of 20 N at any site 15 min and 30 min after LA. No further statistical evaluation of these results was performed as a high proportion of cows revealed no aversive response to mechanical nociceptive stimulation prior to anaesthesia. No skin alterations were apparent macroscopically after mechanical stimulation.

Stress responses to changes in body position

Changing the cow's body position from standing to LR on a surgical tipping table resulted in significantly increased blood concentrations of mean cortisol ($p < 0.001$), lactate ($p < 0.001$), NEFA ($p = 0.038$), and in the HR ($p = 0.002$), while blood glucose concentrations, RR, and MAP remained statistically unaffected (Table 1). No statistically significant group differences or group*time interactions were found.

Stress response to local anaesthesia

After administering LA, a significant increase occurred in the mean blood concentrations of cortisol (time effects: $p < 0.001$), lactate ($p = 0.013$), glucose ($p < 0.001$), and NEFA ($p = 0.001$), and in HR ($p = 0.005$), RR ($p < 0.001$), and MAP ($p = 0.016$; Table 2). No statistically significant group effects were found, but the RR revealed a group*time interaction of $p = 0.016$. Results revealed a significant increase in mean RR after IVRA while RR remained almost unchanged after NBA.

Stress response to removal of tourniquet

The IVRA tourniquet was removed from the metatarsus after cows were returned from LR to standing. In NBA cows no tourniquet was used. Mean blood concentrations of cortisol, lactate, and NEFA as well as HR, RR, and MAP significantly decreased over time, while blood glucose concentrations continued to increase (Table 3). Statistical analysis revealed significant group*time interactions for NEFA ($p = 0.015$), and MAP ($p = 0.010$), and a group effect for MAP ($p = 0.004$). Before removal of the tourniquet, mean serum NEFA concentrations ($p < 0.05$), and before and after tourniquet removal MAP ($p < 0.05$) were greater in IVRA compared to NBA cows.

Study 2

Electrical nociceptive stimulation

Before LA, mean thresholds of electrical nociceptive stimulation at the dorso-lateral coronary band was 30.4 ± 4.3 V (range: 25–40 V). At the heel bulb skin, aversive responses to electrical nociceptive stimulation were registered after exceeding a mean threshold of 41.2 ± 8.7 V (range: 25–60 V; $p = 0.015$; Figs. 2 and 3). Mean thresholds of electrical stimulation at the dorso-lateral coronary band in cows from study 1 (Fig. 1) and study 2 (Fig. 2) were not significantly different ($p = 0.32$) before LA.

At the dorso-lateral coronary band (time effect: $p < 0.001$), as well as at the heel bulb skin, a steady increase in mean electrical stimulation thresholds was recorded after LA (time effect: $p < 0.001$). After NBA, the recorded increase in mean thresholds at the coronary band (ANOVA group effect: $p = 0.018$, group x time effect: $p < 0.001$) and at the heel skin (ANOVA group effect: $p = 0.014$, group x time effect: $p = 0.001$) were faster than those of IVRA (Figs. 2 and 3).

At the coronary band, no cows in the NBA group showed an aversive reaction 10 min after anaesthesia when stimulated at the cut-off level of 150 V, while 4 of

Table 1 Hormonal-metabolic (blood concentrations of cortisol, lactate, glucose, and NEFA) and cardio-respiratory stress responses (HR, RR, MAP; mean ± SEM) to lateral recumbency (LR) restraint in dairy cows (before local anaesthesia in a distal hind distal limb; IVRA: intra venous regional anaesthesia, *N* = 8; NBA: hindfoot nerve block anaesthesia, N = 8; t-20: 20 min before, t-5: five min before, t5: five min after restraining the cows)

		Time	Relative	To	LR		(Minutes)	Statistics		(p-values)
		t-20		t-5		t5				
Parameter	Group	Mean	SEM	Mean	SEM	Mean	SEM	Group	Time	Group*time
Cortisol	IVRA	20.7	3.0	18.0	2.2	**43.2**	5.2	0.19	< 0.001	0.70
(ng/ml)	NBA	17.9		10.7		**37.2**				
Lactate	IVRA	2.04	0.20	2.04	0.23	**2.61**	0.30	0.31	< 0.001	0.98
(mmol/l)	NBA	2.34		2.36		**2.93**				
Glucose	IVRA	3.90	0.13	3.91	0.15	3.81	0.09	0.14	0.30	0.91
(mmol/l)	NBA	4.16		4.13		3.99				
NEFA	IVRA	262	45	275	49	**318**	70	0.74	0.038	0.89
(µmol/l)	NBA	231		250		**298**				
HR	IVRA	62	4	63	5	**70**	6	0.12	0.002	0.54
(1/min)	NBA	71		71		**83**				
RR	IVRA	28	3	30	3	31	3	0.59	0.51	0.46
(1/min)	NBA	27		30		26				
MAP	IVRA	142	5	145	5	141	7	0.76	0.75	0.37
(mmHg)	NBA	143		143		149				

Bold mean values are significantly different ($p < 0.05$) from the baseline values within groups
NEFA non-esterified fatty acids, *HR* heart rate, *RR* respiratory rate, *MAP* mean arterial blood pressure

Table 2 Hormonal-metabolic (blood concentrations of cortisol, lactate, glucose, and NEFA) and cardio-respiratory stress responses (HR, RR, MAP; mean ± SEM) to local anaesthesia in the distal hind limb (IVRA: intravenous regional anaesthesia, N = 8; NBA: hindfoot nerve block anaesthesia, N = 8; t-5: 5 min before, t2, t14, t29, t40: 2, 14, 29, and 40 min after anaesthesia)

		Time	Relative	To	Anesthesia		(Minutes)					Statistics		(p-values)
		t-5		t2		t14		t29		t40				
Parameter	Group	Mean	SEM	Mean	SEM	Mean	SEM	Mean	SEM	Mean	SEM	Group	Time	Group*time
Cortisol	IVRA	48.7	6.6	53.5	6.9	**65.4**	7.0	**73.3**	7.0	**77.4**	6.3	0.90	< 0.001	0.17
(ng/ml)	NBA	43.1		**61.9**		**62.8**		**69.3**		**81.6**				
Lactate	IVRA	2.99	0.54	**4.36**	0.76	**5.02**	0.78	**5.57**	0.99	**5.60**	1.02	0.80	0.013	0.34
(mmol/l)	NBA	3.76		4.05		4.01		4.77		5.65				
Glucose	IVRA	3.81	0.14	4.01	0.11	4.06	0.16	4.48	0.25	**4.84**	0.30	0.66	< 0.001	0.78
(mmol/l)	NBA	3.98		**4.21**		4.18		**4.46**		**4.86**				
NEFA	IVRA	324	69	321	72	346	91	**397**	81	**451**	84	0.70	0.001	0.14
(µmol/l)	NBA	308		320		308		328		358				
HR	IVRA	73	6	**83**	7	68	6	**84**	8	**84**	7	0.53	0.019	0.44
(1/min)	NBA	76		**88**		82		**83**		**88**				
RR	IVRA	23	3	**32**	4	**29.8**	3.1	**33.0**	3.2	**36.3**	4	0.90	< 0.001	0.016
(1/min)	NBA	29		29		27.8		**35.3**		31.8				
MAP	IVRA	137	5	**148**	5	**148**	4	145	4	**146**	4	0.35	0.016	0.47
(mmHg)	NBA	146		148		151		149		154				

Bold mean values indicate significant differences ($p < 0.05$) in mean values within groups compared to baseline values
NEFA non-esterified fatty acids, *HR* heart rate, *RR* respiratory rate, *MAP* mean arterial blood pressure

Table 3 Hormonal-metabolic (blood concentrations of cortisol, lactate, glucose, and NEFA) and cardio-respiratory stress responses (HR, RR, MAP; mean ± SEM) in standing cows to removal of the metatarsal tourniquet used for IVRA in a hind limb (intravenous regional anaesthesia, N = 8)

		Time	Relative	To	Tourniquet	Removal	(Min)	Statistics		(p-values)
		t-5		t10		t20				
Parameter	Group	Mean	SEM	Mean	SEM	Mean	SEM	Group	Time	Group*time
Cortisol	IVRA	76.2	6.8	72.7	7.0	**62.9**	7.2	0.83	0.002	0.84
(ng/ml)	NBA	72.9		71.7		61.0				
Lactate	IVRA	5.71	0.94	**5.02**	0.85	**3.52**	0.61	0.82	< 0.001	0.27
(mmol/l)	NBA	5.14		**4.70**		**3.63**				
Glucose	IVRA	5.17	0.34	5.17	0.32	5.35	0.35	0.49	0.017	0.32
(mmol/l)	NBA	5.34		**5.53**		5.81				
NEFA	IVRA	481[a]	62	435	59	**315**	44	0.11	< 0.001	0.015
(μmol/l)	NBA	297[b]		293		**246**				
HR	IVRA	84	8	86	6	80	6	0.97	0.088	0.16
(1/min)	NBA	95		**79**		**76**				
RR	IVRA	39.5	4.1	34.8	5.2	**30.0**	3.2	0.34	0.007	0.71
(1/min)	NBA	32.0		30.8		25.3				
MAP	IVRA	168[a]	4	**151[a]**	3	**146[a]**	3	0.004	< 0.001	0.010
(mmHg)	NBA	142[b]		142[b]		135[b]				

Recordings at corresponding time points in cows after NBA in which no tourniquet was used (hindfoot nerve block anaesthesia, N = 8; t-5: 5 min before, and t10, t20: 10 and 20 min after tourniquet removal in IVRA cows and at corresponding time points in NBA cows). Significant differences (p < 0.05) between corresponding group mean values at specific time points are indicated by different superscript letters (a,b). Bold mean values indicate significant differences (p < 0.05) in mean values within groups compared to baseline values
NEFA non-esterified fatty acids, *HR* heart rate, *RR* respiratory rate, *MAP* mean arterial blood pressure

the 6 cows in the IVRA group responded (responses at 65 V, 120 V, 120 V, and 140 V). At 20 min after anaesthesia, one IVRA cow responded to electrical stimulation at 80 V (Fisher's EXACT test: group effect: $p = 0.008$).

At the heel bulb skin 10 min after NBA, no cows responded to the cut-off stimulation of 150 V, except one who responded at 140 V. The last cow in the NBA group reached the cut-off level at 12.5 min after LA. Ten minutes after IVRA, three of six cows responded to

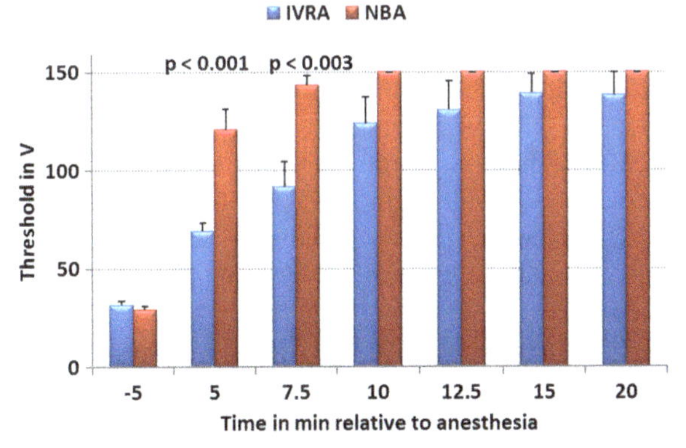

Fig. 2 Mean thresholds (mean ± SEM) of electrical nociceptive stimulation at the lateral coronary band of the lateral claw of the right hind limb until an aversion response was noticed in dairy cows before (t-5 min) and after intravenous regional anaesthesia (IVRA; $N = 6$) and hindfoot nerve block anaesthesia (NBA; N = 6; study 2). Safety cut-off of electrical stimulation was 150 V (ANOVA effects: group: $p = 0.018$, time: $p < 0.001$, group x time: $p < 0.001$; significant group differences in mean values at specific time points are indicated by *p*-values within the graph)

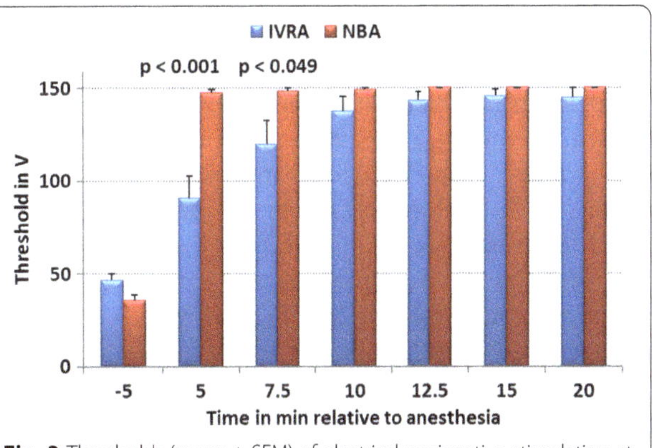

Fig. 3 Thresholds (mean ± SEM) of electrical nociceptive stimulation at the heel bulb skin of the lateral claw of the right hind limb until an aversion response was noticed in dairy cows before (t-5 min) and after intravenous regional anaesthesia (IVRA; N = 6) and hindfoot nerve block anaesthesia (NBA; N = 6; study 2). Safety cut-off of electrical stimulation was 150 V (ANOVA effects: group: $p = 0.014$, time: $p < 0.001$, group x time: $p < 0.001$; significant group differences in mean values at specific time points are indicated by p-values within the graph)

100 V, 130 V, and 140 V stimulation. Twenty min after IVRA, one cow responded to 120 V stimulation (Fisher's EXACT test: group effect: $p < 0.001$).

Discussion

Intravenous regional anaesthesia [9, 18, 19, 36] and hindfoot nerve blocks [28] are long-standing recommended techniques for anaesthesia of the bovine digit. Onset of anaesthesia occurs approximately five minutes after IVRA and is optimal after ten minutes for major surgeries [20, 21, 26, 27, 37]. However, our study found that not all cows were completely desensitized to electrical stimulation of the distal limb at 15 min (study 1) and 20 min (study 2) after IVRA. In contrast, all cows in the NBA group were completely desensitized after approximately 12 min based on electrical stimulation results. This relatively slow onset of anaesthesia after IVRA contradicts previous reports and may be relevant in the field when major surgeries are performed under time constraints. In this study, only healthy animals were used. However, when anaesthesia of the distal limb is required for major surgeries of the claw, the affected distal limb frequently exhibits infectious cellulitis [9]. Onset and quality of IVRA and NBA in infectious, hyperaemic, and inflamed limbs may deviate from these results. In this study, 20 ml of procaine 2% was used for IVRA. Higher injection volumes [21, 27, 38] of procaine may expedite the desensitization process after IVRA. Other local anaesthetics, or concentrations higher than 2%, may also result in faster desensitization of the distal limb. This study used the common dorsal digital vein for procaine injections; however, use of the medial or lateral common plantar digital veins may reduce anaesthesia onset times. Before IVRA is used in humans, exsanguination is achieved using pneumatic tourniquets and lifting the treated extremity [24, 39, 40]. In this study, blood was drained from the punctured vein under gentle flexion of the fetlock joint, which is likely less effective in clearing space for local anaesthetics within the vascular system than the use of pneumatic tourniquets. However, options for improving IVRA efficacy have not been tested in cattle.

A valid instrument for demonstrating pain should produce a consistent response, show a correlation between stimulus intensity and pain and be easily interpreted by the observer. It should be easy to apply and have a clear cut-off point, while not damaging the tissue [41]. Electrical, mechanical and thermal nociceptive tests are validated in various species and commonly applied [41–46]. Electrical stimuli activate nonspecific peripheral fibres, including large-diameter fibres not directly involved in nociception [42]. Mechanical stimuli generate action potentials due to ionic influx in channels activated by stretch, and thermal stimuli involve membrane proteins or intracellular effector molecules with high temperature coefficients [43]. Because electrical stimulation is seen as a nonspecific nociceptive test, thermal and mechanical stimulation were also used to evaluate efficacy of distal limb desensitization after LA; however, a high proportion of cows did not respond to thermal or mechanical stimuli at the chosen cut-off levels before LA. Accordingly, mechanical and thermal stimulation appears unsuitable to detect nociception in cows restrained on a tipping table in LR.

Luna et al. [41] validated electrical, mechanical and thermal stimulation and found all techniques were valid for assessing nociception in horses. Mechanical and thermal stimulation were performed using wireless devices while the horses were free and undisturbed in their stalls. The authors reported a thermal stimulation response at an average of 51 °C at the limb and 55 °C at the thorax using a safety cut-off level of 60 °C. Most horses responded to thermal stimulation prior to the cut-off. The reported thermal thresholds were close to or at the safety cut-off level of 55 °C used in this study. Ambient temperature influences the outcome of thermal nociceptive testing [44]. In horses, thermal stimulation at higher ambient temperatures (> 20 °C: 53 ± 4 °C) resulted in greater thresholds until a response occurred compared to low ambient temperatures (10 °C; 49 ± 6 °C) [31]. Performing thermal nociceptive stimulation at room temperature (21 ° C) on the cows in this study may have contributed to the high frequency of non-responders at the cut-off level. It appears that a cut-off level of 60 °C would have reduced the number of non-responders to thermal stimulation in this study. However, it is important to consider that all cows exhibited superficial skin alterations at the coronary band after thermal stimulation using the 55 °C cut-off level. Before higher cut-off levels for thermal stimulation are used, more studies are needed to evaluate the degree of tissue damage from heat stimulation at different cut-off levels and anatomic locations after single and repeated stimulations.

In this study, using a probe tip of 1.7 mm diameter and a safety cut-off of 20 N, most cows did not respond to mechanical stimulation at the distal limb even before LA. In contrast, unrestrained horses always responded to mechanical stimulation at the metacarpus using a wireless device with a 1 mm probe. The average nociceptive threshold at which the horses responded was 3–4 N [41]. Recently, Raundal et al. [45] compared two handheld devices for mechanical stimulation, a 0.8 mm vs. 6.5 mm diameter probe, at the metatarsus of unrestrained cattle. On average cows reacted aversively to both the 0.8 mm and 6.5 mm probes at 4 N and at 45 N, respectively. Other authors have reported mean mechanical nociceptive stimulation thresholds at the coronary band in cattle to be 14 N using a 2.5 mm probe [46], and 9 N using a 2 mm probe, at the dorsal aspect of the metatarsus [47]. Thus, it is unlikely that the stoic nature of cows [13] explains the high frequency of non-responders to mechanical nociceptive stimulation in this study.

Restraining cows in LR on a hydraulic surgical tipping table allowed safe LA application. Artificial isolation from the herd, close human-cow contact, a novel environment and restraint induce stress response [15, 34] as indicated by increased blood concentrations of cortisol, lactate, and NEFA, and by an increased HR. LR also impairs respiration in cows leading to a moderate increase in arterial pCO_2 and a decrease in pO_2. This impairment, together with the adrenergic stress response, is reflected here and in previous studies [15, 34] by an increase in blood lactate levels caused by anaerobic glycolysis in peripheral tissues. Clinical pain is a complex mechanism involving sensory-discriminative, cognitive and emotional aspects [41]. An increase in nociceptive thresholds after exposure to acute stressors (stress induced hypo- or analgesia: SIA) [44] has been found in various species [48, 49] including cattle [50, 51]. It is likely that LR-induced SIA contributed to the cows' unresponsiveness to mechanical and thermal nociceptive stimulation. Accordingly, needle pricks to the interdigital space as a common nociceptive test for foot desensitization [38] after LA and before surgical interventions should be reconsidered. Major claw surgeries are common on restrained and sick cows. Thus, SIA may also modulate needle prick results, providing inaccurate information on the degree of desensitization after LA.

Administering IVRA requires placing a tourniquet. Human patients commonly experience tourniquet pain within 30 min after placement [39, 52–54]. In this study, mean hormonal-metabolic and cardiac stress responses to IVRA and NBA in cows, were of similar magnitude (except the respiratory rate), giving no clear indication of specific pain-related stress responses by either technique. However, pain-induced stress responses may have been masked by stress from the LR. In the IVRA group the tourniquet was removed from the metatarsus after returning the cows back to standing position. The significantly higher mean blood NEFA concentration and MAP in the IVRA cows before and after removing the tourniquet compared to those of the NBA group may indicate a stress response induced by tourniquet pain.

Conclusion

Despite the small sample number in this study, the results from the electrical nociceptive stimulation indicate that intravenous regional anaesthesia and hindfoot four-point nerve blocks completely desensitize the distal limb in dairy cows. However, complete desensitization developed significantly faster after NBA (12 min) than after IVRA (≥ 20 min).

This study found no specific stress re-sponses to placement of the tourniquet, but they may have been masked by the responses of the animals to re-straint in lateral recumbency. The time difference until complete desensitization between both LA techniques in this study appears clinically relevant, as distal limb an-aesthesia is often necessary for major claw surgeries under time constraints in the field.

Abbreviations

HF: Holstein Friesian; HR: Heart rate; IVRA: Intravenous regional anaesthesia; LA: Local anaesthesia; LR: Lateral recumbency; MAP: Mean arterial blood pressure; NBA: hindfoot four-point nerve block anaesthesia; NEFA: Non-esterified fatty acids; RR: Respiratory rate

Acknowledgements

The authors appreciate the support of the animal keeper of the Clinic for Cattle during the practical part of the study, particularly the help of Marian Overmann, Michael Sachs, Lars Klütz, and Rolf Schwarze.

Funding

This study was self-funded, no external resources were used.

Authors' contributions

JR, SBRK, AS, MH, SY and WK conceived the project. SY, NK, GS and JR accomplished the practical part of the study. SY, EE, and CP performed dissections of hind limbs to verify appropriate injection sites. MS performed laboratory analysis. SY and JR carried out statistical analysis. All authors participated in interpretation of results. SY, SBRK and JR drafted the paper, which was revised by all co-authors through contributions to the content of the paper. All authors read and approved the final manuscript.

Consent for publication

Not applicable.

Competing interests

The authors declare that they have no competing interests.

Author details

[1]Clinic for Cattle, University of Veterinary Medicine Hannover, Foundation, Hannover, Germany. [2]Department for Veterinary Surgery, College of Veterinary Medicine, An-Najah National University, Nablus, Israel. [3]Clinic for Ruminants, University of Leipzig, Leipzig, Germany. [4]Institute for Anatomy, University of Veterinary Medicine Hannover, Foundation, Hannover, Germany. [5]Clinic for Horses, University of Veterinary Medicine Hannover, Foundation,

Hannover, Germany. [6]Clinic for Small Animals, University of Veterinary Medicine Hannover, Foundation, Hannover, Germany.

References

1. Murray RD, Downham DY, Clarkson MJ, Faull WB, Hughes JW, Manson FJ, Merritt JB, Russell WB, Sutherst JE, Ward WR. Epidemiology of lameness in dairy cattle: description and analysis of foot lesions. Vet Rec. 1996;138(24): 586–91.
2. Barker ZE, Leach KA, Whay HR, Bell NJ, Main DC. Assessment of lameness prevalence and associated risk factors in dairy herds in England and Wales. J Dairy Sci. 2010;93(3):932–41.
3. Main DC, Barker ZE, Leach KA, Bell NJ, Whay HR, Browne WJ. Sampling strategies for monitoring lameness in dairy cattle. J Dairy Sci. 2010;93(5): 1970–8.
4. Solano L, Barkema HW, Mason S, Pajor EA, LeBlanc SJ, Orsel K. Prevalence and distribution of foot lesions in dairy cattle in Alberta Canada. J Dairy Sci. 2016;99(8):6828–41.
5. Cook NB, Hess JP, Foy MR, Bennett TB, Brotzman RL. Management characteristics, lameness, and body injuries of dairy cattle housed in high-performance dairy herds in Wisconsin. J Dairy Sci. 2016;99(7):5879–91.
6. Thomas HJ, Miguel-Pacheco GG, Bollard NJ, Archer SC, Bell NJ, Mason C, Maxwell OJ, Remnant JG, Sleeman P, Whay HR, et al. Evaluation of treatments for claw horn lesions in dairy cows in a randomized controlled trial. J Dairy Sci. 2015;98(7):4477–86.
7. Thomas HJ, Remnant JG, Bollard NJ, Burrows A, Whay HR, Bell NJ, Mason C, Huxley JN. Recovery of chronically lame dairy cows following treatment for claw horn lesions: a randomised controlled trial. Vet Rec. 2016;178(5):116.
8. Toussaint Raven E. Cattle Footcare and claw trimming. Marlborough: The Cowood Press Ltd; 2003.
9. Heppelmann M, Kofler J, Meyer H, Rehage J, Starke A. Advances in surgical treatment of septic arthritis of the distal interphalangeal joint in cattle: a review. Vet J. 2009;182(2):162–75.
10. Starke A, Heppelmann M, Beyerbach M, Rehage J. Septic arthritis of the distal interphalangeal joint in cattle: comparison of digital amputation and joint resection by solar approach. Vet Surg. 2007;36(4):350–9.
11. Offinger J, Herdtweck S, Rizk A, Starke A, Heppelmann M, Meyer H, Janssen S, Beyerbach M, Rehage J. Postoperative analgesic efficacy of meloxicam in lame dairy cows undergoing resection of the distal interphalangeal joint. J Dairy Sci. 2013;96(2):866–76.
12. Desrochers A, St Jean G. Surgical management of digit disorders in cattle. Vet Clin Food Anim. 1996;12(1):277–98.
13. Anderson DE, Edmondson MA. Prevention and management of surgical pain in cattle. Vet Clin Food Anim. 2013;29(1):157–84.
14. Edmondson MA. Local and regional anesthesia in cattle. Vet Clin Food Anim. 2008;24(2):211–26.
15. Rizk A, Herdtweck S, Offinger J, Meyer H, Zaghloul A, Rehage J. The use of xylazine hydrochloride in an analgesic protocol for claw treatment of lame dairy cows in lateral recumbency on a surgical tipping table. Vet J. 2012; 192(2):193–8.
16. Janssen S, Wunderlich C, Heppelmann M, Palme R, Starke A, Kehler W, Steiner A, Rizk A, Meyer U, Daenicke S, et al. Short communication: pilot study on hormonal, metabolic, and behavioral stress response to treatment of claw horn lesions in acutely lame dairy cows. J Dairy Sci. 2016;99(9):7481–8.
17. Bier A. A new way to induce local anesthesia in the distal limb (Über einen neuen Weg Lokalanästhesie an den Gliedmaßen zu erzeugen). Arc Klin Chir. 1908;86:1007–16.
18. Antalovsky A. Technika mistni nitrozilni anestezie na distalnich castech koncetin u skotu (technique of intravenous local anesthesia in the distal limb in cattle). Vet Med. 1965;7:413–20.
19. Avemann M. Prüfung des von ANTALOVSKY angegebenen Verfahrens zur intravenösen regionalen Betäubung im Zehenbereich des Rindes auf seine Brauchbarkeit (usefulness of intravenenous regional anaesthesia according to ANTALOVSKY in the distal limb of cattle). Doctoral-Thesis: Univ Vet Med Hannover, Hannover, Germany; 1974.
20. Maierl J, Nuss K. Anatomische Grundlagen und Lokalanästhesie. In: Fiedler A, Maierl J, editors. Nuss K: Erkankungen der Klauen und der Zehen des Rindes. Stuttgart, New York: Schattauer; 2004. p. 29–43.
21. Greenough PR: Regional anesthesia, regional antibiotic perfusion, and nerve blocks. In: Geenough PR: Bovine laminitis and lameness. A hands-on approach. London: Saunders Elsevier; 2007. p.252.
22. Potterton SL, Bell NJ, Whay HR, Berry EA, Atkinson OC, Dean RS, Main DC, Huxley JN. A descriptive review of the peer and non-peer reviewed literature on the treatment and prevention of foot lameness in cattle published between 2000 and 2011. Vet J. 2012;193(3):612–6.
23. Halladin NL, Zahle FV, Rosenberg J, Gogenur I. Interventions to reduce tourniquet-related ischaemic damage in orthopaedic surgery: a qualitative systematic review of randomised trials. Anaesthesia. 2014;69(9):1033–50.
24. Estebe JP, Davies JM, Richebe P. The pneumatic tourniquet: mechanical, ischaemia-reperfusion and systemic effects. Europ J Anaes. 2011;28(6): 404–11.
25. Copland VS, Hildebrand SV, Hill T 3rd, Wong P, Brock N. Blood pressure response to tourniquet use in anesthetized horses. JAVMA. 1989;195(8): 1097–103.
26. Skarda R: Local and Regional Anesthetic Techniques: Ruminants and Swine. In Thurmon JC, Tranquilli WJ, Benson GJ, editors. Lumb & Jones' Veterinary Anesthesia. *Baltimore: Williams & Wilkins, A Waverly Company; 1996. p. 479–514.*
27. Skarda RT: Local and regional analgesia. In: Short CE, editor. Principles & practice of veterinary anesthesia. William & Wilkens, Baltimore, USA; 1987. *P.* 91–133, ISBN 0-683-07702-3.
28. Raker CW. Regional anesthesia of the bovine foot. JAVMA. 1956;128(5):238–9.
29. Levionnois OL, Spadavecchia C, Kronen PW, Schatzmann U. Determination of the minimum alveolar concentration of isoflurane in Shetland ponies using constant current or constant voltage electrical stimulation. Vet Anaesth Analg. 2009;36(1):9–17.
30. Hopster K, Muller C, Hopster-Iversen C, Stahl J, Rohn K, Kastner S. Effects of dexmedetomidine and xylazine on cardiovascular function during total intravenous anaesthesia with midazolam and ketamine and recovery quality and duration in horses. Vet Anaesth Analg. 2014;41(1):25–35.
31. Poller C, Hopster K, Rohn K, Kastner SB. Evaluation of contact heat thermal threshold testing for standardized assessment of cutaneous nociception in horses - comparison of different locations and environmental conditions. BMC Vet Res. 2013;9:4.
32. Poller C, Hopster K, Rohn K, Kastner SB. Nociceptive thermal threshold testing in horses - effect of neuroleptic sedation and neuroleptanalgesia at different stimulation sites. BMC Vet Res. 2013;9:135.
33. Schütter AF, Tunsmeyer J, Kastner SB: Influence of metamizole on 1) minimal alveolar concentration of sevoflurane in dogs and 2) on thermal and mechanical nociception in conscious dogs. Vet Anaesth Analg 2016, 43: 215–226.
34. Rizk A, Herdtweck S, Meyer H, Offinger J, Zaghloul A, Rehage J. Effects of xylazine hydrochloride on hormonal, metabolic, and cardiorespiratory stress responses to lateral recumbency and claw trimming in dairy cows. JAVMA. 2012;240(10):1223–30.
35. Offinger J, Meyer H, Fischer J, Kastner SB, Piechotta M, Rehage J. Comparison of isoflurane inhalation anaesthesia, injection anaesthesia and high volume caudal epidural anaesthesia for umbilical surgery in calves; metabolic, endocrine and cardiopulmonary effects. Vet Anaesth Analg. 2012; 39(2):123–36.
36. Prentice DE, Wyn-Jones GW, Jones RS, Jagger DW. Intravenous regional anaesthesia of the bovine foot. Vet Rec. 1974;94(13):293–5.
37. Trim CM: Special Anesthesia Considerations in the Ruminant. *in Short CE, editor: Principles and Practice of Veterinary Anesthesia. Baltimore: Williams & Williams;* 1987. p.285–308. ISBN 0-683-07702-3:285–308.
38. Hudson C, Whay H, Huxley J: Recognition and management of pain in cattle. *In Practice (0263841X)* 2008, 30(3):126–134.
39. Rodola F, Vagnoni S, Ingletti S. An update on intravenous regional Anaesthesia of the arm. Eur Rev Med Pharmaco. 2003;7(5):131–8.
40. Flamer D, Peng PW. Intravenous regional anesthesia: a review of common local anesthetic options and the use of opioids and muscle relaxants as adjuncts. Local Reg Anesth. 2011;4:57–76.
41. Luna SP, Lopes C, Rosa AC, Oliveira FA, Crosignani N, Taylor PM, Pantoja JC. Validation of mechanical, electrical and thermal nociceptive stimulation methods in horses. Equine Vet J. 2015;47(5):609–14.
42. Le Bars D, Gozariu M, Cadden SW. Animal models of nociception. Pharmacol Rev. 2001;53(4):597–652.
43. Neddermeyer TJ, Fluhr K, Lotsch J. Principle components analysis of pain thresholds to thermal, electrical, and mechanical stimuli suggests a predominant common source of variance. Pain. 2008;138(2):286–91.
44. Love EJ, Murrell J, Whay HR. Thermal and mechanical nociceptive threshold testing in horses: a review. Vet Anaesth Analg. 2011;38(1):3–14.

45. Raundal PM, Andersen PH, Toft N, Forkman B, Munksgaard L, Herskin MS. Handheld mechanical nociceptive threshold testing in dairy cows - intra-individual variation, inter-observer agreement and variation over time. Vet Anaesth Analg. 2014;41(6):660–9.
46. Tadich N, Tejeda C, Bastias S, Rosenfeld C, Green LE. Nociceptive threshold, blood constituents and physiological values in 213 cows with locomotion scores ranging from normal to severely lame. Vet J. 2013;197(2):401–5.
47. Whay HR, Webster AJ, Waterman-Pearson AE. Role of ketoprofen in the modulation of hyperalgesia associated with lameness in dairy cattle. Vet Rec. 2005;157(23):729–33.
48. Rushen J, Ladewig J. Stress-induced hypoalgesia and opioid inhibition of pigs' responses to restraint. Physiol Behav. 1991;50(6):1093–6.
49. Cook CJ, Maasland SA, Devine CE. Social behaviour in sheep relates to behaviour and neurotransmitter responses to nociceptive stimuli. Physiol Behav. 1996;60(3):741–51.
50. Rushen J, Boissy A, Terlouw EM, de Passille AM. Opioid peptides and behavioral and physiological responses of dairy cows to social isolation in unfamiliar surroundings. J Anim Sci. 1999;77(11):2918–24.
51. Herskin MS, Munksgaard L, Ladewig J. Effects of acute stressors on nociception, adrenocortical responses and behavior of dairy cows. Physiol Behav. 2004;83(3):411–20.
52. Chiao FB, Chen J, Lesser JB, Resta-Flarer F, Bennett H. Single-cuff forearm tourniquet in intravenous regional anaesthesia results in less pain and fewer sedation requirements than upper arm tourniquet. Brit J Anaesth. 2013; 111(2):271–5.
53. Kam PC, Kavanagh R, Yoong FF. The arterial tourniquet: pathophysiological consequences and anaesthetic implications. Anaesthesia. 2001;56(6):534–45.
54. Choyce A, Peng P. A systematic review of adjuncts for intravenous regional anesthesia for surgical procedures. Can J Anaesth. 2002;49(1):32–45.

6

Effects of glucose on lactose synthesis in mammary epithelial cells from dairy cow

Ye Lin, Xiaoxu Sun, Xiaoming Hou, Bo Qu, Xuejun Gao and Qingzhang Li*

Abstract

Background: Lactose, as the primary osmotic component in milk, is the major determinant of milk volume. Glucose is the primary precursor of lactose. However, the effect of glucose on lactose synthesis in dairy cow mammary glands and the mechanism governing this process are poorly understood.

Results: Here we showed that glucose has the ability to induce lactose synthesis in dairy cow mammary epithelial cells, as well as increase cell viability and proliferation. A concentration of 12 m*M* glucose was the optimum concentration to induce cell growth and lactose synthesis in cultured dairy cow mammary epithelial cells. In vitro, 12 m*M* glucose enhanced lactose content, along with the expression of genes involved in glucose transportation and the lactose biosynthesis pathway, including GLUT1, SLC35A2, SLC35B1, HK2, β4GalT-I, and AKT1. In addition, we found that *AKT1* knockdown inhibited cell growth and lactose synthesis as well as expression of GLUT1, SLC35A2, SLC35B1, HK2, and β4GalT-I.

Conclusions: Glucose induces cell growth and lactose synthesis in dairy cow mammary epithelial cells. Protein kinase B alpha acts as a regulator of metabolism in dairy cow mammary gland to mediate the effects of glucose on lactose synthesis.

Keywords: Dairy cow, Mammary epithelial cell, Glucose, Lactose synthesis, AKT1

Background

The basic function of the lactating mammary gland is to produce milk, providing nutrients for growth and development of the offspring. The main constituents of bovine milk are lactose, proteins, and lipids [1]. Among these, lactose is the major carbohydrate of milk. It plays a primary role in milk production because it represents the main osmotic constituent in milk and draws water into the mammary epithelial cells [2].

Lactose is synthesized from free glucose and uridine diphosphate (UDP)-galactose. In lactating ruminants, the mammary gland can consume up to 85 % of circulating glucose [3]. Uridine diphosphate-galactose is also derived from glucose. Analysis of lactose synthesis in rodent and bovine models of secretory activation has demonstrated increases in a number of enzyme activities and carbohydrate metabolites that together form the fundamental framework of the lactose synthesis pathway [4, 5]. The lactose synthesis reaction is catalyzed by lactose synthase (UDP-galactose : D-glucose l-galactosyl transferase, EC 2.4.1.22), a complex of β-1, 4-galactosyltransferase (β4GalT-I) and the essential cofactor α-lactalbumin (α-LA), in the Golgi compartment [6, 7]. The expression of α-LA is very low during pregnancy and increases significantly with parturition [8]. Beginning in late pregnancy, the β4GalT-I level in the mammary gland is estimated to increase about 50-fold in preparation for lactose biosynthesis [8]. As the substrate of lactose synthase, the glucose concentration in the lactating mammary epithelial cells is higher than usual [1]. Generally, animal nutrition has a substantial effect on milk composition [9]. Glucose is a dietary factor that affects milk synthesis [10], but whether glucose supplementation can affect lactose synthesis in dairy cow mammary gland is not well understood.

Lactose synthesis and secretion by the mammary gland involve the expression of a large number of genes [9]. However, few studies have focused on the effects of glucose supplementation on the expression of genes involved in lactose biosynthesis. The protein kinase B (PKB, also known as AKT) is a critical downstream

* Correspondence: qzli@neau.edu.cn
Key Laboratory of Dairy Science of Education Ministry, Northeast Agricultural University, Harbin 150030, China

effector in multiple signal transduction pathways and regulates cellular proliferation, survival, and metabolism [11]. Among the members of the AKT family, only AKT1 is upregulated in the mammary gland during pregnancy and lactation [12]. Knockout studies have demonstrated that AKT1 has roles in the functional differentiation of the secretory epithelium and in metabolic pathways that regulate milk synthesis [13]. Loss of AKT1 results in failure of the coordinated metabolic responses required for the establishment of lactation at parturition, including increased glucose uptake and lipid synthesis, which in turn results in decreased milk production [13]. The lactation defect observed in $AKT1^{-/-}$ mice is thus due to a metabolic defect that results from a failure to upregulate glucose transporter 1 (GLUT1) and other AKT1-specific target genes [11]. It is unknown whether glucose supplementation modulation of lactose synthesis could be related to *AKT1* expression in dairy cow mammary gland.

In this study, we hypothesized glucose supplementation could affect lactose synthesis in lactating mammary gland of dairy cow. Additionally, glucose induced lactose synthesis is related to AKT1 expression in lactating cow mammary epithelial cells. To meet these objectives, we evaluated the effects of different concentrations of glucose on mammary epithelial cell survival, proliferation, and lactose synthesis. The expression of genes known to be involved in glucose transportation and lactose synthesis was examined by quantitative real-time PCR (qPCR) and western blot when cells were cultured with DMEM containing 12 m*M* glucose. To evaluate if glucose modulates lactose synthesis via AKT1 activation, siRNA-mediated knockdown of *AKT1* in cultured mammary epithelia cells was performed.

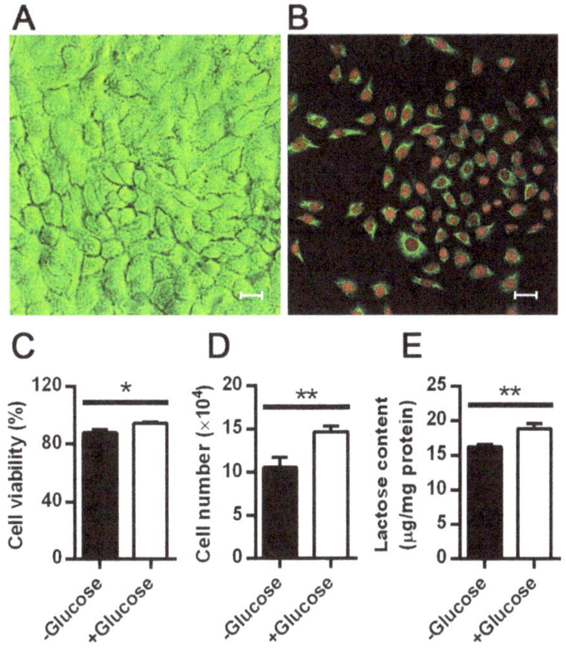

Fig. 1 Glucose induces dairy cow mammary epithelial cell growth and lactose synthesis. **a** Mammary epithelial cells isolated from mid-lactating mammary tissues of dairy cows were acquired using a phase-contrast light microscopy with a Leica L 40 × 0.5 PH2 objective. **b** Immunofluorescence staining for cytokeratin 18 in dairy cow mammary epithelial cells was acquired using a confocal microscopy with a Leica HCX PL Apo CS 40 × 1.25 oil objective. Cytokeratin 18 was stained with FITC (green), and nuclei were stained with propidium iodide (red). For A and B, Scale bar, 75 µm. **c**, **d** The effect of glucose on cell viability (**c**) and proliferation (**d**) in dairy cow mammary epithelial cells. **e** Lactose secretion from dairy cow mammary epithelial cells cultured with or without glucose. Lactose content in medium was measured with the Lactose/D-Glucose (Rapid) Assay kit (K-LACGAR, Megazyme, Ireland, UK). For **c**, **d**, and **e**, mammary epithelial cells were cultured in DMEM with high glucose (+Glucose, 25 mM) or without glucose (-Glucose) for 24 h. Data are shown as the mean ± SEM from 3 independent replicates. *$P < 0.05$, **$P < 0.01$

Results

Effect of glucose on lactose synthesis in dairy cow mammary epithelial cells

Glucose is the primary precursor of lactose in lactating mammary glands. In animal and human models, plasma glucose gives rise to the vast majority of the monosaccharides of lactose [14, 15]. To investigate if glucose supplementation has the ability to induce lactose synthesis in lactating dairy cow mammary gland, we generated mammary epithelial cells from mid-lactating mammary tissues of dairy cows (Fig. 1a). Immunofluorescence staining of cytokeratin 18 was observed in the cytoplasm (Fig. 1b), indicating that the cells we cultured were purified mammary epithelial cells [16].

To detect the effects of glucose on mammary epithelial cell survival, proliferation, and lactose biosynthesis, mammary epithelial cells were cultured in DMEM with or without glucose for 24 h. Cell viability (Fig. 1c) and proliferation (Fig. 1d) were both markedly increased in mammary epithelial cells cultured with high glucose compared with those in glucose-free medium ($P < 0.05$). We also measured the lactose secreted from mammary epithelial cells into the medium. The amount of lactose in medium collected from cells cultured with high glucose for 24 h was significantly higher than that from glucose free-cultured cells ($P < 0.01$, Fig. 1e). These data suggested that glucose induced mammary epithelial cell growth and lactose biosynthesis.

Next, we measured the effects of different concentrations of glucose on lactose synthesis in dairy cow mammary epithelial cells. Mammary epithelial cells were cultured in DMEM with glucose at concentrations ranging from 0 to 20 m*M*. The effects of glucose concentration on cell viability and proliferation were examined after 12, 24, 36, and 48 h of treatment. Cell viability and proliferation both increased for the first 24 h, followed by a decrease (Fig. 2a, b). However, the viability and proliferation of only those cells cultured with 12 m*M* glucose were upregulated significantly, peaking at 24 h, compared with the other glucose concentration groups (at 24 h). The lactose content in the medium increased for the first 24 h, followed by a plateau when cells were cultured with 8, 12, and 16 m*M* glucose. Similarly, lactose content reached highest in the medium when cells were cultured with 12 m*M* glucose for the first 24 h (Fig. 2c). As a result, a concentration of 12 m*M* glucose was determined to be the optimum concentration to induce lactose synthesis in cultured dairy cow mammary epithelial cells.

Effect of glucose on expression of genes involved in lactose synthesis in dairy cow mammary epithelial cells

In lactating mammary gland, lactose synthase catalyzes the conversion of glucose and UDP-galactose to lactose in the Golgi [8]. Glucose is passed across the plasma membrane and Golgi membrane into the Golgi lumen by GLUTs [17]. Uridine diphosphate-galactose is actively transported into the Golgi lumen by solute carrier family 35 member A2 (SLC35A2) and solute carrier family 35 member B1 (SLC35B1) [18]. To explore the molecular process by which glucose induces lactose synthesis, we first examined the expression of *GLUT1*, *GLUT4*, *GLUT8*, *GLUT12*, *SLC35A2*, and *SLC35B1*, which mediate glucose and UDP-galactose transportation in mammary gland [19–21]. As shown in Fig. 3a, *GLUT1*, *SLC35A2*, and *SLC35B1* mRNA levels were significantly increased in mammary epithelial cells cultured in DMEM with 12 m*M* glucose for 24 h ($P < 0.05$), whereas *GLUT4*, *GLUT8*, and *GLUT12* mRNA levels did not change compared to the control. Next, we examined the expression of hexokinase 2 (*HK2*), *α-LA*, *β4GalT-I*, and *AKT1*, which are involved in catalyzing and regulating lactose biosynthesis in lactating mammary gland [22]. The results showed that *HK2*, *β4GalT-I*, and *AKT1* mRNA levels were significantly increased in mammary epithelial cells cultured in DMEM with 12 m*M* glucose for 24 h ($P < 0.05$), whereas expression of *α-LA* remained relatively unchanged (Fig. 3b).

On the basis of the mRNA expression data, we analyzed protein expression of GLUT1, SLC35A2, SLC35B1, HK2, β4GalT-I, AKT1, and pAKT1 in mammary epithelial cells by western blot. Compared with the control, the levels of GLUT1, SLC35A2, and SLC35B1 were significantly increased ($P < 0.05$) in mammary epithelial cells cultured in DMEM with 12 m*M* glucose for 24 h. Consistent with this, HK2 and β4GalT-I levels were also increased ($P < 0.05$) (Fig. 3c, d). Protein kinase B alpha and its active form, pAKT1, were also upregulated in mammary epithelial cells cultured in DMEM with 12 m*M* glucose compared with the control ($P < 0.05$; Fig. 3e, f). Overall, these results strongly suggested that glucose induced lactose synthesis in mammary epithelial cells by upregulating GLUT1, SLC35A2, SLC35B1, HK2, β4GalT-I, and AKT1.

Relationship between AKT1 and glucose modulation of lactose synthesis in dairy cow mammary gland

Next we examined the regulatory mechanism underlying the effects of glucose on lactose synthesis. Protein kinase B alpha has a critical role in the proliferation, survival, and metabolism of numerous cell types [23, 24]. The above findings demonstrated that AKT1 and pAKT1 expression were both upregulated by glucose in vitro cow mammary epithelial cells. To determine whether glucose modulates lactose synthesis via AKT1 activation in dairy cow mammary gland, siRNA-mediated knockdown of

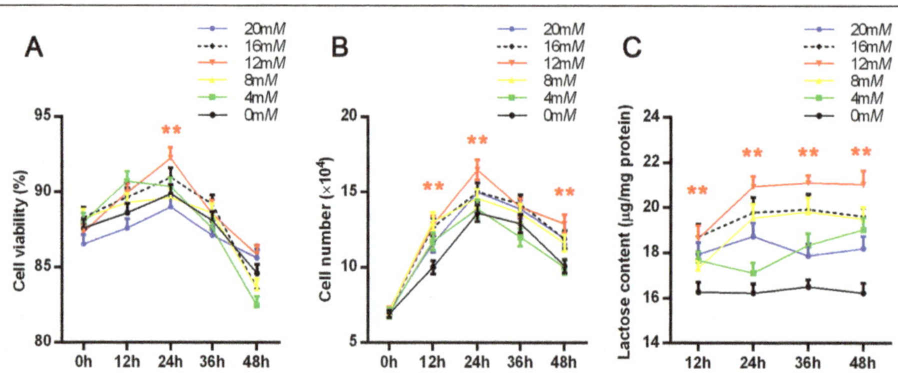

Fig. 2 The effect of glucose concentration on cell growth and lactose synthesis. **a**, **b** The effect of glucose concentration on cell viability (**a**) and proliferation (**b**) in dairy cow mammary epithelial cells. **c** The effect of glucose concentration on lactose synthesis in dairy cow mammary epithelial cells. Dairy cow mammary epithelial cells were cultured in DMEM with glucose at concentrations ranging from 0 to 20 m*M*. The effects of glucose concentration on cell viability, proliferation, and lactose content were examined after 12, 24, 36, and 48 h of treatment. Data are shown as the mean ± SEM from 3 independent replicates. *$P < 0.05$; **$P < 0.01$ versus corresponding value for cells cultured in DMEM with 0 m*M* glucose

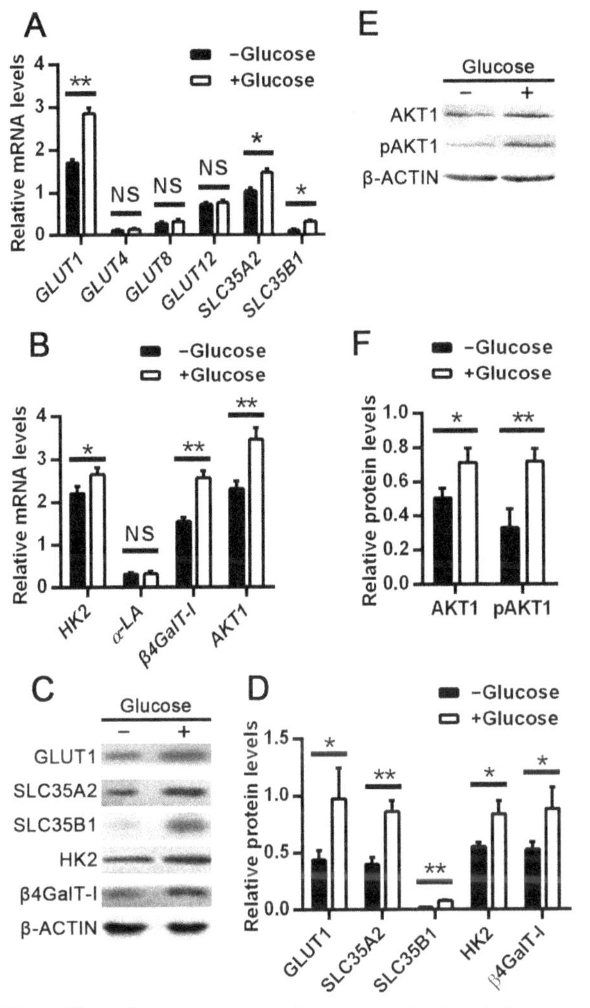

Fig. 3 Effect of glucose on expression of genes involved in lactose synthesis. **a** Quantitative real-time PCR analysis of the mRNAs encoding glucose and UDP-galactose transporters, including glucose transporter 1 (GLUT1), glucose transporter 4 (GLUT4), glucose transporter 8 (GLUT8), glucose transporter 12 (GLUT12), solute carrier family 35 member A2 (SLC35A2), and solute carrier family 35 member B1 (SLC35B1). **b** Quantitative real-time PCR analysis of the mRNAs encoding key enzymes and regulators in the lactose synthesis pathway, including hexokinase-2 (HK2), α-lactalbumin (α-LA), β-1,4-galactosyltransferase 1 (β4GalT-I), and protein kinase B alpha (AKT1). For **a** and **b**, total RNA was isolated from dairy cow mammary epithelial cells treated with or without 12 m*M* glucose for 24 h. The mRNA levels were measured by qPCR and normalized to *β-ACTIN* mRNA. **c**, **d** Western blot analysis (**c**) and quantification (**d**) of proteins involved in lactose synthesis. **e**, **f** Western blot analysis (**e**) and quantification (**f**) of AKT1 and pAKT1. For **c** and **e**, protein lysates from dairy cow mammary epithelial cells treated with or without 12 m*M* glucose for 24 h were collected for western blot analysis. Data are shown as the mean ± SEM from 3 independent replicates. $^{*}P < 0.05$; $^{**}P < 0.01$; NS, no statistical significance

AKT1 in dairy cow mammary epithelial cells was performed, followed by measurement of *AKT1* mRNA and protein 48 h later. The expression of *AKT1* mRNA was suppressed significantly by transfection with *AKT1* siRNA ($P < 0.01$) but not by the scrambled siRNA (Fig. 4a). Consistent with the qPCR results, knockdown of *AKT1* in mammary epithelial cells also led to substantially lower levels of AKT1 ($P < 0.01$; Fig. 4b, c).

We then investigated the effect of *AKT1* knockdown on lactose synthesis in mammary epithelial cells and measured lactose production under both basal (without glucose) and stimulated (glucose concentrations ranging from 4 to 20 m*M*) conditions. Compared with the control, *AKT1* knockdown decreased lactose synthesis in both basal and glucose-stimulated conditions. The effect at 0 m*M* (basal) was not significant, and the inhibition effect was significant at 8 *mM* and 12 m*M* glucose (Fig. 4d). We further examined the effect of *AKT1* knockdown on cell viability and proliferation when cells were cultured in medium with different concentrations of glucose (from 0 to 20 m*M*). Compared with the control, cell viability and proliferation were both significantly decreased when *AKT1* was knocked down (Fig. 4e, f). However, cell viability and proliferation were increased by treatment with 12 m*M* glucose compared to other glucose concentrations when *AKT1* was knocked down.

We next examined whether *AKT1* knockdown in dairy cow mammary epithelial cells cultured in medium with 12 m*M* glucose downregulated the expression of a glucose transporter and genes encoding enzymes involved in lactose synthesis. Quantitative real-time PCR analysis showed that the mRNA levels of *GLUT1*, *HK2*, and *β4GalT-I* in *AKT1* knockdown mammary epithelial cells were significantly reduced ($P < 0.01$), whereas the expression of *α-LA* remained relatively unchanged (Fig. 5a). Consistent with these qPCR results, knockdown of *AKT1* in mammary epithelial cells also led to significantly lower levels of GLUT1, HK2, and β4GalT-I, as well as SLC35A2 and SLC35B1 ($P < 0.05$; Fig. 5b, c). Overall, these data strongly support the hypothesis that glucose modulation of lactose synthesis is related to *AKT1* expression in dairy cow mammary gland.

Discussion

Lactose synthesis is related to milk volume. Glucose is the primary precursor of lactose. Once taken up by the lactating mammary epithelial cell, glucose is either used in the synthesis of lactose or processed by glycolysis to provide energy [25]. Compared with non-lactating dairy cows, lactating dairy cows have an approximately 4-fold higher requirement for glucose [26]. Because lactating dairy cows need more glucose than non-lactating dairy cows, one might reasonably expect that glucose availability could affect lactose synthesis in lactating dairy cow mammary gland. In this study, we observed that glucose increased lactose synthesis in cultured dairy cow mammary epithelial cells, as well as cell viability and proliferation. A concentration of 12 m*M* glucose was the

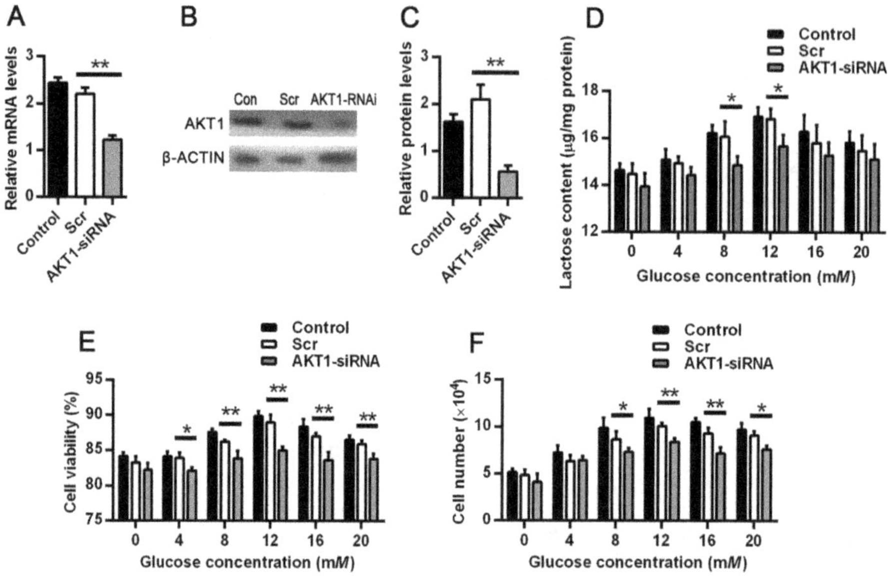

Fig. 4 Glucose modulates cell growth and lactose synthesis via AKT1 activation. Nonspecific scrambled siRNA (Scr) was used as a negative control in all experiments. **a** Relative *AKT1* mRNA levels in siRNA-transfected dairy cow mammary epithelial cells. The mRNA levels were measured by qPCR and normalized to *β-ACTIN* mRNA. **b**, **c** Western blot analysis (**b**) and quantification (**c**) of AKT1 in siRNA-transfected dairy cow mammary epithelial cells. The protein levels were normalized to β-ACTIN. **d**, **e**, **f** The effect of glucose concentration on lactose synthesis (**d**), cell viability (**e**), and proliferation (**f**) in dairy cow mammary epithelial cells when cells were transiently transfected with *AKT1* siRNA. siRNA-transfected dairy cow mammary epithelial cells were cultured in DMEM with glucose at concentrations ranging from 0 to 20 m*M* for 24 h. The effect of glucose concentration on lactose content, cell viability, and proliferation were examined. Data are shown as the mean ± SEM from 3 independent replicates. $^{*}P < 0.05$; $^{**}P < 0.01$

optimum concentration to induce lactose synthesis in cultured dairy cow mammary epithelial cells within 24 h. Protein kinase B alpha was activated by glucose in lactose synthesis process. Glucose transporter 1, SLC35A2, SLC35B1, HK2, and β4GalT-I expression were also increased as a result of AKT1 signaling in dairy cow mammary gland.

In mammals, glucose transport across the basolateral membrane of epithelial cells is carried out by glucose transporters [17]. In the current research, we examined the expression of *GLUT1*, *GLUT4*, *GLUT8*, and *GLUT12*, which has been previously reported in mammary gland [19–21]. Glucose induced significant increases in *GLUT1* mRNA and protein levels in dairy cow mammary epithelial cells but did not affect *GLUT4*, *GLUT8*, and *GLUT12* mRNA, suggesting that GLUT1 is one of the glucose transporters induced by glucose in dairy cow mammary gland for lactose synthesis. Glucose transporter 1 is abundantly expressed in bovine mammary gland and increases during pregnancy and lactation [27]. Our result is in agreement with a report that GLUT1 is the main extracellular and intercellular (Golgi membrane) glucose transporter in mammary epithelial cells [28]. In mouse and dairy goat mammary epithelial cells, GLUT1 is mainly distributed along the plasma membrane and Golgi apparatus [28, 29]. The presence of GLUT1 on the Golgi membrane appears to be specific to mammary epithelial cells, as most cells do not have this glucose transporter on the Golgi membrane [28].

In rat and mouse mammary glands, GLUT4 expression markedly decreases during pregnancy and practically disappears during lactation [30, 31]. This is because GLUT4 is mainly located in the adipose compartment, not in the plasma membrane [31]. In bovine mammary gland, the expression of GLUT8 increases 10-fold from prepartum to postpartum, and fairly strong GLUT8 staining is observed at both the apical and basolateral membrane of lactating bovine mammary epithelial cells [17]. But our results showed that *GLUT8* expression in cells cultured with 12 m*M* glucose did not changed compared to the control (without glucose), which is consistent with the report that *GLUT8* mRNA is not affected by various concentration of glucose in cultured bovine mammary epithelial cells [32]. Although GLUT12 is detected in the apical plasma membrane of epithelial cells during lactation in rats, forced weaning results in decreased cytoplasmic GLUT1 staining intensity but no change in GLUT12 staining [20]. Our data, along with previous evidence, demonstrate that GLUT1 is the main glucose transporter in dairy cow mammary epithelial cells for lactose synthesis, and its levels are regulated by glucose.

The synthesis of lactose takes place in the Golgi compartment of mammary epithelial cells. Lactose synthesis

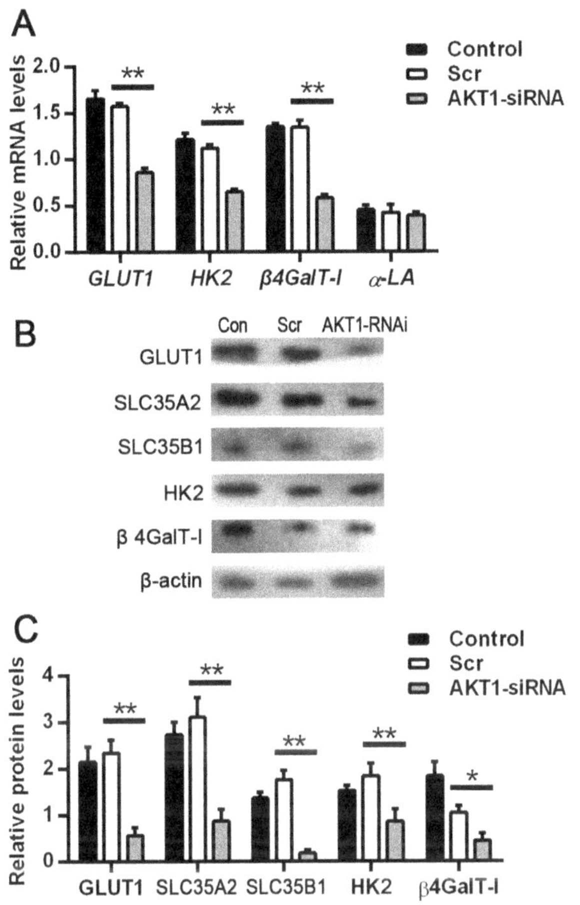

Fig. 5 Protein kinase B knockdown decreased the expression of genes involved glucose transportation and lactose synthesis. Nonspecific scrambled siRNA (Scr) was used as a negative control in all experiments. **a** Relative mRNA levels of glucose transporter 1 (GLUT1), hexokinase-2 (HK2), β-1,4-galactosyltransferase 1 (β4GalT-I), and α-lactalbumin (α-LA) in siRNA-transfected dairy cow mammary epithelial cells. siRNA-transfected dairy cow mammary epithelial cells were cultured in DMEM with 12 m*M* glucose for 24 h. Cells were collected and the total RNA was isolated for gene expression analysis. The mRNA levels were measured by qPCR and normalized to *β-ACTIN* mRNA. **b**, **c** Western blot analysis (**b**) and quantification (**c**) of protein expression levels of GLUT1, solute carrier family 35 member A2 (SLC35A2), solute carrier family 35 member B1 (SLC35B1), HK2, and β4GalT-I in siRNA-transfected dairy cow mammary epithelial cells. siRNA-transfected dairy cow mammary epithelial cells were cultured in DMEM with 12 m*M* glucose for 24 h. Cells were collected for western blot analysis, and the protein levels were normalized to β-ACTIN. Data are shown as the mean ± SEM from 3 independent replicates. $^{*}P < 0.05$; $^{**}P < 0.01$

paralleled the induction of expression of proteins involved in UDP-galactose synthesis and its transport across the Golgi membrane. The initial step in the synthesis of UDP-galactose is phosphorylation of glucose by HK. Bovine mammary epithelial cells grown in higher glucose concentrations have higher HK activity [32]. In rodents, HK1 is expressed in the mammary gland throughout pregnancy and lactation, whereas HK2 is expressed only following parturition [33]. In the current study, we detected *HK2* mRNA and protein expression in cultured dairy cow mammary epithelial cells, and this expression was upregulated when cells were treated with 12 m*M* glucose for 24 h. These expression data are positively correlated with lactose synthesis, indicating that *HK2* expression was affected by glucose during lactation. Glucose has the ability to induce *HK2* expression in mammary epithelial cells, but various concentrations of glucose do not elevate *HK2* mRNA expression levels [32]. Uridine diphosphate-galactose is actively transported into the Golgi lumen by SLC35A2 and SLC35B1. We found that the levels of SLC35A2 and SLC35B1 were markedly increased in mammary epithelial cells treated with 12 m*M* glucose for 24 h. Similarly, the expression of SLC35A2 and SLC35B1 strongly correlated with lactose concentration.

In the Golgi compartment, the biosynthesis of lactose involves the combination of free glucose and UDP-galactose, which is catalyzed by the lactose synthase system. This enzyme complex is composed of β4GalT-I and α-LA [7]. β-1, 4-galactosyltransferase is a constitutively expressed, trans-Golgi resident type II membrane-bound glycoprotein [34]. In mammals, β4GalT-I is required for the tissue-specific production of lactose, which takes place exclusively in the lactating mammary gland. α-lactalbumin is a noncatalytic mammalian protein expressed de novo exclusively in the mammary gland during lactation [35, 36]. The physiological function of α-LA is to lower the glucose K_m of β4GalT-I so that it may be used maximally for the synthesis of lactose and it changes the specificity of β4GalT-I from N-acetylglucosamine to glucose [37]. Mice with a null mutation of the *α-LA* gene fail to lactate [38]. In this study, we found that *β4GalT-I* levels were upregulated in the presence of additional glucose, but the expression of *α-LA* in dairy cow mammary epithelial cells was not affected by glucose. This result is in agreement with the previous report that elevated level of glucose increases *β4GalT-I* expression, but not *α-LA* in bovine mammary epithelial cells [39]. Our data, along with previous evidence, indicate that glucose supplementation may stimulate lactose synthesis partly by altering the expression of *β4GalT-I*, but not *α-LA*.

In this study, we also demonstrate that glucose modulation of lactose synthesis is likely related to AKT1 in lactating mammary glands of dairy cows. Protein kinase B alpha serves as a potent survival factor for secretory epithelial cells [40]. Protein kinase B alpha is upregulated in the mammary gland during pregnancy and lactation, and knockout studies have demonstrated that this particular serine-threonine protein kinase plays a role in the functional differentiation of the secretory epithelium and in metabolic pathways that regulate milk synthesis [41].

In our experiments, *AKT1* knockdown decreased both basal and glucose-stimulated lactose synthesis, as well as cell viability and proliferation, indicating that *AKT1* is required for lactose synthesis in response to glucose.

The protein kinase B family stimulates transport and metabolism of glucose [42]. In hematopoietic cell line, AKT1 activation increases the localization of GLUT1 to the cell surface and maintenance of HK function [43]. In mammary gland, previous studies have shown that mice lacking *AKT1* exhibit a pronounced metabolic defect during late pregnancy and lactation that results from a failure to upregulate GLUT1 [11, 13]. Our data showed that knockdown of *AKT1* in dairy cow mammary epithelial cells decreased both the mRNA and protein expression of *GLUT1* in response to glucose. These data, along with previous evidence, demonstrate that AKT1 is required for upregulation of glucose uptake in dairy cow mammary epithelial cells. Consistent with a role in upregulation of GLUT1, *AKT1* knockdown also resulted in deceased expression of SLC35A2 and SLC35B1, as well as HK2 and β4GalT-I, in dairy cow mammary epithelial cells.

Conclusions

Glucose supplementation has the ability to induce lactose synthesis in in-vitro bovine mammary epithelial cells. Glucose modulation of lactose synthesis is related to AKT1 activation in lactating mammary glands of dairy cows. Additionally, GLUT1, SLC35A2, SLC35B1, HK2, and β4GalT-I expression were increased as a result of AKT1 signaling in dairy cow mammary gland.

Methods

Mammary epithelial cell cultures

Primary mammary epithelial cells were prepared from mid-lactating mammary tissues of dairy cows as described previously [44]. Briefly, lactating cows were slaughtered and several pieces of mammary parenchyma tissue were aseptically removed from the midregion of the mammary glands. Mammary epithelial cells were isolated from mammary tissues by collagenase digestion and cultured in DMEM/F12 (12400–024, Life Technologies, Carlsbad, CA, USA) containing 10 % (v/v) fetal bovine serum (16000–044, Life Technologies, Carlsbad, CA, USA), 100 U/mL penicillin, 100 μg/mL streptomycin, and lactating hormones (5 μg/mL insulin, 1 μg/mL prolactin, and 1 μg/mL hydrocortisone; Sigma-Aldrich, St Louis, MO, USA) on cell culture dishes coated with 0.5 % rat tail collagen. Cells were maintained at 37 °C in an atmosphere containing 5 % CO_2. When cells grew to 80 % confluency, the primary mammary epithelial cells were trypsinized with 0.25 % trypsin plus 0.02 % EDTA and passaged. The pure mammary epithelial cells were isolated after 3 to 4 passages.

To confirm that the cells we isolated were pure mammary epithelial cells, cells plated on glass coverslips were washed twice with PBS and fixed in ice-cold methanol for 10 min. Cells were blocked with 10 % normal rabbit serum (ZLI-9025, ZSGB-BIO, Beijing, China) in PBS at room temperature for 1 h and then treated with cytokeratin 18 antibody (diluted 1:200 in PBS containing 10 % normal rabbit serum; sc-31700, Santa Cruz, CA, USA) at 4 °C overnight. Following three washes with PBS, cells were incubated with fluorescein isothiocyanate (FITC)-conjugated AffiniPure rabbit anti-goat IgG (diluted 1:50 in PBS containing 10 % normal rabbit serum; ZF-0314, ZSGB-BIO, Beijing, China) for 1 h at room temperature. To stain the nuclei, cells were treated with propidium iodide (Sigma-Aldrich, St Louis, MO, USA) for 10 min at room temperature, followed by three washes with PBS, and mounted with DABCO (Sigma-Aldrich, St Louis, MO, USA). Fluorescence was assessed by confocal microscopy (TCS SP2, Leica Microsystems, Wetzlar, GmbH, Germany).

To investigate if glucose has the ability to induce cell growth and lactose synthesis, mammary epithelial cells were plated at 6×10^4 cells/well in 12-well culture plates. When cells grew to 80 % confluency, the medium was changed to DMEM with high glucose (25 mM, 11965–092, Life Technologies, Carlsbad, CA, USA) or without glucose (11966–025, Life Technologies, Carlsbad, CA, USA). After 24 h of treatment, cells in triplicate wells were collected for viability and proliferation analyses, and medium in triplicate wells was collected for lactose content analysis.

To detect the effects of different concentrations of glucose on cell growth and lactose synthesis, mammary epithelial cells were plated at 6×10^4 cells/well in 12-well culture plates. When cells grew to 80 % confluency, the medium was changed to DMEM with different concentrations of glucose (ranging from 0 to 20 m*M*). After 12, 24, 36, and 48 h of treatment, cells in triplicate wells were collected for viability and proliferation analyses, and medium was collected for lactose content analysis.

To explore the molecular process by which glucose induces lactose synthesis, mammary epithelial cells were plated at 2×10^5 cells/well in 6-well culture plates. When cells grew to 80 % confluency, the medium was changed to DMEM with 12 m*M* glucose or without glucose. After 24 h of treatment, cells in triplicate wells were collected for indicated assays.

Cell viability and proliferation analyses

Cell viability and proliferation were assayed with the CASY-TT Analyzer System (Schärfe System GmbH, Reutlingen, Germany) as described previously [45]. Briefly, 100 μL mammary epithelial cell-suspension aliquots were transferred to a CASY cup containing 10 mL

CASY ton, mixed by inverting three times, and placed in the CASY cell counter. All experiments were performed in triplicate.

Lactose analysis

Lactose content in medium was measured with the Lactose/D-Glucose (Rapid) Assay kit (K-LACGAR, Megazyme, Ireland, UK) as described [46]. All experiments were performed in triplicate.

qPCR

Total RNA was extracted from each well using TRIzol reagent (15596–026, Life Technologies, Carlsbad, CA, USA). M-MLV reverse transcriptase (28025–013, Life Technologies, Carlsbad, CA, USA) was used with oligo (dT) primers to perform first-strand synthesis. Quantitative real-time PCR was performed using mRNA-specific primers for *GLUT1, GLUT4, GLUT8, GLUT12, SLC35A2, SLC35B1, AKT1, HK2, α-LA*, and *β4GalT-I* (Table 1) in the Applied Biosystems 7300 Real-time PCR system (Applied Biosystems, NY, USA). Primers were designed using Primer Premier 5.0 (PREMIER Biosoft, Palo Alto, CA) with amplicon size more than 100 bp and limited 3' G + C. Major parts of the primers were designed to span the junctions of 2 exons when possible to avoid amplification of genomic DNA. Prior to qPCR the primers were tested by a 20 μL PCR reaction using the same protocol described for qPCR and the PCR products were analyzed by gel electrophoresis on 2 % agarose gels (Additional file 1). The primers that presented a single band at the expected size were used for qPCR. The qPCR was performed in a 20-μL volume containing 10 μL of SYBR Premix Ex Taq II (RR820A, TaKaRa, China, Dalian), 0.4 μL ROX Reference Dye II (RR820A, TaKaRa, China, Dalian), 0.8 μL each of 10 μM forward and reverse primers, 2 μL cDNA and 6 μL DNase/RNase free water. Reactions were performed using the following conditions: 30 s of predenaturalization at 95 °C, followed by 40 cycles of 5 s denaturation at 95 °C and 34 s annealing and extension at 60 °C. The presence of a single PCR product was verified by the dissociation protocol as follows: 15 s at 95 °C, followed by 1 min at 65 °C, and 15 s at 95 °C. Each gene was amplified in a separated reaction and triplicate PCRs were performed on each sample of cDNA. Repeatability

Table 1 Primers sequences used for qPCR

Gene symbol[1]	GenBank accession	Primer sequence (5′ to 3′)[2]	Melting temperature	Product size
SLC2A1	NM_174602.2	F: GACACTTGCCTTCTTTGCCA	58.97	164 bp
		R: AACCTAATGGAGCCTGACCC	59.08	
SLC2A4	NM_174604.1	F: TCATTCTTGGACGGTTCTTC	55.40	159 bp
		R: CTAGCACCTGGGCGATTA	56.08	
SLC2A8	NM_201528.1	F: CGGCTCAGAACCTGTGGAT	59.40	165 bp
		R: GGAGGATGCCTGTGACTACC	59.53	
SLC2A12	NM_001011683.2	F: ATTGTCATCGGCATTCTT	51.80	165 bp
		R: ATGTCCCTTCATCACCAG	53.43	
SLC35A2	NM_176640.2	F: GTGGTCCAGAATGCTTCCCTC	60.68	163 bp
		R: CCAGGTGCTTCACGTTACCC	60.95	
SLC35B1	NM_181338.2	F: GACCTGCTCCATCATCACCAC	60.75	126 bp
		R: AGACCGAGACCCAAGAACACC	61.71	
AKT1	NM_173986.2	F: TCATGCAGCACCGATTCTT	57.44	159 bp
		R: CTTGGTCAGGTGGCGTAAT	57.45	
HK2	XM_015473383.1	F: AAGATGCTGCCCACCTACG	59.78	123 bp
		R: TCGCTTCCCATTCCTCACA	58.63	
LALBA	NM_174378.2	F: AGACTTGAAGGGCTACGGA	57.62	177 bp
		R: TAGTTGCTTGAGTGAGGGTT	56.38	
B4GALT-I	NM_177512.2	F: TGCCCTGAGGAGTCCCC	59.92	151 bp
		R: GGCCACCTTGTGAGGAGAG	59.70	
ACTB	NM_173979.3	F: AGGACCTCTACGCCAACACG	61.87	249 bp
		R: TTTGCGGTGGACGATGGAG	60.38	

[1]SLC2A1 = glucose transporter 1(GLUT1); SLC2A4 = glucose transporter 4 (GLUT4); SLC2A8 = glucose transporter 8 (GLUT8); SLC2A12 = glucose transporter 12 (GLUT12); SLC35A2 = solute carrier family 35 member A2; SLC35B1 = solute carrier family 35 member B1; AKT1 = PKB alpha; HK2 = hexokinase 2; LALBA = α-lactalbumin (α-LA); B4GALT1 = β-1,4-galactosyltransferase 1 (β4GalT-I); ACTB = β-ACTIN

[2]F: forward; R: reverse

(intra-assay) and reproducibility (inter-assay) were assessed by computing the coefficient of variation under different conditions. The intra-assay variability was tested in triplicate in the same run. The inter-assay variability was evaluated in three independent runs, performed in different days. The intra-assay and inter-assay coefficients of variation were both less than 2 %. The comparative Ct method was used to analyze the relative changes of each gene [47]. The abundance of *β-ACTIN* was not significantly different between treatments, therefore, *β-ACTIN* was chosen for the purpose of normalization.

Small interfering RNA (siRNA)-mediated gene knockdown

The siRNA against *AKT1* and scrambled siRNA (Scr) were chemically synthesized by GenePharma Co., Ltd, Shanghai, China. Gene knockdown was achieved by transient transfection with *AKT1* siRNA or scrambled siRNA as negative control using Lipofectamine 2000 Reagent (11668–019, Life Technologies, Carlsbad, CA, USA) as described previously [16]. Transfection efficiency was detected by fluorescence microscope (DM LB2, Leica Microsystems, Wetzlar, GmbH, Germany) after 6 h of transfection (Additional file 2). After 48 h of transfection, cells were extracted for the *AKT1* mRNA and protein expression detection.

To investigate the effect of *AKT1* knockdown on lactose synthesis in mammary epithelial cells, after 48 h of transfection, the medium was changed to DMEM with different concentrations of glucose (ranging from 0 to 20 m*M*). After 24 h of treatment, cells and medium in triplicate wells were collected for indicated assays.

Western blot analysis

For western blot analysis, dairy cow mammary epithelial cells were washed twice with ice-cold PBS and lysed with RIPA buffer (P0013C, Beyotime Biotechnology, Shanghai, China) containing 0.5 m*M* PMSF (ST506, Beyotime Biotechnology, Shanghai, China),5 μg/mL aprotinin (A1153, Sigma-Aldrich, St Louis, MO, USA) and 5 μg/mL leupeptin (L2884, Sigma-Aldrich, St Louis, MO, USA) on ice. The cell lysates were centrifuged at 17,800 × g for 10 min at 4 °C. Protein quantification was performed using the BCA Protein Assay kit (23227, Thermo Fisher Scientific Inc., Waltham, MA, USA). Equal amounts of protein (40 μg) were separated by 10 % SDS-PAGE and transferred onto nitrocellulose membranes. Membranes were blocked for 1 h with 5 % nonfat milk in Tris-buffered saline with 0.1 % Tween 20 (TBST) at room temperature and incubated overnight at 4 °C with primary antibodies specific to GLUT1 (1:200 dilution; sc-7903, Santa Cruz, CA, USA), SLC35A2 (1:200 dilution; sc-82032, Santa Cruz, CA, USA), SLC35B1 (1:200 dilution; sc-132864, Santa Cruz, CA, USA), HK2 (1:200 dilution; sc-6521, Santa Cruz, CA, USA), β4GalT-I (1:200 dilution; sc-22279, Santa Cruz, CA, USA), AKT1 (1:200 dilution; sc-1618, Santa Cruz, CA, USA), pAKT1 (1:1000 dilution, Abcam, Cambridge, MA, USA), and β-ACTIN (1:1000 dilution; 4970, Cell Signaling Technology, Beverly, MA, USA). The primary antibodies used were all diluted in 5 % BSA in TBST. The blots were then treated with horseradish peroxidase (HRP)–conjugated anti-goat IgG (1:500 dilution; ZB-2306, ZSGB-BIO, Beijing, China) or HRP-conjugated anti-rabbit IgG (1:500 dilution; ZB-2301, ZSGB-BIO, Beijing, China) for 1 h at room temperature. Proteins were visualized using the SignalFire ECL Reagent (6683, Cell Signaling Technology, Beverly, MA, USA). Three samples were assessed per time point or treatment. The protein bands were used for densitometric analysis using Image-Pro Plus 6.0 (Media Cybernetics, Inc., Warrendale, PA, USA) to obtain relative protein levels expressed as integrated density values. The integrated density, representing the mean density of pixels multiplied by the area was determined from equal sized rectangles drawn around the bands of interest minus the background to remove non-specific antibody staining. The protein levels were normalized to β-ACTIN.

Statistical analysis

Experimental data were analyzed using GraphPad Prism 6 (GraphPad Software, La Jolla, CA). Quantitative data is presented as the mean ± SEM. A two-tailed unpaired *t*-test was used for two condition comparison. Two-way or one-way ANOVA with Bonferroni multiple comparison was used for multiple conditions comparison. $P < 0.05$ was considered to be statistically significant.

Abbreviations

UDP, uridine diphosphate; β4GalT-I, β-1, 4-galactosyltransferase, α-LA: α-lactalbumin; PKB, the protein kinase B, also known as AKT; GLUT, glucose transporter; qPCR, quantitative real-time PCR; SLC35A2, solute carrier family 35 member A2; SLC35B1, solute carrier family 35 member B1; HK2, hexokinase 2; Scr, scrambled siRNA; FITC, fluorescein isothiocyanate.

Acknowledgements

Not applicable.

Funding

This work was supported by grants from National Basic Research Program (2011CB100804) from the Ministry of Science and Technology of China and the National Natural Science Foundation of China (31401109 to YL and 31200984 to XH).

Authors' contributions

YL devised the hypothesis, designed the experiments and wrote the manuscript. XS performed the experiments. XH performed the data analysis and made the figures. BQ helped with the confocal microscopy. XG helped with experimental design. QL is also responsible for the writing, revision and finalization of the manuscript as well as for the decision to submit the manuscript for publication. All authors read and approved the final manuscript.

Author's information
Not applicable.

Competing interests
The authors declare that they have no competing interests.

Consent for publication
Not applicable.

References

1. Anderson SM, Rudolph MC, McManaman JL, Neville MC. Key stages in mammary gland development. Secretory activation in the mammary gland: it's not just about milk protein synthesis! Breast Cancer Res. 2007;9(1):204.
2. Holt C. Swelling of Golgi vesicles in mammary secretory cells and its relation to the yield and quantitative composition of milk. J Theor Biol. 1983;101(2):247–61.
3. Annison EF, Jl L. The oxidation and utilization of glucose and acetate by the mammary gland of the goat in relation to their over-all metabolism and to milk formation. J Physiol. 1964;175:372–85.
4. Bartley JC, Abraham S, Chaikoff IL. Biosynthesis of lactose by mammary gland slices from the lactating rat. J Biol Chem. 1966;241:1132–7.
5. Kuhn NJ, Carrick DT, Wilde CJ. Lactose synthesis: the possibilities of regulation. J Dairy Sci. 1980;63:328–36.
6. Keenan TW, Huang CM, Morre DJ. Membranes of mammary gland. V. Isolation of Golgi apparatus and rough endoplasmic reticulum from bovine mammary gland. J Dairy Sci. 1972;55(11):1577–85.
7. Watkins WM, Hassid WZ. Synthesis of lactose by particulate enzyme preparations from guinea pig and bovine mammary glands. Science. 1962; 136(3513):329.
8. Turkington RW, Brew K, Vanaman TC, Hill RL. The hormonal control of lactose synthetase in the developing mouse mammary gland. J Biol Chem. 1968;243(12):3382–7.
9. Ollier S, Robert-Granié C, Bernard L, Chilliard Y, Leroux C. Mammary transcriptome analysis of food-deprived lactating goats highlights genes involved in milk secretion and programmed cell death. J Nutr. 2007;137(3):560–7.
10. Xiao CT, Cant JP. Relationship between glucose transport and metabolism in isolated bovine mammary epithelial cells. J Dairy Sci. 2005;88(8):2794–805.
11. Chen CC, Boxer RB, Stairs DB, Portocarrero CP, Horton RH, Alvarez JV, et al. Akt is required for Stat5 activation and mammary differentiation. Breast Cancer Res. 2010;12(5):R72.
12. Chodosh LA, Gardner HP, Rajan JV, Stairs DB, Marquis ST, Leder PA. Protein kinase expression during murine mammary development. Dev Biol. 2000; 219(2):259–76.
13. Boxer RB, Stairs DB, Dugan KD, Notarfrancesco KL, Portocarrero CP, Keister BA, et al. Isoform-specific requirement for Akt1 in the developmental regulation of cellular metabolism during lactation. Cell Metab. 2006;4(6):475–90.
14. Cherepanov GG, Danfaer A, Cant JP. Simulation analysis of substrate utilization in the mammary gland of lactating cows. J Dairy Res. 2000;67(2):171–88.
15. Kuhn NJ, White A. Milk glucose as an index of the intracellular glucose concentration of rat mammary gland. Biochem J. 1975;152(1):153–5.
16. Wang L, Lin Y, Bian Y, Liu L, Shao L, Lin L, et al. Leucyl-tRNA synthetase regulates lactation and cell proliferation via mTOR signaling in dairy cow mammary epithelial cells. Int J Mol Sci. 2014;15(4):5952–69.
17. Zhao FQ. Biology of glucose transport in the mammary gland. J Mammary Gland Biol Neoplasia. 2014;19(1):3–17.
18. Mohammad MA, Hadsell DL, Haymond MW. Gene regulation of UDP-galactose synthesis and transport: potential rate-limiting processes in initiation of milk production in humans. Am J Physiol Endocrinol Metab. 2012;303(3):E365–76.
19. Zhao FQ, Keating AF. Expression and regulation of glucose transporters in the bovine mammary gland. J Dairy Sci. 2007;90 Suppl 1:E76–86.
20. Macheda ML, Williams E, Best JD, Wlodek ME, Rogers S. Expression and localisation of GLUT1 and GLUT12 glucose transporters in the pregnant and lactating rat mammary gland. Cell Tissue Res. 2003;311(1):91–7.
21. Sharma A, Aggarwal J, Sodhi M, Kishore A, Mishra BP, Mohanty AK, et al. Stage specific expression of ATP-binding cassette and solute carrier superfamily of transporter genes inmammary gland of riverine buffalo (Bubalus bubalis). Anim Biotechnol. 2014;25(3):200–9.
22. Charron M, Shaper JH, Shaper NL. The increased level of beta1,4-galactosyltransferase required for lactose biosynthesis is achieved in part by translational control. Proc Natl Acad Sci U S A. 1998;95(25):14805–10.
23. Tan SX, Ng Y, James DE. Akt inhibitors reduce glucose uptake independently of their effects on Akt. Biochem J. 2010;432(1):191–7.
24. Whiteman EL, Cho H, Birnbaum MJ. Role of Akt/protein kinase B in metabolism. Trends Endocrinol Metab. 2002;13(10):444–51.
25. Kleiber M, Black AL, Brown MA, Baxter CF, Luick JR, Stadtman FH. Glucose as a precursor of milk constituents in the intact dairy cow. Biochim Biophys Acta. 1955;17(2):252–60.
26. Bell AW, Bauman DE. Adaptations of glucose metabolism during pregnancy and lactation. J Mammary Gland Biol Neoplasia. 1997;2(3):265–78.
27. Zhao FQ, Glimm DR, Kennelly JJ. Distribution of mammalian facilitative glucose transporter messenger RNA in bovine tissues. Int J Biochem. 1993;25(12):1897–903.
28. Nemeth BA, Tsang SW, Geske RS, Haney PM. Golgi targeting of the GLUT1 glucose transporter in lactating mouse mammary gland. Pediatr Res. 2000; 47(4 Pt 1):444–50.
29. Yu Q, Zhu L, Lin J, Zhang Q, Tian Q, Hu W, et al. Functional analyse of GLUT1 and GLUT12 in glucose uptake in goat mammary gland epithelial cells. PLoS One. 2013;8(5):e65013.
30. Camps M, Vilaro S, Testar X, Palacín M, Zorzano A. High and polarized expression of GLUT1 glucose transporters in epithelial cells from mammary gland: acute down-regulation of GLUT1 carriers by weaning. Endocrinology. 1994;134(2):924–34.
31. Rudolph MC, McManaman JL, Phang T, Russell T, Kominsky DJ, Serkova NJ, et al. Metabolic regulation in the lactating mammary gland: a lipid synthesizing machine. Physiol Genomics. 2007;28(3):323–36.
32. Zhao K, Liu HY, Wang HF, Zhou MM, Liu JX. Effect of glucose availability on glucose transport in bovine mammary epithelial cells. Animal. 2012;6(3):488–93.
33. Kaselonis GL, McCabe ER, Gray SM. Expression of hexokinase 1 and hexokinase 2 in mammary tissue of nonlactating and lactating rats: evaluation by RT-PCR. Mol Genet Metab. 1999;68(3):371–4.
34. Qasba PK, Ramakrishnan B, Boeggeman E. Structure and function of beta -1,4-galactosyltransferase. Curr Drug Targets. 2008;9(4):292–309.
35. Hill RL, Brew K, Vanaman TC, Trayer IP, Mattock P. The structure, function, and evolution of alpha-lactalbumin. Brookhaven Symp Biol. 1968;21(1):139–54.
36. Brew K, Vanaman TC, Hill RL. The role of alpha-lactalbumin and the A protein in lactose synthetase: a unique mechanism for the control of a biological reaction. Proc Natl Acad Sci U S A. 1968;59(2):491–7.
37. Fitzgerald DK, Brodbeck U, Kiyosawa I, Mawal R, Colvin B, Ebner KE. Alpha-lactalbumin and the lactose synthetase reaction. J Biol Chem. 1970;245(8):2013–8.
38. Stacey A, Schnieke A, Kerr M, Scott A, McKee C, Cottingham I, et al. Lactation is disrupted by alpha-lactalbumin deficiency and can be restored by human alpha-lactalbumin gene replacement in mice. Proc Natl Acad Sci U S A. 1995;92(7):2835–9.
39. Liu H, Zhao K, Liu J. Effects of glucose availability on expression of the key genes involved in synthesis of milk fat, lactose and glucose metabolism in bovine mammary epithelial cells. PLoS One. 2013;8(6):e66092.
40. Kennedy SG, Wanger AJ, Conzen SD, Jordán J, Bellacosa A, Tsichlis PN, et al. The PI 3-kinase/Akt signaling pathway delivers an anti-apoptotic signal. Genes Dev. 1997;11(6):701–13.
41. Maroulakou IG, Oemler W, Naber SP, Klebba I, Kuperwasser C, Tsichlis PN. Distinct roles of the three Akt isoforms in lactogenic differentiation and involution. J Cell Physiol. 2008;217(2):468–77.
42. Plas DR, Thompson CB. Akt-dependent transformation: there is more to growth than just surviving. Oncogene. 2005;24(50):7435–42.
43. Rathmell JC, Fox CJ, Plas DR, Hammerman PS, Cinalli RM, Thompson CB. Akt-directed glucose metabolism can prevent Bax conformation change and promote growth factor-independent survival. Mol Cell Biol. 2003;23(20):7315–28.
44. Hou X, Lin L, Xing W, Yang Y, Duan X, Li Q, et al. Spleen tyrosine kinase regulates mammary epithelial cell proliferation in mammary glands of dairy cows. J Dairy Sci. 2016;99(5):3858–68.
45. Li HM, Wang CM, Li QZ, Gao XJ. MiR-15a decreases bovine mammary epithelial cell viability and lactation and regulates growth hormone receptor expression. Molecules. 2012;17(10):12037–48.
46. Wang Z, Hou X, Qu B, Wang J, Gao X, Li Q. Pten Regulates Development and Lactation in the Mammary Glands of Dairy Cows. PLoS ONE. 2014;9(7):e102118.
47. Livak KJ, Schmittgen TD. Analysis of relative gene expression data using real-time quantitative PCR and the 2(-Delta Delta C(T)) Method. Methods. 2001;25(4):402–8.

7

Herd-level seroprevalence of *Fasciola hepatica* and *Ostertagia ostertagi* infection in dairy cattle population in the central and northeastern Poland

Sławomir J. Kowalczyk[1], Michał Czopowicz[1*], Corinna N. Weber[2], Elisabeth Müller[2], Tomasz Nalbert[1], Andrzej Bereznowski[1] and Jarosław Kaba[1]

Abstract

Background: *Fasciola hepatica* and *Ostertagia ostertagi* infections are widespread in cattle population of Europe, however data on their prevalence in Poland are only fragmentary. Therefore, the cross-sectional study was carried out to determine the herd-level seroprevalence of *F. hepatica* and *O. ostertagi* infection in dairy cattle population in the central and north-eastern provinces Poland, and to identify basic local risk factors for these infections. In total, 598 herds were enrolled, 394 (65.9%) in the north-eastern province and 204 (34.1%) in the central province. In each herd the questionnaire survey was conducted and bulk-tank milk (BTM) sample was collected and screened using two indirect immunoenzymatic tests. Optical density ratio (ODR) was regarded as the quantitative proxy of exposure to either of the two parasites.

Results: Both *Fasciola* and *Ostertagia* ELISA ODR in the north-eastern province was significantly higher than ODR in the central province. At the cut-off value of ODR = 0.27 the true herd-level seroprevalence of *F. hepatica* was 79.6% (95% CI: 74.0%, 84.3%) in the north-eastern province and 13.0% (95% CI: 5.3%, 21.7%) in the central province. At the cut-off of ODR = 0.50151 of 188 herds (80.3%, 95% CI: 74.1%, 85.4%) were seropositive for *O. ostertagi* in the north-eastern province and only 70 of 136 herds (51.5%, 95% CI: 43.1%, 59.7%) were seropositive in the central province. Location of a herd in the north-eastern province, longer grazing period practiced in a herd and > 50%-share of grazing grass in roughage were all positively related to the increase in exposure to both parasites. Moreover, the use of hay or haylage as main roughage proved to be positively related to the increase in exposure to *F. hepatica*.

Conclusions: *F. hepatica* and *O. ostertagi* are widespread in cattle population in Poland, however their occurrence at a herd-level varies between different regions of Poland. This diversity can only partly be explained by different herd management, and appears linked to environmental and climate conditions typical for these regions.

Keywords: Bulk-tank milk, ELISA, Grazing, Hay, Haylage

* Correspondence: mczopowicz@gmail.com
[1]Laboratory of Veterinary Epidemiology and Economics, Faculty of Veterinary Medicine, Warsaw University of Life Sciences, Nowoursynowska 159c, 02-776 Warsaw, Poland
Full list of author information is available at the end of the article

Background

Parasitic diseases have great impact on cattle productivity and welfare in all regions of the world [1]. Infections with two groups of internal parasites – liver flukes and gastrointestinal nematodes – are commonly regarded as most detrimental for dairy cattle [2, 3]. The former group is represented in Europe by an ubiquitous fluke *Fasciola hepatica*, responsible for fascioliasis – a disease usually following a chronic course in cattle and known to reduce both the milk yield and quality [4]. Gastrointestinal nematodes are a very diverse group of roundworms of which *Ostertagia ostertagi*, responsible for pathologic lesions in the abomasum, appears to be of main importance in the temperate climate zone [5, 6].

Both *F. hepatica* and *O. ostertagi* infections are widespread in cattle population of Europe. Herd-level prevalence of *F. hepatica* infection ranges from below 10% in Sweden to almost 90% in Wales [3, 7, 8]. Intensity of exposure to *O. ostertagi* expressed as a quantitative result of bulk tank milk ELISA shows similar spatial trends [9–11] with over 90% of affected herds in Great Britain [12]. Similar risk factors have so far been identified for both infections with access to pastures, grazing season length and climate conditions, especially frequent rainfalls, being the most commonly mentioned [9, 11, 13–17].

Latest spatial analysis indicated Poland as the region with low-to-moderate risk of exposure to *F. hepatica* with the north-eastern part of the country as potentially the high-risk area [14]. Observations from the second half of the twentieth century appear to corroborate this result [18], however, recent large-scale studies allowing to verify this model are lacking. The snail *Galba truncatula*, an intermediate host of *F. hepatica*, is widespread in Poland and appears to be quite often a source of cercariae [19]. *F. hepatica* infection has been diagnosed at individual-animal level in 4–20% cows in north-eastern regions using fecal egg count [20], in 5–65% cows in the south-eastern province and 28% in the southern province of Poland on the basis of post-slaughter examinations [21–24] and in 20–50% cows in central and eastern regions using PCR for fecal samples [25]. However, no herd-level disease survey based on bulk-tank milk testing has so far been published. On the other hand, two studies carried out in cattle herds in the southern Poland [26, 27] showed presence of *O. ostertagi* antibodies in bulk-tank milk of all herds enrolled, with 30–56% herds considered as economically affected by the infection. No data on *O. ostertagi* prevalence from other parts of Poland are available.

Therefore, we decided to carry out the cross-sectional study to determine the herd-level seroprevalence of *F. hepatica* and *O. ostertagi* infection in dairy cattle population in the central and north-eastern Poland, and to identify basic local risk factors for these infections.

Methods

Herds and questionnaires

The study was carried out in summer 2012 and 2013 in 2 of 16 Polish provinces – Podlaskie province located in the north-eastern part of the country and Łódzkie province situated in the central Poland. Herds were selected from the lists of farmers who used to sell milk to five large dairies located in the north-eastern (3 dairies) and the central province (2 dairies) and they were enrolled in the study provided that they counted at least 10 adult cows (> 24 month-old) and their owners granted informed verbal consent for participation in the study. Given that the estimated parameter in this study was the quantitative result of an immunoenzymatic test (optical density ratio, ODR), only the lower limit of the sample size required (n) was determined at the 95% level of confidence according to the following formula: $n = (1.96 \times \sigma)^2 / M^2$, where σ^2 stood for the variance in the population and M was the desired size of the 95% margin of error (precision). The minimum required sample size was determined so that the desired precision of the estimation (M) was not lower than 20% of the dispersion of observed ODRs (σ^2) and equaled 97 herds.

Each herd was visited once in 2012 or 2013. During the visit bulk-tank milk (BTM) samples were collected and the questionnaire survey was conducted personally by the author (SJK). The questionnaire had been developed specifically for this study (Additional file 1) and it included following information: herd size (only adult dairy cows), average milk yield per lactation, grazing policy (recorded as no grazing, grazing roughly half a day, roughly whole day or whole 24 h), length of grazing period (in months), main roughage used (corn silage, haylage or hay), proportion of grazing grass in diet (no grazing grass, < 50% of all roughage, > 50% of all roughage), and treatment against worms used in the herd. The study complied with the Directive 2010/63/EU and the Act of Polish Parliament of 15 January 2015 on protection of animals used for scientific purposes. According to Polish legal regulations no formal approval of the ethical commission regarding participation of cattle herd owners in the study (completing questionnaires) was required since questionnaires applied to animal not human health (verbal information from the National Ethical Commission).

Bulk-tank milk sampling and serological tests

Bulk-tank milk (BTM) samples were collected into 50 ml Falcon with 1 sodium azide tablet per falcon and centrifuged for 15 min. at 1200×g in a cooling centrifuge set to 4 °C. Then, the fat fraction was carefully moved to the side with the plastic spatula, and the whey fraction was collected with a plastic Pasteur pipette, aliquoted into 2 ml Eppendorf tubes and stored at – 20 °C until testing.

Then, BTM samples were screened using two semi-quantitative indirect immunoenzymatic tests from Boehringer Ingelheim Svanova, Sweden: SVANOVIR®F. hepatica-Ab (all samples), based on excretory and secretory antigens (E/S ELISA), and SVANOVIR®O.ostertagi-Ab (only samples collected in 2013). ELISAs were performed according to manufacturer's manuals. The results were expressed as optical density (OD) ratio (ODR) calculated with the following formula:

$$ODR = (OD_{BTM\ sample} - OD_{negative\ control})/(OD_{positive\ control} - OD_{negative\ control})$$

ODR was regarded as the quantitative proxy of exposure to either of the two parasites.

To report herd-level seroprevalences for the infections some cut-offs must have been assumed. For *F. hepatica* we set the cut-off value at ODR = 0.27 as for this figure sensitivity and specificity for identifying herds in which more than 25% of the cows were infected was known and equaled 96% (95 CI: 89%, 100%) and 80% (95 CI: 66%, 94%), respectively [8]. Within-herd prevalence of *F. hepatica* of at least 25% had been in turn previously proposed as a threshold for significant production losses [28]. Furthermore, we used the cut-off values recommended by the manufacturer to identify herds affected by *F. hepatica* at low (economically insignificant) or high (economically significant) level, which equaled ODR = 0. 30 and ODR = 0.60, respectively.

For *O. ostertagi* we assumed the cut-off recommended by the manufacturer, which equaled ODR = 0.50, and allowed to identify herds economically affected by the parasite.

Nevertheless, all risk analyses were performed using the crude ODRs instead of categorical results based on the aforementioned cut-off values, in order to retain as much information as possible.

Statistical analysis

Numerical variables were presented as the median, interquartile range (IQR) and range, categorical variables as the count and percentage in the group. Ninety five per cent confidence intervals (95% CI) for the apparent and true prevalence were determined using the Wilson's score method. Mean ODRs (standard deviation, SD) were also reported to allow comparisons with other studies, however all statistical analyses were non-parametric as ODRs were non-normally distributed according to the Shapiro-Wilk test ($p < 0.001$ for ODRs of both ELISAs except for ODR of *Ostertagia* ELISA in the central province where $p = 0.201$). In the univariable analysis ODR was correlated with numerical variables using the Spearman's rank-ordered correlation coefficient (r_s), and compared between two groups of a categorical variable using the Mann-Whitney U test, or between more than 2 groups of a categorical variable using the Kruskal-Wallis test with Dunn's post-hoc test. Variables identified as linked with ODR at p-value below 0.1 in the univariable analysis were offered to the multivariable linear regression analysis on the basis of the backward stepwise procedure. As correlations between explanatory variables were observed, only the one showing stronger link with ODR was entered into the multivariable analysis. Proportion of ODR variation explained by the model was determined by the coefficient of determination adjusted by the number of explanatory variables included in the model (adjusted-R^2). All statistical tests were two-sided. A significance level (α) was set at 0. 05. Statistical analyses were performed in Statistica 12 (StatSoft Inc.) and the figure in the Microsoft Office Excel.

Results

In total, 598 herds were visited, 274 (45.8%) in 2012 and 324 (54.2%) in 2013. Of them 394 (65.9%) herds were located in the north-eastern province and 204 (34.1%) in the central province. Herd size ranged from 10 to 109 adult dairy cows and herds were larger in the north-eastern (median of 24, IQR from 17 to 32 heads) than in the central province (median of 18, IQR from 14 to 25 heads) ($p < 0.001$). Of herds tested in 2013 188 (58.0%) were located in the north-eastern province and 136 (42. 0%) in the central province. Detailed characteristics of herds from both provinces are given in Table 1 and Additional file 2. In any herd neither regular deworming of adult cows was practiced nor any medicine against parasites had been given for at least 3 months preceding sample collection.

Fasciola ELISA ODR in the north-eastern province ranged from 0.062 to 1.516, with the median of 0.596 and IQR from 0.332 to 0.764, and was significantly higher than ODR in the central province which ranged from 0.002 to 0.989, with the median of 0.146 and IQR from 0.098 to 0.334 ($p < 0.001$). Mean ODRs (SD) were 0.553 (0.266) and 0.235 (0.206), respectively.

At the cut-off of ODR = 0.27378 herds (63.2%) tested positive for *F. hepatica*, however there was a substantial discrepancy between provinces: as many as 317 of 394 herds (80.5%) were positive in the north-eastern province and only 61 of 204 (29.9%) were positive in the central province. The overall (common for both provinces) true herd-level seroprevalence was 56.9% (95% CI: 52.9%, 60.8%) , but it was 79.6% (95% CI: 74.0%, 84.3%) in the north-eastern province and 13.0% (95% CI: 5.3%, 21.7%) in the central province. Also much higher percentage of herds were economically affected by *F. hepatica* infection according to the manufacturer's cut-offs in the north-eastern province (Fig. 1).

Ostertagia ELISA ODR in the north-eastern province ranged from 0.010 to 1.080, with the median of 0.725 and IQR from 0.570 to 0.845, and was significantly higher than ODR in the central province, which ranged from 0.010 to 1.040, with the median of 0.505 and IQR

Table 1 Characteristics of herds included in the study given as the median, interquartile range and range or as the count and percentage of the entire group

	Entire study population (n = 598)	Province		p
		north-eastern (Podlaskie) (n = 394)	central (Łódzkie) (n = 204)	
Herd size	21, 15–30 (10–109)	24, 17–32 (10–109)	18, 14–25 (10–100)	< 0.001
Average milk yield	5500, 460–6400 (1700–10,200)	5500, 4500–6400 (1700–10,200)	5200, 4800–6150 (3000–9360)	0.982
Grazing policy				< 0.001
No grazing	264 (44.2%)	118 (30.0%)	146 (71.6%)	
6 h a day	66 (11.0%)	46 (11.7%)	20 (9.8%)	
12 h a day	179 (29.9%)	156 (39.6%)	23 (11.3%)	
24 h a day	89 (14.9%)	74 (18.7%)	15 (7.3%)	
Length of grazing period in months (only grazed herds)	5 (1–8)	5 (1–8)	5 (1–7)	0.023
Main roughage				< 0.001
Haylage	311 (52.0%)	273 (69.3%)	38 (18.6%)	
Corn silage	263 (44.0%)	103 (26.1%)	160 (78.4%)	
Hay	24 (4.0%)	18 (4.6%)	6 (3.0%)	
Proportion of grazing grass in diet				< 0.001
No grazing grass	203 (33.9%)	90 (22.8%)	113 (55.4%)	
Grazing grass < 50% of all roughage	204 (34.1%)	135 (34.3%)	69 (33.1%)	
Grazing grass > 50% of all roughage	191 (32.0%)	169 (42.9%)	22 (11.5%)	

from 0.335 to 0.710 ($p < 0.001$). Mean ODRs (SD) were 0.680 (0.242) and 0.513 (0.248), respectively.

At the cut-off of ODR = 0.50 221 of 324 tested herds (68.2%, 95% CI: 63.0%, 73.0%) were positive for *O. ostertagi*, however there was also a great discrepancy between provinces: as many as 151 of 188 herds (80.3%, 95% CI: 74.1%, 85.4%) were positive in the north-eastern province and only 70 of 136 herds (51.5%, 95% CI: 43.1%, 59.7%) were positive in the central province.

Seven and 6 variables were initially offered to the multivariable model for *F. hepatica* (Additional file 3) and *O. ostertagi* exposure (Additional file 4), respectively. Location of a herd in the north-eastern province, longer grazing period practiced in a herd and over 50%-share of grazing grass in roughage were all positively related to the increase in exposure to both *F. hepatica* and *O. ostertagi* (Tables 2 and 3). Moreover, the use of hay or haylage as main roughage was positively related to the increase in exposure to *F. hepatica* (Table 2). The models including the aforementioned variables explained only 38.1% and 33.6% of variation of exposure to *F. hepatica* and *O. ostertagi*, respectively.

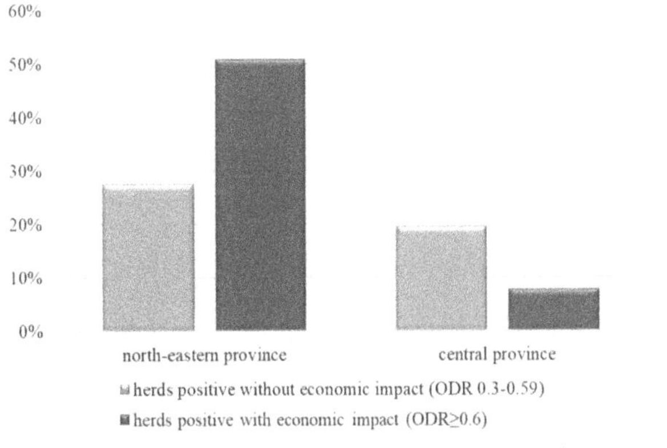

Fig. 1 Percentage of herds affected by *Fasciola hepatica* infection in two provinces of Poland. ODR stands for optical density ratio of the ELISA used for screening bulk-tank milk samples

Fasciola ELISA ODR and *Ostertagia* ELISA ODR were moderately positively correlated with each other in both provinces (the north-eastern province $r_s = 0.45$, $p < 0.001$, $n = 188$; the central province $r_s = 0.58$, $p < 0.001$, $n = 136$).

Discussion

Our study shows that both *F. hepatica* and *O. ostertagi* infection is widespread in Polish cattle population, and the north-eastern part of the country appears a high-risk area for both infections.

It is difficult to compare herd-level seroprevalence of *F. hepatica* from our study with the results of studies from other countries since various methods such as coproantigen ELISA [29], f2-antigen- or cathepsin L1 antigen-based serum ELISA [17, 30, 31], post-slaughter examinations [32], or combinations of the aforementioned [33] have been used, and marked differences in

Table 2 Multivariable linear regression model for herd-level exposure to *Fasciola hepatica*

Risk factor	Regression coefficient (B) and standard error (SE)	95% confidence interval (95% CI) for regression coefficient	t-statistics	*p*-value
Intercept	0.18 (0.02)	–	–	–
The north-eastern province	0.21 (0.02)	0.17, 0.26	9.05	< 0.001[a]
Length of grazing period (months)	0.03 (0.01)	0.02, 0.04	6.35	< 0.001 [a]
Grazing grass > 50% of all roughage	0.08 (0.02)	0.04, 0.13	3.38	0.001 [a]
Hay or haylage as main roughage	0.05 (0.02)	0.01, 0.09	2.03	0.043 [a]
Grazing on pasture > 6 h	0.06 (0.03)	−0.01, 0.12	1.67	0.095
Any grazing grass in roughage	−0.03 (0.03)	−0.08, 0.03	− 0.89	0.373
Grazing on pasture	0.03 (0.07)	−0.11, 0.17	0.38	0.706

[a] significant risk factor for herd-level exposure to *Fasciola hepatica* at a significance level of 0.05

their diagnostic performance have recently been demonstrated [34, 35]. Even if the same E/S ELISA had been used the authors assumed different cut-off values [7, 15, 35, 36]. We based our interpretation of *Fasciola* ELISA ODR on logical argumentation of Vercruysse and Claerebout [28] and Salimi-Bejestani et al. [8] – the cut-off of ODR = 0.27 allows to detect herds in which at least one fourth of cows is infected by *F. hepatica*, which is considered an economically significant level. Following this interpretation the north-eastern part of our country seems to be as much affected as England, Scotland and Wales (76–86%; [37]). In the Belgian and German studies [36, 38] much higher cut-off of ODR = 0.8 was used and the herd-level seroprevalences were 37% and 50–57%, respectively. If the cut-off of ODR = 0.27 had been assumed these figures perhaps would have been much higher, which implies that situation in the central part of Poland is better than in other countries of the western Europe. It seems to be more like in low-risk Scandinavian countries [7].

Median *Ostertagia* ELISA ODR in the central province of Poland (0.51, IQR from 0.34 to 0.71) was similar or slightly higher than the figures presented in two other studies carried out in the southern province [26, 27] which were 0.59 (IQR from 0.37 to 0.66) and 0.29 (IQR from 0. 12 to 0.59), respectively. In the north-east median *Ostertagia* ELISA ODR (0.73, IQR from 0.57 to 0.85) was much higher than in the southern and central provinces. The same trend can be observed for the herd-level seroprevalence at the cut-off set to ODR of 0.5–30-56% in the south, 50% in the center and 80% in the north-east of the country. These results confirm that the north-east of Poland is a high-risk area. In general, situation in Poland corresponds to other European countries with the central part of the country closer to countries with moderate exposure like France, Germany, Denmark, Sweden and Netherlands, and the north-eastern part more like high-risk countries such as Belgium, UK, Ireland and Switzerland [9, 10, 17, 39]. Nevertheless, the conclusions on the economic role of *O. ostertagi* should be drawn with caution as the antigen used for the *Ostertagia* ELISA is considered to cross-react with antibodies to several other parasites including *Cooperia* spp. [40], *F. hepatica* and lungworm *Dictyocaulus viviparus* [35]. Given that these parasites differ considerably from *O. ostertagi, Ostertagia* ELISA is likely to cross-react with antibodies against any members of *Trichostrongyloidea* superfamily or against *Paramphistomidae* family (rumen flukes). In our study quantitative results of ELISAs for *F. hepatica* and *O. ostertagi* were also mutually correlated which may at least partly be explained by overlapping risk factors for both infections, however the role of cross-reactions can neither

Table 3 Multivariable linear regression model for herd-level exposure to *Ostertagia ostertagi*

Risk factor	Regression coefficient (B) and standard error (SE)	95% confidence interval (95% CI) for regression coefficient	t-statistics	*p*-value
Intercept	0.42 (0.02)	–	–	–
Length of grazing period (months)	0.05 (0.01)	0.03, 0.06	8.40	< 0.001 [b]
Grazing grass > 50% of all roughage	0.08 (0.03)	0.02, 0.13	2.60	0.010 [b]
The north-eastern province	0.06 (0.03)	0.01, 0.11	2.24	0.026[b]
Grazing on pasture	−0.16 (0.08)	−0.31, − 0.01	−1.98	0.048[a]
Hay or haylage as main roughage	−0.02 (0.03)	−0.07, 0.04	− 0.50	0.620
Herd size	0.01 (0.01)	−0.01, 0.01	−0.11	0.912

[a] excluded for very strong correlation with the length of grazing period.
[b] significant risk factor for herd-level exposure to *Ostertagia ostertagi* at a significance level of 0.05.

be excluded nor reliably assessed. Similar doubts have also been raised by other authors [17, 36]. Therefore, *Ostertagia* ELISA may have overestimated the true level of infestation with *O. ostertagi* in Polish cattle herds. Nevertheless, our results are comparable with studies carried out in other regions as the same ELISA has been used worldwide.

All included risk factors were more often encountered in the north-eastern province, which might have accounted for the higher prevalence of both infections. Nevertheless, the location of a cattle herd in this region still proved to be an independent risk factor for both infections when the model was adjusted by all the other risk factors. This likely rises from climatic differences between central and north-eastern Poland, mainly higher average monthly rainfalls in the spring and early summer in the north Poland [41], which has been previously shown to explain a considerable part of variation of *F. hepatica* occurrence [42–44].

Risk factors identified in our study generally match to these already known. Length of grazing period and high proportion of grazing grass in roughage have been evidenced to increase the risk of exposure to *F. hepatica* and *O. ostertagi* [9–11, 15, 17] simply because they are pasture-borne parasites. The univariable analysis indicated that herd size was negatively linked to the risk of exposure to *O. ostertagi*, which was consistent with many previous studies [7, 9–11], however, this factor did not hold in the multivariable analysis. This may result from the fact that Polish herds are small compared to those included in most of other studies. On the other hand, in the recent Swiss study [17], where most of herds were similarly small, herd size proved to increase the risk of exposure to *O. ostertagi*, which indicates that the role of herd size in the epidemiology of *O. ostertagi* is unclear and any conclusion should be drawn very cautiously. Contrary to some previous studies [11, 17] we did not include milk yield in our risk analysis as reduced milk yield is a consequence of more intensive exposure to the parasites [2, 37, 45] rather than a factor affecting the exposure. The only thing not mentioned before is the fact that using hay or haylage as main roughage increases risk of exposure to *F. hepatica*, when compared to corn silage, even when no pasture or fresh grass is available. This likely results from the presence of infective encysted metacercariae in forages harvested from meadows, and their resistance to the process of drying or fermentation involved in production of hay or silages. While it is known that metacercariae remain viable and infective in drying hay for at least several months, haylage has been considered as an unfavorable environment for metacercariae, which sustains their viability for less than 3 weeks [46]. Our observation seems to contradict this statement. Most likely explanation is that the viability of metacercariae depends on the quality of fermentation process. The process of haylage preparation in field conditions simply may not be as careful as it was in the experimental study [46], which allows metacercariae to remain viable and infective. Corn silage is safer perhaps because environmental conditions on corn fields are less favorable for metacerkariae survival than on grass pastures [47]. Moreover, corn for silage is harvested at considerably higher cutting height (usually more than 20 cm from the ground) than hay for haylage (8–10 cm). This may reduce the risk of inclusion of metacercariae in the roughage, as they have been shown to encyst mainly on the bottom parts of plants [48, 49].

The main shortcoming of our study is the lack of knowledge of climate (e.g. rainfalls, temperatures) and environmental conditions (e.g. soil humidity on pastures) on farms included in the study, as well as low refinement of farm management factors (i.e. type of pasture and water source utilized, exact share of grazing grass in diet, turnover of calves and heifers, stocking density etc.), which make the risk analysis quite superficial (which is highlighted by the low amount of variation explained by the models), in particular when compared with thorough risk analyzes published in recent years [30, 32, 37]. However, we mainly aimed to characterize epidemiological situation of *F. hepatica* and *O. ostertagi* in our country, especially in view of scarcity of currently available data not only from Poland but also from our eastern neighbors. Risk analysis was rather about to confirm that factors commonly considered as linked to the occurrence of these infections play their role in Poland as well, than to find any new, so far uncovered, ones. Furthermore, herds were not randomly enrolled in this study so results should be extrapolated with caution. We hope that using lists of collaborators of main dairies and enrolling quite high proportion of herds from each province (5–10%) at least partially compensates for lack of randomness.

Conclusions

Both *F. hepatica* and *O. ostertagi* are widespread in cattle population in Poland, however their occurrence at a herd-level varies between different regions of Poland. This diversity can only partly be explained by different herd management, and appears linked to environmental and climate conditions typical for these regions. Our study provides basic knowledge of the epidemiological situation in the country, which is essential for effective control of those two parasitic infections in Polish cattle population.

Additional files

Additional file 1: Clean copy of the questionnaire used in this study. The questionnaire was developed specifically for this study.

Additional file 2: Detailed data regarding cattle herds enrolled in the study. Detailed data regarding cattle herds enrolled in the study including province, optical density ratio (ODR) in *Fasciola hepatica* ELISA, result of *Fasciola hepatica* ELISA according to the cut-off of 0.270, optical density ratio (ODR) in *Ostertagia ostertagi* ELISA, result of *Ostertagia ostertagi* ELISA according to the cut-off of 0.50, herd size given as the number of adult cows (> 24 month-old), average milk yield per lactation in kg, grazing policy classed into four categories (no pasture, grazing for 6 h a day, for 12 h a day, for 24 h), grazing period length in months, main roughage used in the herd (corn silage, haylage, hay), and proportion of grazing grass in diet classed into three categories (no grazing grass, < 50%, > 50%).

Additional file 3: Univariable analysis of herd-level risk factors for *Fasciola hepatica* exposure. Descriptive statistics and results of univariable statistical analyses comparing *Fasciola hepatica* bulk-tank milk (BTM) optical density ratio (ODR) between cattle herds with different characteristics.

Additional file 4: Univariable analysis of herd-level risk factors for *Ostertagia ostertagi* exposure. Descriptive statistics and results of univariable statistical analyses comparing *Ostertagia ostertagi* bulk-tank milk (BTM) optical density ratio (ODR) between cattle herds with different characteristics.

Abbreviations

BTM: Bulk-tank milk; CI: Confidence interval; IQR: Interquartile range; M: Desired precision of the estimation; n: Sample size required; OD: Optical density; ODR: Optical density ratio; r_S: Spearman's rank-ordered correlation coefficient; σ^2: Variance in the population

Funding

This study was financed by the European Union Seventh Framework Programme for the Research and Technological Development "GLOWORM" no. FP7-KBBE-2011-5 288975. The costs of article publication were covered by KNOW (Leading National Research Centre) Scientific Consortium "Healthy Animal - Safe Food", decision of Ministry of Science and Higher Education No. 05-1/KNOW2/2015. The funders had no role in study design, data collection and analysis, decision to publish, conclusions and preparation of the manuscript.

Authors' contributions

SJK, MC, CNW, EM, and JK designed research; SJK, TN, CNW, EM, and AB performed research; MC, SJK, and JK analyzed data and wrote the paper. All authors read and approved the final manuscript.

Ethics approval

The study complied with the Directive 2010/63/EU and the Act of Polish Parliament of 15 January 2015 on protection of animals used for scientific purposes (Journal of Laws 2015, item 266).
Cattle herd owners granted informed verbal consent for participation in the study.
According to Polish legal regulations no formal approval of the bioethical commission regarding participation of cattle herd owners in the study (completing questionnaires) was required since questionnaires applied to animal not human health (verbal information from the National Ethical Commission on the basis of the Act of Polish Parliament of 6 September 2001 Pharmaceutical Law, Journal of Laws 2001, No. 126, item 1381, which regulates participation in medical experiments).

Competing interests

The authors declare that they have no competing interests.

Author details

[1]Laboratory of Veterinary Epidemiology and Economics, Faculty of Veterinary Medicine, Warsaw University of Life Sciences, Nowoursynowska 159c, 02-776 Warsaw, Poland. [2]Laboklin GmbH & Co. KG, Steubenstrasse 4, 97688 Bad Kissingen, Germany.

References

1. Fitzpatrick JL. Global food security: the impact of veterinary parasites and parasitologists. Vet Parasitol. 2013;195:233–48.
2. Charlier J, van der Voort M, Kenyon F, Skuce P, Vercruysse J. Chasing helminths and their economic impact on farmed ruminants. Trends Parasitol. 2014;30:361–7.
3. Mehmood K, Zhang H, Sabir AJ, Abbas RZ, Ijaz M, Durrani AZ, Saleem MH, Ur Rehman M, Iqbal MK, Wang Y, Ahmad HI, Abbas T, Hussain R, Ghori MT, Ali S, Khan AU, Li J. A review on epidemiology, global prevalence and economical losses of fasciolosis in ruminants. Microb Pathog. 2017;109:253–62.
4. Mezo M, González-Warleta M, Castro-Hermida JA, Muiño L, Ubeira FM. Association between anti-F. hepatica antibody levels in milk and production losses in dairy cows. Vet Parasitol. 2011;180:237–42.
5. Bellet C, Green MJ, Vickers M, Forbes A, Berry E, Kaler J. *Ostertagia* spp., rumen fluke and liver fluke single- and poly-infections in cattle: an abattoir study of prevalence and production impacts in England and Wales. Prev Vet Med. 2016;132:98–106.
6. Fanke J, Charlier J, Steppin T, von Samson-Himmelstjerna G, Vercruysse J, Demeler J. Economic assessment of *Ostertagia ostertagi* and *Fasciola hepatica* infections in dairy cattle herds in Germany using Paracalc®. Vet Parasitol. 2017;240:39–48.
7. Höglund J, Dahlstrom F, Engstrom A, Hessle A, Jakubek E-B, Schnieder T, Strube C, Sollenberg S. Antibodies to major pasture borne helminth infections in bulk-tank milk samples from organic and nearby conventional dairy herds in south-central Sweden. Vet Parasitol. 2010;171:293–9.
8. Salimi-Bejestani MR, Daniel RG, Felstead SM, Cripps PJ, Mahmoody H, Williams DJ. Prevalence of *Fasciola hepatica* in dairy herds in England and Wales measured with an ELISA applied to bulk-tank milk. Vet Rec. 2005;156:729–31.
9. Forbes AB, Vercruysse J, Charlier J. A survey of the exposure to *Ostertagia ostertagi* in dairy cow herds in Europe through the measurement of antibodies in milk samples from the bulk tank. Vet Parasitol. 2008;157:100–7.
10. Bennema SC, Vercruysse J, Morgan E, Stafford K, Höglund J, Demeler J, von Samson-Himmelstjerna G, Charlier J. Epidemiology and risk factors for exposure to gastrointestinal nematodes in dairy herds in northwestern Europe. Vet Parasitol. 2010;173:247–54.
11. Pablos-Tanarro A, Pérez-Cabal MÁ, Ortega-Mora LM, Ferre I. Presence of Ostertagia ostertagi antibodies in bulk tank milk from cattle herds in northern Spain. Vet Parasitol. 2013;197:388–92.
12. Velasova M, Damaso A, Prakashbabu BC, Gibbons J, Wheelhouse N, Longbottom D, Van Winden S, Green M, Guitian J. Herd-level prevalence of selected endemic infectious diseases of dairy cows in Great Britain. J Dairy Sci. 2017;100:9215–33.
13. Charlier J, Ghebretinsae AH, Levecke B, Ducheyne E, Claerebout E, Vercruysse J. Climate-driven longitudinal trends in pasture-borne helminth infections of dairy cattle. Int J Parasitol. 2016;46:881–8.
14. Ducheyne E, Charlier J, Vercruysse J, Rinaldi L, Biggeri A, Demeler J, Brandt C, De Waal T, Selemetas N, Höglund J, Kaba J, Kowalczyk SJ, Hendrickx G. Modelling the spatial distribution of *Fasciola hepatica* in dairy cattle in Europe. Geospat Health. 2015;9:261–70.
15. Novobilský A, Sollenberg S, Höglund J. Distribution of *Fasciola hepatica* in Swedish dairy cattle and associations with pasture management factors. Geospat Health. 2015;9:293–300.
16. Okulewicz A. The impact of global climate change on the spread of parasitic nematodes. Ann Parasitol. 2017;63:15–20.
17. Frey CF, Eicher R, Raue K, Strube C, Bodmer M, Hentrich B, Gottstein B, Marreros N. Apparent prevalence of and risk factors for infection with Ostertagia ostertagi, Fasciola hepatica and Dictyocaulus viviparus in Swiss dairy herds. Vet Parasitol. 2018;250:52–9.
18. Deryło A, Szilman P. Exensivity of invasion *Fasciola hepatica* L. in cattle in the period 1957-1998 and valuation of losses caused by staggers disease among cattle in Poland, 1995-1997 (in Polish). Wiad Parazytol. 1999;45:169–79.
19. Kozak M, Wędrychowicz H. The performance of a PCR assay for field studies on the prevalence of *Fasciola hepatica* infection in *Galba truncatula* intermediate host snails. Vet Parasitol. 2010;168:25–30.
20. Michalski M, Romaniuk K. Liver fluke (*Fasciola hepatica* L) in dairy cows in north-east Poland (in Polish). Med Weter. 2000;56:182–4.
21. Deryło A, Kozłowska-Łój J, Szilman P, Najda N, Seniuk A, Wasilewski K. Extensivity of infection of *Fasciola hepatica* L. in cattle in Lublin's voivodship in the years 1993-2001 (in Polish). Wiad Parazytol. 2001;47:775–8.

22. Kornaś S, Nowosad B, Skalska M, Wróbel A. Dairy cattle infection with *Fasciola hepatica* in small farms. Med Weter. 2005;61:1368–9.
23. Kozłowska-Łój J. Prevalence of *Fasciola hepatica* L. infection in cattle in the Lublin province (Poland) in the years 2005-2008 (in Polish). Wiad Parazytol. 2011;57:127–8.
24. Kozłowska-Łój J, Łój-Maczulska A. The prevalence of *Fasciola hepatica* L. infection in cattle in the Lublin province in the years 2009-2012. Ann Parasitol. 2013;59:207–8.
25. Kozak-Cieszczyk M. The occurrence of *Fasciola hepatica* in chosen regions of Poland based on molecular and serological methods (in Polish). Wiad Parazytol. 2006;52:137–9.
26. Płoneczka-Janeczko K, Piekarska J, Rypuła K, Mazurkiewicz M. A survey of anti-*Ostertagia ostertagii* antibody levels in bulk tank milk samples (BTM) in dairy herds in lower Silesia region (Poland). Pol J Vet Sci. 2011;14:135–6.
27. Piekarska J, Płoneczka-Janeczko K, Kantyka M, Kuczaj M, Gorczykowski M, Janeczko K. Gastrointestinal nematodes in grazing dairy cattle from small and medium-sized farms in southern Poland. Vet Parasitol. 2013;198:250–3.
28. Vercruysse J, Claerebout E. Treatment vs. non-treatment of helminth infections in cattle: defining the threshold. Vet Parasitol. 2001;98:195–214.
29. Cringoli G, Rinaldi L, Veneziano V, Capelli G, Malone JB. A cross-sectional coprological survey of liver flukes in cattle and sheep from an area of the southern Italian Apennines. Vet Parasitol. 2002;108:137–43.
30. Bloemhoff Y, Forbes A, Danaher M, Good B, Morgan E, Mulcahy G, Sekiya M, Sayers R. Determining the prevalence and seasonality of *Fasciola hepatica* in pasture-based dairy herds in Ireland using a bulk tank milk ELISA. Ir Vet J. 2015;68:16.
31. Munita MP, Rea R, Bloemhoff Y, Byrne N, Martinez-Ibeas AM, Sayers RG. Six-year longitudinal study of *Fasciola hepatica* bulk milk antibody ELISA in the dairy dense region of the Republic Ireland. Prev Vet Med. 2016;134:16–25.
32. Olsen A, Frankena K, Bødker R, Toft N, Thamsborg SM, Enemark HL, Halasa T. Prevalence, risk factors and spatial analysis of liver fluke infections in Danish cattle herds. Parasit Vectors. 2015;8:160.
33. Innocent GT, Gilbert L, Jones EO, McLeod JE, Gunn G, McKendrick IJ, Albon SD. Combining slaughterhouse surveillance data with cattle tracing scheme and environmental data to quantify environmental risk factors for liver fluke in cattle. Front Vet Sci. 2017;4:65.
34. Arifin MI, Höglund J, Novobilský A. Comparison of molecular and conventional methods for the diagnosis of *Fasciola hepatica* infection in the field. Vet Parasitol. 2016;232:8–11.
35. Mazeri S, Sargison N, Kelly RF, Bronsvoort BM, Handel I. Evaluation of the performance of five diagnostic tests for *Fasciola hepatica* infection in naturally infected cattle using a Bayesian no gold standard approach. PLoS One. 2016;11:e0161621.
36. Bennema S, Vercruysse J, Claerebout E, Schnieder T, Strube C, Ducheyne E, Hendrickx G, Charlier J. The use of bulk-tank milk ELISAs to assess the spatial distribution of *Fasciola hepatica*, *Ostertagia ostertagi* and *Dictyocaulus viviparus* in dairy cattle in Flanders (Belgium). Vet Parasitol. 2009;165:51–7.
37. Howell A, Baylis M, Smith R, Pinchbeck G, Williams D. Epidemiology and impact of *Fasciola hepatica* exposure in high-yielding dairy herds. Prev Vet Med. 2015;121:41–8.
38. Kuerpick B, Schnieder T, Strube C. Seasonal pattern of *Fasciola hepatica* antibodies in dairy herds in northern Germany. Parasitol Res. 2012;111:1085–92.
39. Charlier J, Claerebout E, Duchateau L, Vercruysse J. A survey to determine relationships between bulk tank milk antibodies against *Ostertagia ostertagi* and milk production parameters. Vet Parasitol. 2005;129:67–75.
40. Keus A, Kloosterman A, Van Den Brink R. Detection of antibodies to *Cooperia* spp. and *Ostertagia* spp. in calves with the enzyme linked immunosorbent assay (ELISA). Vet Parasitol. 1981;8:229–36.
41. Czarnecka M, Nidzgorska-Lencewicz J. Multiannual variability of seasonal precipitation in Poland (in Polish). Water Environ Rural Areas. 2012;12:45–60.
42. Bennema SC, Ducheyne E, Vercruysse J, Claerebout E, Hendrickx G, Charlier J. Relative importance of management, meteorological and environmental factors in the spatial distribution of *Fasciola hepatica* in dairy cattle in a temperate climate zone. Int J Parasitol. 2011;41:225–33.
43. Mas-Coma S, Valero MA, Bargues MD. Climate change effects on trematodiases, with emphasis on zoonotic fascioliasis and schistosomiasis. Vet Parasitol. 2009;163:264–80.
44. McCann CM, Baylis M, Williams DJL. The development of linear regression models using environmental variables to explain the spatial distribution of *Fasciola hepatica* infection in dairy herds in England and Wales. Int J Parasitol. 2010;40:1021–8.
45. Köstenberger K, Tichy A, Bauer K, Pless P, Wittek T. Associations between fasciolosis and milk production, and the impact of anthelmintic treatment in dairy herds. Parasitol Res. 2017;116:1981–7.
46. Tarczyński S, Podkówka W. Effect of ensilage of green roughage and drying hay on viability and infectivity of *Fasciola hepatica* metacercariae. Acta Parasitol. 1964;20:201–8.
47. Meek AH, Morris RS. The longevity of *Fasciola hepatica* metacercariae encysted on herbage. Aust Vet J. 1979;55:58–60.
48. Dreyfuss G, Abrous M, Vignoles P, Rondelaud D. Fasciola hepatica and Paramphistomum daubneyi: vertical distribution of metacercariae on plants under natural conditions. Parasitol Res. 2004;94:70–3.
49. Suhardono, Roberts JA, Copeman DB. Distribution of Metacercariae of Fasciola gigantica on rice straw. Trop Anim Health Prod. 2006;38:117–9.

8

Evaluation of arterial digital blood flow using Doppler ultrasonography in healthy dairy cows

H. Müller[1*], M. Heinrich[1], N. Mielenz[2], S. Reese[3], A. Steiner[4] and A. Starke[1]

Abstract

Background: Local circulatory disturbances have been implicated in the development of foot disorders in cattle. The goals of this study were to evaluate the suitability of the interdigital artery in the pastern region in both hind limbs using pulsed-wave (PW) Doppler ultrasonography and to investigate quantitative arterial blood flow variables at that site in dairy cows. An Esaote MyLabOne ultrasound machine with a 10-MHz linear transducer was used to assess blood flow in the interdigital artery in the pastern region in both hind limbs of 22 healthy German Holstein cows. The cows originated from three commercial farms and were restrained in a standing hoof trimming chute without sedation.

Results: A PW Doppler signal suitable for analysis was obtained in 17 of 22 cows. The blood flow profiles were categorised into four curve types, and the following quantitative variables were measured in three uniform cardiac cycles: vessel diameter, pulse rate, maximum systolic velocity, maximum diastolic velocity, end-diastolic velocity, reverse velocity, maximum time-averaged mean velocity, blood flow rate, resistance index and persistence index. The measurements did not differ among cows from the three farms. Maximum systolic velocity, vessel diameter and pulse rate did not differ but other variables differed significantly among blood flow profiles.

Conclusions: Differences in weight-bearing are thought to be responsible for the normal variability of blood flow profiles in healthy cows. The scanning technique used in this report for evaluation of blood flow in the interdigital artery appears suitable for further investigations in healthy and in lame cows.

Keywords: Arterial blood flow, Cattle, Foot, Pulsed wave Doppler, Ultrasonography

Background

Diseases of the locomotor system are of major welfare and economic importance in dairy cattle. The prevalence of lameness can range from 21 to 55%, depending on the geographical region, farm management, and observer [1–4]. Nearly 90% of lameness is caused by diseases of the foot [5], which have a multifactorial aetiology; the structure of the stable flooring, the design of cubicles as well as standing and lying times all play a role [4, 6, 7]. Currently, the most common foot disorders in dairy cows are claw horn lesions associated with laminitis (sole ulcer and white line disease) and digital dermatitis [3, 8, 9].

Scanning electron microscopy studies have revealed vascular alterations in feet collected post mortem from cows with laminitis and digital dermatitis [10]. In-vitro studies have shown that ischaemia results in a reduction in oxygen and glucose transfer from the perioplic and coronary epidermis to keratinocytes of the bovine claw leading to reduced cell viability and abnormal cell differentiation [11]. It has been suggested that blood circulation plays an important role in the development of foot disorders in cattle [12]. Therefore, the effects of the type of barn flooring on digital blood flow should be investigated to better understand pathophysiological implications of vascular changes in the feet of dairy cows [13].

A variety of techniques have been used for assessing blood vessels in the bovine foot. Angiography was used to show that vascular changes play a role in the pathogenesis of laminitis in bovine feet collected post mortem

* Correspondence: hendrik.mueller@vetmed.uni-leipzig.de
[1]Clinic for Ruminants and Swine, Faculty of Veterinary Medicine, University of Leipzig, An den Tierkliniken 11, 04103 Leipzig, Germany
Full list of author information is available at the end of the article

[14, 15] and in the feet of anaesthetised cattle [16, 17]. Brightness (B)-mode ultrasonography is a reliable diagnostic tool for the identification of abnormalities in bones, tendons, ligaments and synovial structures in the bovine foot [18, 19] and has been used to examine blood vessels in the feet of healthy cows [20]. Pulsed-wave (PW) Doppler is a non-invasive technique for the evaluation of blood perfusion of an organ. Doppler ultrasonography has been used in cattle to study uterine blood flow [21–23], fatty liver syndrome [24], the musculophrenic vein [25] and as a tool for echocardiography [26]. A literature search (pubmed, October 2016, keywords: PW Doppler limb cattle) failed to identify publications describing the use of PW Doppler ultrasonography in the bovine foot. In contrast, Doppler ultrasonography has been used widely in horses for the examination of blood flow in the distal limb, although it is not yet a routine diagnostic procedure. Doppler ultrasonography has been used to monitor vascular blood flow dynamics in the feet of horses with laminitis and septic pododermatitis [27], to study the temporal relationship between lipopolysaccharide-induced decrease in digital blood flow and the appearance of vasoconstrictor mediators in plasma [28], to determine the repeatability of ultrasonographic measurements and the variability in blood flow in weight-bearing and non-weight-bearing limbs [29], and to determine the effects of acetylpromazine and xylazine on digital blood flow in standing horses [30].

The goal of this study was to evaluate the suitability of the interdigital artery in the pastern region in both hind limbs for PW Doppler ultrasonographic examination in cattle. A secondary goal was to investigate quantitative arterial blood flow variables relative to the blood flow profiles in the hind feet of healthy dairy cows.

Methods

The study was approved under the guidelines of the Research Animal Act (research permit number 24–9168.21/4/30; A 30/12) by the Regional Council of Saxony, Department of Veterinary Public Health Service and Foodstuffs Control, Dresden, Germany.

Animals and study design

Twenty-two healthy German Holstein cows (age 2.2 to 5.6 (mean 4.0) years, lactation number 1 to 3 (mean 2.3), days in milk 52 to 147 (mean 95), milk yield 36.2 ± 7.1 kg / day (mean ± standard deviation), body mass 652 ± 84 kg, withers height 143 ± 5 cm, body condition score 2.5 to 3.75 / 5 (mean 3.0)) were enrolled. The cows were from three commercial free-stall farms in Saxony, Germany (farm 1, n = 11; farm 2, n = 5; farm 3, n = 6). They underwent an on-farm clinical examination, which included lameness scoring [31], and all feet were trimmed [32] 1 day before ultrasonographic examination of both hind limbs. Blood samples were collected from the external jugular vein into heparinised and EDTA tubes for the determination of haematocrit and the concentrations of urea, creatinine and total protein. All measured blood variables were in the respective reference intervals [33, 34].

Ultrasonographic examination

The unsedated cows were placed in a standing hoof-trimming chute and scanned by an experienced examiner (HM) using a MyLabOne ultrasound machine (Esaote Deutschland GmbH, Germany) with a linear transducer (linear array 33 mm; 10 MHz) without a standoff pad. The hair of both hind limbs was clipped from the tarsal region to the coronary band 1 day before the scan. Immediately before the scan, the clipped lower legs were washed with soap (Baktolin, Bode Chemie GmbH, Germany), warm water and a soft brush. Ultrasound gel (Henry Schein GmbH, Germany) was applied with a paintbrush, and both hind limbs, referred to as locations, were scanned in random order. The behaviour of the cow, leg position [31], duration of examination and the ambient temperature during the examination were recorded. An attempt was made to examine the cows when they were standing squarely, but forceful correction of the stance was not done to avoid undue excitement of the cows. The following scoring system was used for behaviour:

0 - calm - occasional movements of the hind limbs
1 - agitated - frequent movements of the hind limbs
2 - excited - frequent movements of the hind limbs accompanied by swishing of the tail and turning of the head toward the hind limbs.

The ultrasonographic examination started with B-mode using a protocol described by Heppelmann et al. [35] for examination of the soft tissues, joints and bone surfaces, and a protocol by Kofler [20] for examination of the blood vessels. Pulsed-wave Doppler mode was then used to measure blood flow in a longitudinal section of the interdigital artery in the pastern region distal to the convergence of the dorsal metatarsal artery III with the dorsal common digital artery III [36] (Fig. 1). In this region, the interdigital artery runs from the dorsal aspect of the foot in a plantar direction between the digits. The suitability of the interdigital artery for PW Doppler ultrasonography was determined in a preliminary trial. The penetration depth was set at 3 cm. When the widest section of the artery was clearly visualised, beam steering (10°) was used to ensure that the angle between blood flow direction and ultrasound waves was less than 60°. The B-mode focus zone was set at the level of the vessel (1.25 or 1.75 cm image depth), and a low level of gain was used

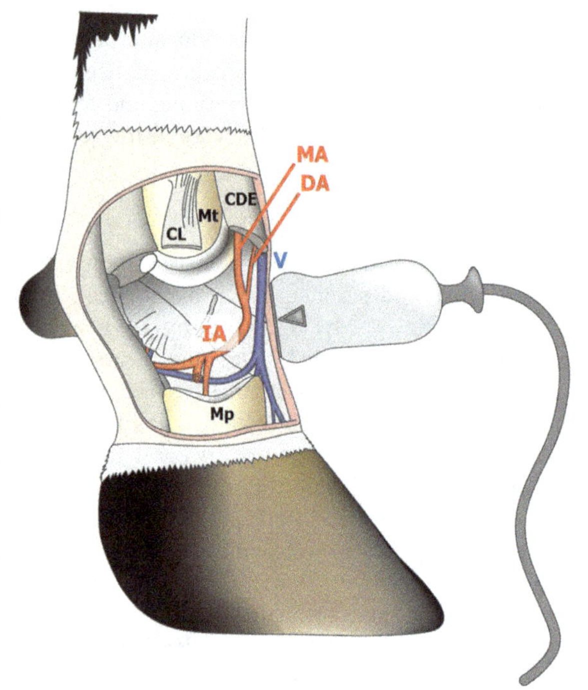

Fig. 1 Schematic longitudinal view of a bovine foot showing the position of the linear transducer for pulsed-wave Doppler examination of the interdigital artery in the pastern region distal to the convergence of the dorsal metatarsal artery III with the dorsal common digital artery III. The digit closest to the viewer has been removed to show the plantar course of the artery between the digits. MA = dorsal metatarsal artery III, DA = dorsal common digital artery III, IA = interdigital artery, V = dorsal common digital vein III, CDE = common digital extensor tendon, CL = collateral ligament, Mt = metatarsal bone IV, Mp = middle phalanx

to obtain a slightly dark B-mode picture. Pulsed-wave ultrasonography in the duplex mode was then started while placing the transducer on the artery with as little pressure as possible. The Doppler volume sample cursor was placed in the middle of the vessel and occupied half to two-thirds of the vessel diameter. The Doppler angle was set in the vessel course. The pulse repetition frequency and level of the zero line were adapted to the blood flow velocity. The PW gain was adjusted to obtain optimum contrast between the Doppler wave form and the background [37]. Self-prepared pre-sets were used to minimise the time required to obtain an image of recordable quality. One image recording a PW Doppler signal of at least 5 s was stored for each measurement.

Analysis of PW Doppler flow characteristics

The PW Doppler ultrasound images and data were transferred to the analysis program MyLabDesk (Esaote Biomedica Deutschland GmbH, Germany) on a laptop. Three similar cardiac cycles of the recorded measurement series were analysed by the same examiner (HM). The profile of the cardiac cycles was evaluated qualitatively and assigned to one of four types.

Type 1 - *biphasic blood flow profile without reverse blood flow*; the maximum diastolic velocity (Vdiast) is at least half of the maximum systolic velocity (Vsyst; Fig. 2a)
Type 2 - *biphasic blood flow profile without reverse blood flow*; the Vdiast is between one fourth and one third of Vsyst (Fig. 2b)
Type 3 - *biphasic blood flow profile without reverse blood flow*; blood flow to the zero line between the Vsyst and Vdiast (Fig. 2c)
Type 4 - *triphasic blood flow profile with one or two episodes of reverse diastolic blood flow* (Fig. 2d).

The following quantitative variables were measured at each location: vessel diameter (D), pulse rate (PR), maximum systolic velocity, maximum diastolic velocity, end-diastolic (Venddiast) and reverse velocity (Vrev). The maximum time-averaged mean velocity (Vmean = FVI × t; t = duration of one pulse cycle, FVI = flow velocity integral, FVI = ΣVi × ΔT, ΔT = time interval, Vi = instant velocity), blood flow rate (BF = Vmean × π × $(D / 2)^2$, resistance index (biphasic waveform RI = Vsyst – Venddiast / Vsyst or triphasic waveform RI = Vsyst – Vrev / Vsyst) and persistence index (biphasic waveform PI = Vsyst – Venddiast / Vmean or triphasic waveform PI = Vsyst – Vrev / Vmean) were calculated.

Statistical analyses

The blood flow variables were statistically analysed using a linear model with correlated residual effects. The analysis was based on the following model with regard to farm and location:

$$\text{(A) } y_{ijk} = \mu + \alpha_i + \beta_k + (\alpha\beta)_{ik} + e_{ijk} \text{ with cov } (e_{ij1}, e_{ij2}) = \rho \times \sigma_{e1} \times \sigma_{e2}$$

where y_{ijk} is the response variable of the j-th animal from the i-th farm (i = 1,2,3) at the k-th location (k = 1,2), μ is an intercept, α_i is the fixed effect of the i-th farm, β_k is the fixed effect of k-th location, $(\alpha\beta)_{ik}$ is the interaction between farm and location, and e_{ijk} is the random residual effect, which is assumed to have a normal distribution. The term "location" refers to left or right hind limb. A non-significant interaction term was dropped from the model. The residual effects of the locations on the same animal were assumed to be dependent with correlation ρ and with heterogeneous residual variances. The parameter ρ can be interpreted as the phenotypic correlation between the observations of the two locations on the same animal.

Phenotypic correlations between different traits were estimated by using two-trait analyses within locations.

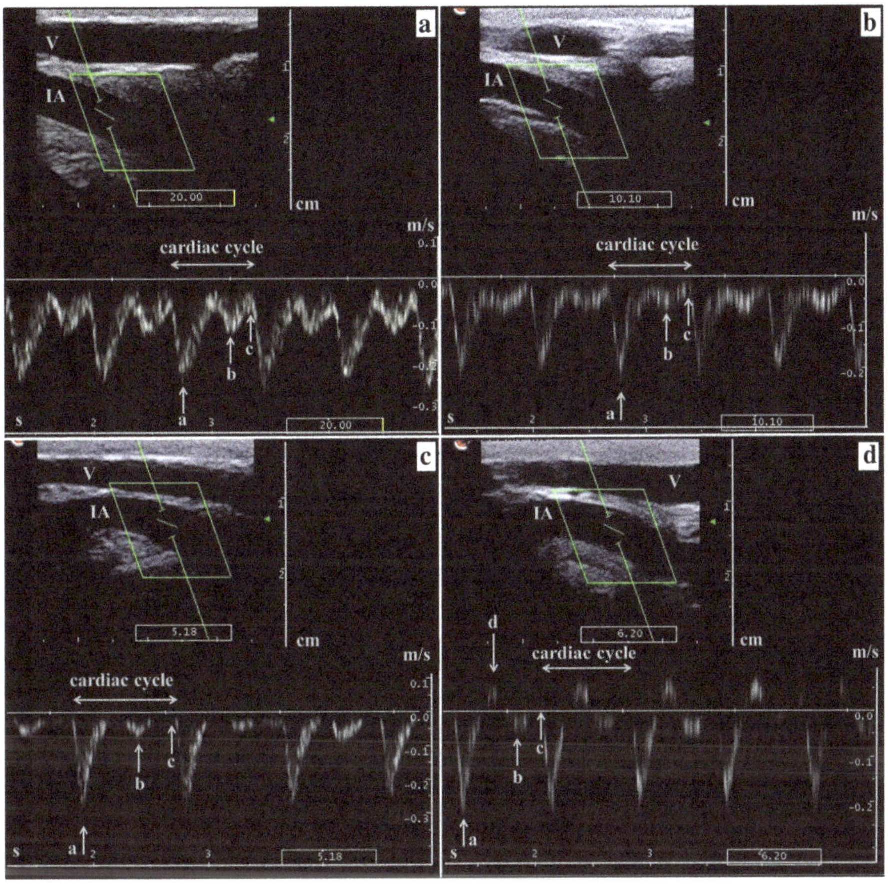

Fig. 2 The four different blood flow profiles of the interdigital artery in the pastern region distal to the convergence of the dorsal metatarsal artery III with the dorsal common digital artery III. The image on top shows the interdigital artery in sagittal section and brightness mode. Below is the pulsed wave Doppler curve. **a** Type 1. The blood flow profile is biphasic without reverse blood flow, and the maximum diastolic velocity is at least half of the maximum systolic velocity. **b** Type 2. The blood flow profile is biphasic without reverse blood flow, and the maximum diastolic velocity is between one fourth and one third of the maximum systolic velocity. **c** Type 3. The blood flow profile is biphasic without reverse blood flow, and the blood flow velocity approaches the zero line between the maximum systolic and the maximum diastolic velocities. **d** Type 4. The blood flow profile is characterised by a triphasic waveform with one episode of reverse blood flow. IA = interdigital artery, V = dorsal common digital vein III, a = maximum systolic velocity (Vsyst), b = maximum diastolic velocity (Vdiast), c = end-diastolic velocity (Venddiast), d = reverse velocity (Vrev)

Because of the analysis within the locations, the linear models contain only the fixed effects of the farms and random residual effects.

Because of the small sample size, simultaneous analysis of the factors farm, location and curve type in one complex model was omitted. With model A, most variables did not differ significantly among farms and between locations and therefore an alternative model B was used for the analysis of the curve types:

$$\text{(B)}\quad y_{ijk} = \mu + \alpha_i + e_{ijk} \text{ with } \operatorname{cov}(e_{ij1}, e_{ij2}) = \rho \times \sigma_{e1} \times \sigma_{e2}$$

where y_{ijk} is the response variable of the j-th animal with the i-th curve type (i = 1,2,3,4) at the k-th location (k = 1,2), μ is an intercept, α_i is the fixed effect of the i-th curve type and e_{ijk} is the random residual effect, which is assumed to have a normal distribution. As with model A, correlations of residual effects were considered because of repeated measurements within cows.

The MIXED procedure (SAS, version 9.4, SAS Institute Inc., Cary, N.C., USA) was used for estimation of the model parameters. Differences between means were analysed using the F-test or the Tukey-Kramer test with Kenward-Roger approximation to calculate the degrees of freedom. The Shapiro-Wilk test was used to examine the studentized residuals for normality. Results were presented as least square means. Differences were considered significant at $p < 0.05$.

Results and discussion

Measurability of blood flow in the interdigital artery and selection of the measuring point

A PW Doppler signal suitable for analysis could not be obtained in either of the hind limbs in five of the 22 cows. In these ten hind limbs, the following B-mode conditions accompanied the lack of a suitable PW Doppler signal:

- good B-mode image of the artery, no PW Doppler signal (n = 2)
- moderate B-mode image of the artery, no PW Doppler signal (n = 5)
- moderate B-mode image of the artery, PW Doppler signal obtained but not suitable for analysis (n = 3).

The lateral palmar digital artery at the level of the fetlock joint [38–40] and the medial palmar digital artery [27] have been used for PW Doppler ultrasonography in horses. The plantar common digital arteries II, III and IV have a diameter of only 1 to 2 mm in cattle and cannot be imaged ultrasonographically in all cows [20]. By contrast, the dorsal metatarsal artery III has a diameter of 4 mm [20] and can be consistently visualised ultrasonographically at the level of the fetlock joint in cattle. Visualisation of this artery further distally (interdigital artery) is difficult because the vessel courses between the digits toward the plantar aspect of the foot [20]. Therefore we had hoped to identify a suitable measuring site for the dorsal metatarsal artery III at the level of the metatarsus in preliminary trials. PW Doppler ultrasonography of dorsal metatarsal artery III in the metatarsal region was only possible when a make-shift standoff pad was used to obtain a Doppler angle smaller than 60°; suitable reusable diagonal standoff pads for the linear transducer that we used were not commercially available. To achieve a Doppler angle smaller than 60°, we selected the measuring site in a longitudinal section of the interdigital artery in the pastern region distal to the convergence of the dorsal metatarsal artery III with the dorsal common digital artery III. The less-than-ideal visualisation of the artery by B-mode ultrasonography in this region and the low occlusive pressure of the transducer (in an attempt to minimise compression of the artery) were possible reasons for the lack of suitable Doppler signals in five cows.

Ambient temperature during ultrasonography

The ambient temperature ranged from 4 to 24 °C (mean 11 °C) and the examination room was draught-free. It was not possible to maintain a constant ambient temperature for all cows, similar to studies in horses [40, 41]. However, the barn and the examination room had a common air space and therefore the cows did not encounter major changes in ambient temperature when moved from the barn to the examination room.

Behaviour and leg position during ultrasonography

Thirteen of the 22 cows were calm, three were agitated and six were excited during the examination. Several attempts were necessary to obtain a PW Doppler signal in agitated and excited cows. A PW Doppler signal suitable for analysis could not be obtained in three calm and two agitated cows. There were no significant associations between behaviour and quality of B-mode images and behaviour and the generation of PW Doppler images suitable for analysis.

The position of the hind limbs of the 17 cows from which a useful PW Doppler signal was obtained was normal (n = 20 hind limbs) or slightly forward (n = 14 hind limbs). To maintain normal blood circulation, the limbs were not fixed, and forceful manipulation of the leg position was avoided. Although the administration of xylazine did not change blood flow velocity or the diameter of the musculophrenic vein in cattle [25], it reduced the blood flow rate in the toe of healthy horses [30]. Therefore, we did not sedate the cows and instead dealt with a certain amount of shifting of weight between the limbs.

Results of the B-mode ultrasonography

There were no abnormal findings on B-mode ultrasonograms of the larger blood vessels, synovial structures, tendons, bone surfaces, and skin.

Results of the PW Doppler ultrasonography

Qualitative results - blood flow profiles

The distribution of the curve types of the different blood flow profiles is shown in Table 1. Six cows had the same curve type in both locations. Curve type 4 was only seen in images of the right hind limb. There were no significant associations between the order of measurement and curve type and the duration of examination and curve type.

Biphasic as well as triphasic arterial blood flow profiles have been described in humans; the former occurs with low peripheral resistance in an artery [37]. Diastolic blood flow decreases with increasing peripheral vascular resistance as for example with an increased perfusion demand or ischaemia [37]. In humans, triphasic blood flow profiles are typical of peripheral arteries and they occur when peripheral resistance is high. This flow pattern can change to a biphasic pattern when vasodilatory second messengers are released, e.g. physical activity [37]. Healthy horses had biphasic blood flow profiles with one systolic peak, two or

Table 1 Distribution of four types of blood flow profiles seen on PW Doppler ultrasound images of the interdigital artery in the pastern region of the left and right hind limb in 17 healthy German Holstein cows

Curve type	Left hind limb	Right hind limb	Total
Type 1	3	2	5
Type 2	11	7	18
Type 3	3	3	6
Type 4	0	5	5

three diastolic peaks and an end-diastolic plateau [39–41]. Biphasic blood flow profiles also occur in horses with laminitis, pododermatitis [27, 40] and systemic inflammatory response syndrome [40]. Triphasic blood flow profiles have been documented in healthy horses [27, 42] and in horses with laminitis [40]. Furthermore, the end-diastolic plateau can approach the zero line on PW Doppler ultrasound images from healthy horses when the examined limb bears more weight [29]. Blood flow profiles and various quantitative PW Doppler variables obtained from restless healthy horses had a higher variability than those obtained from relaxed healthy horses [29]. Degree of weight-bearing and blood flow rate in the limb were inversely related in healthy horses because weight-bearing increases peripheral resistance in the toe [29, 30] and causes the diastolic plateau to be lower. Therefore non-weight bearing limbs of healthy horses showed a low or non-resistance blood flow profile and weight bearing limbs a high resistance blood flow profile [29, 39]. The blood flow profile is prone to change particularly in diastole. We suspect that the different curve types observed in the present study were caused by variability in weight-bearing.

Quantitative results – Doppler parameters

The quantitative PW Doppler variables for both limbs are shown in Table 2 according to the statistical model A. The differences in LSMeans of the variable Vsyst (Δ 0.03 m / s), Vmean (Δ 0.02 m / s), and RI (Δ 0.11) between the hind limbs found in this study were significant but the amount seemed to be small. The diameter, Vdiast, Venddiast, BF, PR and PI did not differ significantly between locations. Quantitative PW Doppler variables measured at the various farms did not differ.

Table 2 Least square means (LSM) and standard error (SE) of the brightness mode and pulsed-wave Doppler variables of the interdigital artery in the pastern region in both hind limbs in 17 healthy German Holstein cows using statistical model A, which considered farm and location

Variable	n	LSM ± SE HL	LSM ± SE HR	p location	p farm
D (cm)	17	0.54 ± 0.02	0.54 ± 0.02	1.00	0.08
Vsyst (m/s)	17	0.35 ± 0.02	0.32 ± 0.02	0.04*	0.18
Vdiast (m/s)	17	0.12 ± 0.01	0.09 ± 0.01	0.06	0.35
Venddiast (m/s)	17	0.05 ± 0.01	0.03 ± 0.01	0.34	0.82
Vrev (m/s)	5		0.10 ± 0.02		
Vmean (m/s)	17	0.13 ± 0.01	0.11 ± 0.01	0.05*	0.63
PI[a]	16	2.18 ± 1.08	3.12 ± 1.38	0.10	0.09
RI	17	0.84 ± 0.02	0.95 ± 0.06	0.03*	0.24
BF[a] (ml/min)	17	177.40 ± 1.10	131.10 ± 1.20	0.14	0.31
PR (1/min)	17	77 ± 2	77 ± 3	0.84	0.06

D vessel diameter, *Vsyst* maximum systolic velocity, *Vdiast* maximum diastolic velocity, *Venddiast* end-diastolic velocity, *Vrev* reverse velocity, *Vmean* maximum time-averaged mean velocity, *PI* persistence Index, *RI* resistance Index, *BF* blood flow rate, *PR* pulse frequency, *HR* right hind limb, *HL* left hind limb, *p location* differences between LSM of right and left hind limb, *p farm* F-test of differences between farms

*significance ($p < 0.05$)

[a]median in the original scale, calculated from the LSM estimated in the log transformed scale

The diameter of the interdigital artery was in general agreement with a mean diameter of 4 mm obtained at a similar measuring site further proximally in mature Simmental and Brown Swiss cows aged 6.1 ± 1.7 years and weighing 510 – 637 kg. The same study did not find a difference in vessel diameter between the limbs [20]. Horses with acute laminitis or pododermatitis (4.3 mm) had a larger diameter of the medial palmar digital artery than healthy horses (4.0 mm) [27]. The blood flow velocities and the resistance and persistence indices found in this study are largely comparable to the values measured in healthy horses [27, 39]. Blood flow velocities obtained from healthy horses were greater when the leg was lifted than when it was weight-bearing, but the resistance index did not differ [29]. In another study of healthy horses, diastolic and mean velocities decreased and resistance index increased with an increase in weight-bearing, whereas the maximum systolic velocity did not differ between weight-bearing and non-weight-bearing legs [39]. A possible reason for the discrepancy between these two studies with respect to maximum systolic velocity of weight-bearing and non-weight-bearing limbs could be a difference in the measuring site; in one study [39], the site was closer to the hoof capsule, where the effect of weight-bearing was more easily recognized, than in the other study [29]. The measuring point in our study was also close to the claw capsule. There was a remarkable difference in the Vsyst compared with that found by Pietra et al. [39]. However the difference in the mean of Vsyst between weight bearing (0.32 m / s) and non-weight bearing limbs (0.37 m / s) [29] were small. The difference in the diastolic velocities in healthy horses between weight bearing and non-weight bearing had a range from 0.03 to 0.1 m / s [29] or 0.02 to 0.04 m / s [39]. Also the RI had a similar range (0.09 [29], 0.07 [39]) like in this study. The described blood flow profiles in this two horse studies were consistent in the group of non-weight bearing and weight bearing limbs. Thus the comparison of the quantitative PW Doppler variables was between non- or low resistance and high resistance blood flow profiles. In our study the distribution of blood flow profiles differed considerably between locations, which could account for the difference in Vsyst, Vmean and RI. Horses with septic pododermatitis showed much higher amounts in blood flow velocities and BF than healthy horses [27] (e.g. Vsyst (mean) 0.52 vs. 0.31 m / s, Vmean 0.15 vs. 0.08 m / s, BF 90 ml / min vs. 40 ml / min).

The PR measured in the right and left hind limbs did not differ and the two rates were strongly correlated ($r = 0.9$, $p = 0.001$). This means that the degree of

Table 3 Correlations of pulsed-wave Doppler variables of both hind limbs and between hind limbs in 17 healthy German Holstein cows using statistical model A, which considered farm and location

		HL						
		Vsyst	Venddiast	Vdiast	Vmean	PI	RI	BF
HR	**Vsyst**	**0.77**[c]	0.19[b]	0.44[b]	**0.67**[b]	0.26[b]	0.25[b]	0.42[b]
	Venddiast	**−0.45**[a]	−0.29[c]	**0.9**[b]	**0.74**[b]	**−0.79**[b]	**−0.87**[b]	**0.67**[b]
	Vdiast	−0.28[a]	**0.75**[a]	−0.26[c]	**0.86**[b]	**−0.66**[b]	**−0.67**[b]	**0.80**[b]
	Vmean	−0.27[a]	**0.79**[a]	**0.87**[a]	−0.45[c]	**−0.52**[b]	−0.43[b]	**0.84**[b]
	PI	**0.72**[a]	**−0.69**[a]	**−0.76**[a]	**−0.79**[a]	0.09[c]	**0.91**[b]	**−0.55**[b]
	RI	**0.64**[a]	**−0.86**[a]	**−0.61**[a]	**−0.73**[a]	**0.82**[a]	0.27[c]	−0.41[b]
	BF	−0.37[a]	**0.60**[a]	**0.82**[a]	**0.89**[a]	**−0.87**[a]	**−0.59**[a]	−0.26[c]

Vsyst maximum systolic velocity, *Vdiast* maximum diastolic velocity, *Venddiast* end-diastolic velocity, *Vmean* maximum time-averaged mean velocity, *PI* log transformed persistence index, *RI* resistance index, *BF* log transformed blood flow rate, *HR* right hind limb, *HL* left hind limb
bold text = different from zero ($p < 0.05$)
[a]correlations between pulsed-wave Doppler variables of the right hind limb
[b]correlations between pulsed-wave Doppler variables of the left hind limb
[c]correlations of the pulsed-wave Doppler variables between right and left hind limb

agitation or excitement did not change markedly between the examinations at the two locations.

In the left hind limb, Vdiast was positively correlated with Venddiast, Vmean and BF, and in the right hind limb, Vdiast was negatively correlated with the same variables (Table 3). Furthermore, the correlations between Vsyst and RI and between Vsyst and PI were stronger in the right than in the left hind limb. We suspect that these correlations depend on the distribution of the blood flow profiles between the hind limbs; types 3 and 4 are more resistive than types 1 and 2 and were more common in the right limb than in the left. Correlations between Vdiast and Venddiast, between Vdiast and Vmean, RI, PI and BF, and between Vendiast and Vmean, RI, PI and BF were similar for the two locations. The maximum systolic velocities measured at the two locations were positively correlated and Vdiast, Venddiast, Vmean and BF were negatively correlated. Even though the distribution of blood flow profiles differed between locations, maximum systolic velocities within cows were similar. In contrast, diastolic velocities varied considerably between locations. The hind limbs were not scanned simultaneously and therefore changes in the overall circulation or local factors could account for the lack of correlation between locations. However, major circulatory changes are unlikely because pulse rate and vessel diameter did not differ between locations, and therefore, local factors (weight-bearing, temperature) are more likely to be responsible.

Therefore quantitative PW Doppler variables were also evaluated according to the statistical model B, where the blood flow profile was included without consideration of location and farm (Table 4). Vessel diameter, PR and Vsyst did not differ significantly among curve types but the remaining variables did in spite of the small sample

Table 4 Least square means (LSM) and standard error (SE) of the brightness mode and pulsed-wave Doppler variables of the interdigital artery in the pastern region in both hind limbs in 17 healthy German Holstein cows using statistical model B, which considered curve type (CT)

Variable	LSM ± SE CT1	LSM ± SE CT2	LSM ± SE CT3	LSM ± SE CT4
D (cm)	0.56[a] ± 0.03	0.55[a] ± 0.02	0.51[a] ± 0.03	0.55[a] ± 0.03
Vsyst (m/s)	0.36[a] ± 0.03	0.31[a] ± 0.02	0.35[a] ± 0.03	0.32[a] ± 0.04
Vdiast (m/s)	0.16[a] ± 0.01	0.10[b] ± 0.01	0.09[b] ± 0.01	0.06[b] ± 0.01
Venddiast (m/s)	0.08[a] ± 0.01	0.05[b] ± 0.00	0.03[b] ± 0.01	0.00[c] ± 0.01
Vmean (m/s)	0.17[a] ± 0.02	0.12[ab] ± 0.01	0.11[bc] ± 0.02	0.06[c] ± 0.02
PI[e]	1.41[a] ± 1.15	1.98[ab] ± 1.08	3.04[b] ± 1.15	13.38[c] ± 1.37
RI	0.74[a] ± 0.03	0.83[b] ± 0.01	0.91[c] ± 0.03	1.29[d] ± 0.03
BF[e] (ml/min)	267.86[a] ± 1.21	164.23[a] ± 1.10	134.02[ab] ± 1.20	63.05[b] ± 1.38
PR (1/min)	78[a] ± 3	77[a] ± 2	76[a] ± 3	75[a] ± 3

D vessel diameter, *Vsyst* maximum systolic velocity, *Vdiast* maximum diastolic velocity, *Venddiast* end-diastolic velocity, *Vmean* maximum time-averaged mean velocity, *PI* persistence index, *RI* resistance index, *BF* blood flow rate, *PR* pulse frequency
[a,b,c,d] within rows, means with different superscripts are different ($p < 0.05$)
[e]median in the original scale, calculated from the LSM estimated in the log transformed scale

size in curve type 3 and 4. Maximum systolic velocity, Venddiast, Vmean and BF decreased progressively from curve type 1 to curve type 4. The differences in LSMeans in the diastolic velocities (Vdiast Δ 0.1 m / s, Venddiast Δ 0.08 m / s) as well as in the Vmean (Δ 0.11 m / s) had a higher range between the curve types than in the Vsyst (Δ 0.05 m / s). Also the BF (Δ 200 ml / min) and the RI (Δ 0.55) differed strongly between the curve types.

These findings support the view that a blood flow profile is primarily defined by diastole and Vsyst is independent of the curve type in healthy cows. The RI increased from type 1 to type 4, reflecting a parallel increase in peripheral resistance possibly related to an increase in weight-bearing as described before. Horses with triphasic blood flow profiles also had higher peripheral vascular resistance, which was negatively correlated with blood flow rate [43].

Conclusions

It is possible to measure blood flow in the hind limbs of healthy dairy cows by using PW-Doppler ultrasonography. There appears a considerable variability of blood flow profiles, possibly related to shifting of weight during the measurements. The differences in the quantitative PW Doppler variables are of higher importance in the comparison of the curve types than in the comparison of the hind limbs despite of the small sample size. The blood flow velocities measured in this study are similar to velocities obtained at the lower legs of healthy horses. The technique and measuring site in the interdigital artery is well suited for further studies on the effects of different degrees of weight-bearing on blood circulation and development of claw disorders.

Abbreviations

a: Maximum systolic velocity; b: Maximum diastolic velocity; BF: Blood flow rate; B-mode: Brightness mode; c: End-diastolic velocity; CDE: Common digital extensor tendon; CL: Collateral ligament; CT: Curve type; d: Reverse velocity; D: Vessel diameter; DA: Dorsal common digital artery III; EDTA: Ethylenediaminetetraacetic acid; FVI: Flow velocity integral; HL: Left hind limb; HR: Right hind limb; IA: Interdigital artery; LSM: Least square means; MA: Dorsal metatarsal artery III; Mp: Middle phalanx; Mt.: Metatarsal bone IV; PI: Persistence index; PR: Pulse rate; PW: Pulsed-wave; RI: Resistance index; SD: Standard deviation; SE: Standard error; t: Duration of one pulse cycle; V: Dorsal common digital vein III; Vdiast: Maximum diastolic velocity; Venddiast: End-diastolic velocity; Vi: Instant velocity; Vmean: Maximum time-averaged mean velocity; Vrev: Reverse velocity; Vsyst: Maximum systolic velocity; Δ: Difference; ΔT: Time interval

Acknowledgements

This work was generously supported by grants from the H. WILHELM SCHAUMANN STIFTUNG, Esaote Biomedica Deutschland GmbH, Boehringer Ingelheim Pharma GmbH & Co. KG and Selectavet Dr. Otto Fischer GmbH.

The authors thank the farm Wirtschaftshof Sachsenland Röhrsdorf/Wittgensdorf e.G., Kathrin Nechanitzky, Helena Fieseler, Anne Kretschmar, Tanja Gärtner, Halina Masur and hoof trimmers Andreas Busch, Silvio Wiesner and Andre Wissel from the Genossenschaft Klauenpfleger e.G. Sachsen for their help and support. Special thanks go to Dr. Stefan Nüske from the Lehr- und Versuchsgut Oberschleißheim, Ludwig-Maximilians-University, Munich.

Funding

This study was fund by H. WILHELM SCHAUMANN STIFTUNG, Esaote Biomedica Deutschland GmbH, Boehringer Ingelheim Pharma GmbH & Co. KG and Selectavet Dr. Otto Fischer GmbH. They had no role in the design of the study and collection, analysis, and interpretation of data and in writing the manuscript.

Authors' contributions

HM and MH carried out the practical work with the study animals and managed the data. HM did the ultrasonographic examination and analysed the blood flow profiles. Furthermore he reviewed the literature and drafted the manuscript. NM performed the statistical analyses. SR helped with the identification of a suitable measuring site for the PW Doppler in the dorsal artery in preliminary trials and gave advice in the interpretation of the data. AS participated in the study design and gave practical advice. AA was the project leader, coordinated the project, gave practical advice for all tasks and substantially helped to draft the manuscript. All authors read, revised and approved the final manuscript.

Competing interests

The authors declare that they have no competing interests.

Consent for publication

Not applicable.

Author details

[1]Clinic for Ruminants and Swine, Faculty of Veterinary Medicine, University of Leipzig, An den Tierkliniken 11, 04103 Leipzig, Germany. [2]Institute for Agriculture and Nutrition, Biometrics and Informatics in Agriculture Group, Martin-Luther-University, Karl-Freiherr-von-Fritsch-Straße 4, 06120 Halle (Saale), Germany. [3]Institute of Anatomy, Histology, and Embryology, Department of Veterinary Sciences, Faculty of Veterinary Medicine, Ludwig-Maximilians-University, Veterinaerstraße 13, 80539 Munich, Germany. [4]Clinic for Ruminants, Vetsuisse-Faculty, University of Bern, Bremgartenstrasse 109a, 3012 Bern, Switzerland.

References

1. Barker ZE, Leach KA, Whay HR, Bell NJ, Main DCJ. Assessment of lameness prevalence and associated risk factors in dairy herds in England and Wales. J Dairy Sci. 2010;93:932–41.
2. Von Keyserlingk MAG, Barrientos A, Ito K, Galo E, Weary DM. Benchmarking cow comfort on North American freestall dairies: lameness, leg injuries, lying time, facility design, and management for high-producing Holstein dairy cows. J Dairy Sci. 2012;95:7399–408.
3. Becker J, Steiner A, Kohler S, Koller-Bähler A, Wüthrich M, Reist M. Lameness and foot lesions in Swiss dairy cows: I. Prevalence. Schweiz Arch für Tierheilkd. 2014;156:71–8.
4. Solano L, Barkema HW, Pajor EA, Mason S, LeBlanc SJ, Zaffino Heyerhoff JC, et al. Prevalence of lameness and associated risk factors in Canadian Holstein-Friesian cows housed in freestall barns. J Dairy Sci. 2015;98:6978–91.
5. Shearer JK, Van Amstel SR, Brodersen BW. Clinical diagnosis of foot and leg lameness in cattle. Vet Clin North Am Food Anim Pract. 2012;28:535–56.

6. Galindo F, Broom DM. The relationships between social behaviour of dairy cows and the occurrence of lameness in three herds. Res Vet Sci. 2000;69:75–9.
7. De Vries M, Bokkers EAM, van Reenen CG, Engel B, van Schaik G, Dijkstra T, et al. Housing and management factors associated with indicators of dairy cattle welfare. Prev Vet Med. 2015;118:80–92.
8. De Frain JM, Socha MT, Tomlinson DJ. Analysis of foot health records from 17 confinement dairies. J Dairy Sci. 2013;96:7329–39.
9. Schulz T, Gundelach Y, Feldmann M, Hoedemarker M. Early detection and treatment of lame cows – effect on duration and prevalence of lesion-specific lameness. Tierarztl Prax Ausg G Grosstiere Nutztiere. 2016; doi:10.15653/TPG-150534.
10. Hirschberg RM, Plendl J. Pododermal angiogenesis and angioadaptation in the bovine claw. Microsc Res Tech. 2005;66:145–55.
11. Lübbe K. Entwicklung und Einsatz eines In-vitro-Ischämiemodells zur Untersuchung zellulärer Pathomechanismen der Klauenrehe des Rindes. Inaugural dissertation, Leipzig, Germany, 2015.
12. Hirschberg RM, Mülling CKW. Preferential pathways and hemodynamic bottlenecks in the vascular system of the healthy and diseased bovine claw. In: Proceedings of the 12th International Symposium on Lameness in Ruminants. Orlando, Florida, USA; 2002. p. 223–6.
13. Rajapaksha E, Winkler C, Tucker CB. Effect of rubber flooring on dairy cattle stepping behavior and muscle activity. J Dairy Sci. 2015;98:2462–71.
14. Boosman R, Nemeth F, Gruys E, Klarenbeek A. Arteriographical and pathological changes in chronic laminitis in dairy cattle. Vet Q. 1989;11:144–55.
15. Singh SS, Ward WR, Murray RD. An angiographic evaluation of vascular changes in sole lesions in the hooves of cattle. Br Vet J. 1994;150:41–52.
16. Prentice DE, Wyn-Jones G. A technique for angiography of the bovine foot. Res Vet Sci. 1973;14:86–90.
17. Gogoi SN, Nigam JM, Singh AP. Angiographic evaluation of bovine foot abnormalities. Vet Radiol. 1982;23:171–4.
18. Tyron KA, Clark CR. Ultrasonographic examination of the distal limb of cattle. Vet Clin North Am Food Anim Pract. 1999;15:275–300.
19. Kofler J. Diagnostic imaging in bovine orthopedics. Vet Clin North Am Food Anim Pract. 2014;30:11–53.
20. Kofler J. Darstellung und Durchmesserbestimmung der Arterien und Venen am Hinterfuß des Rindes mittels B-Mode. J Vet Med A Physiol Pathol Clin Med. 1995;42:253–66.
21. Herzog K, Koerte J, Flachowsky G, Bollwein H. Variability of uterine blood flow in lactating cows during the second half of gestation. Theriogenology. 2011;75:1688–94.
22. Heppelmann M, Weinert M, Brömmling A, Piechotta M, Hoedemaker M, Bollwein H. The effect of puerperal uterine disease on uterine involution in cows assessed by Doppler sonography of the uterine arteries. Anim Reprod Sci. 2013;143:1–7.
23. Magata F, Hartmann D, Ishii M, Miura R, Takahashi H, Matsui M, et al. Effects of exogenous oxytocin on uterine blood flow in puerperal dairy cows: the impact of days after parturition and retained fetal membranes. Vet J. 2013; 196:76–80.
24. Starke A, Schmidt S, Haudum A, Scholbach T, Wohlsein P, Beyersbach M, et al. Evaluation of portal blood flow using transcutanous and intraoperative Doppler ultrasonography in dairy cows with fatty liver. J Dairy Sci. 2011;94: 2964–71.
25. Braun U, Hoegger R, Haessig M. Colour-Doppler sonography of the musculophrenic vein in cows. Vet J. 2009;179:451–4.
26. Hallowell GD, Potter TJ, Bowen IM. Methods and normal values for echocardiography in adult dairy cattle. J Vet Cardiol. 2007;9:91–8.
27. Wongaumnuakyl S, Siedler C, Schobesberger H, Stanek C. Doppler sonographic evaluation of the digital blood flow in horses with laminitis or septic pododermatitis. Vet Radiol Ultrasound. 2006;47:199–205.
28. Menzies-Gow NJ, Bailey SR, Katz LM, Marr CM, Elliott J. Endotoxin-induced digital vasoconstriction in horses: associated changes in plasma concentrations of vasoconstrictor mediators. Equine Vet J. 2004;36:273–8.
29. Hoffmann KL, Wood AKW, Griffiths KA, Evans DL, Gill RW, Kirby AC. Doppler sonographic measurements of arterial blood flow and their repeatability in the equine foot during weight bearing and non-weight bearing. Res Vet Sci. 2001;70:199–204.
30. Hunt RJ, Brandon CI, McCann ME. Effects of acetylpromazine, xylazine and vertical load on digital arterial blood flow in horses. Am J Vet Res. 1994;55:375–8.
31. Dirksen G, Gründer H-D, Stöber M. Die klinische Untersuchung des Rindes. 4th ed. Germany: Enke; 2012.
32. Toussaint Raven E. In: Toussaint Raven E, editor. Klauenpflege beim Rind: über die Entstehung und Vorbeuge von Sohlengeschwüren. Netherlands: Brouwer Uithof; 1998. p. 75–108.
33. Fürll M. Stoffwechselkontrollen und Stoffwechselüberwachung bei Rindern. Nutztierpraxis aktuell. 2004;9:1–10.
34. George JW, Snipes J, Lane M. Comparison of bovine hematology reference intervals from 1957 to 2006. Vet Clin Pathol. 2010;39:138–48.
35. Heppelmann M. Eitrige Klauengelenksentzündung beim Rind: Vergleich des Heilungsverlaufes nach Klauengelenksresektion mit plantarem Zugang und nach hoher Klauenamputation. Inaugural dissertation, Hannover, Germany, 2004.
36. Nickel R, Schummer A, Seiferle E. Lehrbuch der Anatomie der Haustiere – Band III Kreislaufsystem, Haut und Hautorgane. 4th ed. Germany: Parey; 2005. p. 158–62.
37. Hofer M. In: FKDS-Trainer –Ein Arbeitsbuch für den Einstieg in die Farbkodierte Duplexsonographie und Echokardiographie. 3rd ed. Germany: Didamed; 2009. p. 10–8.
38. Bailey SR, Menzies-Gow NJ, Marr CM, Elliot J. The effects of vasoactive amines found in the equine hindgut on digital blood flow in the normal horse. Equine Vet J. 2004;36:267–72.
39. Pietra M, Guglielmini C, Nardi S, Gandini G, Cipone M. Influence of weight bearing and hoof position on Doppler evaluation of lateral digital arteries in healthy horses. Am J Vet Res. 2004;65:1211–5.
40. Aguirre C, Talavera J, Fernandez del Palacio MJ. Usefulness of doppler ultrasonography to assess digital vascular dynamics in horses with systemic inflammatory response syndrome or laminitis. J Am Vet Med Assoc. 2013;243:1756–61.
41. Cochard T, Toal RL, Saxton AM. Doppler ultrasonographic features of thoracic limb arteries in clinically normal horses. Am J Vet Res. 2000;61:183–90.
42. Raisis AL, Young LE, Meire HB, Taylor PM, Walsh K, Lekeux P. Variability of Doppler ultrasound measurements of hindlimb blood flow in conscious horses. Equine Vet J. 2000;32:125–32.
43. Raisis AL, Young LE, Meire HB, Taylor PM, Blissitt KJ, Marlin D, et al. Measurements of hindlimb blood flow recorded using Doppler ultrasound during administration of vasoactive agents in halothane-anesthetized horses. Vet Radiol Ultrasound. 2000;41:64–72.

9

Inflammatory mechanism of Rumenitis in dairy cows with subacute ruminal acidosis

Chenxu Zhao[1†], Guowen Liu[1†], Xiaobing Li[1†], Yuan Guan[2], Yazhou Wang[1], Xue Yuan[3], Guoquan Sun[3], Zhe Wang[1] and Xinwei Li[1*]

Abstract

Background: Subacute ruminal acidosis (SARA) is a metabolic disease in high-producing dairy cattle, and is accompanied by rumenitis. However, the mechanism of rumenitis remains unclear. Therefore, the aim of this study was to investigate the molecular mechanism of rumenitis in dairy cows with SARA.

Results: The results showed that SARA cows displayed high concentrations of ruminal volatile fatty acids, lactic acid and lipopolysaccharide (LPS). Furthermore, the blood concentrations of LPS and acute phase proteins haptoglobin, serum amyloid-A, and LPS binding protein were significantly higher in SARA cows than in control cows. Importantly, the phosphorylation levels of nuclear factor-kappaB (NF-κB) p65, inhibitor of NF-κB (IκB), c-Jun N-terminal kinase (JNK), and extracellular signal-regulated kinase 1/2 (ERK1/2) were significantly higher in the rumen epithelium of SARA cows than those of control cows. The ruminal mRNA and protein levels of NF-κB- and mitogen-activated protein kinase (MAPK)s -regulated inflammatory cytokines, tumor necrosis factor α (TNF-α), interleukin 6 (IL-6) and interleukin 1β (IL-1β), were markedly higher in SARA cows than in control cows. Similarly, serum concentrations of TNF-α and IL-6 were also significantly higher in SARA cows.

Conclusions: These results indicate that SARA results in high concentration of ruminal LPS, which over activates the NF-κB and MAPKs inflammatory pathways and then significantly increases the expression and synthesis of pro-inflammation cytokines in the rumen epithelium, thereby partly inducing rumenitis.

Keywords: Subacute ruminal acidosis, LPS, Rumenitis, Inflammatory pathway

Background

Subacute ruminal acidosis (SARA) is a metabolic disease in high-producing dairy cattle. This disease is caused by feeding high concentrate diets and is defined as a depression of ruminal pH below 5.6 at least 3 h/day [1]. Study in the United States has been indicated that up to 19% of early lactation dairy cows as well as 26% of mid-lactation cows have SARA [2]. This disease affects feed intake, milk production, rumen microflora, rumen digestion, and can cause diarrhea, rumen mucosal damage, laminitis, rumenitis, and liver abscesses in dairy cows [3–6].

Rumen acidosis can induce rumenitis, but the inflammatory pathogenesis is not well characterized. It has been suggested that low rumen pH could result in death and lysis of gram-negative bacteria and hence increase free lipopolysaccharide (LPS) in the rumen [7]. This increase accompanies the augmented LPS binding protein (LBP) concentration [8]. LBP is a specific acute phase protein, and is considered as a marker to evaluate inflammatory response in cattle [8]. Furthermore, haptoglobin (Hp) and serum amyloid-A (SAA) is also considered as inflammatory markers in cattle. Their concentrations in blood have been shown to increase as a result of SARA [1, 9].

Ruminal LPS is an inducer for inflammation and is associated with the development of rumenitis. Among the many signaling pathways that respond to inflammatory response, the nuclear factor-kappaB (NF-κB) pathway and mitogen-activated protein kinase (MAPK) family members are crucial signal transduction mediators that respond to LPS [10, 11]. Normally, the transcription factor NF-κB is sequestered in the cytoplasm bound to its inhibitor IκBα. Once stimulated by some stimuli, NF-κB

* Correspondence: lixinwei100@126.com
†Equal contributors
[1]Key Laboratory of Zoonosis, Ministry of Education, College of Veterinary Medicine, Jilin University, 5333 Xi'an Road, Changchun, Jilin 130062, China
Full list of author information is available at the end of the article

unit p65 separates from inhibitor of NF-κB (IκB) and translocates into the nucleus where NF-κB can regulate the transcription of several inflammatory cytokines genes, such as tumor necrosis factor alpha (TNF-α), interleukin 6 (IL-6), and interleukin 1 beta (IL-1β). Three subfamilies of MAPKs have been identified: c-Jun N-terminal kinase (JNK), extracellular signal-regulated kinase 1/2 (ERK1/2), and p38 mitogen-activated protein kinase (p38MAPK), and are deemed as stress responsive and thus involved in inflammation [12]. Nevertheless, the expression changes of these inflammatory pathways were not investigated in the rumen epithelium of cows with SARA. The ruminal LPS content was significantly increased in cows with SARA [13]. Accordingly, we hypothesized that over activation of NF-κB and MAPKs inflammatory pathways induced by LPS resulted in the inflammation response of rumen epithelium in SARA cows.

Methods

Animals

The Ethics Committee on the Use and Care of Animals at Jilin University approved the study protocol (Changchun, China). Ten multiparous mid-lactating Holstein cows fitted with ruminal fistula were randomly assigned into SARA group (5 cows) and control group (5 cows). These Holstein cows were selected from a 3000-cow dairy farm located in Suihua City, Heilongjiang Province, China. Cows averaged 539 ± 12.6 kg of body weight and with an average milk yield of 27 ± 1.6 kg/d (mean ± SD) at the beginning of the experiment. The SARA model was prepared for induced by Khafipour et al. (2009b) [14] and Jin et al. (2016) [15]. The component of ingredients, nutrient composition, and forage to concentrate ratio (F:C) of the total mixed ratio (TMR) and wheat-barley pellets (WBP) were also according to the study of Khafipour et al. (2009b) [14]. The experiment was continued 8 wks. In control group, cows were fed with TMR with an F:C of 50:50 during the 8 wks. In SARA group, cows were fed with TMR with an F:C of 50:50 during wk. 1 to 3. A SARA challenge was conducted by increased the ratio of concentrate to an F:C of 40:60 from wk. 4 to 8. Cows were fed two times daily and had unlimited access to fresh water throughout the experiment. After this study, these cows have been kept for our other studies about SARA. During the experimental work, the cows were housed in a climate-controlled barn in individual tie stalls to reduce environmental effects.

Rumen pH measurement

Rumen pH was monitored continuously during the experiment in 5 ruminally cannulated cows (SARA group) using indwelling pH probes as described by Gozho et al. (2006) [12]. The pH probes were placed in the ventral sac of the rumen. Rumen pH was measured at 1-min intervals and data stored in a data logger for subsequent analysis as described by Gozho et al. (2006) [12]. The purpose of ruminal pH measurement is to evaluate the success of high concentrate-induced SARA models. The pH data were summarized as average pH and time spent below pH 5.6 for each 24 h period. SARA was defined that rumen pH was below 5.6 for at least 3 h/day. The average pH in SARA group was 5.95 and the duration with pH below 5.6 was 6.15 h/day. However, the average pH in control group was 6.51 and the duration was 0. 6 h/day. The rumen fluid, rumen epithelium and blood samples were collected at 8w.

Rumen fluid sampling and analysis

Rumen fluid samples were collected from all cows at 15 min before feeding and 6 h after feeding. We opened the rumen fistula cap. A 50-mL centrifuge tube was invaded into the rumen liquid phase and the rumen fluid was collected. In ruminally cannulated cows, rumen fluid was sampled from the ventral sac of the rumen. Ruminal contents were filtered through four layers of cheesecloth. Briefly, rumen fluid samples were centrifuged at 10,000×g for 45 min at 4 °C. The supernatant was aspirated gently to prevent its mixing with the pellet and passed through a disposable 0.22-μm LPS-free filter. The filtrate was collected in a sterile tube for subsequent LPS measurement. LPS concentration was determined using a chromogenic Limulus amoebocyte lysate (LAL) endpoint assay (CE64406; Xiamen BioEndo Technology, Co. Ltd., Xiamen, China) according to the manufacturer's instructions.

The rumen fluid sample was immediately centrifuged at 2500×g for 15 min. The supernatant was used to determine the concentrations of volatile fatty acids (VFA; acetate, propionate, butyrate, valerate, isobutyrate, and isovalerate) and lactate by gas chromatography (Model 3400 Star, Varian, Walnut Creek, CA) as described by Khafipour et al. (2009a) [4].

The determination of blood parameters

Blood samples were collected by tail venipuncture from each cow at 15 min before feeding and 6 h after feeding. The blood samples were collected in blank and heparinized 10-mL evacuated tubes for serum and plasma collection, respectively. The concentration of LPS in plasma was determined by a chromogenic kinetic LAL assay (CE32545; Xiamen BioEndo Technology, Co. Ltd.) with a minimum detection limit of 0.005 EU/mL according to the manufacturer's instructions. The plasma levels of β-hydroxybutyrate (BHB) and glucose were determined using an automatic biochemistry analyzer with commercially available kits (Randox Laboratories, Beijing, China).

The serum concentrations of Hp and plasma concentrations of SAA and LBP were determined using enzyme-linked immunosorbent assay (ELISA) kits (Hp: ml002480; SAA: ml002466; LBP: ml024581; Shanghai Enzyme-linked Biotechnology Co., Ltd., Shanghai, China), respectively. Samples were initially diluted 1:1 for Hp and SAA and 1:5 for LBP and assayed according to the manufacturer's instructions. Samples were analyzed and absorbance values were read at 450 nm for Hp, SAA, and LBP using a spectrophotometer (Thermo Scientific Instrument Inc., Shanghai, China). Furthermore, the serum concentrations of TNF-α, IL-6, and IL-1β were also determined using ELISA kits (TNF-α: ml024586; IL-6: ml023756; IL-1β: ml023753; Shanghai Enzyme-linked Biotechnology Co., Ltd.) according to the manufacturer's instructions, respectively.

Quantitative real-time polymerase chain reaction (qRT-PCR) assay

A part of rumen content was moved out through rumen fistula to facilitate the retraction of the ventral sac. Approximately 1 g rumen papillae were excised at the ventral sac by an experienced surgical veterinarian.. The ruminal epithelium samples were washed 10 times by ice-cold saline, and were stored in the − 80 °C freezer and for the RNA and protein extraction. The ruminal incision was closed immediately. The total RNA was extracted from 0.2 g rumen epithelium using RNAiso Plus (TaKaRa Biotechnology Co., Ltd., Dalian, China) according to the manufacturer's instructions. RNA concentration and quality was detected using an RNA/DNA calculator (Cambridge, UK) and electrophoresis (1% agarose gels), respectively. Then, total RNA in each sample was reverse transcribed to cDNA using a reverse transcription kit (TaKaRa Biotechnology Co. Ltd.), according to the supplier's protocol. mRNA expression levels were evaluated using qRT-PCR with the SYBR Green QuantiTect RT-PCR Kit (TaKaRa Biotechnology Co., Ltd.) and a 7500 Real-Time PCR System (Applied Biosystems Inc.). The relative expression of each gene was normalized to β-actin. The primers for each gene were shown in Table 1.

Western blotting assay

The total protein was extracted from 0.5 g ruminal epithelium according to the manufacturer's instructions (Sangon Biotech Co., Ltd., Shanghai, China). Thus, protein concentration was determined using a protein assay reagent (Beyotime Biotechnology Inc., China). Target proteins were separated by sodium dodecyl sulfate-polyacrylamide gel electrophoresis (SDS-PAGE) and then electrophoretically transferred to a polyvinylidene difluoride (PVDF) membrane. Next, the membranes were blocked in 3% bovine serum albumin (BSA)-Tris-buffered saline-Tween buffer for 4 h. The blocked membranes were hybridized overnight at 4 °C with primary antibodies (Abcam, Cambridge, UK; Cell Signaling Technology, Danvers, MA, USA), respectively. After being washed 4 times, the blocked membranes were incubated in a horseradish peroxidase (HRP)-conjugated secondary antibody at room temperature for 45 min. Immunoreactive bands were visualized by enhanced chemiluminescence solution (ECL, Beyotime Biotechnology Inc., Shanghai, China). Finally, the bands were imaged using a protein simple imager (ProteinSimple, Santa Clara, CA, USA).

Statistical analysis

Data are expressed as the means ± standard deviation (SD) and analyzed using SPSS (Statistical Package for the Social Sciences) 16.0 software (SPSS Incorporated, Chicago, IL, USA). Differences among groups were compared with Student's test. A P value < 0.05 was considered statistically significant and P value < 0.01 was marked significant compared to control group.

Results

The rumen levels of VFA, lactic acid, and LPS

As shown in the Table 2, the rumen content of total VFA ($P < 0.01$), acetate ($P < 0.01$), propionate ($P < 0.05$), butyrate ($P < 0.05$), isobutyrate ($P < 0.05$), and valerate ($P < 0.05$) were significantly higher in cows with SARA than in control cows. Importantly, the ruminal lactic acid and LPS concentrations in SARA cows was 3.6 times and 4.2 times as great as the control cows ($P < 0.01$), respectively. These

Table 1 The primers sequences used for cDNA generation

Gene	Sequence number	Primer sequences (5′-3′)	Length (bp)
TNFα	NW_003104557.1	For CTGCCGGACTACCTGGACTAT Rev. CCTCACTTCCCTACATCCCTAA	234
IL6	NW_00310889.1	For AACGAGTGGGTAAAGAACGC Rev. CTGACCAGAGGAGGGAATGC	144
IL1β	NW_003104294.1	For CTGAACCCATCAACGAAA Rev. ATGACCGACACCACCTGC	190
β-actin	BC 142413.1	For GCCCTGAGGCTCTCTTCCA Rev. GCGGATGTCGACGTCACA	101

Table 2 The concentration of ruminal and blood parameters in the control and SARA cows

Item	Control ($n = 5$)	SARA ($n = 5$)	P-value
Total Ruminal VFA, *mM*	93.6 ± 5.6	130.9 ± 6.2**	0.001
Acetate, *mM*	66.2 ± 4.1	88.3 ± 4.5**	0.008
Propionate, *mM*	15.5 ± 2.9	23.5 ± 2.6*	0.021
Butyrate, *mM*	10.9 ± 1.7	17.1 ± 2.6*	0.024
Isobutyrate, *mM*	2.0 ± 0.3	2.8 ± 0.2*	0.036
Valerate, *mM*	1.7 ± 0.2	2.4 ± 0.3*	0.033
LPS, EU/mL	30,768 ± 1035	130,589 ± 1675**	0.004
Lactic acid, *mM*	0.95 ± 0.2	3.38 ± 0.4**	0.000
Plasma BHBA, *mM*	0.62 ± 0.03	0.66 ± 0.02	0.129
Plasma glucose, *mM*	3.87 ± 0.11	3.94 ± 0.12	0.632
Plasma LPS, EU/mL	< 0.005	0.21 ± 0.06**	0.000

* $P < 0.05$, ** $P < 0.01$ versus the control group

results indicate that cows with SARA display ruminal VFA, lactic acid, and LPS accumulation.

Blood parameters

As shown in Table 2, the blood concentrations of glucose and BHB were slightly higher in the SARA cows than in the control cows, but there were no significant statistical difference ($P > 0.05$). Plasma LPS concentration in control cows was below the minimum detection level of 0.005 EU/mL for the method. Following the SARA challenge, the average LPS in plasma was 0.21 EU/mL. The blood concentrations of the acute phase proteins LBP, Hp, and SAA in SARA group were 40.9 μg/mL, 0.300 mg/mL, and 313.9 μg/mL, and were significantly higher than those of 18.6 μg/mL, 0.168 mg/mL, and 114. 5 μg/mL in control group, respectively (Fig. 1, $P < 0.01$), which suggested that cows with SARA displayed severe inflammation response.

MAPKs and NF-κB inflammation pathways were over induced in the ruminal epithelium of cows with SARA

To investigate the mechanism of rumenitis, the inflammatory pathways MAPKs and NF-κB were detected using Western Blotting. As shown in Fig. 2, the phosphorylation levels of IκBα (p-IκBα/IκBα) and NF-κB p65 (p-NF-κB p65/NF-κB p65) were significantly higher in cows with SARA than in control cows ($P < 0.01$). Furthermore, the phosphorylation levels of JNK (p-JNK/JNK) and ERK1/2 (p-ERK1/2/ERK1/2) were significantly increased in cows with SARA ($P < 0.05$). In addition, p38MAPK (p-p38MAPK/p38MAPK) phosphorylation level was slightly higher in SARA cows than in control cows ($P > 0.05$). Taken together, these data indicate that MAPKs and NF-κB inflammation pathways are over induced in the ruminal epithelium of cows with SARA.

The mRNA and protein expression levels of inflammatory cytokines

MAPKs and NF-κB pathways over activation could promote the transcription expression of genes coded

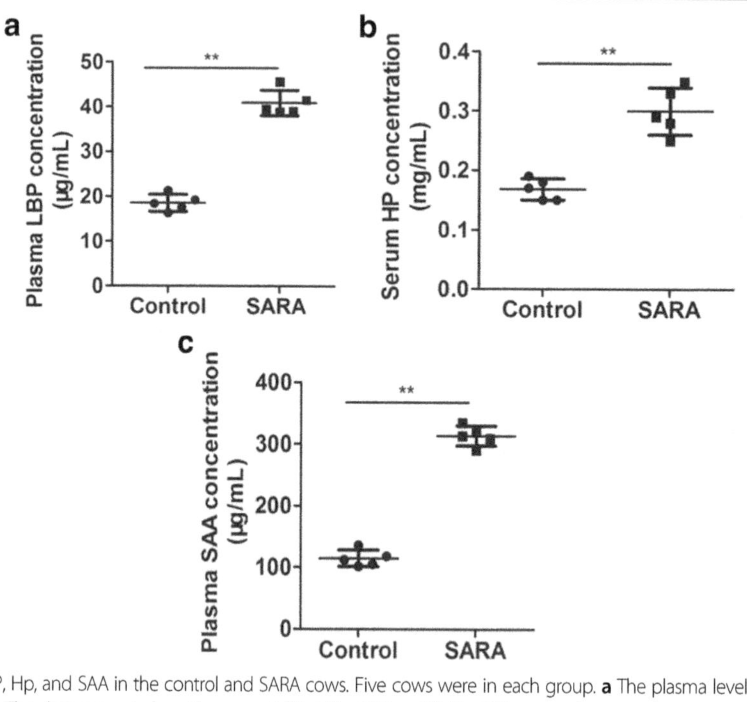

Fig. 1 The blood levels of LBP, Hp, and SAA in the control and SARA cows. Five cows were in each group. **a** The plasma level of LBP; **b** The serum level of Hp; **c** The plasma level of SAA. The data presented are the mean ± SD. * $P < 0.05$ and ** $P < 0.01$ versus the control group

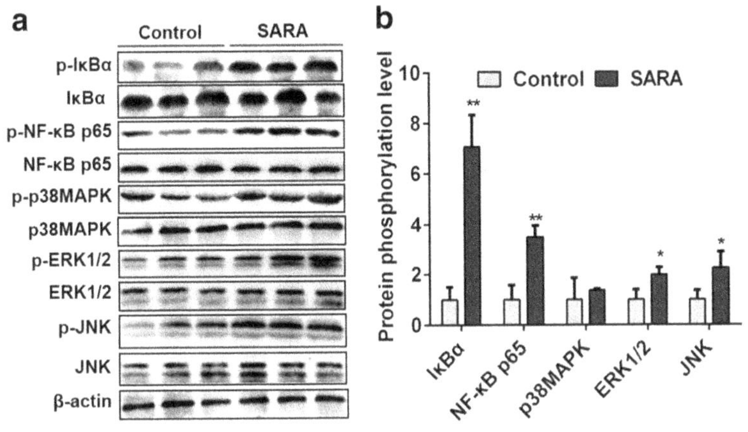

Fig. 2 The phosphorylation levels of NF-κB and MAPKs in the ruminal epithelium of control and SARA cows. Five cows were in each group. **a** The western blotting results of p-IκBα, IκBα, p-NF-κB p65, NF-κB p65, p-p38MAPK, p38MAPK, p-ERK, ERK, p-JNK, JNK, and β-actin; B: The phosphorylation levels of IκBα, NF-κB p65, p38MAPK, ERK1/2 and JNK. The data presented are the mean ± SD. * $P < 0.05$, ** $P < 0.01$ versus the control group

inflammatory cytokines [16]. In this study, the mRNA and protein expression levels of TNF-α, IL-6, and IL-1β in the ruminal epithelium of cows with SARA were significantly higher than those of control cows (Fig. 3, $P < 0.01$; Fig. 4, $P < 0.05$).

The blood concentrations of inflammatory cytokines

As shown in Fig. 5, the blood concentrations of TNF-α and IL-6 was significantly higher in SARA cows than in control cows ($P < 0.01$, $P < 0.05$). Furthermore, the IL-1β concentrations were higher in SARA cows than in control cows, but there were no significant statistical difference ($P > 0.05$). These results further demonstrated that high expression of TNF-α, IL-6, and IL-1β further mediated the inflammation response of ruminal epithelium in SARA cows.

Discussion

SARA is a common health problem in dairy cattle. Suddenly switching cattle from a high-forage to a high-starch diet result in a decrease of ruminal pH that is characteristic of SARA because VFA accumulates in the rumen [17]. Furthermore, the increase of high-starch diet of cattle during gradual grain adaptation results in microbiological changes in the rumen. If the rate of concentrate inclusion in the diet is higher than the rate at which lactate-utilizing bacteria can increase, then lactic acid accumulates in the rumen and depresses rumen pH more drastically than similar amounts of VFA [18]. In this study, we found the content of total VFA, acetate, propionate, butyrate, and lactic acid in rumen fluid were significantly higher in SARA cows than in control cows. Our data further indicated that SARA appeared to be caused by an elevation in total VFA and lactic acid, which further resulted in a significant decrease in ruminal pH.

During low rumen pH, gram-negative bacteria lyse more rapidly, increasing the concentration of LPS in the rumen [7]. In our experiment, the ruminal content of LPS was 130,589 EU/mL in grain-induced SARA cows and significantly higher than that of 30,768 EU/mL in control cows signi. This change was greater than the

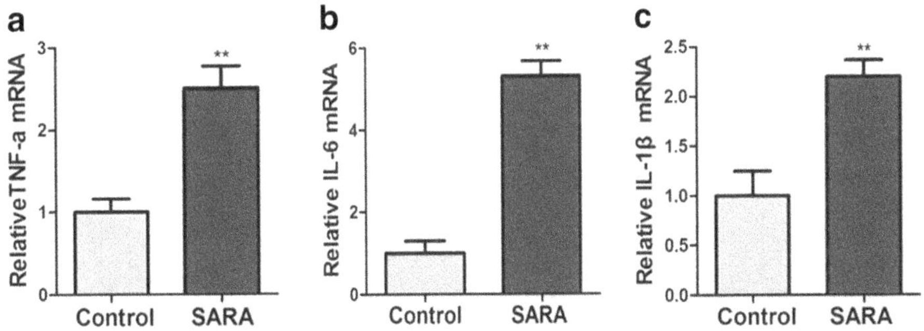

Fig. 3 The mRNA expression levels of inflammatory cytokines in the ruminal epithelium of control and SARA cows. Five cows were in each group. **a** The mRNA expression levels of *TNF-α*; **b** The mRNA expression levels of *IL-6*; (C) The mRNA expression levels of *IL-1β*. The data presented are the mean ± SD. * $P < 0.05$, ** $P < 0.01$ versus the control group

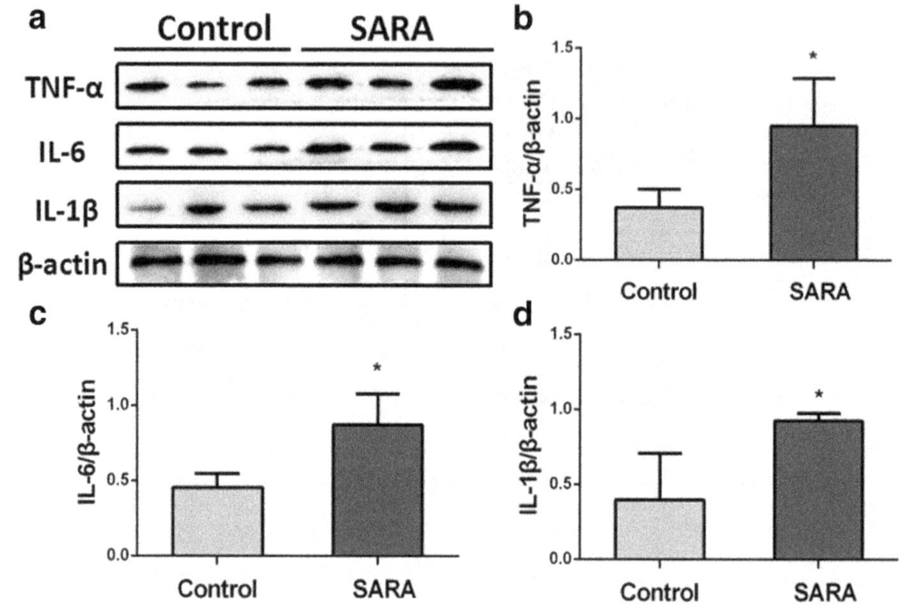

Fig. 4 The protein expression levels of inflammatory cytokines in the ruminal epithelium of control and SARA cows. Five cows were in each group. **a** The western blotting results of TNF-α, IL-6, and IL-1β; **b** The protein expression levels of TNF-α; **c** The protein expression levels of IL-6; **d** The protein expression levels of IL-1β. The data presented are the mean ± SD. * $P < 0.05$, ** $P < 0.01$ versus the control group

reported ranges for grain-induced SARA, which was from 3715 to 12,589 EU/mL, 24,547 to 128,820 EU/mL, from 28,184 to 107,150 EU/mL, reported by Gozho et al. (2005) [1], Gozho et al. (2007) [3] and Khafipour et al. (2009a) [4], respectively. Furthermore, Rodríguez-Lecompte et al. (2014) reported that ruminal LPS content of cows with grain-based SARA was up to 168,391 EU/mL [6], which was greater than our data. The discrepancy between our study and previous studies is probably because of differences in the body

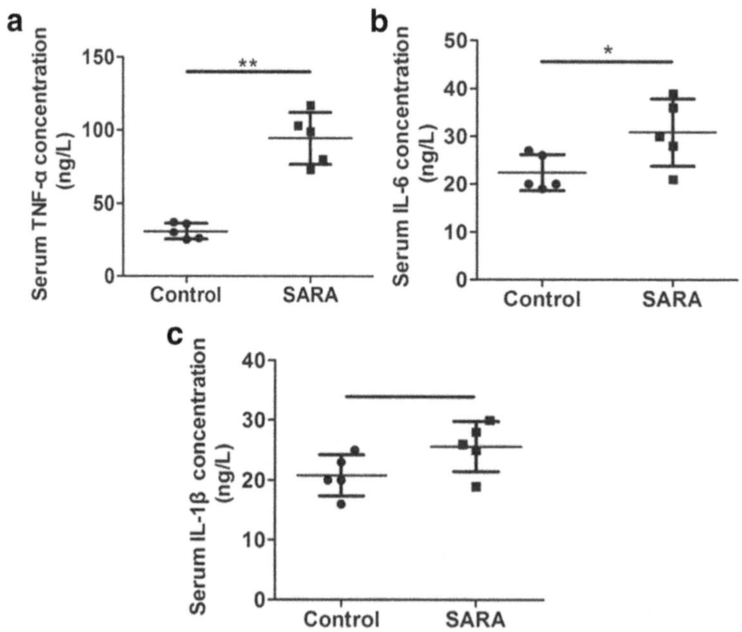

Fig. 5 The blood concentrations of inflammatory cytokines in the control and SARA cows. Five cows were in each group. **a**The blood concentration of TNF-α; **b** The blood concentration of IL-6; **c** The blood concentration of IL-1β. The data presented are the mean ± SD. * $P < 0.05$, ** $P < 0.01$ versus the control group

condition, nutrient composition, the method of LPS determination, which affects the ruminal LPS production. The presence of LPS could result in the production of multiple proinflammatory cytokines [15] and cause massive disruption of ruminal epithelial tight junctions [19]. Rumen wall damage associated with SARA further increased ruminal LPS translocation into the bloodstream [4, 5, 14]. Khafipour et al. (2009b) reported for the first time that a grain-based SARA challenge increased the LPS from < 0.05 (Control) to 0.52 EU/mL (SARA) [14]. In addition, Jin et al. (2016) also reported that the plasma LPS concentration of lacteal artery and vein in cows with long-term high-concentrate diet feeding was 0.86 and 0.27 EU/mL, respectively [15]. In our study, the plasma LPS concentration was 0.21 EU/mL in SARA cows. However, many studies did not detect the LPS in peripheral circulation during experimentally induced ruminal acidosis [6, 20]. This discrepancy may be due to the difference of damage degree of rumen wall in SARA cows.

When LPS released in large quantities, these mediators induced an acute phase response. Hp, SAA, and LBP are such proteins that are used as inflammatory markers in cattle [8, 21]. In the present study, the blood Hp, SAA and LBP concentrations were markedly increased in cows with SARA, which indicated that cows with SARA displayed high inflammatory levels. Many previous studies also demonstrated that blood Hp, SAA and LBP concentrations were significantly increased in cows with SARA or grain-induced SARA modal [1, 4, 14, 22, 23], which further supported our data.

Rumenitis is the direct result of SARA in dairy cows. As the degradation of ruminal environment, ruminal epithelium was continuously exposed to high levels of LPS in SARA cows. LPS is an inflammatory inducer and may induce the development of rumenitis. Zhang et al. (2016) also reported that LPS involved in the development of inflammation of cows with SARA [24]. MAPKs and NF-κB pathways are the main inflammatory pathways that mediate the LPS challenge. Over activation of NF-κB, JNK, p38MAPK and ERK1/2 were also observed in the mastitis induced by LPS [25]. In this study, the phosphorylation levels of IκBα and NF-κB were significantly increased in rumen epithelium of cows with SARA than in control cows. This indicated that NF-κB pathway was involved in the development of rumenitis. Interestingly, Fan et al. (2016) found that NF-κB was involved in the LPS-mediated proliferation and apoptosis of MAC-T epithelial cells as part of the SARA response in cows, which further supported our results in vitro [26]. In addition, we found that JNK and ERK1/2 phosphorylation levels were markedly increased in the rumen epithelium of SARA cows. Over activation of MAPKs and NF-κB inflammatory pathways could significantly increase the expression of inflammatory cytokines [25]. As expected, our results showed that the mRNA and protein levels of TNF-α, IL-6 and IL-1β in ruminal epithelium and the blood concentrations of TNF-α and IL-6 were significantly higher in SARA cows than in control cows. These results indicate that MAPKs and NF-κB inflammatory pathways are over-activated and further induce the overproduction of inflammatory cytokines in ruminal epithelium. These inflammatory cytokines further mediated the inflammation response in ruminal epithelium through a paracrine fashion, thereby partly inducing rumenitis in cows with SARA. Zhang et al. (2016) reported that high-concentrate feeding upregulated the expression of inflammation-related genes IL-1β, IL-2 and IL-22 in the ruminal epithelium of dairy cattle [24]. They also found that LPS treatment significantly increased the mRNA expression of IL-1β, IL-2, IL-6 and IL-8 in ruminal epithelial cells of Holstein cows in vitro. These studies further demonstrated that over activation of inflammatory pathways mediated the development of rumenitis in SARA cows.

Conclusion

In summary, our results suggest that SARA-induced high levels of LPS over activate the MAPKs and NF-κB inflammatory pathways and increase the production of inflammatory cytokines in ruminal epithelium, thereby inducing rumenitis at least in part. These findings will promote new exploration into the prophylaxis and treatment of SARA in dairy cows.

Abbreviations

ERK1/2: Extracellular signal-regulated kinase 1/2; Hp: Haptoglobin; IL-1β: Interleukin 1β; IL-6: Interleukin 6; JNK: c-Jun N-terminal kinase; LAL: Limulus amoebocyte lysate; LBP: LPS binding protein; LPS: Lipopolysaccharide; MAPK: Mitogen-activated protein kinase; NF-κB: Nuclear factor-kappaB; p38MAPK: p38 mitogen-activated protein kinase; PVDF: Polyvinylidene difluoride; qRT-PCR: Quantitative real-time polymerase chain reaction; SAA: Serum amyloid-A; SARA: Subacute ruminal acidosis; SDS-PAGE: Sodium dodecyl sulfate-polyacrylamide gel electrophoresis; TNF-α: Tumor necrosis factor α

Funding

This work was supported by the National key research and development program (Beijing, China; grant no. 2016YFD0501206), National High Technology Research and Development Program 863 (Beijing, China; grant no. 2013AA102806), the National Natural Science Foundation of China (Beijing, China; grant nos. 31360630, 31460681, and 31372494), and the Jilin Province Science Foundation for Youths (Changchun, China; grant no.20160520063JH).

Authors' contributions

CXZ, GWL and XBL conceived the study and participated in its design. CXZ performed the experimental work and wrote the manuscript. XBL, XY, and YG assisted in the phenotyping and cytokine analysis. YZW, GQS, ZW and

XWL participated in the writing of the manuscript and its critical review. All co-authors revised the manuscript and approved the final submitted version.

Competing interests
The authors declare that they have no competing interests.

Author details
[1]Key Laboratory of Zoonosis, Ministry of Education, College of Veterinary Medicine, Jilin University, 5333 Xi'an Road, Changchun, Jilin 130062, China. [2]College of Animal Science, Jilin University, 5333 Xi'an Road, Changchun, Jilin 130062, China. [3]College of Animal Science and Technology, Inner Mongolia National University, Tongliao 028000, Inner Mongolia, China.

References
1. Gozho GN, Plaizier JC, Krause DO, Kennedy AD, Wittenberg KM. Subacute ruminal acidosis induces ruminal lipopolysaccharide endotoxin release and triggers an inflammatory response. J Dairy Sci. 2005;88(4):1399–403.
2. Enemark JM. The monitoring, prevention and treatment of sub-acute ruminal acidosis (SARA): a review. Vet J. 2008;176(1):32.
3. Gozho GN, Krause DO, Plaizier JC. Ruminal lipopolysaccharide concentration and inflammatory response during grain-induced subacute ruminal acidosis in dairy cows. J Dairy Sci. 2007;90(2):856–66.
4. Khafipour E, Krause DO, Plaizier JC. Alfalfa pellet-induced subacute ruminal acidosis in dairy cows increases bacterial endotoxin in the rumen without causing inflammation. J Dairy Sci. 2009a;92(4):1712–24.
5. Plaizier JC, Khafipour E, Li S, Gozho GN, Krause DO. Subacute ruminal acidosis (SARA), endotoxins and health consequences. Anim Feed Sci Technol. 2012;172(1–2):9–21.
6. Rodríguez-Lecompte JC, Kroeker AD, Ceballos-Márquez A, Li S, Plaizier JC, Gomez DE. Evaluation of the systemic innate immune response and metabolic alterations of nonlactating cows with diet-induced subacute ruminal acidosis. J Dairy Sci. 2014;97(12):7777–87.
7. Khafipour E, Plaizier JC, Aikman PC, Krause DO. Population structure of rumen Escherichia coli associated with subacute ruminal acidosis (SARA) in dairy cattle. J Dairy Sci. 2011;94(1):351.
8. Zebeli Q, Ametaj BN. Relationships between rumen lipopolysaccharide and mediators of inflammatory response with milk fat production and efficiency in dairy cows. J Dairy Sci. 2009;92(8):3800.
9. Tothova C, Nagy O, Kovac G. Acute phase proteins and their use in the diagnosis of diseases in ruminants: a review. Vet Med. 2014;59(59):163–80.
10. Shi X, Li D, Deng Q, Yu L, Sun G, Xue Y, Song Y, Zhe W, Li X, Li X, Liu G. NEFAs activate the oxidative stress-mediated NF-κB signaling pathway to induce inflammatory response in calf hepatocytes. Journal of Steroid Biochemistry & Molecular Biology. 2015;145:103–12.
11. Park J, Min JS, Kim B, Chae UB, Yun JW, Choi MS, Kong IK, Chang KT, Lee DS. Mitochondrial ROS govern the LPS-induced pro-inflammatory response in microglia cells by regulating MAPK and NF-κB pathways. Neurosci Lett. 2015;584:191.
12. Ki YW, Park JH, Lee JE, Shin IC, Koh HC. JNK and p38 MAPK regulate oxidative stress and the inflammatory response in chlorpyrifos-induced apoptosis. Toxicol Lett. 2013;218(3):235–45.
13. Gozho GN, Krause DO, Plaizier JC. Rumen lipopolysaccharide and inflammation during grain adaptation and subacute ruminal acidosis in steers. J Dairy Sci. 2006;89(11):4404–13.
14. Khafipour E, Krause DO, Plaizier JC. A grain-based subacute ruminal acidosis challenge causes translocation of lipopolysaccharide and triggers inflammation. J Dairy Sci. 2009;92(3):1060.
15. Jin D, Chang G, Zhang K, Guo J, Xu T, Shen X. Rumen-derived lipopolysaccharide enhances the expression of lingual antimicrobial peptide in mammary glands of dairy cows fed a high-concentrate diet. BMC Vet Res. 2016;12(1):128.
16. Fu Y, Liu B, Zhang N, Liu Z, Liang D, Li F, Cao Y, Feng X, Zhang X, Yang Z. Magnolol inhibits lipopolysaccharide-induced inflammatory response by interfering with TLR4 mediated NF-κB and MAPKs signaling pathways. J Ethnopharmacol. 2013;145(1):193–9.
17. Humer E, Ghareeb K, Harder H, Mickdam E, Kholparisini A, Zebeli Q. Peripartal changes in reticuloruminal pH and temperature in dairy cows differing in the susceptibility to subacute rumen acidosis. J Dairy Sci. 2015; 98(12):8788–99.
18. Owens FN, Secrist DS, Hill WJ, Gill DR. Acidosis in cattle: a review. J Anim Sci. 1998;76(1):275–86.
19. Liu JH, Xu TT, Liu YJ, Zhu WY, Mao SY. A high-grain diet causes massive disruption of ruminal epithelial tight junctions in goats. American Journal of Physiology Regulatory Integrative & Comparative Physiology. 2013;305(3):232.
20. Li S, Khafipour E, Krause DO, Kroeker A, Rodriguez-Lecompte JC, Gozho GN, Plaizier JC. Effects of subacute ruminal acidosis challenges on fermentation and endotoxins in the rumen and hindgut of dairy cows. J Dairy Sci. 2012; 95(1):294–303.
21. Eckersall PD, Bell R. Acute phase proteins: biomarkers of infection and inflammation in veterinary medicine. Vet J. 2010;185(1):23.
22. Heegaard PM, Godson DL, Toussaint MJ, Tjørnehøj K, Larsen LE, Viuff B, Rønsholt L. The acute phase response of haptoglobin and serum amyloid a (SAA) in cattle undergoing experimental infection with bovine respiratory syncytial virus. Veterinary Immunology & Immunopathology. 2000;77(1–2):151.
23. Li S, Danscher AM, Andersen PH, Khafipour E, Kristensen NB, Plaizier JC. The effects of grain-induced subactue ruminal acidosis on interleukin-6 and acute phase response in dairy cows. J Anim Sci. 2014;92(Issue E suppl):2.
24. Zhang R, Zhu W, Mao S. High-concentrate feeding upregulates the expression of inflammation-related genes in the ruminal epithelium of dairy cattle. Journal of Animal Science Biotechnology. 2016;7(1):42.
25. Gao XJ, Guo MY, Zhang ZC, Wang TC, Cao YG, Zhang NS. Bergenin plays an anti-inflammatory role via the modulation of MAPK and NF-κB signaling pathways in a mouse model of LPS-induced mastitis. Inflammation. 2015; 38(3):1142–50.
26. Fan WJ, Li HP, Zhu HS, Sui SP, Chen PG, Deng Y, Sui TM, Wang YY. NF-κB is involved in the LPS-mediated proliferation and apoptosis of MAC-T epithelial cells as part of the subacute ruminal acidosis response in cows. Biotechnol Lett. 2016;38(11):1–11.

10

Evaluation of fecal culture and fecal RT-PCR to detect *Mycobacterium avium* ssp. *paratuberculosis* fecal shedding in dairy goats and dairy sheep using latent class Bayesian modeling

Cathy A. Bauman[1*], Andria Jones-Bitton[1], Jocelyn Jansen[2], David Kelton[1] and Paula Menzies[1]

Abstract

Background: The study's objective was to evaluate the ability of fecal culture (FCUL) and fecal PCR (FPCR) to identify dairy goat and dairy sheep shedding *Mycobacterium avium* ssp. *paratuberculosis*. A cross-sectional study of the small ruminant populations was performed in Ontario, Canada between October 2010 and August 2011. Twenty-nine dairy goat herds and 21 dairy sheep flocks were visited, and 20 lactating females > two years of age were randomly selected from each farm resulting in 580 goats and 397 sheep participating in the study. Feces were collected per rectum and cultured using the BD BACTEC™ MGIT™ 960 system using a standard (49 days) and an extended (240 days) incubation time, and underwent RT-PCR based on the *hsp*-X gene (Tetracore®). Statistical analysis was performed using a 2-test latent class Bayesian hierarchical model for each species fitted in WinBUGS.

Results: Extending the fecal culture incubation time statistically improved FCUL sensitivity from 23.1 % (95 % PI: 15.9-34.1) to 42.7 % (95 % PI: 33.0-54.5) in dairy goats and from 5.8 % (95 % PI: 2.3-12.4) to 19.0 % (95 % PI: 11.9-28.9) in dairy sheep. FPCR demonstrated statistically higher sensitivity than FCUL (49 day incubation) with a sensitivity of 31.9 % (95 % PI: 22.4-43.1) in goats and 42.6 % (95 % PI: 28.8-63.3) in sheep.

Conclusions: Fecal culture demonstrates such low sensitivity at the standard incubation time it cannot be recommended as a screening test to detect shedding of MAP in either goats or sheep. Extending the incubation time resulted in improved sensitivity; however, it is still disappointingly low for screening purposes. Fecal PCR should be the screening test of choice in both species; however, it is important to recognize that control programs should not be based on testing alone when they demonstrate such low sensitivity.

Keywords: Dairy goats, Dairy sheep, Fecal PCR, Fecal culture, Paratuberculosis

Background

In Canada, the province of Ontario has the largest dairy small ruminant industries in the country. Over the last 10 years the industry has grown rapidly, but has been impacted by production-limiting diseases such as paratuberculosis, similar in scope to other dairy sectors. The true herd- and flock-level prevalences of paratuberculosis have been estimated to be 83.0 % (95 % PI: 62.6-98.1) in dairy goats and 66.8 % (95 % PI: 41.6-91.4) in dairy sheep [1]. This widely-recognized chronic enteric wasting disease of domestic ruminants is caused by infection with *Mycobacterium avium* ssp. *paratuberculosis* (MAP). Transmission is mainly fecal-oral, although infectious animals can also shed bacteria in colostrum and milk, and in utero transmission has been shown to occur [2]. Infectious animals are the primary source of infection as this facultative intracellular bacterium has not demonstrated the ability to replicate in the environment [3]. Detection and culling of

* Correspondence: cbauman@uoguelph.ca
[1]Department of Population Medicine, University of Guelph, Guelph, Ontario N1G 2W1, Canada
Full list of author information is available at the end of the article

infectious animals in the early stages of infection is one component of control programs in countries such as Canada where the paratuberculosis vaccine has not been approved for use. The presence of tuberculosis in North America means there is an inherent concern that vaccinated animals could not be differentiated from animals exposed to tuberculosis.

The general consensus is that fecal tests such as fecal culture (FCUL) and PCR are the most sensitive MAP tests and able to detect shedding animals first [4]. While the strength of fecal culture is that it consistently demonstrates almost perfect specificity [4], sensitivity estimates vary widely and are dependent on the stage of infection in the study population and on the MAP strain circulating in a given geographic region and animal species [5]. The S-strain (sheep strain) of MAP has been previously shown to be more sensitive to particular decontamination antibiotics and detergents [6], more fastidious about the components of the culture media it is grown in [6, 7], and takes longer to grow in vitro [8]. The current recommendation for incubating fecal samples from a previously untested population where the circulating strain is unknown is at least 8 months [7]. Since automated fecal culture systems often incubate samples for less than 60 days, the sensitivity of fecal culture may be underestimated in populations where the S-strain predominates.

In theory, fecal PCR performance should not be influenced by MAP strain and has the potential to demonstrate higher sensitivity [9, 10]. However, the test can exhibit lower than expected specificity if the specific gene sequence cross-reacts with other pathogens [11].

Neither fecal culture nor fecal PCR (FPCR) is a suitable MAP reference test with which to evaluate the other test. In the absence of a gold standard, test evaluations often use a latent class (LC) approach where no test is considered the gold standard and both tests are interpreted in parallel [12]. One LC model commonly in use is the Hui-Walter (HW) 2-test maximum likelihood estimates model. However, in order for this HW model to be identifiable, wherein there is sufficient degrees of freedom in the data to calculate the parameters of interest, it must fulfill three assumptions: the tests must be conditionally independent, the tested animals must be divided into two populations with different prevalences, and the two tests must demonstrate the same test sensitivity and specificity in each of the two populations [13, 14]. Fulfilling this last assumption is an inherent challenge with paratuberculosis tests as it is generally acknowledged that they demonstrate higher sensitivity in populations with higher prevalences [5, 15]. An alternative 2-test method is to use a hierarchical approach incorporating prior information in a Bayesian framework [16]. The main assumptions with this model are that prevalences may differ within subpopulations or herds, but that there is a fixed test sensitivity and specificity in the population analyzed as one group, the population is obtained randomly, and the animals tested represent the population or field conditions in which the test will be utilized [16].

As yet no MAP test evaluations have been undertaken in either of the dairy goat or dairy sheep populations in Canada. Therefore, the objective of this study was to evaluate the performance of fecal culture and fecal PCR. To account for the potential presence of the slow-growing S-strain of MAP, an additional objective was to assess the impact of extending the culture incubation time on the performance of the same 2 tests.

Methods

Herd and animal selection

This study was part of a larger cross-sectional study conducted between October 2010 and August 2011. The purpose of the original study was to determine prevalence, identify the risk factors for herds/flocks to test positive, and to evaluate the performance of 7 individual- and herd-level diagnostic tests. Greater detail on herd and animal sampling are available [1].

Briefly, 29 dairy goat herds and 21 sheep flocks in the province of Ontario were visited once during milking time, at which time 20 lactating does or ewes over the age of 2 years were selected as they came through the milking parlour using a formal systematic random sampling procedure. The inclusion criteria at the animal-level were: milking doe/ewe greater than 2 years of age and greater than 3 days post-kidding/lambing. Animals were excluded if no feces could be detected intra-rectum at the time of sampling; in such cases, the next animal fulfilling the inclusion requirements was substituted.

No farm had concurrent or recent cohabitation with other domestic ruminant species, and no farm had ever used a vaccine for the prevention of paratuberculosis in their animals. MAP-infection status on all farms was unknown to the researchers at the time of enrolment in the study.

Sample collection and handling

A minimum of 20 g of feces was collected by the same handler from the rectum of each selected animal, and placed into a sterile plastic vial (1 per animal). Single-use polyethylene gloves lubricated with 0.3mLs of sterile lubricant (Lubricating Jelly, Healthcare Plus, Canadian Packaging Company, Toronto, CA) were changed between each animal to reduce risk of cross-contamination. Fecal samples were transported in a Styrofoam cooler with ice packs, and stored at 4-8 °C, until submission to the Animal Health Laboratory (AHL, Laboratory Services

Division, University of Guelph, Guelph, ON) within 12-18 h of collection. Two tests were performed: FCUL and FPCR. Generally, feces were processed at the AHL while fresh; however, samples from 7 goat and 10 sheep farms were frozen at -80 °C due to high work-volume at the laboratory, and were processed within 7-21 days.

Laboratory testing

All technicians processing samples did not have access to the results of the other fecal test.

Culture of Mycobacterium avium ssp. paratuberculosis

Fecal samples underwent a 3-day decontamination procedure prior to being cultured using the liquid culture BD BACTEC™ MGIT™ 960 paraTB non-radiometric automated system (Becton Dickinson and Company, Franklin Lakes, NJ, US). In brief, 2.0 g of feces were mechanically broken up with a tongue depressor and vortexed with sterile, purified water prior to incubation with Bacto yeast extract and sodium pyruvate. After overnight incubation with 0.9 % hexadecylpyridinium chloride (HPC), the samples were decanted and incubated a second night with an antibiotic brew of amphotericin B, vancomycin HCl, and nalidixic acid sodium salt. On the third day, 0.1 mL of sample was inoculated into a BACTEC™ MGIT™ ParaTB medium tube supplemented with BACTEC™ MGIT™ ParaTB supplement (casein, catalase, oleic acid and bovine albumin), antibiotic brew and egg yolk. Each 7 mL tube of medium contains modified Middlebrook 7H9 Broth base with mycobactin J and a fluorescent indicator compound embedded in silicone located in the bottom. Also present in the broth base is dissolved oxygen that upon being consumed by multiplying organisms causes the indicator to fluoresce. The system monitors the samples every 60 min for signs of fluorescence. The standard incubation length for samples in the system is 49 days; however, this was extended to 240 days in order to give the opportunity for slower-growing strains of MAP (S-strain), which are more common in sheep than other domestic ruminants [17], to be detected.

The time in days that each sample took to fluoresce positive was recorded. Fluorescence positive colonies underwent acid-fast staining and PCR confirmation with the *hsp*X gene (Culture Confirmation Protocol, MAP Extraction System, Tetracore®, Rockville, MD, US) according to the manufacturer's instructions. Briefly, 1 mL of broth was placed in a tube with glass beads and bead-beated for 5 min at 4800 rpm, centrifuged for 10 min at 16 000 × g and the supernatant processed using the VetAlert™ Johne's Real-Time PCR (Tetracore®) according to manufacturer's instructions with modifications for the RocheLightCycler® 2.0 thermocycler (Roche Applied Science, Laval, QC). The cycling program of this thermocycler is 10s @ 95 °C for enzyme activation and a 2-step PCR for 50 cycles (95 °C × 5 s and 95 °C x 30s). Samples that reached fluorescence with a cycle count (Ct) ≤ 42.0 were considered positive. Therefore, for a fecal sample to be designated culture positive it had to fluoresce, have a positive PCR confirmation test, and stain positive using acid-fast stain.

Direct real-time polymerase chain reaction test (RT-PCR)

Prior to the fecal culture decontamination procedure feces underwent DNA extraction using the Two Gram MAP Extraction and Mini Beadbeater protocols (Tetracore®) according to manufacturer's instructions. In summary, 2.0 g of feces were manually broken apart as described above, vortexed with 35 mL of DNase free water in a conical tube, and then allowed to incubate at room temperature for 30 min to allow the settling of larger particulate matter. The top 20 mL were removed and centrifuged in a new conical tube at 2500 × g for 10 min after which the pellet was removed and re-suspended in 1 mL of 1 × TE (Tris-HCl/EDTA) and placed in a disruption tube containing glass beads. The sample was briefly vortexed and then beadbeated at 4800 rpm for 5 min. DNA purification occurred according to manufacturer's instructions and the sample underwent PCR using the VetAlert™ Johne's Real-Time PCR (Tetracore®) with modifications for the Roche-LightCycler® 2.0 thermocycler (Roche Applied Science, Laval, QC). Samples that reached fluorescence with a cycle count (Ct) ≤ 42.0 were considered positive.

Even though the *hsp*X gene is thought to be exclusive to MAP [18], there was a concern regarding possible cross-reaction with antigens of *Corynebacterium pseudotuberculosis*, the causal agent of caseous lymphadenitis (CL), a common disease in Ontario sheep and goats. Antibodies to *C. pseudotuberculosis* had earlier been shown to cause false positive reactions in a previously available ELISA [19] and no literature could be found indicating *hsp*X had been evaluated against *C. pseudotuberculosis* antigens. Therefore, ten samples of *C. pseudotuberculosis* from a stock of pure culture obtained from the AHL were processed as described above and evaluated using the same real-time PCR test.

Purpose, target conditions, and case definitions

The purpose of this evaluation was to provide an accurate appraisal of test sensitivities and specificities for both fecal tests to evaluate if fecal PCR could be substituted for fecal culture in future test evaluations and to see if these results would be impacted by extending the culture period. The target condition for this test evaluation was MAP infectious animals which as a case definition, was animals sufficiently infected with MAP that they shed enough bacteria in their feces to potentially test positive on fecal culture, fecal PCR or both.

A 'contaminated' fecal sample was defined as a sample that signalled positive in the MGIT fecal culture system by demonstrating fluorescence, but was negative on one or more of the confirmation tests (acid-fast and/or PCR).

Statistical analyses

All data were entered and managed in Microsoft® Excel (2007). Descriptive statistics and contingency tables were generated in StataIC® 11.1 (StataCorp LP, College Station, TX, US). The tests were evaluated separately in goats and sheep as it was expected that there would be species-specific differences in test performance [20]. Test sensitivities and specificities were determined using a 2-test multiple populations latent class Bayesian model with a random effect as described previously [16, 21]. This model allowed for dependence between the 2 fecal tests [22] as they are based on a similar biological process. Non-informative priors were used for test sensitivity. Specificity priors were based on previously published literature [4, 19, 22, 23, 24] that evaluated tests in sheep and goats when available and other species when not. These priors were reviewed for appropriateness by a researcher with small ruminant MAP experience (MT Collins, University of Madison, WI). The final priors were represented by beta distributions in the format (a,b) and generated using the software BetaBuster (http://betabuster.software.informer.com). The same specificity priors were used for both the goat and sheep models. Priors for the prevalence parameters were based on the data acquired during an earlier study [1]. Separate herd-level (HLP) and within-herd prevalence (WiHP) priors were used for each species. The herd-level prior was represented by a beta distribution (also generated through BetaBuster) while the prior for the estimated prevalence within an infected herd was represented by a normal distribution in the format (μ, precision of μ). The 'true' within-herd prevalence of any sampled herd or flock was represented by a zero-inflated logit-normal variable that was a function of the probability of the herd/flock to be infected (FP), the WiHP given the farm was infected and a random effect (U). The random effect represented the potential variability in prevalence between farms and was modeled by a normal distribution with a mean of 0 and a precision (prec) represented by a non-informative gamma distribution [21]. All prior data is available in Table 1.

The sheep and goat models were each fitted using WinBUGS software (Available at: http://www.mrc-bsu.cam.ac.uk/software/bugs/the-bugs-project-winbugs/). Posterior estimates for test sensitivity and specificity were generated using the Markov Chain Monte Carlo (MCMC) sampling method and the Gibbs algorithm. Model convergence was assessed visually by examining the traces, histories, Monte Carlo (MC) errors and autocorrelations plots. Diagnostically, CODA outputs were generated from WinBUGS and evaluated using the coda package in [25] (R) and the Raftery-Lewis diagnostic [26]. In this package the I statistic is examined for variations exceeding 5 and the M (minimum burnin period) and N (minimum iterations needed) are checked for values exceeding those used in the primary model [27]. To assess the influence of the initial values, the Gelman-Rubin [28] statistic in WinBUGS was generated using 3 different sets of starting points. To assess the need to account for dependence between the 2 fecal tests 4 models were run: fully independent, fully dependent, 1 with only a covariance in the diseased state and 1 with only the covariance on the non-diseased state. After each model was run, any covariances with a 95 % probability interval that contained zero were removed and the model was re-run.

The final estimates of the test and prevalence parameters generated from the model were represented by the median and its 95 % probability intervals (PI) after the first 10 000 iterations were dropped as the burnin period and 60 000 iterations were run. To determine the probability of one test demonstrating a statistically important difference from another test the Bayesian posterior probability (POPR) [29] was estimated. The POPR is calculated by first creating a Boolean or indicator variable using the step function in WinBUGS and a non-specific node X [30]. The value of 1 is then assigned to step(X) when the function is 'true' (e.g. Se FPCR- Se

Table 1 Prior information used in the 2-test latent class Bayesian hierarchical model developed to estimate test sensitivity and specificity for 2 paratuberculosis fecal tests in dairy goats and dairy sheep in Ontario, Canada

Parameter	Distribution	Dairy Goats			Dairy Sheep		
		(a, b)	95 % certain	Mode	(a, b)	95 % certain	Mode
Fecal culture specificity	Beta	560.72, 6.65	> 0.98	0.99	560.72, 6.65	> 0.98	0.99
Fecal PCR specificity	Beta	99.70, 6.19	> 0.90	0.95	99.70, 6.19	> 0.90	0.95
Herd-level prevalence	Beta	19.48, 5.62	> 0.63	0.80	8.35, 4.62	> 0.42	0.67
		precision μ	95 % certain	Median	precision μ	95 % certain	Median
Within-herd prevalence	Normal	175.41	< 0.50	0.35	55.62	< 0.74	0.48

FCUL ≥ 0) and the value of 0 when the function is 'false' (e.g. Se FPCR- Se FCUL < 0) thus acting as a counter of the number of iterations where this statement is true. The mean value of X subsequently represents the Monte-Carlo estimate of the posterior probability that there was a statistically important difference between the 2 tests; a POPR threshold of ≥0.95 was used as the cut-off [20].

Sensitivity analysis was performed to assess the influence of the priors on the sensitivity and specificity estimates. A table of the values uses in this analysis is available in the Additional file 1. The model was also run using 3 different chains of starting points to assess the influence that these points had on the model results and assessing the Gelman-Rubin statistic [28].

Results

Test population

Fecal samples were collected on 29 dairy goat farms from 580 dairy goats representing Saanen, Alpine, Toggenberg, Nubian, La Mancha breeds, and their crosses. Fecal samples were also collected on 21 dairy sheep farms from 400 dairy sheep representing East Friesian, British Milk Sheep breeds and their crosses. However, samples from 3 sheep were later discarded, as they were determined to be less than 2 years of age resulting in a sample size of 397 ewes.

Laboratory results

Culture of Mycobacterium avium ssp. paratuberculosis

When the standard incubation time (49 days) was used, the proportion of fecal positive dairy goats was 9.8 % (57/580) and fecal positive dairy sheep was 2.8 % (11/397) (Tables 2 and 3 respectively).

The proportions of fecal samples that were positive after the extended incubation time (240 days) were 18.3 % (95 % PI: 15.1–21.4 %) in dairy goats and 7.6 % (95 % PI: 4.9-10.2 %) in dairy sheep. Using extended incubation, the mean times for fecal samples to culture positive (machine-signalled and confirmed by PCR and acid-fast staining) was 66.4 days (median: 36.0 days) for goat samples and 104.2 days (median: 54.5 days) for sheep samples.

The proportion of contaminated fecal samples at the 240-day cut-off was 41.6 % in goats (241/580) and 34.3 % (136/397) in sheep. Of these, 38/241 goats (15.8 %) and 21/136 sheep (15.4 %) samples were positive on the direct PCR test performed prior to the samples undergoing decontamination and culture. Contamination proportions were not significantly different between frozen and fresh samples (goats $\chi^2 = 3.131$, $p < 0.0777$; sheep $\chi^2 = 0.1412$, $p < 0.7077$).

Direct real-time polymerase chain reaction test (RT-PCR)

The proportion of FPCR-positive samples was 16.6 % (96/580) in dairy goats and 15.1 % (60/297) in dairy sheep. The Ct values for FPCR-positive samples ranged from 23.96–41.66 (mean positive Ct =36.04; median Ct =36.63) for goats and 21.88-41.19 (mean positive Ct =35.67; median Ct = 36.68) for sheep. FPCR was negative on all 10 samples of *C. pseudotuberculosis* tested.

Test performance

Dairy goats

Sensitivity and specificity results for both tests are presented in Table 4. Extending the culture period statistically increased the test sensitivity for FCUL from 23.1–42.7 % ($p = 0.9954$). Specificity was not statistically different ($p = 0.5735$) when estimated at the extended time. In addition, the sensitivity ($p = 0.5754$) and specificity ($p = 0.4832$) for FPCR estimated in each of the 2 different models did not demonstrate a statistically important difference from one another.

Sensitivity of fecal culture was not statistically lower ($p = 0.9488$) than fecal PCR when evaluated at the standard time, but was statistically higher ($p = 0.9902$) than the sensitivity of PCR at the extended time. At both the 49-day ($p = 0.9996$) and 240-day ($p = 0.9998$) incubation periods the specificity of fecal culture was statistically higher than the specificity of fecal PCR.

Dairy sheep

Test sensitivities and specificities determined using the standard and extended incubation time are listed in Table 5. Increasing the incubation time of cultured samples, statistically improved the sensitivity of FCUL from 6.0–19.7 % ($p = 0.9944$), but had no impact on specificity ($p = 0.7762$). FPCR test sensitivity ($p = 0.7829$) and specificity ($p = 0.3359$) were not statistically impacted by the

Table 2 Comparison of *Mycobacterium avium* ssp. *paratuberculosis* fecal shedding status as determined by fecal culture using standard and extended incubation times, and fecal PCR, in dairy goats sampled in Ontario, Canada in 2010–2011

		Fecal PCR positive	Fecal PCR negative	Total
Standard Incubation Time (49 days)	Fecal culture positive	30 (5.2 %)	54 (9.3 %)	57 (9.8 %)
	Fecal culture negative	66 (11.4 %)	457 (78.8 %)	523 (90.2 %)
	Total	96 (16.6 %)	484 (83.4 %)	580
Extended Incubation Time (240 days)	Fecal culture positive	35 (6.0 %)	71 (12.2 %)	106 (18.3 %)
	Fecal culture negative	61 (10.5 %)	413 (71.2 %)	474 (81.7 %)
	Total	96 (16.6 %)	484 (83.4 %)	580

Table 3 Comparison of *Mycobacterium avium* ssp. *paratuberculosis* fecal shedding status as determined by fecal culture using standard and extended incubation times, and fecal PCR, in 397 dairy sheep sampled in Ontario, Canada in 2010–2011

		Fecal PCR positive	Fecal PCR negative	Total
Standard Incubation Time (49 days)	Fecal culture positive	4 (1.0 %)	7 (1.8 %)	11 (2.8 %)
	Fecal culture negative	56 (14.1 %)	330 (83.1 %)	386 (97.2 %)
	Total	60 (15.1 %)	337 (84.9 %)	397
Extended Incubation Time (240 days)	Fecal culture positive	10 (2.5 %)	20 (5.0 %)	30 (7.6 %)
	Fecal culture negative	50 (12.6 %)	317 (79.8 %)	367 (92.4 %)
	Total	60 (15.1 %)	337 (84.9 %)	397

extended incubation time. At the standard incubation time, FPCR was statistically more sensitive than FCUL ($p = 1.0$), but FCUL was statistically more specific than FPCR ($p = 0.9967$). At extended incubation times, the FPCR test was still statistically more sensitive than FCUL ($p = 0.9940$), and FCUL was still statistically more specific than FPCR ($p = 0.9931$).

Latent class Bayesian models

Based on visual observation of the model traces, histories, and densities there was no evidence that convergence of the model had not occurred. The autocorrelations did not drop off to 0 before 10 on the sensitivity parameters therefore the model was thinned by a value of 10. This was further corroborated when the Raftery-Lewis diagnostic was performed in R and some of the I values were between 12–15 and decreased to less than 5 after thinning occurred. The starting values had little impact on the posterior estimates as the BGR diagnostic approached 1.0 for all parameters. Sensitivity analysis indicated that the specificity priors for FCUL and FPCR most influenced the goat models. When the prior for FCUL was made the same as the FPCR prior or lower, the sensitivity estimate for FCUL was increased to 22.5 % (standard time) and 43.8 % (extended time) in goats, with the other parameters remaining unchanged (<2 % change); the sheep estimates were unaffected. The specificity prior for fecal PCR only impacted the goat model results when it was increased to the same confidence level as FCUL. Again, the impact of increasing this prior was only on the sensitivity estimates for FCUL, increasing it to 21.1 % (standard time) and 36.3 % (extended time) in goats, but leaving all other parameters unchanged; again, increase of the prior did not affect the sheep results. All other priors had limited impact on the posterior estimates.

When assessing dependence of all models the only covariance interval that did not contain 0 was the covariance in the diseased state for the goat tests at the standard incubation time (covDFS). Therefore, covDFS was included in the primary analysis goat standard incubation model. All covariances evaluated in the sheep models contained 0, therefore the completely independent model was used for the primary analysis.

Discussion

Test population and study design

The objective of this study was to evaluate 2 paratuberculosis fecal tests to determine which test should be used for potentially screening dairy does and ewes for fecal shedding of MAP. To have external validity, reduce selection bias, and fulfill the model assumptions, the study population must be representative of the target population within which the tests will be applied and is best accomplished through formal random sampling at the herd and animal level [16, 31]. For all herds and animals in the dairy goat population this was successfully achieved; however, the ability to randomly sample the sheep flocks was hampered by the fact that dairy sheep are not licensed under the provincial Milk Act (Milk Act, 1990, Available at https://www.ontario.ca/laws/statute/90m12) and therefore, no complete sampling frame exists. Every attempt was made to contact producers from all the milk processors across Ontario and the participating farms represented a variety of flock sizes, regions, and management styles. In addition,

Table 4 Performance of two paratuberculosis tests (fecal culture and fecal PCR) at two different culture incubation times in 580 dairy goats in Ontario, Canada

	Standard Incubation Time		Extended Incubation Time	
	Fecal culture (95 % PI)	Fecal PCR (95 % PI)	Fecal culture (95 % PI)	Fecal PCR (95 % PI)
Sensitivity	23.1 % (15.9-34.1)	31.9 % (22.4-43.1)	42.7 % (33.0-54.5)	30.5 % (23.3-38.8)
Specificity	98.9 % (98.0-99.5)	93.3 % (89.0-97.1)	98.8 % (97.8-99.5)	93.4 % (90.0-96.5)

PI = probability interval

Table 5 Performance of two paratuberculosis tests (fecal culture and fecal PCR) at two different culture incubation times in 397 dairy sheep in Ontario, Canada

	Standard Incubation Time		Extended Incubation Time	
	Fecal culture (95 % PI)	Fecal PCR (95 % PI)	Fecal culture (95 % PI)	Fecal PCR (95 % PI)
Sensitivity	5.8 % (2.3-12.4)	42.6 % (28.8-63.3)	19.0 % (11.9-28.9)	35.3 % (24.7-49.8)
Specificity	98.8 % (97.8-99.4)	97.8 % (93.0-99.9)	98.9 % (97.8-99.5)	97.4 % (93.5-99.7)

PI = probability interval

random selection of ewes occurred on-farm, therefore the researchers involved are confident that the dairy sheep sample population is reflective of the target population. Random sampling also helps to minimize spectrum bias which occurs when animals from one stage of infection are more likely to be sampled than another stage. Since paratuberculosis tests demonstrate higher sensitivity in the later stages of infection it is important that the spectrum of infection in the study population match the range of infection stages in the target population. No pre-selection of animals based on weight, thriftiness or the presence/absence of diarrhea occurred so as to mimic the conditions experienced when screening animals.

Sensitivity analysis of model

The high level of confidence placed on the specificity prior for FCUL had a large influence on the posterior estimates for the goat models; however, due to the large number of bacteria that need to be present in a sample to culture positive, and the double-confirmation step there is strong support for this prior to be as high as it is. The specificity prior for FPCR only impacted the goat estimates when it was increased to the same prior as FCUL. There is not enough evidence to date to warrant weighting the specificity prior of FPCR equal to FCUL. It is unclear at the moment why these priors had little impact on the sheep estimates. Otherwise, all the models converged well.

Laboratory test performance

Fecal culture

Overall, the average length of time for cultures to signal positive in this test evaluation (goats: 66 days, sheep: 100 days) fell outside of the standard length of time that cultures would normally be kept (49 days). Thirty per cent of positive goat cultures and 60 % of positive sheep cultures would have been classified as negative had the standard time been observed. Increasing the incubation time of cultured samples dramatically improved test sensitivity of FCUL from 23.1 % (95 % PI: 15.9-34.1) to 42.7 % (95 % PI: 33.0-54.5) in the dairy goat population and from 5.8 % (95 % PI: 2.3-12.4) to 19.0 % (95 % PI: 11.9-28.9) in the dairy sheep population. Even though the sensitivity is statistically higher in both species, increasing incubation times would not be practical for screening animals in the general population. FCUL at the standard incubation time is already an expensive, labor-intensive procedure with an impractical delay between sample submission and acquisition of results for producers and veterinarians. However, extending the incubation time to 240 days should be considered in test evaluation and prevalence studies where the objective is to maximize the sensitivity of the test. Small ruminant prevalence studies are likely to be significantly underestimating the level of infection when using FCUL at the standard time. It must be noted, that even if incubation time is extended, under-reporting would likely still occur as the fluorescent indicator present in the media has a limited life-expectancy. The actual life-expectancy has not been published, however there is the possibility that some of the samples in this study did not 'signal' positive, even with the extra incubation time, because the indicator was no longer active.

Fecal culture using the BACTEC™ MGIT™ system demonstrated such low sensitivity in sheep that it cannot be recommended in this species in this region for any purpose. Gumber and Whittington [24] previously expressed concern that the liquid culture media (modified Middlebrook 7H9) in this culture system may not support the growth of S-strains of MAP either due to the lower concentration of egg yolk than is present in other media or the presence of vancomycin in the antibiotic brew used to prevent contamination. While the S-strain has been most commonly detected in sheep in other countries [17, 32] it has also been isolated from cattle, goats and other species [8]. Conversely the other strains, C- (cattle) and I- (intermediate) strains, have also been detected in sheep [32]. Prior to this study, the only other previous literature available on the MAP strains present in Canadian sheep was a study in 1990 [32] that used hybridization and restriction analysis to identify strain types. Seven different Canadian sheep fecal/intestinal tissue samples were processed and identified as 6 cattle strains and 1 intermediate strain. However, given the current test results it is suspected that the S-strain may be the predominant strain in this population and requires further investigation.

Fecal PCR

The use of FPCR has the potential to be a rapid, relatively inexpensive, and sensitive method of MAP diagnosis, especially in sheep where FCUL performed so poorly. However, PCR-based tests are only as useful as the uniqueness of the DNA primers used. PCR MAP tests based on the IS900 gene sequence had previously cross-reacted with other Mycobacterium species [11, 33] and therefore, had limited usefulness. Evaluations of the Tetracore® *hsp*X sequence to date report no cross-reaction with common bacterial pathogens [18], (written communication, Tracy Fecteau, Tetracore®). However, prior to this study, the test had not been evaluated against *C. pseudotuberculosis* the causative agent of caseous lymphadenitis. Antibodies to this bacteria demonstrated cross-reactivity with a previous MAP ELISA test used in goats and sheep [19]. While antibody cross-reactivity does not necessarily infer DNA cross-reactivity, it was reassuring that the Tetracore® test did not identify any of the *C. pseudotuberculosis* samples as positive.

Overall, the performance of FPCR was disappointing in this study with a sensitivity of 30–40 %. While it performed significantly higher than FCUL (at the standard incubation time) in both species, previous literature had indicated a potential sensitivity of 60 % [23] could be achieved. The detection limit for the Tetracore® test, in theory, is a single gene from one bacterium, and much lower than the detection limit of FCUL (2.3×10^1–2.3×10^3 bacteria) [24] therefore the test should be able to detect animals shedding at a low level. However, it is difficult to isolate DNA from feces as they contain compounds that can inhibit the PCR reaction: bile salts, ionic detergents, hemoglobin degradation products, polyphenol-based substances, and humic-like acid [34]. Additionally, MAP aggregates in clusters [34], so even when a substantial amount of bacteria may be present in a sample it may be missed when only a 2.0 g amount is processed as the protocol requires. The Alinovi study likely demonstrated higher sensitivity because they studied cattle [23]. While the influence of animal species is not referred to in the manufacturer's instructions for the Tetracore® kit, it specifically states that sampled feces should be fresh and moist. Small ruminant feces are inherently dry in comparison to cattle. Goat and sheep feces are excreted in formed, firm pellets that are difficult to manually break apart. This hampers the ability to properly prepare the sample for FPCR testing. Alinovi's study also used culture-positive clinically affected animals. The cattle tested were shedding MAP bacteria at a high enough level to culture positive [23]; therefore, there would be more bacteria to potentially detect via PCR thus improving sensitivity. MAP tests typically exhibit lower test sensitivity when evaluated in populations of sub-clinically rather than clinically diseased animals [35]. In this study, where the MAP infection status was unknown, animals would potentially be in the early stages of infection, thereby shedding organisms at very low levels, which would reduce the perceived sensitivity, but better represent what is encountered when screening animals in the general population. Further work is needed to identify if the dry feces of small ruminants can be processed in a way to further improve the recovery of DNA and possibly improve sensitivity.

Conclusions

Fecal culture is extremely useful in test comparison and prevalence studies as the 2-step confirmation process almost ensures 100 % specificity, and as demonstrated in this study, its sensitivity can be statistically increased by extending the incubation time. However, compared to FCUL, FPCR is more economical, has a shorter turnaround time, and is thus more likely to be accepted by veterinarians and producers. In this study, FPCR also demonstrated statistically higher sensitivity than FCUL (incubated at the standard time) in both species, and with a specificity exceeding 93 %, it should be the screening test of choice in both these species. However, a sensitivity of 30–40 %, means that for every animal identified as positive there are 2–3 animals undetected. As such, a successful paratuberculosis control program cannot be based solely on animal testing alone, and must incorporate other management and biosecurity measures.

Abbreviations

MAP: Mycobacterium avium ssp. paratuberculosis; PI: Probability interval

Acknowledgements

The lead author received a graduate student stipend through the Ontario Veterinary College (OVC) Fellowship program. The support and participation of the small ruminant producers in this study is gratefully acknowledged.

Funding

This study was funded by the Animal Health Strategic Initiative, the Ontario Ministry of Agriculture, Food and Rural Affairs, and the Ontario Ministry of Agriculture, Food and Rural Affairs (OMAFRA) -University of Guelph Research program. The funding bodies did not have any direct role in study design or sample collection. The role of Dr. Jansen, who works for OMAFRA, is outlined below.

Authors' contributions

PM, AJB and JJ developed the study and participated in the design and coordination of the study. CAB carried out the animal sampling, sample handling, data analysis and drafted the initial manuscript. DK provided additional assistance in coordination of the study. CAB, PM, and AJB drafted early versions of the manuscript; JJ and DK contributed to later versions. All authors read and approved the final manuscript.

Competing interests

The authors declare that they have no competing interests.

Consent for publication

Not applicable.

Author details

[1]Department of Population Medicine, University of Guelph, Guelph, Ontario N1G 2W1, Canada. [2]Veterinary Science & Policy Unit, Animal Health and Welfare Branch, Ontario Ministry of Agriculture, Food & Rural Affairs, Elora, Ontario N0B 1S0, Canada.

References

1. Bauman C, Jones-Bitton A, Menzies P, Toft N, Jansen J, Kelton D. Prevalence of paratuberculosis in the dairy goat and dairy sheep industries in Ontario, Canada. Can Vet J. 2016;57:169–75.
2. Chiodini R, Van Kruiningen H, Merkal R. Ruminant paratuberculosis (Johne's disease): the current status and future prospects. Cornell Vet. 1984;74:218–62.
3. National Research Council. Diagnosis and Control of Johne's Disease. 1st ed. 2003.
4. Whitlock R, Wells S, Sweeney R, Van Tiem J. ELISA and fecal culture for paratuberculosis (Johne's disease): sensitivity and specificity of each method. Vet Microbiol. 2000;77:387–98.
5. Whittington R, Sergeant E. Progress towards understanding the spread, detection and control of Mycobacterium avium subsp paratuberculosis in animal populations. Aust Vet J. 2001;79:267–78.
6. Whittington RJ. Factors affecting isolation and identification of *Mycobacterium avium* subsp. *paratuberculosis* from fecal and tissue samples in a liquid culture system. J Clin Microbiol. 2009;47:614–22.
7. de Juan L, Álvarez J, Romero B, Bezos J, Castellanos E, Aranaz A, Mateos A, Domínguez L. Comparison of Four Different Culture Media for Isolation and Growth of Type II and Type I/III *Mycobacterium avium* subsp. *paratuberculosis* Strains Isolated from Cattle and Goats. Appl Environ Microbiol. 2006;72:5927–32.
8. Stevenson K. Genetic diversity of *Mycobacterium avium* subspecies *paratuberculosis* and the influence of strain type on infection and pathogenesis: a review. Vet Res. 2015;46:64–77.
9. Stabel J, Bannantine J. Development of a Nested PCR Method Targeting a Unique Multicopy Element, ISMap02, for Detection of *Mycobacterium avium* subsp. *paratuberculosis* in Fecal Samples. J Clin Microbiol. 2005;43:4744–50.
10. Douarre P, Cashman W, Buckley J, Coffey A, O'Mahony J. Isolation and detection of *Mycobacterium avium* subsp. *paratuberculosis* (MAP) from cattle in Ireland using both traditional culture and molecular based methods. Gut Pathog. 2010;2:11–8.
11. Cousins D, Whittington R, Marsh I, Masters A, Evans RJ, Kluver P. Mycobacteria distenct from *Mycobacterium avium* subsp. *paratuberculosis* isolated from the faeces of ruminants possess IS900-like sequences detectable IS900 polymerase chain reaction: implications for diagnosis. Mol Cell Probes. 1999;13:431–42.
12. Branscum AJ, Gardner IA, Johnson WO. Estimation of diagnostic-test sensitivity and specificity through Bayesian modeling. Prev Vet Med. 2005;68:145–63.
13. Hui SL, Walter SD. Estimating the Error Rates of Diagnostic Tests. Biometrics. 1980;36:167–71.
14. Toft N, Jørgensen E, Højsgaard S. Diagnosing diagnostic tests: Evaluating the assumptions underlying the estimation of sensitivity and specificity in the absence of a gold standard. Prev Vet Med. 2005;68:19–33.
15. Gardner I, Nielsen S, Whittington R, Collins M, Bakker D, Harris B, Sreevatsan S, Lombard J, Sweeney R, Smith D, Gavalchin J, Eda S. Consensus-based reporting standards for diagnostic test accuracy studies for paratuberculosis in ruminants. Prev Vet Med. 2011;101:18–34.
16. Hanson T, Johnson WO, Gardner IA. Hierarchical Models for Estimating Herd Prevalence and Test Accuracy in the Absence of a Gold Standard. J Agric Biol Environ Stat. 2003;8:223–39.
17. Whittington RJ, Hope AF, Marshall DJ, Taragel CA, Marsh I. 2000. Molecular epidemiology of Mycobacterium avium subsp. paratuberculosis: IS900 restriction fragment length polymorphism and IS1311 polymorphism analyses of isolates from animals and a human in Australia. J Clin Microbiol. 2000;38:3240–3248.
18. Ellingson JL, Bolin CA, Stabel JR. Identification of a gene unique to *Mycobacterium avium* subspecies *paratuberculosis* and application to diagnosis of paratuberculosis. Mol Cell Probes. 1998;12:133–42.
19. Manning EJB, Cushing HF, Hietala S, Wolf CB. Impact of *Corynebacterium pseudotuberculosis* infection on serologic surveillance for Johne's disease in goats. J Vet Diagn Invest. 2007;19:187–90.
20. Kostoulas P, Leontides L, Enøe C, Billinis C, Florou M, Sofia M. Bayesian estimation of sensitivity and specificity of serum ELISA and faecal culture for diagnosis of paratuberculosis in Greek dairy sheep and goats. Prev Vet Med. 2006;76:56–73.
21. Stringer LA, Jones G, Jewell CP, Noble AD, Heuer C, Wilson PR, Johnson WO. Bayesian estimation of the sensitivity and specificity of individual fecal culture and Paralisa to detect *Mycobacterium avium* subspecies *paratuberculosis* infection in young farmed deer. J Vet diagnostic Investig. 2013;25:759–64.
22. Dendukuri N, Joseph L. Bayesian approaches to modeling the conditional dependence between multiple diagnostic tests. Biometrics. 2001;57:158–67.
23. Alinovi CA, Ward MP, Lin TL, Moore GE, Wu CC. Real-time PCR, compared to liquid and solid culture media and ELISA, for the detection of *Mycobacterium avium* ssp. *paratuberculosis*. Vet Microbiol. 2009;136:177–9.
24. Gumber S, Whittington RJ. Comparison of BACTEC 460 and MGIT 960 systems for the culture of *Mycobacterium avium* subsp. *paratuberculosis* S strain and observations on the effect of inclusion of ampicillin in culture media to reduce contamination. Vet Microbiol. 2007;119:42–52.
25. R: A language and environment for statistical computing. [http://www.r-project.org]. Accessed 31 Aug 2016.
26. Raftery A, Lewis S. How many iterations in the Gibbs sampler? In Bayesian Statistics. 1992;763–774.
27. Ntzoufras I. Bayesian Modeling Using WinBUGS. 1st ed. Hoboken: John Wiley & Sons Inc; 2009.
28. Gelman A, Rubin D. Inference from iterative simulation using multiple sequences. Stats Sci. 1992;7:457–511.
29. Okura H, Nielsen SS, Toft N. Prevalence of *Mycobacterium avium* subsp. *paratuberculosis* infection in adult Danish non-dairy cattle sampled at slaughter. Prev Vet Med. 2010;94:185–90.
30. Appendix B Winbugs. :316.
31. Dohoo IR, Martin W, Stryhn H. Veterinary Epidemiologic Research. 2nd ed. Charlottetown: VER Inc; 2009.
32. Collins DM, Gabric DM, de Lisle GW. Identification of two groups of Mycobacterium paratuberculosis strains by restriction endonuclease analysis and DNA hybridization. J Clin Microbiol. 1990;28:1591–6.
33. Englund S, Bolske G, Johansson K. An IS 900 -like sequence found in a Mycobacterium sp.other than *Mycobacterium avium* subsp. *paratuberculosis*. FEMS Microbiol Lett. 2002;209:267–71.
34. Christopher-Hennings J, Dammen MA, Weeks SR, Epperson WB, Singh SN, Steinlicht GL, Fang Y, Skaare JL, Larsen JL, Payeur JB, Nelson EA. Comparison of two DNA extractions and nested PCR, real-time PCR, a new commercial PCR assay, and bacterial culture for detection of *Mycobacterium avium* subsp. *paratuberculosis* in bovine feces. J Vet Diagn Invest. 2003;15:87–93.
35. Gilardoni LR, Paolicchi FA, Mundo SL. Bovine paratuberculosis: a review of the advantages and disadvantages of different diagnostic tests. Rev Argent Microbiol. 2012;44:201–15.

11

The association between serum ß-hydroxybutyrate and milk fatty acid profile with special emphasis on conjugated linoleic acid in postpartum Holstein cows

Pedro Melendez[1], Pablo Pinedo[2,5*], José Bastias[3], Maria Paz Marin[3], Carolina Rios[3], Consuelo Bustamante[3], Natalia Adaro[3] and Mario Duchens[4]

Abstract

Background: Ketogenesis is a secondary metabolic pathway to provide energy to dairy cows during early lactation; however when the production of ketone bodies (acetoacetate, acetone, ß- hydroxybutyrate) is above certain levels a subclinical disorder may appear. The aim of the present study was to investigate the association between serum concentrations of ß- hydroxybutyrate (BHBA) and fatty acid (FA) profile of milk with emphasis in conjugated linoleic acid (CLA) in a population of early lactation Holstein cows. Fifty cows between parity 1 and 5, ranging from 14 to 21 days in milk, were randomly selected from 3 farms of the central area of Chile for determination of serum BHBA concentrations, milk fat content, and milk FA profiles.

Results: Cows were divided in low ($n = 26$) and high ($n = 24$) BHBA groups considering the median value of the serum concentration of BHBA (0.7 mmol/L) (SEM = 0.094). Mean milk fat % was 3.45 % and 3.60 % for cows in the low and high BHBA groups, respectively ($P = 0.15$). Concentrations of several FA were significantly different between both groups. Specifically, mean CLA concentrations were 0.40 % (4 ± 0.03 g/kg) and 0.33 % (3.3 ± 0.03 g/kg) for the low and high BHBA groups, respectively ($P = 0.05$).

Conclusions: It is concluded that early postpartum cows with serum BHBA > 0.7 mmol/L tended to have higher milk fat % and had significantly lower concentrations of CLA than early postpartum cows with BHBA ≤ 0.7 mmol/L.

Keywords: Ketone bodies, Conjugated linoleic acid, ß- hydroxybutyrate

Background

Ketogenesis is a common metabolic pathway to provide energy to dairy cows during early lactation; however the synthesis of large amounts of ketone bodies (acetoacetate, acetone, and beta hydroxyl butyrate [BHBA]) may result in a typical metabolic disorder denominated ketosis [4]. Several studies have shown that high levels of BHBA are associated with increased risk of periparturient diseases, losses in milk yield and impaired fertility [4, 9, 11, 12]. In addition, increased concentrations of BHBA were associated with greater milk fat and lower milk protein percentages in the first Dairy Herd Improvement test day of lactation [4].

Conjugated linoleic acid (CLA) is one component of milk fat that might be beneficial to human health due to anticarcinogenic properties [15]. However, this potential benefit has been described for only few isomers of CLA, especially for those called CLA cis-9, tans-11. Strategies such as commercial sources of CLA added into dairy cattle diets is one potential means of increasing CLA content of milk [5]. Conversely, there might be several factors decreasing the levels of CLA in milk such as feeding excessive grains (starch), corn silage and products rich in polyunsaturated fatty acids (e.g. vegetable

* Correspondence: pinedop@colostate.edu
[2]Texas A&M AgriLife Research & Extension Center, College of Veterinary Medicine and Biomedical Sciences, Texas A&M University System, Amarillo, TX 79106-1769, USA
[5]Present Address: Department of Animal Sciences, Colorado State University, Fort Collins, CO 80523, USA
Full list of author information is available at the end of the article

oils) [8]. As stated previously, the association of high levels of ketone bodies and greater milk fat percentage in dairy cattle has been reported [3, 4, 10]. However, these studies did not evaluate fatty acid (FA) profile in milk. The hypothesis of this study was that there is a relationship between BHBA levels and FA profile of milk in early lactation Holstein cows. The objective of this study was to determine the association of plasma concentrations of BHBA and FA profile of milk with special emphasis in CLA in early lactation Holstein cows.

Results and discussion

The median value of BHBA concentration was 0.7 ± 0.094 mmol/L. Twenty six cows were classified as having a plasma concentration of BHBA of ≤ a 0,7 mmol/L and 24 cows as having a plasma concentration of BHBA of > 0,7 mmol/L. The reason for using the median value as cut-off point was to develop two groups with comparable number of animals. When the cut-off for subclinical ketosis of 1.2 mmol/L of BHBA [14] was considered, only 7 cows (14 %) were above this concentration, and 43 cows were below this value; consequently there was not sufficient statistical power to conduct the analysis.

For all study variables (fat % and fatty acids) the interaction of group by parity was not significant; therefore only main effects are reported. The milk fat content (g/kg) of cows with high and low BHBA was 3.60 % and 3.45 %, respectively ($P = 0.15$). Consequently, there was a tendency for milk fat content to be higher in the group of cows with BHBA > 0.7 mmol/L. This is in agreement with the study conducted by Duffield et al [4] where cows with higher levels of BHBA had higher concentration of milk fat than cows with low levels of BHBA. This association might be reasonable since BHBA is used as precursor for milk fat synthesis in the mammary gland. Short- and medium-chain FA (4 to 14 carbons) and a portion of the 16-carbon FA are derived from de novo synthesis from acetate and to a lesser extent butyrate [1]. Conversely, preformed FA that account for the remaining 16-carbon and all of the longer-chain FA (>16 carbons) are taken up from the circulating plasma pool. These long FA originate from absorption from the digestive tract or mobilization from body reserves. Adipose tissue mobilization accounts for less than 10 % of preformed FA in milk fat, except during periods of negative energy balance when their proportion increases substantially [1].

In Table 1, the least squared mean concentration and SEM of milk FA (g/kg of fat) of both group is presented. An interesting finding of this study was that the concentration of CLA in milk in cows with BHBA ≤ 0.7 mmol/L was higher than in cows with levels of BHBA > 0.7 mmol/L ($P \leq 0.05$). Differences were also found for 18:1 (all isomers). In addition, results from the multivariate regression model (Table 2) show there was a tendency that for each 0.1 mmol/L of BHBA in plasma, CLA (all isomers) decreased in 0.2 g/kg of total fat ($P = 0.15$). This association was established correcting for milk yield, solid contents, somatic cell counts and milk urea nitrogen. The total r^2 for the predicting equation was 0.59, indicating that 59 % of the variation of CLA in milk was explained by the current model. The partial slope for BHBA was essentially linear, without evidence of polynomial effects. Fatty acid metabolism in the rumen has a major influence on the FA profile in milk [7]. The CLA are present at higher concentrations in ruminant products because they are formed in the rumen from dietary linoleic and linolenic acid. This transformation is due to via three major processes carried out by rumen microbes: lipolysis, biohydrogenation and isomerization. These mechanisms are dependent on the type and amount of fat entering the rumen [2] and ruminal pH [14]. The question arising is why levels of blood BHBA > 0.7 mmol/L might be negatively related to CLA in milk. BHBA is synthesized in the liver through the process of ketogenesis [6] and in the rumen through butyrate-producing bacteria [7]. In addition, all bacteria that produced substantial quantities of *cis*-9, *trans*-11-CLA, *trans*-11-18:1, or both from linoleic acid were butyrate producers (mostly *Butyrivibrio fibrisolvens*). However,

Table 1 Least squared mean concentration and SEM of milk FA in cows with BHBA plasma concentrations ≤ 0,7 mmol/L ($n = 26$) and > 0,7 mmol/L ($n = 24$)[a]

Fatty acid (%)	BHB ≤ 0,7 mmol/L	BHB > 0,7 mmol/L	SEM	*p*-value
4:0	2.43	2.35	0.1	0.63
6:0	1.75	1.13	0.1	0.0007
8:0	1.77	0.77	0.32	0.0001
10:0	0.95	0.4	0.35	0.0001
12:0	2.09	1.29	0.13	0.001
14:0	6.23	7.87	0.32	0.003
14:1	0.3	0.21	0.04	0.66
15:0	0.78	0.63	0.03	0.01
16:0	28.68	28.67	0.5	0.92
16:1	1.12	1.68	0.08	0.0001
17:0	0.82	0.96	0.03	0.0008
18:0	17.96	18.25	0.56	0.42
18:1 (all isomers)	30.69	32.67	0.43	0.04
18:2 (all isomers CLA)	0.4	0.33	0.03	0.048
18:2 (all isomers no CLA)	0.52	0.43	0.04	0.28
20:0	0.51	0.47	0.03	0.43

[a]All models included the effect of parity (primiparous, multiparous) and farm (1, 2, and 3); consequently reported least squared means for all fatty acids are adjusted for these two variables. For all FA, the interaction between group and parity was not significant

Table 2 Multivariate regression model for CLA (g/kg) in milk. Coefficient of determination (r^2) = 0.594

Independent variables	Coefficients	SEM	*P*-value
Constant	6.68	4.76	0.06
BHBA (per 0.1 mmol/L)	−0.22	0.14	0.15
Fat (%)	−0.34	0.16	0.05
Protein (%)	−0.56	0.98	0.58
Milk urea nitrogen (per 1 mg/dl)	0.30	0.18	0.14
Somatic cells count (per 1,000 cells/ml)	−0.00024	0.0001	0.05
Average milk yield up to 60 days in milk (per 1 kg/day)	0.0041	0.048	0.25

not all butyrate producers yield *cis*-9, *trans*-11-CLA or *trans*-11-18:1. Other source of CLA is cell membrane from protozoa. Protozoa contained at least 2 to 3 times more unsaturated FA than bacteria. These unsaturated FA included CLA and *trans*-11-18:1, which were more than 8- and 3-fold more abundant in protozoa than in bacteria, respectively. As a result, the flow of FA from protozoa accounted for between 30 and 43 % of the CLA and 40 % of the *trans*-11-18:1 reaching the duodenum [17]. The contribution of protozoa to the flows of 16:0 and 18:0 to the duodenum was less than 20 and 10 %, respectively. Although protozoa do not themselves produce CLA and *trans*-11-18:1 by their own metabolism; they might be expected to have a significant influence on CLA and *trans*-11-18:1 available to the host animal [17]. In this sense, the content of CLA in milk is explained by several highly complex mechanisms. Since butyrate is precursor for short and medium chain milk FA it is reasonable to suggest that higher concentrations of blood butyrate would increase milk fat through fatty acids < 16 C; and would decrease proportionally long chain fatty acids. This is partially true, because in the present study most of short and medium chain FA, and 18:1 were higher in cows with BHBA > 0.7 mmol/L than in cows with BHBA ≤ 0.7 mmol/L. Accordingly, even though, *trans-11* C18:1 and *cis-9, trans-11* CLA are the predominant trans FA intermediates produced from the ruminal metabolism of linoleic acid; ruminal biohydrogenation pathways are dynamic, allowing the production of a wide range of positional and geometric isomers as well as modified FA such as hydroxy and keto derivatives [7, 13].

Finally, if the negative association between levels of BHBA in blood and proportion of CLA in milk remain consistent in further studies, especially in cows with high levels of BHBA (>1.2 mmol/L), it would be another reason to consider essential the prevention of high levels of ketone bodies, if the dairy industry is looking for milk and its derivatives with high levels of CLA.

Conclusions

Early postpartum cows with plasma BHBA > 0.7 mmol/L tended to have higher milk fat content and significantly lower proportion of CLA than early postpartum cows with BHBA ≤ 0.7 mmol/L. For each 0.1 mmol/L of BHBA in plasma, CLA (all isomers) decreased in 0.2 g/kg of total fat.

Methods

Dairy farms

The study was conducted during 2010 in three dairy farms from the central area of Chile. The study was performed in accordance with the recommendations of the University Santo Tomas animal care guidelines. No procedure causing more than momentary discomfort was performed during this study. In addition, all the animal related actions had the consent of the University Animal Care Committee and the dairy owners. Average herd size was 350 lactating Holstein cows. Rolling herd average milk yield was 10,000 to 12,000 kg ME 305 and selected farms followed similar management practices. Farms housed cows in sand-bedded free-stalls and milked 3 times a day. Feed management consisted of a unique total mixed ration based on alfalfa hay, corn silage, corn grain, soybean meal, wet brewer grain, wheat bran, distiller derivative grain solubles, minerals and vitamins. Farms used computerized record systems; consequently individual milk yield was recorded daily based on computerized milking parlors. During early lactation cows were subjected to a postpartum health management protocol. After calving cows were checked for retained fetal membranes and the development of milk fever. Daily clinical examination and rectal temperature were conducted until 14 days postpartum. Milk BHBA colorimetric testing (Ketotest, Nagoya, Japan) was conducted at once between 14 to 21 days postpartum. Sick cows were treated as needed.

Study design

In order to find a difference in 0.1 ± 0.1 g of CLA per kg of milk fat, with a 95 % of confidence and 80 % of power, a sample size of 13 cows per group was required. Weekly visits for the 3 farms were conducted between May and August of 2010 until 50 cows within 14 and 21 days in milk were obtained. Blood and milk samples were collected from each cow the same day of assortment. Blood sample for plasma collection was obtained from the coccygeal vessels using a vacuatainer system and centrifuged at 4000 × g for 10 min. Plasma was separated and stored in plastic tubes and frozen at - 20 °C until analysis. Plasma samples were submitted to the laboratory of clinical pathology of University of Santo Tomas, Viña del Mar, Chile. The determination of plasma BHBA concentrations was conducted using a commercial kit (Pointe Scientific, Inc. BHA Set., Lincoln

Park, MI, USA), based on an enzymatic-colorimetric method [16]. Samples were assayed in duplicate with a coefficient of variation of 4.8 %.

A milk composite sample was obtained during the morning milking in a plastic tube with preservative (2-bromo-2-nitropropane- 1-2-diol).and immediately after collection submitted to the laboratory of biochemistry at the Catholic University of Valparaiso, Chile, for milk fat content and milk FA profile analysis. The determination of milk fat (g/kg) was determined using infrared spectroscopy (MilkoScan™ Minor, DK-3400, Hilleroed, Denmark). Milk samples were analyzed for individual FA by Gas Chromatograph-Mass Spectrometry (GCMS) of butyl esters. The identification of fatty acids was based on the comparison of retention times of standard methyl esters ranging from C4 to C24 (Sigma-Aldrich, St. Louis, MO, USA). Calibration standards were achieved from seven different concentrations, depending on the particular FA. Quality control samples were prepared at four different concentrations.

The methodology consisted as follow: 0.5 ml of milk was placed into test tubes with teflon-lined screw caps, followed by the addition of 750 μl of n-butanol. Samples were vortexed at low speed while 75 μl of acetyl chloride was added. Then, samples were gassed with N_2, capped tightly, and heated at 100 °C for 1.5 h. After samples cooled to room temperature, 5 ml of 6 % K_2CO_3 and 1 ml of hexane were added, and the samples were vortexed for 30 s. Samples were centrifuged for 20 min at 2500 *g*, and the bottom layer was aspirated and discarded. The remaining layer was washed four times (20 min at 2500 *g*) with distilled, deionized water. The upper layer was removed and placed in injection vials for analysis. Fatty acid analysis of butyl esters was conducted with a GCMS model QP5050A, series Splitless (Shimadzu®, Shimadzu Europa GmbH, Albert-Hahn-Str. 6-10 47269, Duisburg, Germany).

Results were expressed as percentage relative to each FA detected in the chromatogram according to its retention time and identification based on pure standard. Subsequently, the median value for BHBA concentrations (0.7 mmol/L) was considered as a cut-off value to obtain balanced numbers of animals in the low and high BHBA groups. Milk fat (g/kg) and FA (g/kg) content between the 2 groups were analyzed by ANOVA, developing a mixed model that also included parity (primiparous, multiparous) and farm (1, 2, 3). Potential interaction between group and parity was also explored. Least squared means were estimated and tested for significant differences. In addition, a multivariate regression model was conducted to predict total CLA (g/kg) in milk from plasma BHBA as a continuous variable, correcting for fat and protein % in milk, milk urea nitrogen (mg/dl), somatic cell count (cells/ml) and average milk yield until 60 days in milk (kg/d). Statistical analysis was performed using the corresponding procedure of SAS (release 9.0; SAS Institute Inc., Cary, NC). Statistical significance was set at $P \leq 0.05$; a tendency was set at $0.05 < P \leq 0.15$.

Competing interests

The authors declare that they have no competing interests.

Authors' contributions

PM: Experimental design, statistical analysis, manuscript preparation. PP: Statistical analysis, manuscript preparation. JB, CB, NA: Sample collection, laboratory analyses; CR, MPM, MD: Experimental design, manuscript preparation. All the authors read and approved the final version of the manuscript.

Acknowledgments

The authors acknowledge University Santo Tomas for financial support and thank dairy farmers that provided access to their dairies for this study.

Author details

College of Veterinary Medicine, University of Missouri, Columbia, MO 65211, USA. [2]Texas A&M AgriLife Research & Extension Center, College of Veterinary Medicine and Biomedical Sciences, Texas A&M University System, Amarillo, TX 79106-1769, USA. [3]College of Veterinary Medicine, University Santo Tomas, Viña del Mar, Chile. [4]College of Veterinary Medicine, University of Chile, Santiago, Chile. [5]Present Address: Department of Animal Sciences, Colorado State University, Fort Collins, CO 80523, USA.

References

1. Bauman DE, Griinari JM. Nutritional regulation of milk fat synthesis. Annu Rev Nutr. 2003;23:203.
2. Beam TM, Jenkins TC, Moate PJ, Kohn RA, Palmquist DL. Effects of amount and source of fat on the rates of lipolysis and biohydrogenation of fatty acids in ruminal contents. J Dairy Sci. 2000;83:2564.
3. Duffield TF, Kelton DF, Leslie KE, Lissemore K, Lumsden JH. Use of test day milk fat and milk protein to predict subclinical ketosis in Ontario dairy cattle. Can Vet J. 1997;38:713.
4. Duffield TF, Lissemore KD, McBride BW, Leslie KE. Impact of hyperketonemia in early lactation dairy cows on health and production. J Dairy Sci. 2009;92: 571.
5. Giesy JG, McGuire MA, Shafii B, Hanson TW. Effect of dose of calcium salts of conjugated linoleic acid (CLA) on percentage and fatty acid content of milk fat in milk lactation Holstein cows. J Dairy Sci. 2002;85:2023.
6. Herdt TH. Ruminant adaptation to negative energy balance. Vet Clin North Am Food Anim Pract. 2000;16:215.
7. Jenkins TC, Harvatine KJ. Lipid feeding and milk fat depression. Vet Clin North Am Food Anim Pract. 2014;30:623.
8. Jenkins TC, Wallace RJ, Moate PJ, Mosley EE. Recent advances in biohydrogenation of unsaturated fatty acids within the rumen microbial ecosystem. J Anim Sci. 2008;86:397.
9. LeBlanc SJ, Leslie KE, Duffield TF. Metabolic predictors of displaced abomasum in dairy cattle. J Dairy Sci. 2005;88:159.
10. Miettinen PVA, Setala JJ. Relationships between subclinical ketosis, milk production and fertility in Finnish dairy cattle. Prev Vet Med. 1993;17:1.
11. Ospina PA, Nydam DV, Stokol T, Overton TR. Associations of elevated nonesterified fatty acids and β-hydroxybutyrate concentrations with early lactation reproductive performance and milk production in transition dairy cattle in the northeastern United States. J Dairy Sci. 2010a;93:1596.
12. Ospina PA, Nydam DV, Stokol T, Overton TR. Association between the proportion of sampled transition cows with increased nonesterified fatty acids and β-hydroxybutyrate and disease incidence, pregnancy rate and milk production at the herd level. J Dairy Sci. 2010b;93:3595.
13. Palmquist DL, Lock AL, Shingfield KJ, Bauman DE. Biosynthesis of conjugated linoleic acid in ruminants and humans. Adv Food Nutr Res. 2005;50:179.

14. Van Nevel CJ, Demeyer DI. Influence of pH on lipolysis and biohydrogenation of soybean oil by rumen contents in vitro. Reprod Nutr Dev. 1996;36:53.
15. Whigham LD, Cook ME, Atkinson RL. Conjugated linoleic acid: Implications for human health. Pharmacol Res. 2000;42:503.
16. Williamson DH, Mellanby J. D-3-hydroxybutyrate. In: Bergmeyer HU, editor. Methods of enzymatic analysis. London, England: Academic; 1974. p. 1836–8.
17. Yanez-Ruiz DR, Scollan ND, Merry RJ, Newbold CJ. Contribution of rumen protozoa to duodenal flow of nitrogen, conjugated linoleic acid and vaccenic acid in steers fed silages differing in their water-soluble carbohydrate content. Br J Nutr. 2006;96:861.

12

Characterization of *Staphylococcus aureus* isolated from milk samples of dairy cows in small holder farms of North-Western Ethiopia

S. A. Mekonnen[1,2*], T. J. G. M. Lam[1,3], J. Hoekstra[1,4], V. P. M. G. Rutten[4,5], T. S. Tessema[6], E. M. Broens[4], A. E. Riesebos[1], M. P. Spaninks[4] and G. Koop[1]

Abstract

Background: *Staphylococcus aureus* is a contagious, opportunistic pathogen that causes clinical or subclinical mastitis in dairy cattle. The genetic background and antimicrobial resistance of isolates from Ethiopian dairy farms has not been studied. Therefore, the aim of this study was to characterize *S. aureus* from Ethiopian hand milked dairy cows, by *spa*, MLST and virulence factor typing, and by assessment of antimicrobial susceptibility. A total of 79 *S. aureus* isolates from intramammary infections was studied. A PCR was used to detect *lukM-lukF'* and *pvl* genes encoding the bovine and human associated bi-component leukocidins, and the toxic shock syndrome toxin gene-1 (*tst*). Antimicrobial susceptibility was determined using the broth microdilution method.

Results: Twenty different *spa* types were identified, most isolates were t042 (58%), and the closely related t15786 (11%). The proportion of isolates positive for *lukM-lukF'*, *tst* and *pvl* was low at 0.04, 0.10 and 0.09 respectively, with *lukM-lukF'* often co-occurring with *tst*, but not with *pvl*. Methicillin-resistance was not found, but resistance to penicillin/ampicillin (86%) and tetracycline (54%) was very common.

Conclusions: We found a high degree of relatedness among bovine *S. aureus* isolates in North-Western Ethiopia, suggesting contagious within and between farm transmission of strains that are often resistant to commonly used antimicrobials. This highlights the need for effective preventive measures that aim at limiting transmission of bacteria rather than using antimicrobials to control *S. aureus* mastitis in Ethiopia.

Keywords: *Staphylococcus aureus*, *Spa*, MLST, Virulence factor, Antimicrobial resistance, Mastitis, Ethiopia

Background

Staphylococcus aureus is a contagious pathogen that causes mastitis in dairy cattle [1, 2], and is an opportunistic pathogen in humans and many other animal species [3–5]. Also in Ethiopia, *S. aureus* is frequently isolated from cows with mastitis [6–8], and from milk or milk products [9].

In dairy cows, *S. aureus* can be isolated from milk as well as from different other body sites [10, 11]. Transmission of *S. aureus* intramammary infections (IMI) is believed to mainly occur during the milking process [12], but this has been poorly researched in situations where hand milking is common, as is the case in Ethiopia. Additionally, hand milking may introduce possibilities for transmission between farmers and their cattle. Typing of *S. aureus* isolates may give insight in their likely origin, because MLST and spa-types and virulence factors often are host-associated [13, 14]. Genetic similarity of isolates within and between farms suggests contagious transmission in the spread of bacteria [4] whereas a greater variety of genotypes within herds or regions is more suggestive of environmental pathogens [15]. Genotyping does not only give information on modes of transmission, but can also be used to identify

* Correspondence: sefiale@yahoo.com
[1]Department of Farm Animal Health, Faculty of Veterinary Medicine, Utrecht University, Yalelaan 7, 3584, CL, Utrecht, The Netherlands
[2]Faculty of Veterinary Medicine, University of Gondar, P.O. Box 196, Gondar, Ethiopia
Full list of author information is available at the end of the article

virulence factors of the bacteria. Depending on their clonal lineage, *S. aureus* may carry the phage-encoded leukocidin genes *lukM-lukF'* [16–18]. LukMF' is a potent toxin specifically killing bovine neutrophils [19] and likely to contribute to the clinical severity of mastitis [20]. The proportion of bovine *S. aureus* isolates encoding this toxin varies largely between countries [21], but to our knowledge this has not been estimated in isolates of IMI from African cattle.

Because the use of antimicrobial drugs, both in humans and animals, is poorly controlled, multidrug resistant *S. aureus* are frequently isolated from animals [6, 7] and humans in Ethiopia [22–24]. In *S. aureus* isolated from humans, there is a trend of increasing antimicrobial resistance (AMR) [25]. Describing AMR patterns in both humans and animals may contribute to the knowledge on the importance of the issue in Ethiopia and to a more prudent and effective antimicrobial use.

In smallholder dairy farming, effective control of *S. aureus* mastitis is important, because mastitis can have a substantial effect on family income. Since only a limited number of studies characterized bovine *S. aureus* isolated from Ethiopia [9, 26], the aim of the present study was to characterize *S. aureus* isolates from milk samples from hand milked dairy cows in North-Western Ethiopia by *spa* and MLST typing, virulence genes and antimicrobial susceptibility.

Methods

Isolate collection

A total of 135 phenotypically isolated *S. aureus* were available from an earlier cross-sectional study conducted from October 2014 to December 2016 in 167 urban and peri-urban smallholder dairy farms in the regions Gondar and Bahir Dar, in the North-West of Ethiopia [8]. Nine isolates were lost. Forty seven isolates were excluded from the study because data of isolates were lost (11), data mismatched (8), and genotypically non *S. aureus* (28). Seventy nine *S. aureus* isolates that had complete data records were available (two from clinical mastitis (CM), 53 from California mastitis test (CMT) positive and 24 from CMT negative quarters) for *spa* and MLST typing, virulence genes detection and antimicrobial susceptibility testing. The 79 isolates were derived from 60 dairy cows in 42 dairy herds and were kept in glycerol stocks at – 80 °C until further study.

DNA extraction

Single colonies from an overnight culture grown on blood agar plates were suspended in 1 mL of distilled water, homogenized using Vortex and centrifuged for 1 min at 13,000 rpm. Supernatant was removed and the bacteria were re-suspended in 200 μL of distilled water, boiled for 10 min at 100 °C and centrifuged at 13,000 rpm for 1 min. The extracted DNA was diluted 1:10 in distilled water and stored at – 20 °C until further analysis.

Virulence genes detection

Polymerase chain reactions were performed on 10 μL of the DNA of all 79 isolates to confirm the isolates were *S. aureus* (*femA*), to test for presence of the *mecA* gene, encoding resistance to methicillin, and to amplify the polymorphic X region of the surface protein A (*spa*) gene. We also used PCR to detect *lukM-lukF'* and *pvl* genes encoding the bovine and human associated bi-component leukocidins, respectively, and the toxic shock syndrome toxin gene-1 (*tst*). Primers and protocols are summarized in Table 1. Amplifications were carried out in final volumes of 25 μL using a T100™ Thermal Cycler (Bio-Rad, USA). The presence of appropriate amplicon sizes was examined by electrophoresis of 10 μL PCR product on 1.5% agarose at 100 V for 40 min. PCR products were visualized with a molecular Imager Gel Doc XR+ Imaging system (Bio-Rad, USA).

Spa and MLST typing

The product of the *spa* gene PCR was cleaned up from unbound primers and nucleotides using ExoSAP-IT PCR Cleanup Reagent (Affymetrix, USA) according to manufacturer's instructions and was submitted for Sanger sequencing (Baseclear BV, The Netherlands). The *spa* types were assigned using BioNumerics version 7 (Applied Maths, Belgium), using the *spa*-typing plugin and Ridom SpaServer (http://www.spaserver.ridom.de). A minimum spanning tree (MST) was constructed in BioNumerics using the *spa* clustering method of the spa typing plugin. Repeat successions were aligned using a gap creation cost of 100%, gap extension cost of 50%, duplicate creation and duplicate extension costs of both 25% and a maximum duplication number of 3 repeats. The MST was built using the MST cluster method and the default cost matrix with a 1% grouping distance bin. To save costs, multilocus sequence typing (MLST) as described by Enright et al. [27] was performed on a selection of ten of the 79 isolates: on six t042 isolates, and on 4 isolates of *spa* types t273, t17184, t488 and t409. These ten isolates were selected aiming to optimally represent the various spa types and the two regions (Gondar and Bahir Dar). Sequence types were assigned using BioNumerics.

Antimicrobial susceptibility testing

Antimicrobial susceptibility testing was performed by determining minimum inhibitory concentrations (MICs) using the MICRONAUT system (Merlin Diagnostika, Germany) in customized prepared 96-wells microtiter plates containing dilution series of 17 antimicrobials. Inoculum preparation, broth composition and incubation conditions were performed as recommended by the

Table 1 Sequences of primer sets with their corresponding PCR protocols and product sizes

Target gene	Primer sequence (5′-3′)	PCR-protocol	Amplicon size	Reference
femA	F: TGCCTTTACAGATAGCATGCCA	1	142 bp	[52]
	R: AGTAAGTAAGCAAGCTGCAATGACC			
mecA	F: GGCTATCGTGTCACAATCGTT	2	689 bp	[53]
	R: TCACCTTGTCCGTAACCTGA			
spa	F: AGACGATCCTTCGGTGAGC	3	Variable	[54]
	R: GCTTTTGCAATGTCATTTACTG			
lukM	F: TGAGTGGGTATGGCATGAAAGA	4	572 bp	[20]
	R: TGGACATTTTGTGTTACACCCC			
lukF′	F: ACTCAGGCTATACCAACCCA	1	425 bp	[20]
	R: CGAGCTACTCTGTCTGCCAC			
pvl	F: GCTGGACAAAACTTCTTGGAATAT	5	85 bp	[55]
	R: GATAGGACACCAATAAATTCTGGATTG			
tst	F: CAACATACTAGCGAAGGAACT	6	277 bp	[56]
	R: GATATGTGGATCCGTCATTCA			

[1] 95 °C for 2 min, 35× (95 °C for 30 s, 59.5 °C for 30 s, 72 °C for 35 s), 72 °C for 5 min and 12 °C for ∞
[2] 95 °C for 1.5 min, 30× (95 °C for 45 s, 55 °C for 30 s, 72 °C for 45 s), 72 °C for 5 min and 12 °C for ∞
[3] 95 °C for 2 min, 30× (95 °C for 45 s, 60 °C for 45 s, 72 °C for 1:30 min), 72 °C for 5 min and 12 °C for ∞
[4] 95 °C for 2 min, 33× (95 °C for 30 s, 58.2 °C for 30 s, 72 °C for 30 s), 72 °C for 5 min and 12 °C for ∞
[5] 95 °C for 2 min, 35× (95 °C for 30 s, 55.5 °C for 30 s, 72 °C for 35 s), 72 °C for 5 min and 12 °C for ∞
[6] 95 °C for 2 min, 35× (95 °C for 30 s, 59 °C for 30 s, 72 °C for 30 s), 72 °C for 5 min and 12 °C for ∞

manufacturer. Reading of the plates after incubation was done with a photometer (Skan, Merlin Diagnostika, Germany). *Staphylococcus aureus* ATCC 29213 was used as quality control strain. Clinical breakpoints were used according to the Clinical and Laboratory Standards Institute (CLSI) standards [28, 29]. Veterinary breakpoints were used for ceftiofur (cattle), cephalotin (dogs) and enrofloxacin (dogs). Because veterinary breakpoints for several antimicrobials have not been described, human breakpoints were used for penicillin, ampicillin, tetracycline, clindamycin, erythromycin, trimethoprim sulfamethoxazole, rifampicin, amoxicillin/clavulanic acid, cefoxitin, chloramphenicol, gentamicin, kanamycin and neomycin. For fusidic acid, EUCAST-criteria were used (www.eucast.org), because no breakpoints were available in the CLSI standards.

Results

Genetic typing

In total, twenty different *spa* types were found, the majority of isolates belonging to t042 (58%) and the closely related t15786 (11%) and the distribution of *spa* types was similar across the two regions (Fig. 1). The proportion of *lukM-lukF′* positive *S. aureus* in our sample was low (3/79; 4%). Seven isolates carried *pvl* (7/79; 9%) and *tst* was detected in eight isolates (8/79, 10%). All t14061 isolates (3/79, 4%) were positive for *tst*; *pvl* was present in all t355 (3/79, 4%) and t1376 (2/79, 3%) isolates. None of the isolates carried the *mecA* gene. Table 2 summarizes the *spa* typing results and presents the isolates positive for virulence genes.

All six t042 isolates subjected to MLST typing belonged to a new ST, ST4550 which is a double locus variant of ST97. The other isolates subjected to MLST were ST1 (t273), ST22 (t17184), ST88 (t488) and ST848 (t409). All the data in this paper are available in Additional file 1.

Antimicrobial resistance

All 79 isolates were susceptible for cefoxitin and were therefore classified as methicillin-susceptible *S. aureus*. Resistance to penicillin/ampicillin (86%) and tetracycline (54%) was high, while the resistance to other antimicrobials was limited. The MIC values and numbers of isolates resistant to the antimicrobials included in the testing are presented in Table 3. Associations between *spa*-type and AMR patterns were not found. Percentages resistance to antimicrobials by *spa*-type for 79 *S. aureus* isolates is presented in Additional file 2.

Discussion

The aim of this study was to characterize *S. aureus* isolates cultured from milk samples of dairy farms in North-Western Ethiopia based on *spa* typing, virulence genes identification and AMR. The spa types t042 and the related t15786 were dominant, together representing 69% of the isolates. Such a high degree of relatedness suggests contagious transmission [13, 14]. Since our isolates were derived from different farms, contagious transmission is likely to occur not only within, but also between farms, possibly through cattle movements between farms, or from the government breeding and heifer distribution centers. Four of

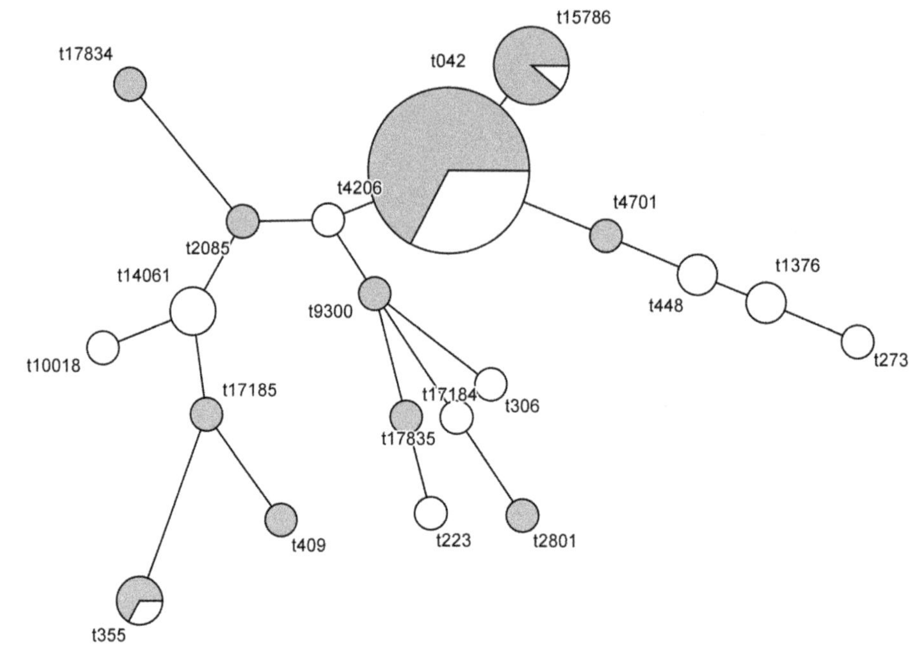

Fig. 1 Minimum spanning tree based on *spa* types of 79 *S. aureus* isolates from Bahir Dar region (grey) and Gondar region (white) in North-Western Ethiopia. The size of the circles is proportional to the number of isolates found

Table 2 Genetic characterization of 79 *Staphylococcus aureus* isolates from quarter milk samples of 60 cows in 42 dairy farms in North-Western Ethiopia

spa type	Total (%)	N isolates positive for			Repeat sequence										
		lukM-lukF'	*tst*	*pvl*											
t042	46 (58)	0	1	0	26	23	12					34	34	33	34
t15786	9 (11)	0	0	0	26	23	12				34	34	34	33	34
t355	3 (4)	0	0	3	07	56	12	17	16	16	33	31	57	12	
t14061	3 (4)	2	3	0	04	31	17								
t1376	2 (3)	0	0	2	07	12	21	17	13	13	34	34	33	13	
t488	2 (3)	0	0	0	07	12	21	17	13	13			34	33	34
t10018	1 (1)	1	1	0	04	31	17	25	17	25	17				
t306	1 (1)	0	1	0	26	23	17	34	17	20	17	12	17	17	16
t17835	1 (1)	0	1	1	14	22	75	16	24	24	24	24	24		
t223	1 (1)	0	1	0	26	23	13	23	05	17	25	17	25	16	28
t273	1 (1)	0	0	1	07	23	21	17	13		34	16	34	33	13
t2085	1 (1)	0	0	0	26									33	34
t4206	1 (1)	0	0	0	26	23						34	34	33	34
t9300	1 (1)	0	0	0	26	23	21								34
t4701	1 (1)	0	0	0	26	12	21	17			34	34	34	33	34
t17184	1 (1)	0	0	0	26	23	13			17	25	17			28
t2801	1 (1)	0	0	0	07	23	34	12	12	23	02	12	23		
t17834	1 (1)	0	0	0	07	23	02	12	23	02	02	02			34
t17185	1 (1)	0	0	0	07	500	22	31	17						
t409	1 (1)	0	0	0	60	61	34	22	34	17					

Table 3 Minimum inhibitory concentrations (MIC) for 79 *S. aureus* isolates from dairy farms in North-Western Ethiopia tested for susceptibility to 17 antimicrobials[1]. The dilution ranges tested are those contained in the white area. Values above this range indicate MICs higher than the highest concentration within the tested range. Vertical lines indicate clinical breakpoints, with values to the left of the line being susceptible or intermediate and those to the right being resistant

Antimicrobial	Number of tested dilutions	Highest tested concentration	MIC (µg/mL)										Resistant n (%)
			0.0625	0.25	0.5	1	2	4	8	16	32	64	
PEN	9	16	11			2	10	27	18	8	3		68 (86)
AMP	8	32		11		2	30	22	7	4	1	2	68 (86)
TET	6	16			36				4	34	5		43 (54)
CLI	4	4			76			3					3 (4)
ERY	6	8		70	6					3			3 (4)
T/S	4	4/76			76	1	1		1				1 (1)
RAM	4	4	78			1							1 (1)
AMC	7	32/16		11	31	35		2					0
CTN	4	16					79						0
CET	4	8					79						0
COX	2	8						79					0
CMP	4	32						42	37				0
FUS	2	2				79							0
ENR	5	4		79									0
GEN	4	16					78	1					0
KAN	3	64								79			0
NEO	3	32							79				0

[1] *PEN* penicillin, *AMP* ampicillin, *TET* tetracycline, *CLI* clindamycin, *ERY* erythromycin, *T/S* Trimethoprim sulfamethoxazole, *RAM* rifampicin, *AMC* amoxicillin/clavulanic acid, *CTN* cephalothin (1st generation cephalosporin), *CET* ceftiofur (3rd generation cephalosporin), *COX* cefoxitin, *CMP* chloramphenicol, *FUS* fusidic acid, *ENR* enrofloxacin, *GEN* gentamicin, *KAN* kanamycin, *NEO* neomycin

the *spa* types (t042, t223, t306 and t355) identified in the current study were previously isolated by Tarekgne et al. [9] in cattle in the Tigray region, Northern Ethiopia. Interestingly, in that study the dominant *spa* type was t314, which has a repeat sequence entirely different from the dominant types in our study. This suggests that, although contagious transmission seems to occur at a local scale, exchange of strains between regions seems to occur less, resulting in dominance of different clones in different regions. The Tigray region has a different cross-breeding and heifer distribution center, possibly explaining different dominant clones between regions. The six t042 isolates subjected to MLST belonged to the new ST4550, which is a double locus variant of ST97, a ST commonly found in bovine mastitis worldwide [30–32].

The proportion of isolates positive for *lukM-lukF'* genes was low. As the majority of isolates in our study belonged to a ST related to ST97, this low *lukM-lukF'* prevalence is in line with previous reports that found ST97 isolates to be negative for *lukM-lukF'* [16, 17]. The isolates that were positive for *lukM-lukF'* belonged to t14061 and t10018, the latter of which has previously been found in nasal swabs from Nigerian abattoir workers [33]. The bi-component leukocidin LukM-LukF' is predominantly produced by strains isolated from bovine *S. aureus* mastitis [34] and specifically kills bovine neutrophils, whereas human neutrophils are not affected [18]. Based on an experimental study, Vrieling et al. [19] reported that LukM-LukF' is produced during the course of infection and that a high levels in milk were associated with increased severity of mastitis. The low proportion of *lukM-lukF'* positive isolates in our samples may be partly due to the fact that they came from a cross-sectional study that included only two cases of clinical mastitis [35]. In our collection the proportion of isolates positive for *pvl* was higher than that of *lukM-lukF',* despite the fact that bovine neutrophils are insensitive to this human-associated leukocidin [21]. Although there are reports of *pvl*-positive cattle *S. aureus* [16, 35], *pvl* is mainly detected in *S. aureus* of human origin [19, 36, 37] whereas cattle isolates are often negative for *pvl* [38–40]. The finding of *pvl* in isolates cultured from cows may suggest human to cow transmission of *S. aureus*. Schmidt et al. [41] reported transmission of a human ST8 isolate to a dairy cow which subsequently caused mastitis and speculated that the close contact between dairy cows and humans, especially in the milking parlor, enables such transmission. Under the Ethiopian circumstances, where cows are hand milked, such transmission seems to be even more likely. Further characterization of the *pvl*-positive *S. aureus* in this study and of *S. aureus* isolates cultured from farmers and their families may help quantifying the importance of human to cattle transmission of *S. aureus*. Likewise, similar studies on bovine isolates that may be found in humans may also show sharing of *S. aureus* isolates between cattle and their farmers, which likely occurs along the same transmission routes. The *tst* gene encodes the toxic shock syndrome toxin, which *S. aureus* uses to facilitate colonization [42]. Although the *tst* gene is documented to be more prevalent in bovine than in human *S. aureus* strains [43], we only found it in a low number of isolates.

In line with the report of Tigabu et al. [44] from the Central highlands of Ethiopia, we did not find *mecA* positive *S. aureus* in our sample. The majority of our isolates were, however, resistant to penicillin, ampicillin and tetracycline. Resistance to penicillin and ampicillin is common among mastitis causing *S. aureus*, but varies largely between geographical regions [45–47]. In Ethiopia, penicillin and ampicillin are commonly used antimicrobials to treat mastitis, and tetracycline is often used to treat other bacterial infections. The high level of resistance to these antimicrobials is in line with previous Ethiopian reports on isolates from bovine mastitis [6, 7, 48]. This consistent picture of widespread AMR in *S. aureus* is alarming and likely results from the high usage of antimicrobials, possibly due to the fact that farmers can obtain antimicrobials over the counter without a prescription. Additionally, farmers are often not instructed on a proper treatment regimen, because guidance by veterinarians is lacking. A final reason for the high level of AMR may be that Ethiopian farmers are not inclined to cull animals, but rather continue treating their cows over and over again for an extended period of time. Treatment of cows with mastitis during lactation is an important component of a standard mastitis control program for contagious pathogens. Because AMR to penicillin and ampicillin is common, amoxicillin-clavulanic acid may be used for treatment of IMI without the need for tetracyclines, macrolides, lincosamides or third-generation cephalosporins. However, antimicrobial treatment is less than desirably effective in eliminating existing *S. aureus* infections [49] and treatment of intramammary infections with penicillin-resistant *S. aureus* strains generally results in a lower cure rate for treatment with either β-lactam or non--β-lactam antibiotics [50]. If cows do not respond favorably to treatment, separation or culling of the infected animals can be used to limit further transmission, but in Ethiopia, dairy cows are not easily culled because cross breed cows are very valuable for reproduction, even if they are chronically infected and have repeated CM cases. Therefore, encouraging dairy farmers to prevent transmission of *S. aureus*, rather than to rely on antimicrobial therapy is a preferable approach for mastitis control in North-Western Ethiopia. For instance, milking hygiene can be improved [8] and post milking teat disinfection [51] is not very expensive, is easy to perform, and is known to be very effective and can be practiced by Ethiopian dairy farmers. Implementation of these and other control measures will likely result in a lower prevalence of mastitis.

Conclusions

We showed a high degree of similarity between the bovine *S. aureus* isolates in North-Western Ethiopia, suggesting contagious transmission within and between farms. Moreover, there was a high level of resistance to antimicrobials that are commonly used to treat mastitis in Ethiopia. Therefore, implementing effective preventive measures aiming to limit transmission, rather than to use antimicrobials to control mastitis, along with guidelines for prudent use of antimicrobials, is important to decrease AMR.

Abbreviations

AMC: Amoxicillin/clavulanic acid; AMP: Ampicillin; AMR: Antimicrobial resistance; CET: Ceftiofur; CLI: Clindamycin; CLSI: Clinical and laboratory standards institute; CM: Clinical mastitis; CMP: Chloramphenicol; CMT: California mastitis test; COX: Cefoxitin; CTN: Cephalothin; ENR: Enrofloxacin; ERY: Erythromycin; FUS: Fusidic acid; GEN: Gentamicin; IMI: Intramammary infections; KAN: Kanamycin; MIC: Minimum inhibitory concentration; MST: Minimum spanning tree; NEO: Neomycin; PEN: Penicillin; RAM: Rifampicin; ST: Sequence type; T/S: Trimethoprim sulfamethoxazole; TET: Tetracycline

Acknowledgements

We acknowledge the dairy farmers who participated in the study. We are thankful to the support of Merijn van den Hout (University of Applied Sciences, Den Bosch) and Lindert Benedictus (Division of Infection and Immunity, The Roslin Institute, The University of Edinburgh, Scotland, UK), for assistance in the lab work. Ato Teklu and Sisay, from Gondar artificial insemination service, are acknowledged for helping selecting participants.

Funding

This study was supported by the Netherlands organization for international cooperation in higher education (Nuffic) and the Gustav Rosenberger Memorial Fund.

Authors' contributions
SAM, TST, GK and TJGML conceived and designed the study. SAM, JH, EMB, AER and MPS performed laboratory work. GK, TJGML, EMB, JH, VPMGR and SAM interpreted the results. SAM and GK: wrote the manuscript. All authors reviewed, edited and approved the manuscript.

Consent for publication
Not applicable.

Competing interests
The authors declare that they have no competing interests.

Author details
[1]Department of Farm Animal Health, Faculty of Veterinary Medicine, Utrecht University, Yalelaan 7, 3584, CL, Utrecht, The Netherlands. [2]Faculty of Veterinary Medicine, University of Gondar, P.O. Box 196, Gondar, Ethiopia. [3]GD Animal Health, P.O. Box 9, 7400 AA Deventer, The Netherlands. [4]Department of Infectious Diseases and Immunology, Faculty of Veterinary Medicine, Utrecht University, Yalelaan 1, 3584 CL Utrecht, The Netherlands. [5]Department of Veterinary Tropical Diseases, Faculty of Veterinary Science, University of Pretoria, Private Bag X04, Onderstepoort, Pretoria 0110, South Africa. [6]Institute of Biotechnology, Addis Ababa University, P.O. Box 1176, Addis Ababa, Ethiopia.

References
1. Cremonesi P, Pozzi F, Raschetti M, Bignoli G, Capra E, Graber HU, Vezzoli F, Piccinini R, Bertasi B, Biffani S, Castiglioni B, Luini M. Genomic characteristics of *Staphylococcus aureus* strains associated with high within-herd prevalence of intramammary infections in dairy cows. J Dairy Sci. 2015;98:6828–38.
2. Boss R, Cosandey A, Luini M, Artursson K, Bardiau M, Breitenwieser F, Hehenberger E, Lam T, Mansfeld M, Michel A, Mösslacher G, Naskova J, Nelson S, Podpečan O, Raemy A, Ryan E, Salat O, Zangerl P, Steiner A, Graber HU. Bovine *Staphylococcus aureus*: subtyping, evolution, and zoonotic transfer. J Dairy Sci. 2016;99:515–28.
3. Nübel U, Strommenger B, Layer F, Witte W. From types to trees: reconstructing the spatial spread of *Staphylococcus aureus* based on DNA variation. Int J Medical Microbiol. 2011;301:614–8.
4. Zadoks RN, Middleton JR, McDougall S, Katholm J, Schukken YH. Molecular epidemiology of mastitis pathogens of dairy cattle and comparative relevance to humans. J Mammary Gland Biol Neoplasia. 2011;16:357–72.
5. Sangvik M, Olsen RS, Olsen K, Simonsen GS, Furberg A, Ericson Sollid JU. Age- and gender-associated *Staphylococcus aureus spa* types found among nasal carriers in a general population: the Tromsø staph and skin study. J Cli Microbiol. 2011;49:4213–8.
6. Getahun K, Kelay B, Bekana M, Lobago F. Bovine mastitis and antibiotic resistance patterns in Selalle smallholder dairy farms, Central Ethiopia. Trop Anim Health Prod. 2008;40:261–8.
7. Haftu R, Taddele H, Gugsa G, Kalayou S. Prevalence, bacterial causes, and antimicrobial susceptibility profile of mastitis isolates from cows in large-scale dairy farms of northern Ethiopia. Trop Anim Health Prod. 2012;44:1765–71.
8. Mekonnen SA, Koop G, Melkie ST, Getahun CD, Hogeveen H, Lam TJGM. Prevalence of subclinical mastitis and associated risk factors at cow and herd level in dairy farms in north-West Ethiopia. Prev Vet Med. 2017;145:23–31.
9. Tarekgne EK, Skjerdal T, Skeie S, Rudi K, Porcellato D, Félix B, Narvhus JA. Enterotoxin gene profile and molecular characterization of *Staphylococcus aureus* isolates from bovine bulk milk and milk products of Tigray region. Northern Ethiopia J Food Protection. 2016;79:1387–95.
10. Joo YS, Fox LK, Davis WC, Bohach GA, Park YH. *Staphylococcus aureus* associated with mammary glands of cows: genotyping to distinguish different strains among herds. Vet Microbiol. 2001;80:131–8.
11. Middleton J, Fox L, Gay J, Tyler J, Besser T. Use of pulsed-field gel electrophoresis for detecting differences in *Staphylococcus aureus* strain populations between dairy herds with different cattle importation practices. Epidemiol Infect. 2002;129:387–95.
12. Keefe G. Update on control of *Staphylococcus aureus* and *Streptococcus agalactiae* for management of mastitis. Vet. Clin. North am. Food Anim. Pract. 2012;28:203–16.
13. Sommerhäuser J, Kloppert B, Wolter W, Zschöck M, Sobiraj A, Failing K. The epidemiology of *Staphylococcus aureus* infections from subclinical mastitis in dairy cows during a control programme. Vet Microbiol. 2003;96:91–102.
14. Gurjar A, Gioia G, Schukken Y, Welcome F, Zadoks R, Moroni P. Molecular diagnostics applied to mastitis problems on dairy farms. Vet Clin Food Anim. 2012;28:565–76.
15. Lam TJGM, Lipman LJA, Schukken YH, Gaastra W, Brand A. Epidemiological characteristics of bovine clinical mastitis caused by *Staphylococcus aureus* and *Escherichia coli* studied by DNA-fingerprinting. Am J Vet Res. 1997;57:39–42.
16. Hata E, Katsuda K, Kobayashi H, Uchida I, Tanaka K, Eguchi M. Genetic variation among *Staphylococcus aureus* strains from bovine milk and their relevance to methicillin-resistant isolates from humans. J. Cli. Microbiol. 2010;48:2130–9.
17. Schlotter K, Ehricht R, Hotzel H, Monecke S, Pfeffer M, Donat K. Leukocidin genes lukF-P83 and lukM are associated with *Staphylococcus aureus* clonal complexes 151, 479 and 133 isolated from bovine udder infections in Thuringia. Germany Vet Res. 2012;43:42.
18. Vrieling M, Koymans KJ, Heesterbeek DAC, Aerts PC, Rutten VPMG, de Haas CJC, van Kessel KPM, Koets AP, Nijland R, van Strijp JAG. Bovine *Staphylococcus aureus* secretes the leukocidin LukMF' to kill migrating neutrophils through CCR1. mBio. 2015;6:3.
19. Vrieling M, Boerhout EM, van Wigcheren GF, Koymans KJ, Mols-Vorstermans TG, de Haas CJC, Aerts PC, Daemen IJJM, van Kessel KPM, Koets AP, Rutten VPMG, Nuijten PJM, van Strijp JAG, Benedictus L. LukMF' is the major secreted leukocidin of bovine *Staphylococcus aureus* and is produced *in vivo* during bovine mastitis. Sci Rep. 2016;6:37759.
20. Hoekstra J, Rutten V, Sommeling L, Werven T, Spaninks M, Duim B, Benedictus L, Koop G. High production of LukMF' in *Staphylococcus* aureus field strains is associated with clinical bovine mastitis. Toxins. 2018;10:200.
21. Koop G, Vrieling M, DML S, LSC L, Monie T, van Wigcheren G, Raisen C, Ba X, Gleadall N, Hadjirin N, Timmerman AJ, Wagenaar JA, Klunder HM, Ross Fitzgerald J, Zadoks R, Paterson GK, Torres C, Waller AS, Loeffler A, Loncaric I, Hoet AE, Bergström K, de Martino L, Pomba C, de Lencastre H, Ben Slama K, Gharsa H, Richardson EJ, Chilvers ER, de Haas C, van Kessel K, van Strijp JAG, Harrison EM, Holmes MA. Identification of LukPQ, a novel, equid-adapted leukocidin of *Staphylococcus aureus*. Sci. Rep. 2017;7:40660.
22. Lemma MT, Zenebe Y, Tulu B, Mekonnen D, Mekonnen Z. Methicillin resistant *Staphylococcus aureus* among HIV infected pediatric patients in Northwest Ethiopia: carriage rates and antibiotic co-resistance profiles. PLoS One. 2015;10:9.
23. Eshetie S, Tarekegn F, Moges F, Amsalu A, Birhan W, Huruy K. Methicillin resistant *Staphylococcus aureus* in Ethiopia: a meta-analysis. Infect Diseases. 2016;16:689.
24. Dilnessa T, Bitew A. Prevalence and antimicrobial susceptibility pattern of methicillin resistant *Staphylococcus aureus* isolated from clinical samples at Yekatit 12 hospital medical college, Addis Ababa. Ethiopia BMC Infect Diseases. 2016;16:398.
25. Moges F, Endris M, Mulu A, Tessema B, Belyhun Y, Shiferaw Y, Huruy K, Unakal C, Kassu A. The growing challenges of antibacterial drug resistance in Ethiopia. J Global Antimicrob Resistance. 2014;2:148–54.
26. Seyoum ET, Mekonene TK, Woldetsadik DA, Zewudie BM, Gebreyes WA. Enterotoxin gene profile of *Staphylococcus aureus* isolates recovered from bovine milk produced in Central Ethiopia. J Infect Dev Ctries. 2016;10:138–42.
27. Enright MC, Day NPJ, Davies CE, Peacock SJ, Spratt BG. Multilocus sequence typing for characterization of methicillin-resistant and methicillin-susceptible clones of *Staphylococcus aureus*. J. Cli. Microbiol. 2000;38:1008–15.
28. CLSI, editor. Performance standards for antimicrobial susceptibility testing. 27th ed. CLSI supplement M100. Clinical and Laboratory Standards Institute: Wayne, PA; 2015.
29. CLSI, editor. Performance standards for antimicrobial disk and dilution

susceptibility tests for bacteria isolated from animals. 3rd ed. CLSI supplement VET01S. Clinical and Laboratory Standards Institute: Wayne, PA; 2017.
30. Smith EM, Green LE, Medley GF, Bird HE, Fox LK, Schukken YH, Kruze JV, Bradley AJ, Zadoks RN, Dowson CG. Multilocus sequence typing of intercontinental bovine *Staphylococcus aureus* isolates. J. Cli. Microbiol. 2005;43:4737–43.
31. Delgado S, García P, Fernández L, Jiméenez E, Mercedes Rodríguez-Baños M, del Campo R, Rodríguez JM. Characterization of *Staphylococcus aureus* strains involved in human and bovine mastitis. FEMS Immunol Med Microbiol. 2011;62:225–35.
32. Rabello RF, Moreira BM, Lopes RMM, Teixeira LM, Riley LW, Castro ACD. Multilocus sequence typing of *Staphylococcus aureus* isolates recovered from cows with mastitis in Brazilian dairy herds. J Medical Microbiol. 2007;56:1505–11.
33. Odetokun IA, Ballhausen B, Adetunji VO, Ghali-Mohammed I, Adelowo MT, Adetunji SA, Fetsch A. *Staphylococcus aureus* in two municipal abattoirs in Nigeria: risk perception, spread and public health implications. Vet Microbiol. 2018;216:52–9.
34. Barrio MB, Rainard P, Prévost G. LukM/LukF0-PV is the most active *Staphylococcus aureus* leukotoxin on bovine neutrophils. Microbes and Infect. 2006;8:2068–74.
35. Haveri M, Roslöf A, Pyörälä S. Virulence genes of bovine *Staphylococcus aureus* from persistent and nonpersistent intramammary infections with different clinical characteristics. J Appl Microbiol. 2007;103:993–1000.
36. Zecconi A, Cesaris L, Liandris E, Daprà V, Piccinini R. Role of several *Staphylococcus aureus* virulence factors on the inflammatory response in bovine mammary gland. Microb Pathog. 2006;40:177–83.
37. Dinges MM, Orwin PM, Schlievert PM. Exotoxins of *Staphylococcus aureus*. Clinical Microbiol Reviews. 2000;13:16–34.
38. Rainard P, Corrales J, Barrio MB, Cochard T, Poutrel B. Leucotoxic activities of *Staphylococcus aureus* strains isolated from cows, ewes, and goats with mastitis: importance of LukM/LukF'-PV leukotoxin. Cli and Diagnostic Lab Immun. 2003;10:272–7.
39. Yamada T, Tochimaru N, Nakasuji S, Hata E, Kobayashi H, Eguchi M, Kaneko J, Kamio Y, Kaidoh T, Takeuchi S. Leukotoxin family genes in *Staphylococcus aureus* isolated from domestic animals and prevalence of lukM-lukF-PV genes by bacteriophages in bovine isolates. Vet Microbiol. 2005;110:97–103.
40. Monecke S, Kuhnert P, Hotzel H, Slickers P, Ehricht R. Microarray based study on virulence-associated genes and resistance determinants of *Staphylococcus aureus* isolates from cattle. Vet Microbiol. 2007;125:128–40.
41. Schmidt T, Kock MM, Ehlers MM. Molecular characterization of *Staphylococcus aureus* isolated from bovine mastitis and close human contacts in south African dairy herds: genetic diversity and inter-species host transmission. Front Microbiol. 2017;8:511.
42. Omoe K, Hu DL, Takahashi-Omoe H, Nakane A, Shinagawa K. Identification and characterization of a new staphylococcal enterotoxin-related putative toxin encoded by two kinds of plasmids. Infect Immun. 2003;71:6088–94.
43. van Leeuwen W, Melles DC, Alaidan A, Al-Ahdal M, Boelens HAM, Snijders SV, Wertheim H, van Duijkeren E, Peeters JK, van der Spek PJ, Gorkink R, Simons G, Verbrugh HA, van Belkum A. Host- and tissue-specific pathogenic traits of *Staphylococcus aureus*. J Bacteriol. 2005;187:4584–91.
44. Tigabu E, Kassa T, Asrat D, Alemayehu H, Sinmegn T, Adkins PRF, Gebreyes W. Phenotypic and genotypic characterization of *Staphylococcus aureus* isolates recovered from bovine milk in central highlands of Ethiopia. African J Microbiol Res. 2015;9:2209–17.
45. Hendriksen RS, Mevius DJ, Schroeter A, Teale C, Meunier D, Butaye P, Franco A, Utinane A, Amado A, Moreno M, Greko C, Stärk K, Berghold C, Myllyniemi AL, Wasyl D, Sunde M, Aarestrup FM. Prevalence of antimicrobial resistance among bacterial pathogens isolated from cattle in different European countries: 2002-2004. Acta Vet Scand. 2008;8:50–28.
46. Jagielski T, Puacz E, Lisowski A, Siedlecki P, Dudziak W, Międzobrodzki J, Krukowski H. Short communication: antimicrobial susceptibility profiling and genotyping of *Staphylococcus aureus* isolates from bovine mastitis in Poland. J Dairy Sci. 2014;97:6122–8.
47. Wang D, Zhang L, Zhou X, He Y, Yong C, Shen M, Szenci O, Han B. Antimicrobial susceptibility, virulence genes, and randomly amplified polymorphic DNA analysis of *Staphylococcus aureus* recovered from bovine mastitis in Ningxia. China J Dairy Sci. 2016;99:9560–9.
48. Mekonnen H, Workineh S, Bayleyegn M, Moges A, Tadele K. Antimicrobial susceptibility profiles of mastitis isolates from cows in three major Ethiopian dairies. Revue Méd Vét. 2005;156:391–4.
49. Philpot WN. Role of therapy in mastitis control. J Dairy Sci. 1969;52:708–13.
50. Barkema HW, Schukken YH, Zadoks RN. Invited review: the role of cow, pathogen, and treatment regimen in the therapeutic success of bovine *Staphylococcus aureus* mastitis. J Dairy Sci. 2006;89:1877–95.
51. Lam TJ, Van Vliet JH, Schukken YH. Udder disinfection and mastitis in cattle: a literature review. Tijdschr Diergeneeskd. 1995;120:392–9.
52. Francois P, Pittet D, Bento M, Pepey B, Vaudaux P, Lew D, Schrenzel J. Rapid detection of methicillin-resistant *Staphylococcus aureus* directly from sterile or nonsterile clinical samples by a new molecular assay. J. Cli. Microbiol. 2003;41:254–60.
53. Melo DA, Coelho IS, Motta CC, Rojas ACCM, Dubenczuk FC, Coelho SMO, Souza MMS. Impairments of *mecA* gene detection in bovine *Staphylococcus* spp. Brazilian J Microbiol. 2014;45:1075–82.
54. Harmsen D, Claus H, Witte W, Rothgänger J, Claus H, Turnwald D, Voge U. Typing of methicillin-resistant *Staphylococcus aureus* in a university hospital setting by using novel software for *spa* repeat determination and database management. J. Cli. Microbiol. 2003;41:5442–8.
55. Pajić MJ, Rašić ZB, Velebit BM, Boboš SF, Mihajlović-Ukropina MM, Radinović MZ, Galfi AL, Petković JM, Trojačanec SI. The prevalence of methicillin resistance and Panton-valentine leukocidin synthesis genes in *Staphylococcus aureus* isolates of bovine and human origin. Vet Arhiv. 2014;84:205–14.
56. Cosandey A, Boss R, Luini M, Artursson K, Bardiau M, Breitenwieser F, Hehenberger E, Lam T, Mansfeld M, Michel A, Mösslacher G, Naskova J, Nelson S, Podpečan O, Raemy A, Ryan E, Salat O, Zangerl P, Steiner A, Graber HU. *Staphylococcus aureus* genotype B and other genotypes isolated from cow milk in european countries. J Dairy Sci. 2016;99:529–40.

13

Estimation of tulathromycin depletion in plasma and milk after subcutaneous injection in lactating goats using a nonlinear mixed-effects pharmacokinetic modeling approach

Zhoumeng Lin[1], Matthew Cuneo[2], Joan D. Rowe[2], Mengjie Li[1,5], Lisa A. Tell[3], Shayna Allison[4], Jan Carlson[4], Jim E. Riviere[1] and Ronette Gehring[1*]

Abstract

Background: Extra-label use of tulathromycin in lactating goats is common and may cause violative residues in milk. The objective of this study was to develop a nonlinear mixed-effects pharmacokinetic (NLME-PK) model to estimate tulathromycin depletion in plasma and milk of lactating goats. Eight lactating goats received two subcutaneous injections of 2.5 mg/kg tulathromycin 7 days apart; blood and milk samples were analyzed for concentrations of tulathromycin and the common fragment of tulathromycin (i.e., the marker residue CP-60,300), respectively, using liquid chromatography mass spectrometry. Based on these new data and related literature data, a NLME-PK compartmental model with first-order absorption and elimination was used to model plasma concentrations and cumulative excreted amount in milk. Monte Carlo simulations with 100 replicates were performed to predict the time when the upper limit of the 95% confidence interval of milk concentrations was below the tolerance.

Results: All animals were healthy throughout the study with normal appetite and milk production levels, and with mild-moderate injection-site reactions that diminished by the end of the study. The measured data showed that milk concentrations of the marker residue of tulathromycin were below the limit of detection (LOD = 1.8 ng/ml) 39 days after the second injection. A 2-compartment model with milk as an excretory compartment best described tulathromycin plasma and CP-60,300 milk pharmacokinetic data. The model-predicted data correlated with the measured data very well. The NLME-PK model estimated that tulathromycin plasma concentrations were below LOD (1.2 ng/ml) 43 days after a single injection, and 62 days after the second injection with a 95% confidence. These estimated times are much longer than the current meat withdrawal time recommendation of 18 days for tulathromycin in non-lactating cattle.

(Continued on next page)

* Correspondence: rgehring@ksu.edu; http://www.k-state.edu/iccm/
[1]Institute of Computational Comparative Medicine (ICCM), Department of Anatomy and Physiology, College of Veterinary Medicine, Kansas State University, 1800 Denison Avenue, P200 Mosier Hall, Manhattan, KS 66506-5802, USA
Full list of author information is available at the end of the article

(Continued from previous page)

Conclusions: The results suggest that twice subcutaneous injections of 2.5 mg/kg tulathromycin are a clinically safe extra-label alternative approach for treating pulmonary infections in lactating goats, but a prolonged withdrawal time of at least 39 days after the second injection should be considered to prevent violative residues in milk and any dairy goat being used for meat should have an extended meat withdrawal time.

Keywords: Tulathromycin, Withdrawal time, Goat, Food safety, Nonlinear mixed-effects (NLME) pharmacokinetic modeling

Background

Drug residues in edible tissues of food-producing animals are a global food safety concern [1, 2]. Tulathromycin is a widely used long-acting triamilide antibiotic labeled for treating respiratory diseases in cattle and swine, and also commonly used in goats in an extra-label manner because it is highly effective against caprine respiratory pathogens [3] and is clinically safe when administered to neonatal dairy and meat goats at 1X, 3X, and 5X the cattle label dose [4]. In the US and Europe, there is no established milk tolerance (equivalent to maximum residue level in Europe) for tulathromycin in goats, thus any detectable residues are considered violative. Therefore, it is important to determine the appropriate withdrawal time needed for tulathromycin concentrations in edible tissues/milk to fall below legally established safe levels (in this case the analytical detection limit) to ensure food safety.

Physiologically based pharmacokinetic (PBPK) and nonlinear mixed-effects pharmacokinetic (NLME-PK) models are helpful tools in the establishment of withdrawal time guidelines for veterinary drugs [5–7]. Currently, there are pharmacokinetic data for tulathromycin in the plasma and tissues of juvenile and market-age goats after single or multiple intravenous (IV), intramuscular (IM), or subcutaneous (SC) injections [3, 8, 9]. A PBPK model has been developed to help estimate meat withdrawal times for tulathromycin in juvenile and market-age goats [10]. For lactating goats, however, there are only pharmacokinetic data in the plasma and milk after single IV, IM, or SC injection [11, 12], but not after repeated administrations. PBPK or NLME-PK models for tulathromycin in lactating goats are not available, either.

Repeated administrations are a common and important dosing strategy in order to maintain therapeutically effective drug concentrations in the plasma and tissues for longer periods of time. A recent study suggested that in order to maintain lung concentrations of tulathromycin higher than minimal inhibitory concentrations (MIC) for respiratory pathogens (e.g., MIC = 2 μg/mL for *M haemolytica*), the SC injection interval in goats should be every 7 days [9]. A major concern of repeated administrations of veterinary drugs in food-producing animals is the potential accumulation of drugs in the body, resulting in prolonged withdrawal times. The objective of this study was to determine the pharmacokinetic characteristics of tulathromycin in lactating goats after two administrations, and to develop a NLME-PK model to estimate tulathromycin depletion in plasma and milk samples after single and two SC injections in lactating goats.

Methods

Pharmacokinetic study design

Eight healthy adult (1–7 years old) lactating goats with a mean body weight of 88 kg (range: 73–109 kg) and a mean milk production of 10.8 lbs/day (range: 7–14 lbs/day) in their first to sixth lactation were used in the present study. Breeds included Saanen ($n = 4$), LaMancha ($n = 3$), and Toggenburg ($n = 1$) goats. Does were housed individually in the Goat Teaching and Research Facility at University of California, Davis, CA. All does met the inclusion criteria, including passing a normal physical examination, lactating for at least 45 days, and no antibiotic treatment in the last 2 months. Prior to tulathromycin injection, composite milk samples representing one entire milking period (instead of foremilk or strippings) were collected using a Dairy Herd Improvement Association-approved milk metering device. Goats were then injected with 2.5 mg/kg tulathromycin (Draxxin®; Zoetis Inc., New York, NY) subcutaneously in the thoracic region caudal to elbow (first injection at left elbow and second injection at right elbow) twice on Day 0 and Day 7 (7-day interval), immediately after the evening milking. Doses were calculated based on body weights collected the day before injection. Health status and injection-site reactions were monitored daily for 64 days after the first injection.

Individual doe milk samples were collected twice daily during regular milking times (morning milking at 6:30 a.m. and evening milking at 5:30 p.m.) for days 1–21 after the first injection, then once every 48 h at the morning milking for days 22–36 after the first injection, and every 72 h at the morning milking for days 37–64 after the first injection. Does were machine-milked; milk was collected using a commercial milk meter/sampler (Waikato Milking Systems, Hamilton, New Zealand) and the milk weights were recorded. Milk samples (~10–13 mL/sample) were transferred into 15 mL conical tubes, frozen at −20 °C within 30 min of collection, and then stored at −80 °C until analysis. Blood was collected twice daily, following each milking, during the first 7 days after the second injection via jugular venipuncture into heparinized blood tubes. Tubes were placed on ice, centrifuged at 1200 g for 5 min at 4 °C, and plasma samples were manually harvested and transferred to storage tubes, which were then immediately frozen at −20 °C and then stored at a −80 °C freezer until analysis. All animal procedures were approved in advance by the Animal Care and Use Committee of the University of California, Davis, CA.

Tulathromycin analysis

Plasma tulathromycin concentrations were determined using the liquid chromatography mass spectrometry (LC-MS) method described in Galer et al. [13], while milk samples were analyzed for the concentrations of CP-60,300, a common fragment of tulathromycin and its metabolites, with an LC-MS method described by Boner et al. [14] because CP-60,300 is the marker residue of tulathromycin and all current US FDA guidelines on tulathromycin tolerances in the milk or target tissues are based on CP-60,300. Briefly, the LC was an Acquity® UPLC from Waters (Milford, MA) and the MS was a Thermo TSQ Quantum Discovery Max (Thermo Fisher Scientific, Waltham, MA) with a heated electrospray ionization source. Sample injections were prepared on an Ace C8, 2.1 × 50 mm, 3 μm column (MacMod, Chadds Ford, PA). Positive ions were monitored in the selected reaction monitoring (SRM) mode for CP-60,300 and an internal standard (a structurally similar compound provided by Pfizer [New York City, NY]) with transitions from 577.2 to 420.2 and from 591.3 to 434.2, respectively. For tulathromycin, the positive ions were monitored in the SRM mode with the doubly charged precursor to product ion pair of 404.1 to 578.0. The limit of detection (LOD) was 1.8 and 1.2 ng/mL for milk and plasma, respectively, and the limit of quantification (LOQ) was 5.0 ng/mL for milk and 4.0 ng/mL for plasma. The average inter-assay variation was measured by relative standard deviation (%RSD) and was 6.6% for milk and 11.3% for plasma. The milk quality control concentrations were 6, 60, and 600 ng/ml, and the plasma quality control concentrations were 15, 225, and 450 ng/ml. The recovery was 100% for milk and 96.7% for plasma.

Pharmacokinetic analysis

Milk and plasma tulathromycin concentration-time data were analyzed for each individual animal using a non-compartmental approach (Phoenix WinNonlin®, version 6.4, Certara Inc., Cary, NC). Furthermore, tulathromycin plasma concentration and cumulative drug amount excreted in milk data were analyzed using the NLME-PK modeling approach in Phoenix NLME (version 1.3, Certara Inc., Cary, NC). The extended least-squares, first-order conditional estimation method (FOCE-ELS) with interaction was employed to fit the data to the model and calculate the pharmacokinetic parameters. Typical population pharmacokinetic parameter values, interindividual variability (IIV), residue error, and 95% confidence interval were calculated [15, 16]. Between-individual and between-occasion variability was assessed with an exponential model. Residual variability was tested using the additive, proportional, and combined error models. Compartment model selection was guided by goodness-of-fit plots (e.g., observed vs. predicted plasma concentrations and milk cumulative amounts, weighted residuals vs. predicted concentrations/amounts, and weighted residuals vs. time), the −2 log-likelihood (−2LL), Akaike information criterion (AIC), as well as the Bayesian information criterion (BIC). The model was selected on the basis of smaller values of AIC, better precision of estimates, and superior goodness-of-fit plots.

In order to determine whether the dosing regimen (single vs. twice injections) has an effect on tulathromycin depletion, plasma concentration and milk amount data of tulathromycin after a single SC injection in lactating goats ($n = 8$) from Grismer et al. [12] were included in the NLME-PK analysis. Briefly, eight adult (2–5 years) lactating dairy goats (body weight range: 72.5-89.0 kg) in their first to fourth lactation that met the inclusion criteria as described above were used in this study. Goats were injected with a single dose of 2.5 mg/kg tulathromycin (Draxxin®; Zoetis Inc., New York, NY) subcutaneously in the left thoracic region caudal to the elbow. Individual milk samples were collected starting at 12 h after injection twice daily during regular milking times (6 a.m. and 5 p.m.) from days 1–14, then once every 48 h (a.m. milking time) from days 14–28, and then every 72 h (a.m. milking time) from days 28–45 post injection. Blood samples were collected twice daily after each milking during the first 7 days after injection via jugular venipuncture into heparinized tubes. In addition, due to lack of absorption phase (<12 h) concentration data in the present study and the study by Grismer et al. [12], raw data from another study by Clothier et al. [3] where tulathromycin plasma concentrations were measured as early as 1 min after a single SC injection (2.5 mg/kg) in market age goats ($n = 10$), was also incorporated into the NLME-PK analysis. In brief, five daily mixed breed and five Boer male (intact and castrated; 6-month-old) goats received one SC injection of 2.5 mg/kg tulathromycin (Draxxin®; Pfizer Animal Health, New York, NY). Blood samples were collected from the jugular vein at 1, 2, 4, 6, 8, and 10 min, as well as at 0.25, 0.5, 1, 2, 4, 8, 12, 24, 48, 72, 96, 120, 144, 168, 192, 216, 240, 264, 288, 312, 336, and 360 h after injection. Plasma tulathromycin and milk CP-60,300 concentrations were analyzed using the same methods as described above.

Simultaneous modeling of plasma and milk data

The structural model for combined plasma and milk data was built sequentially. Firstly, a 1-, 2-, or 3-compartments model with first-order absorption and first-order elimination was fit to the plasma concentration data, and the best structural model was selected. Next, a milk compartment was added, and simultaneous modeling of plasma and milk data was then performed. The final pharmacokinetic structural model developed for tulathromycin plasma concentrations and milk cumulative amount excreted is shown in Fig. 1. Differential equations describing this pharmacokinetic model are shown below:

$$\frac{dA_1}{dt} = A_a \times K_a - (Cl \times C) - [Cl2 \times (C - C2)] - (CL_{milk} \times C) \tag{1}$$

$$\frac{dA_a}{dt} = -A_a \times K_a \tag{2}$$

$$\frac{dA_2}{dt} = Cl2 \times (C - C2) \tag{3}$$

$$C = A_1/V \tag{4}$$

$$C_2 = A_2/V_2 \tag{5}$$

$$\frac{dA_{milk}}{dt} = CL_{milk} \times C \tag{6}$$

where A_1 is the total amount in the central compartment, A_a is the amount in the absorption compartment via SC injection, A_2 is the total amount in peripheral compartment, A_{milk} is the cumulative amount excreted in the milk, C is the concentration in the central compartment, C_2 is the concentration in the peripheral compartment. Definitions of pharmacokinetic parameters are provided in Table 3.

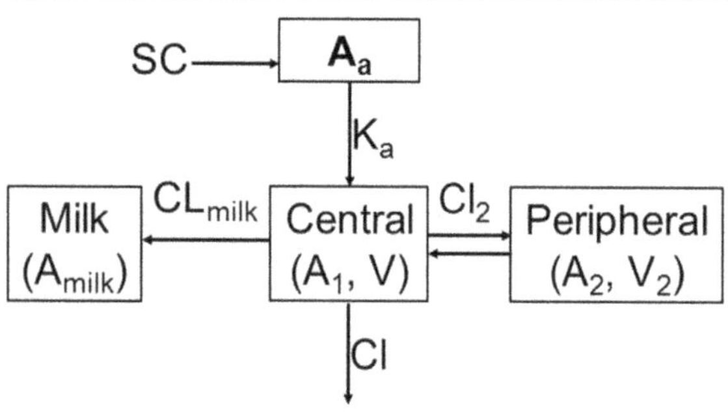

Fig. 1 Schematic representation of the final pharmacokinetic model for plasma tulathromycin concentrations and milk cumulative excreted amounts of tulathromycin marker residue CP-60,300 in lactating goats. A_1 is the total amount in the central compartment (ng); A_a is the amount in the absorption compartment via SC injection (ng); A_2 is the total amount in peripheral compartment (ng); A_{milk} is the cumulative excreted amount in the milk (ng); Cl is central clearance (mL/h); Cl_2 is clearance between the central and the peripheral compartment (mL/h); Cl_{milk} is clearance from the central compartment to the milk (mL/h); K_a is absorption rate constant after SC injection (1/h); V is volume of distribution for the central compartment (mL); V_2 is volume of distribution for the peripheral compartment (mL)

Model evaluation

The performance and stability of the final pharmacokinetic model was evaluated with both graphical method by visually comparing predicted vs. observed data, as well as the bootstrap analysis. The bootstrap samples were collected through random sampling by replacing the original dataset to generate another dataset with the same sample size as the original but with a different combination of subjects. The bootstrap resampling analysis was repeated 100 times and conducted in Phoenix NLME. If the majority of the parameter estimates fell into the 95% confidence intervals (CIs) of the bootstrap values, the model was considered unbiased. However, if any parameter estimate was out of the 95% CI, possible reasons were explored and discussed.

Model simulations

The final model was used in Phoenix NLME to conduct Monte Carlo simulations under dosing regimens of 2.5 mg/kg once or twice with 7-day interval via SC injection(s). For each dosing schedule, the Monte Carlo simulation generated time-concentration (or amount) profiles of tulathromycin in plasma and milk for 100 replicates. The simulated data were used to estimate the times when the upper limits of the 95% confidence intervals of simulated plasma and milk concentrations were below the tolerances. Simulated milk concentrations were calculated by dividing the simulated excreted amounts in milk by the measured milk volumes produced over the period of time between samples. There are no official tolerances for tulathromycin in lactating goats in the US or Europe, so the limits of detection (LODs) in plasma (1.2 ng/ml) and milk (1.8 ng/ml) were used as the most conservative endpoint.

Results

General health and pharmacokinetic data

All goats were healthy throughout the study with normal appetite and milk production levels. The body weight did not change significantly before nor at the end of the study. Four animals exhibited drug administration reactions, including vocalization, extension/flexion of the front limb, and/or looking back at the elbow. Similar to what we observed in our previous single injection study [12], injection-site reactions ranged from mild swelling to multiple skin lesions. All swellings diminished from their maximum size by the end of the study.

The plasma and milk concentration-time data collected during the current study are shown in Fig. 2. The average plasma and milk non-compartmental pharmacokinetic

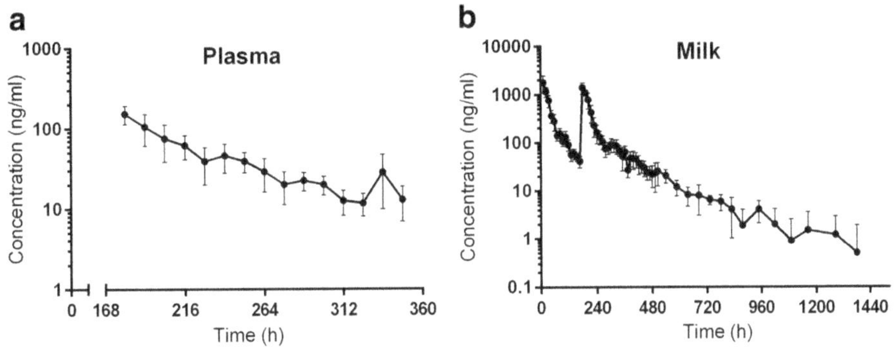

Fig. 2 Plots of tulathromycin concentration versus time in plasma (**a**) and CP-60,300 concentration versus time in milk (**b**)

parameter values are shown in Table 1, while Table 2 shows the average plasma and milk concentration data. The plasma pharmacokinetic profile of tulathromycin after the second injection reported in this study were somewhat different numerically from that observed in the previous study using a different population of lactating goats ($n = 8$) after a single injection [12]. However, the pharmacokinetic parameter values between the first and the second injections could not be compared statistically because the data were from two different studies using two different populations of animals. The milk concentration and pharmacokinetic parameters were analyzed after both the first and the second injections in the present study. The elimination rate constant (λz) was lower and the elimination half-life ($T_{1/2\lambda z}$) longer after the second compared to the first injection; other parameters were not statistically different (Table 1).

Structural NLME-PK model development

Results of model comparison and evaluation between the 3-compartment and 2-compartment models based on the data in lactating goats only are provided in Additional file 1: Figure S1. In addition, comparisons of model evaluation results of the initial 2-compartment model (without the absorption phase data) and the final 2-compartment model (with the absorption phase data from [3]) are shown in Additional file 2: Figure S2. Overall, a 2-compartment model with first-order absorption and first-order elimination best characterized the plasma concentration data. A total of 232 plasma concentrations of tulathromycin in lactating goats and 300 plasma concentrations in market meat goats were simultaneously modeled. The plasma pharmacokinetic model parameters are provided in Table 3. The distribution of residual variability was most adequately characterized with a proportional error model. A total of 411 milk cumulative excreted amounts were added to develop the milk pharmacokinetic model.

Model evaluation

Goodness-of-fit plots from the final pharmacokinetic model are presented in Fig. 3. The plots displayed an excellent correlation between the model-predicted and observed values. Figure 4 shows the comparisons of time-concentration (or amount) profiles vs. the individual predicted data in plasma and milk. The model adequately captured the pharmacokinetic profiles of tulathromycin and CP-60,300 in plasma and milk, respectively.

Table 1 Noncompartmental pharmacokinetic parameters for tulathromycin in plasma and the marker residue CP-60,300 in milk of lactating goats following two subcutaneous injections of 2.5 mg/kg tulathromycin at 7-day interval

Pharmacokinetic parameters (units)	Plasma (injection 2 only)		Milk (injection 1)		Milk (injection 2)		
	mean	SD	mean	SD	mean	SD	p value[#]
$C_{max\ (obs)}$ (ng/mL)	155.211	39.501	1806.280	660.824	1457.238	282.876	0.191
$T_{max\ (obs)}$ (h)	16.828	8.880	15.016	5.240	17.930	9.017	0.574
λz (1/h)	0.013	0.002	0.017	0.006	0.005	0.001	<0.001
$T_{1/2\lambda z}$ (h)	56.421	8.265	51.752	35.768	158.002	40.291	<0.001
AUC_{last} (h*ng/mL)	8050.601	1532.931	61805.331	6351.608	68156.845	12472.633	0.220
AUC_{inf} (h*ng/mL)	9120.588	1599.754	65147.987	7889.685	68978.306	12598.494	0.478

$C_{max\ (obs)}$: observed maximum plasma or milk concentration; $T_{max\ (obs)}$: time to observed maximum plasma or milk concentration after each injection; λz: elimination rate constant; $T_{1/2\lambda z}$: elimination half-life; AUC_{last}: area under the plasma or milk concentration vs. time curve to the last sampled value; AUC_{inf}: area under the plasma or milk concentration vs. time curve extrapolated to infinity

[#] p values of Student's t test of the difference in the milk pharmacokinetic parameters between the first injection and the second injection

Table 2 Average concentrations of tulathromycin in plasma and the marker residue CP-60,300 in milk after twice subcutaneous injections of 2.5 mg/kg tulathromycin in lactating goats

Time	Plasma	SD	Milk	SD	Time	Plasma	SD	Milk	SD
12	NA	NA	1791.4	672.02	360	NA	NA	64.3	20.39
24	NA	NA	1167.0	264.23	372	NA	NA	26.9	8.23
36	NA	NA	763.7	226.64	384	NA	NA	47.7	16.54
48	NA	NA	364.8	109.83	396	NA	NA	46.7	18.93
60	NA	NA	281.1	103.50	408	NA	NA	45.9	16.43
72	NA	NA	140.3	30.71	420	NA	NA	37.0	15.74
84	NA	NA	154.1	45.25	432	NA	NA	32.8	10.11
96	NA	NA	107.1	23.03	444	NA	NA	30.3	10.55
108	NA	NA	131.4	39.94	456	NA	NA	24.0	7.51
120	NA	NA	90.6	23.20	468	NA	NA	23.9	7.56
132	NA	NA	56.7	15.54	480	NA	NA	22.1	10.49
144	NA	NA	59.2	14.50	492	NA	NA	23.3	14.07
156	NA	NA	50.2	9.61	504	NA	NA	25.4	12.87
168	NA	NA	41.0	11.11	540	NA	NA	20.6	6.28
180	152.5	39.03	1392.8	363.32	588	NA	NA	12.0	4.25
192	105.8	44.82	1096.9	151.09	636	NA	NA	8.3	3.55
204	75.1	36.48	789.3	292.69	684	NA	NA	8.0	4.91
216	61.9	20.64	424.9	168.10	732	NA	NA	6.5	1.42
228	39.3	19.24	230.1	101.49	780	NA	NA	6.0	2.21
240	46.2	18.20	164.0	68.47	828	NA	NA	4.1	3.06
252	39.3	10.90	131.2	41.73	876	NA	NA	1.9	2.12
264	29.1	12.87	105.3	33.04	948	NA	NA	4.1	1.98
276	20.1	8.77	74.7	24.81	1020	NA	NA	2.0	2.21
288	22.5	5.94	78.3	21.99	1092	NA	NA	0.9	1.66
300	20.0	5.20	91.6	29.83	1164	NA	NA	1.5	2.11
312	12.6	4.34	88.9	26.16	1284	NA	NA	1.2	1.71
324	11.7	3.66	85.2	28.94	1380	NA	NA	0.5	1.40
336	28.5	18.61	68.4	16.72	1452	NA	NA	0.0	0.00
348	12.9	5.94	52.5	26.43	1524	NA	NA	0.0	0.00

Time time after the first injection (h), *Plasma* average concentration of tulathromycin in the plasma (ng/mL), *Milk* average concentration of CP-60,300 in the milk (ng/mL), *SD* standard deviation, *NA* not applicable as no sample was taken

Table 3 Pharmacokinetic model parameters for plasma tulathromycin and milk CP-60,300 concentrations in lactating goats

Parameter	Description	Population mean	IIV	Bootstrap value	2.5% CI	97.5% CI
V (mL)	Volume of distribution for the central compartment	326627	1.072	260011.65	169213.19	369548.19
Cl (mL/h)	Central clearance	21109.2	0.184	19214.724	16224.881	22463.998
V2 (mL)	Volume of distribution for the peripheral compartment	676758	0.232	721665.92	567400.38	835445.28
Cl2 (mL/h)	Clearance between the central and the peripheral compartment	12550.5	0.710	16958.522	11499.469	24340.243
CLmilk (mL/h)	Clearance from the central compartment to the milk	882.461	0.121	816.285	686.264	957.191
Ka (1/h)	Absorption rate constant after subcutaneous (SC) injection	0.295	6.255	0.938	0.327	2.318
Residue error 1 (%)	Proportional error model of tulathromycin concentrations in plasma	0.485	NA	0.484	0.419	0.549
Residue error 2 (%)	Proportional error model of CP-60,300 cumulative amounts in milk	0.0234	NA	0.0228	0.0157	0.0279

CI confidence interval, *IIV* interindividual variability, *NA* not available

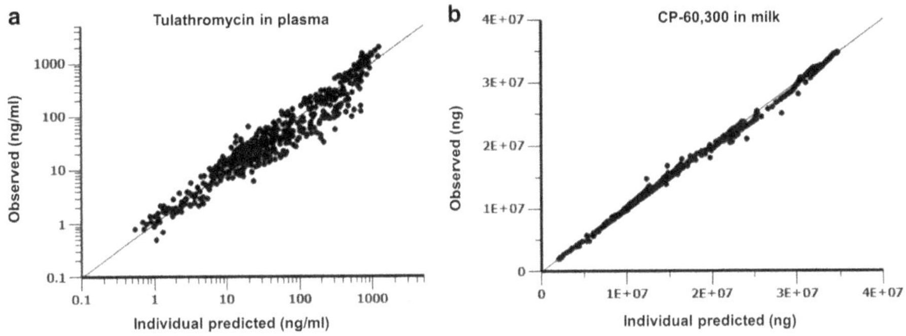

Fig. 3 Goodness-of-fit plots of plasma tulathromycin concentrations and milk cumulative excreted amounts of tulathromycin marker residue CP-60,300 in lactating goats. Comparisons between observed versus individual predicted tulathromycin concentrations or CP-60,300 amounts in the plasma (**a**) and milk (**b**), respectively

Model simulations

The simulated time-concentration (or amount) profiles of tulathromycin in plasma and the marker residue CP-60,300 in milk from the final pharmacokinetic model are shown in Fig. 5. Based on the time-concentration profiles and LOD in plasma, it was estimated with a 95% confidence that the plasma concentration was below LOD 43 days after a single SC injection, and 62 days after the second injection in a 2-injection regimen (7-day interval) (Fig. 5a, b).

The simulated time-cumulative amounts of CP-60,300 in milk displayed a hyperbolic increase at the early phase after injection, started to reach a plateau on ~25 days after the first injection, and continuously increased slowly even up to the last simulation time points, which were day 64 after the single injection (Fig. 5c, d). The simulated time-concentration profiles in milk fluctuated throughout the simulation period (Fig. 5e, f). However, the overall trend of decreasing in the concentrations of CP-60,300 in milk after injection was still observed and the simulated concentrations matched well with the measured concentrations. The fluctuating milk concentrations at the terminal kinetic phase prevented from estimating milk withdrawal time for tulathromycin in goats. Therefore, a definite conclusion on the milk withdrawal time recommendation cannot be made from this study. Nevertheless, based on the experimental data of CP-60,300 in milk, concentrations of the marker residue above the LOD were not measured in the milk beyond 45 days after the single injection [12], and 39 days after the second injection in the 2-injection regimen (7-day interval) for the present studied population of animals.

Discussion

This study reports plasma and milk pharmacokinetic data up to 2 months after twice SC injections of 2.5 mg/kg tulathromycin with a 7-day interval in lactating goats. There were palpable injection-site reactions, but no substantial adverse clinical signs occurred; the body weight and milk production were not affected negatively. These observations suggest that twice SC injections of 2.5 mg/kg tulathromycin do not cause clinical adverse reactions in lactating goats, and could be considered for treating lung infections in lactating goats provided that extra-label drug use requirements are met.

Based on the new data in this study and previously published single SC injection pharmacokinetic data in goats [3, 12], a NLME-PK model for tulathromycin in

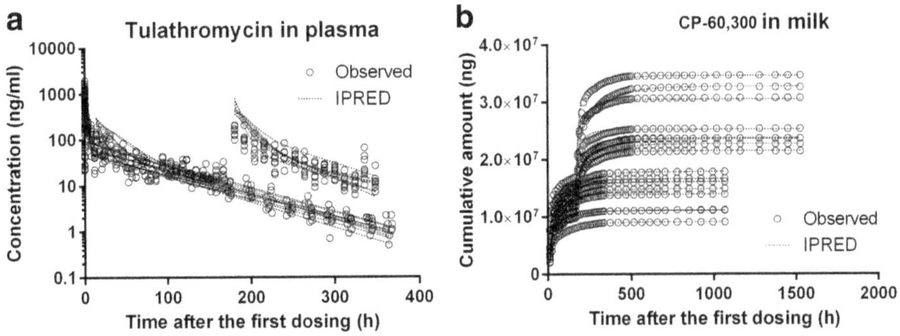

Fig. 4 Comparisons of observed versus individual predicted time-concentration profiles in plasma (**a**) and time-cumulative excreted amount profiles in milk (**b**) of lactating goats. Unfilled circles represent experimentally observed data from Clothier et al. [3], Grismer et al. [12], and the present pharmacokinetic study. Dotted lines represent model simulation results

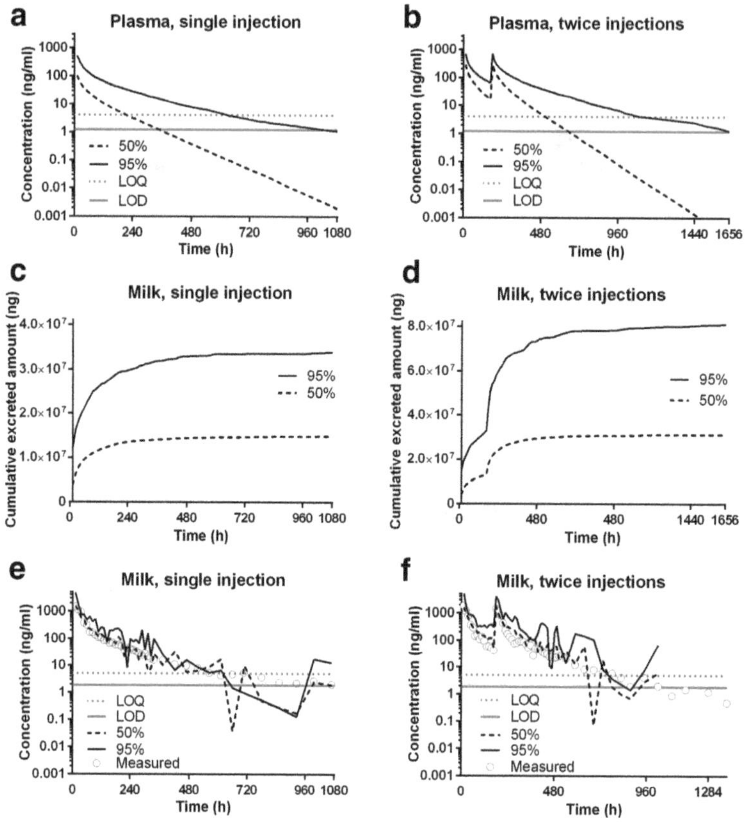

Fig. 5 Simulated data for the plasma tulathromycin concentrations (**a**, **b**), cumulative excreted amounts of marker residue CP-60,300 in milk (**c**, **d**), milk concentrations (**e**, **f**) of CP-60,300 in lactating goats after single or twice injections. The *solid and dashed black lines* represent the 95th and 50th percentiles of the simulated data, respectively. The *solid and dotted red lines* represent limit of detection (LOD) and limit of quantification (LOQ), respectively, in milk (LOD = 1.8 ng/mL, LOQ = 5 ng/ml) and plasma (LOD = 1.2 ng/mL, LOQ = 4 ng/ml). The *empty purple circles* represent the measured data from Grismer et al. [12] and the present pharmacokinetic study

lactating goats was created. The model-predicted plasma concentration and milk cumulative excreted amount data correlated well with the experimental data (Figs. 3 and 4). The model is an excellent tool to estimate tulathromycin plasma and milk depletion profiles (Fig. 5). However, one limitation is that the model can only predict the cumulative excreted amounts in milk and the milk concentrations of tulathromycin marker residue have to be calculated by dividing the measured volumes of milk samples, which were different between different milking periods (range 7–14 lbs/day). This was one of the reasons why the model-predicted milk concentrations of tulathromycin marker residue exhibited a fluctuating pattern at the terminal phase, which was consistent with the measured data (Table 2).

The model predicted that a very long period (62 days) would be needed to ensure that plasma concentrations of tulathromycin and milk concentrations of the marker residue fell below the LOD in 95% of the population following two administrations of tulathromycin 7 days apart. One of the reasons for this is that the model was built using time-concentration data from several studies with variable characteristics such as age, body weight, and milk production. Unfortunately, data for these factors were not reported for all the studies, so explanatory co-variates for the pharmacokinetic variability could not be included in the final model. The pharmacokinetic parameters' variability therefore represents the variability across all the included studies and, as a result, the model predicted a wide range of concentrations to encompass the full 95% of the study population.

Note that the milk concentration analysis was based on the marker residue CP-60,300, which is a common fragment of both the parent compound and metabolites of tulathromycin. In contrast, only the parent tulathromycin was measured in the plasma. This is a limitation of the current study, since it is based on the assumption that the ratio between parent drug concentrations in the plasma and total residues in the milk is constant. However, this approach to the analysis may be a contributing factor to the fluctuating milk concentrations that were observed in the study (i.e., the milk concentration may represent mainly the parent drug at the early kinetic phase and then primarily the metabolites at the later phase). To further optimize the present model, additional studies using more sensitive and specific analytical method are needed to determine milk concentrations of both tulathromycin and the marker residue CP-60,300. In addition, high milk:plasma partition coefficient (~10) and relatively high protein bound percentage (~50%) of tulathromycin in goats may contribute to the fluctuating milk concentrations at the terminal phase [10, 12].

As introduced before, another helpful approach in the estimation of withdrawal times is PBPK modeling, which can be used to predict both the concentrations and the cumulative excreted amounts of drugs in the milk [17, 18]. Therefore, future studies that extend the published tulathromycin PBPK model in juvenile and market-age meat goats [10] to lactating goats are needed to compare the estimated withdrawal intervals from different approaches (NLME-PK vs. PBPK), and then to determine a more accurate estimate to protect food safety.

The estimated times when plasma tulathromycin concentrations were below 1.2 ng/mL LOD were 43 days after a single SC injection and 62 days after the second injection in the 2-injection regimen. This estimation was based on the LOD because the tolerance for tulathromycin in goat milk is not available in the US or Europe. Tulathromycin marker residue LOD in milk is much lower than the tolerance of 5.5 ppm (5500 ng/ml) in cattle liver (the target tissue). However, cattle liver tolerance is not appropriate to be used to estimate milk withdrawal time because milk is consumed on a regular basis and in large quantities compared to cattle liver. Consequently, a conservative estimate based on goat milk-specific data is necessary in order to provide a precautionary withdrawal time estimation.

Our previous study reported that a withdrawal interval of at least 45 days are needed when tulathromycin is administered to lactating goats SC at a dose of 2.5 mg/kg [12]. However, plasma/milk sampling did not extend beyond 45 days after injection, nor was an NLME-PK model developed to estimate the upper limit of the 95% confidence intervals of plasma and milk concentrations. The present study confirms the previous study finding and further demonstrate that for twice SC administrations of tulathromycin a withdrawal interval of 39 days after the second injection (i.e., 46 days after the first injection) is needed for milk concentrations to fall below tolerance (i.e., LOD if official tolerance is not available) for the studied population of animals. Additionally, our NLME-PK model simulations showed that 62 days after the second injection in the 2-injection regimen are needed for plasma concentration to fall below the tolerance for 95% of the simulated population. It should be noted that the longer estimated time for plasma to fall below tolerance than the time for milk was because the former has taken into account the population variability, whereas the later was simply based on the present experiment using a limited number of animals. Other laboratories have reported lower LODs (0.46-0.7 ng/mL in plasma) using more sensitive liquid chromatography tandem mass spectrometry (LC-MS/MS) methods [8, 9, 19]. Based on LOD of 0.7 ng/mL in plasma, the estimated time when plasma concentrations fall below LOD will be >70 days after the second injection in the 2-injection paradigm.

Conclusions

In summary, this study suggests that, although two SC injections of 2.5 mg/kg tulathromycin is a safe treatment for pulmonary infections in lactating goats, there is considerable potential for contamination of the human food supply unless a prolonged and extended withdrawal interval is observed. The experimental data showed that the marker residue CP-60,300 concentrations were below LOD in all milk samples from the studied animals at 39 days after the second injection. However, a much longer withdrawal time of 62 days would be needed to ensure tulathromycin concentrations in both plasma and milk fall below the LOD in 95% of the population. These estimated times are much longer than the current meat withdrawal time recommendation of 18 days in non-lactating cattle and of 5 days for swine. Therefore, tulathromycin use in lactating goats should be cautious and an extended withdrawal period should be considered. Nevertheless, a definite milk withdrawal time recommendation for a large diverse population of lactating goats cannot be made due to the fluctuating milk concentration at the terminal kinetic phase and milk samples should be tested before milk is allowed to enter the human food chain. Additional studies using more sensitive and specific analytical method to determine milk concentrations of both tulathromycin and the marker residue CP-60,300 and using Bayesian population PBPK modeling approaches are needed to provide a comparative, and potentially, more conservative milk withdrawal time estimation.

Abbreviations

CI: Confidence interval; IM: Intramuscular; IV: Intravenous; LC-MS: Liquid chromatography mass spectrometry; LOD: Limit of detection; LOQ: Limit of quantification; MIC: Minimal inhibitory concentration; NLME-PK: Nonlinear mixed-effects pharmacokinetic; PBPK: Physiologically based pharmacokinetic; SC: Subcutaneous; SRM: Selected reaction monitoring

Acknowledgements

Not applicable.

Funding

This study was supported by the United States Department of Agriculture for the Food Animal Residue Avoidance and Depletion Program (USDA 2013-41480-21001), the Kansas Bioscience Authority funds to the Institute of Computational Comparative Medicine at Kansas State University, Certara USA, Inc for software license, and funds from the National Research Support Project-7 (Minor Use Animal Drug Program) and the Center for Food Animal Health (CFAH), School of Veterinary Medicine, University of California, Davis, CA. The authors thank Scott Wetzlich and Dr. Elizabeth Lee for their assistance with sample analysis and Zoetis, Inc. for their provision of analytical standard.

Authors' contributions

LAT, MC, JDR, JC, and SA designed and conducted the experiment. ZL, ML, and RG designed and developed the NLME-PK model. ZL ran the simulations and analyzed the resulting data. ZL, RG, ML, and JER interpreted the simulation data. ZL drafted the manuscript. All authors read, revised, and approved the final manuscript.

Competing interests

The authors declare that there are no conflicts of interest.

Consent for publication

Not applicable.

Author details

[1]Institute of Computational Comparative Medicine (ICCM), Department of Anatomy and Physiology, College of Veterinary Medicine, Kansas State University, 1800 Denison Avenue, P200 Mosier Hall, Manhattan, KS 66506-5802, USA. [2]Department of Population, Health and Reproduction, College of Agricultural and Environmental Sciences, University of California, Davis, CA 95616, USA. [3]Department of Medicine and Epidemiology, College of Agricultural and Environmental Sciences, University of California, Davis, CA 95616, USA. [4]School of Veterinary Medicine, and Department of Animal Science, College of Agricultural and Environmental Sciences, University of California, Davis, CA 95616, USA. [5]Present address: Department of Pharmaceutical Science, College of Pharmacy, University of Oklahoma Health Sciences Center, Oklahoma City, OK 73117, USA.

References

1. Baynes RE, Riviere JE. Strategies for reducing drug and chemical residues in food animals: international approaches to residue avoidance, management, and testing. Hoboken: John Wiley & Sons, Inc; 2014. p. 1–334.
2. Baynes RE, Dedonder K, Kissell L, Mzyk D, Marmulak T, Smith G, Tell L, Gehring R, Davis J, Riviere JE. Health concerns and management of select veterinary drug residues. Food Chem Toxicol. 2016;88:112–22.
3. Clothier KA, Leavens T, Griffith RW, Wetzlich SE, Baynes RE, Riviere JE, Tell LA. Pharmacokinetics of tulathromycin after single and multiple subcutaneous injections in domestic goats (Capra aegagrus hircus). J Vet Pharmacol Ther. 2011;34(5):448–54.
4. Clothier KA, Jordan DM, Loynachan AT, Griffith RW. Safety evaluation of tulathromycin use in the caprine species: tulathromycin toxicity assessment in goats. J Vet Pharmacol Ther. 2010;33(5):499–502.
5. Lin Z, Gehring R, Mochel JP, Lavé T, Riviere JE. Mathematical modeling and simulation in animal health. Part II: principles, methods, applications, and value of physiologically based pharmacokinetic modeling in veterinary medicine and food safety assessment. J Vet Pharmacol Ther. 2016;39(5):421–38.
6. Riviere JE, Gabrielsson J, Fink M, Mochel J. Mathematical modeling and simulation in animal health. Part I: moving beyond pharmacokinetics. J Vet Pharmacol Ther. 2015;39(3):213–23.
7. Lin Z, Vahl CI, Riviere JE. Human food safety implications of variation in food animal drug metabolism. Sci Rep. 2016;6:27907.
8. Young G, Smith GW, Leavens TL, Wetzlich SE, Baynes RE, Mason SE, Riviere JE, Tell LA. Pharmacokinetics of tulathromycin following subcutaneous administration in meat goats. Res Vet Sci. 2011;90(3):477–9.
9. Romanet J, Smith GW, Leavens TL, Baynes RE, Wetzlich SE, Riviere JE, Tell LA. Pharmacokinetics and tissue elimination of tulathromycin following subcutaneous administration in meat goats. Am J Vet Res. 2012;73(10):1634–40.
10. Leavens TL, Tell LA, Clothier KA, Griffith RW, Baynes RE, Riviere JE. Development of a physiologically based pharmacokinetic model to predict tulathromycin distribution in goats. J Vet Pharmacol Ther. 2012;35(2):121–31.
11. Abo-El-Sooud K, Afifi NA, Abd-El-Aty AM. Pharmacokinetics and bioavailability of tulathromycin following intravenous, intramuscular and subcutaneous administrations in healthy rabbits. Vet World. 2012;5(7):424–8.
12. Grismer B, Rowe JD, Carlson J, Wetzlich SE, Tell LA. Pharmacokinetics of tulathromycin in plasma and milk samples after a single subcutaneous injection in lactating goats (Capra hircus). J Vet Pharmacol Ther. 2014;37(2):205–8.
13. Galer D, Hessong S, Beato B, Risk J, Inskeep P, Weerasinghe C, Schneider RP, Langer C, LaPerle J, Renouf D, et al. An analytical method for the analysis of tulathromycin, an equilibrating triamilide, in bovine and porcine plasma and lung. J Agric Food Chem. 2004;52(8):2179–91.
14. Boner PL, Liu DD, Feely WF, Robinson RA, Wu J. Determination of flunixin in edible bovine tissues using liquid chromatography coupled with tandem mass spectrometry. J Agric Food Chem. 2003;51(26):7555–9.
15. Li M, Gehring R, Lin Z, Riviere J. A framework for meta-analysis of veterinary drug pharmacokinetic data using mixed effect modeling. J Pharm Sci. 2015;104(4):1230–9.
16. Sheiner LB, Rosenberg B, Marathe VV. Estimation of population characteristics of pharmacokinetic parameters from routine clinical data. J Pharmacokinet Biopharm. 1977;5(5):445–79.
17. Lin Z, Fisher JW, Wang R, Ross MK, Filipov NM. Estimation of placental and lactational transfer and tissue distribution of atrazine and its main metabolites in rodent dams, fetuses, and neonates with physiologically based pharmacokinetic modeling. Toxicol Appl Pharmacol. 2013;273(1):140–58.
18. Leavens TL, Tell LA, Kissell LW, Smith GW, Smith DJ, Wagner SA, Shelver WL, Wu H, Baynes RE, Riviere JE. Development of a physiologically based pharmacokinetic model for flunixin in cattle (Bos taurus). Food Addit Contam Part A Chem Anal Control Expo Risk Assess. 2014;31(9):1506–21.
19. Martos PA, Lehotay SJ, Shurmer B. Ultratrace analysis of nine macrolides, including tulathromycin A (Draxxin), in edible animal tissues with minicolumn liquid chromatography tandem mass spectrometry. J Agric Food Chem. 2008;56(19):8844–50.

Prevalence and risk factors of bovine tuberculosis in dairy cattle in Eritrea

Michael K. Ghebremariam[1,2*], V. P. M. G Rutten[1,3], J. C. M. Vernooij[4], K. Uqbazghi[5], T. Tesfaalem[6], T. Butsuamlak[7], A. M. Idris[8], M. Nielen[4] and A. L. Michel[3]

Abstract

Background: The prevalence of bovine tuberculosis (BTB) in dairy cattle in the three major milk producing regions of Eritrea was assessed by subjecting 15,354 dairy cattle, 50 % of Eritrea's dairy cattle population, to the single intradermal comparative tuberculin test (SICTT). Skin test results were interpreted according to guidelines of the World Organization for Animal Health (OIE) with >4 mm as cutoff in skin thickness increase. In addition, we studied the relation between 'physiological' variables related to pregnancy and lactation, and the variable 'region' on the probability to be skin test positive.

Results: The BTB prevalences at animal and herd levels were: 21.5 % and 40.9 % in Maekel, 7.3 % and 10 % in Debub, and 0.2 % and 1.6 % in the Anseba region, respectively. Overall, in the regions included, prevalence was 11.3 % (confidence interval (CI) 95 % CI, 11.29 – 11.31 %) and 17.3 % (95 % CI, 17.27–17.33 %), at animal and herd level, respectively. Considering positive herds only, the animal BTB prevalence was 36.8 %, 30.1 %, and 1.8 %, in Maekel, Debub and Anseba, respectively, and the overall animal prevalence within these herds was 32 %. In adult dairy cattle the probability of positive reactivity in the SICTT test was highest in pregnant animals as compared to the other categories.

Conclusion: This study reports persistent prevalence of BTB as defined by positive SICTT in the dairy sector of Eritrea, especially in the regions of Maekel and Debub that are located in the central highlands of the country. To our understanding this is the first report that has encompassed all the major dairy farms in Eritrea and it will be instrumental in advocating future BTB control programs in the dairy sector.

Keywords: Bovine tuberculosis (BTB), Prevalence, Comparative tuberculin test, Dairy, Eritrea

Background

Bovine tuberculosis (BTB) is a chronic, infectious and contagious disease caused by *Mycobacterium bovis (M.bovis)* affecting cattle and other species. The disease in cattle appears both on intensive dairy farms and in extensive pastoral systems [1–5]. The standard diagnostic test used in our study and many others is the 'single intradermal comparative tuberculin test' (SICTT). This test has moderate to high sensitivity (68–95 %) and high specificity (96–99 %) [6–9]. *M. bovis* is zoonotic, and can be transmitted from animals to humans through consumption of raw milk, inhalation, and direct contact with saliva. Beside its zoonotic importance, BTB affects the livelihood of people in developing countries by compromising sustainable food supply, income and social status [10]. In Africa, for example, BTB is ubiquitous: about 85 % of the cattle and 82 % of human populations live in areas where BTB is either not or only partly controlled [11]. In 2004, BTB was reported by 26 of the 51 African countries who filed statistics with OIE [12].

BTB is present in Eritrea [13], where the integrated farming system includes a dairy cattle population of around 16,000 animals of exotic breed (Holstein Friesians (HF)) and crossbreds, 13,000 cattle of the indigenous Barka breed, and 1,500 animals of Sudanese breeds (MOA, Kahsay Negash, personal communication, 2010). BTB, first reported in Eritrea by Pirani in (1929), was reported to be highly prevalent in the capital city Asmara

* Correspondence: michaelvet@yahoo.com
[1]Department of Infectious Diseases and Immunology, Faculty of Veterinary Medicine, Utrecht University, Utrecht, The Netherlands
[2]Department of Veterinary Sciences, Hamelmalo Agricultural College, Keren, Eritrea
Full list of author information is available at the end of the article

(Maekel region) and its surroundings in 2001 [13]. The BTB status in other regions of the country was never addressed.

Thus, the current study focused on the major milk producing regions (Maekel, Debub, Anseba), and its samples included more than 50 % of the total dairy cattle population of the country. It aimed to determine BTB prevalence and its association with various animal 'physiological statuses' related to pregnancy and lactation and 'region' in relation to positive reactivity in the SICTT throughout Eritrea's dairy sector.

Methods

Study area

Eritrea is located in the Horn of Africa and lies north of the equator between latitudes 12°22' N and 18°02' N, and longitudes 36°26' E and 43°13' E. It has an area of 124,400 square kilometres and is divided into six regions (zones) administratively, namely: Maekel, Debub, Anseba, Gash Barka, Southern and Northern Red Sea (Fig. 1), each of them comprising of several sub-regions. The country has diverse climatic zones. The mean annual temperature ranges from 16 °C in the highlands (around the capital Asmara) to extremely high temperatures (31 °C) in the lowlands. In the highlands, where the Maekel and Debub regions are located, the hottest months are usually May to June with the highest temperature reaching around 27 °C to 30 °C. Whereas in the western lowlands, where the Anseba region is located, the temperature ranges from 28 °C to 46 °C, except in December month when the temperature falls as low as 15 °C. The altitude ranges from 120 meters below sea level to 2400 meters above sea level in the central highlands. Dairy farms in the highlands of Eritrea are mainly concentrated in the Maekel and Debub regions [14], and in the lowlands the major dairy region is Anseba.

Fig. 1 Map of Eritrea and the study areas. The map shows the six administrative regions of Eritrea and the three major milk producing regions of the country where the study was conducted: Maekel (1), Debub (2) and Anseba (3). Adapted from https://commons.wikimedia.org/wiki/File:Eritrea_regions_blank.png (website re- visited: 06/05/2016)

The livestock production system in the country is divided into three parts: intensive urban and peri-urban dairy husbandry system, extensive mixed crop-livestock system, and extensive pastoral system. The dairy farms are mainly located in and around the major towns of the regions. In Maekel the dairy farms are located in the capital Asmara and its environs. Dairy farmers in this region have shortage of land for housing, thus keep their dairy herds in confined areas in small houses. Most of the houses lack windows for adequate ventilation. In Debub, most of the dairy farms are located outside the major towns with few farms within the towns. The dairy farms outside the towns have adequate space for housing and grazing. Those dairy houses located in the towns have small sizes but with windows for ventilation. In this region dairy cattle are allowed to partially graze within the compounds of the farms. In Anseba, most of the dairy farms are located outside the major towns and have adequate spaces for partial grazing and housing. The dairy houses have adequate spaces and bigger windows. During the day the dairy cattle spend most of their time under shades outside the buildings.

In the mixed crop-livestock system, livestock are sedentary, kept in villages where grazing is communal. This production system is practiced mainly in the highlands (Maekel and Debub) and partially in Anseba (in the highland areas). In the pastoral system, large mobile herds are kept in the lowlands. Such husbandry system is mainly practiced in the western and eastern lowlands of the country, namely, in Gash Barka, Northern and Southern Red Sea regions, and partially in Anseba (Fig. 1). The herds in this production system come in contact with other herds during their movement. The largest proportion of the livestock (about 60 %) is kept in the lowlands under traditional system. The cattle population and its distribution in the six regions is shown in Table 1. The livestock species in these regions are predominantly small ruminants, cattle and camels.

Study population and skin testing

In 2011 a cross-sectional survey was conducted to assess BTB prevalence in the three major milk producing regions, Maekel, Debub, and Anseba, in Eritrea (Fig. 1). All herds and all individual animals above six weeks of age were included in the study within the selected regions. The inclusion criteria were the same in all the regions. In total 15,354 dairy cattle in 3,149 herds were tested for BTB using the single intradermal comparative tuberculin test (SICTT). The survey was organized and coordinated by the Ministry of Agriculture, Eritrea.

Table 1 Official cattle population in 10^3 in the six regions of Eritrea (Anonymous, 1997)

Region	Cattle population
Anseba	219
Debub	490
Gash Barka	917
Maekel	40
Northern Red Sea	178
Southern Red Sea	82
Total	1,927

The SICTT was used based on the guidelines of the OIE [6]. Two sites on the left side of the mid-neck, 12 to15cm apart, were shaved. The skin thickness was measured with a 'Vernier caliper' and recorded. The upper site was injected with 0.1 ml containing 2,500 IU/ml avian PPD (Instituto Zooprofilatico Sperimentale dell' Umbria e delle Marche, Italia) using McLintock pre-set automatic syringes. Likewise, 0.1 ml of 2,500 UI/ml bovine PPD (Instituto Zooprofilattico Sperimentale deli Abruzzo e del Molise G. Caporale®, Italia) was injected into the lower site. Bovine PPD (PPD-B) and avian PPD (PPD-A) are defined as Purified Protein Derivatives of *M. bovis* and *M. avium*, respectively. All animals were ear tagged and their identities were confirmed at the time of reading. After 72 h, the skinfold thickness at the injection sites was re-measured by the same operator and with the same calliper used before. A skin reaction was considered positive when the skin thickness increase at the bovine site of injection was more than 4 mm greater than the reaction at the avian injection site. The reaction was considered inconclusive when the increase at the bovine site was between 1and 4 mm greater than the avian reaction. The reaction was considered negative when the increase in skin thickness at the bovine site was less than or equal to the increase in the skin reaction at the avian site of injection in the absence of clinical signs at the bovine injection sites. Moreover, information relating to age, sex, physiological status (respectively, bull (castrated/entire), heifer empty, heifer pregnant, lactating empty, lactating pregnant, dry pregnant) and region, was recorded for all the tested cattle (Table 2).

Data entry and analysis

The data was first entered in an Excel spreadsheet and checked for correctness by computing frequencies (pivot tables) using SPSS IBM version 20 software. Analyses of the association of the animals' 'physiological statuses' on the prevalence of BTB as assessed by SICTT were performed using the statistical package R version 3.1.0 [15]. A 'Generalized linear mixed model fit by maximum likelihood' [16] was used with BTB as binary outcome (BTB present: yes or no) and with herd as random effect to account for correlated observations within herd. Explanatory variables were 'region (forced in the model), 'physiological statuses' and 'age category' (the latter variable was included in the model on data of adult animals as 'age' is highly correlated with 'physiological status'). In both models (on all animals and adult animals respectively) the final model contains 'region and 'physiological status.

The AIC (Akaike's Information Criteria) selection criterion was used to select the 'best model'. The analysis was performed in positive herds (at least one reactor animal) with ≥5 animals only (Model 1 & 2) (Tables 3 and 4). The variable 'physiological status' included calf, bull, heifer empty, heifer pregnant, lactating empty, dry pregnant, lactating pregnant. The variable 'age' included different age groups, namely; <3 years old, 3 to ≤5 years old, and >5 years old (Table 2).

Results

Descriptive analysis

Overall, 15,354 individual animals in 3,149 herds were tested in the three regions. The highest proportion of

Table 2 Groups of cattle in different 'Physiological statuses' and different age categories in the study (Eritrea, 2011)

Variables	Description
Calf	Young female/ male animal <2 years old
Bull	Male animal (entire/ castrated) ≥2 years old
Heifer empty	Young female animal ≥2 years old that hasn't calved and is not pregnant
Heifer pregnant	Young female animal ≥2 years old that is pregnant and hasn't calved before
Lactating empty	Mature ≥3 years old, lactating cow that is not pregnant
Lactating pregnant	Mature ≥3 years old, lactating and pregnant cow
Dry pregnant	Mature >3 years old pregnant cow that is approximately 4 to 8 weeks from calving, and not lactating
Age (three categories)	<3 years old
	3 to ≤5 years old
	>5 years old

Table 3 Model 1 'Physiological status' and 'region' as potential risk factors for reactivity in the comparative tuberculin test on positive herds with ≥5 tested animals (4,776 observations within 344 positive herds) and animals tested, number and proportion of positive reactors in all positive farms (5269 observation). Estimated variance for herd: 1.118 on the logit scale

Physiological status and region	OR	95 % confidence interval		Number and % tested and % positive reactors in positive farms		
		Lower bound	Upper bound	Tested	% tested	% positive reactors
Calf (reference)	1.0			1613	30.6	14.6
Bull	5.2	3.3	8.3	147	2.8	30.6
Heifer empty	2.5	1.8	3.5	533	10.1	22.1
Heifer pregnant	5.8	4.2	8.1	460	8.7	34.4
Lactating empty	8.2	6.3	10.6	996	18.9	43.9
Lactating pregnant	10.8	8.4	13.9	1191	22.6	46.9
Dry pregnant	10.2	7.0	14.8	329	6.2	42.3
Anseba region (reference)	1.0			397	7.5	1.76
Debub region	8.5	1.9	37.1	1647	31.30	30.12
Maekel region	13.0	3.0	46.1	3225	61.21	36.84

the cattle population under study was that of young stock <3 years (60 %) followed by mature animals 3 - ≤ 5 years (24 %). The prevalence of SICTT reactors was 11.3 % (95 % CI, 11.29 – 11.31 %) at individual animal level and 17.3 % 95 % CI, 0.1727-0.1733) at herd level. The individual animal prevalence would have increased by 0.4 % had we used single intradermal tuberculin test instead of SICTT. The number of individual animals and herds tested in each of the regions were: Maekel (n = 5,667 and 927), Debub (n = 6,827 and 1,839), Anseba (n = 2,860 and 383), respectively. The region with the highest animal and herd BTB prevalence was Maekel (21.5 % and 40.9 %), followed by Debub (7.3 % and 10 %) and Anseba (0.2 % and 1.6 %). Considering only the positive herds (n = 545) in the three regions, the total number of animals tested was 5,272. These comprised 3,226 animals in Maekel, 1,647 in Debub and 399 in Anseba. The BTB animal prevalence in the positive herds was 36.8 % (n = 1,188), 30.1 % (n = 496), and 1.8 % (n = 7) in Maekel, Debub, and Anseba, respectively. The reaction to bovine tuberculin test was strong with differences in skin thickness (PPD-B minus PPD-A) reaching 4.5 mm to 73 mm (Additional file 1). The additional file shows the bovine versus avian immune reactivity in the skin test in dairy cattle in Eritrea. This file will be useful as a baseline document for further studies in the area.

Table 4 Model 2'Physiological status' as potential risk factor for reacting to tuberculin testing in adult dairy cattle in positive herds with ≥5 tested animals (2,356 observations in 326 herds) in Eritrea-2011, corrected for region. Estimated variance for herd: 1.245 on the logit scale

Physiological status	OR	95 % confidence interval	
		Lower	Upper
Bull (reference)	1.00		
Lactating empty	1.55	0.99	2.43
Lactating pregnant	2.24	1.43	3.50
Dry pregnant	2.01	1.19	3.40

Risk factors analysis

The highest prevalence of SICTT reactors was found in 'lactating-pregnant cows (Table 3). In Table 3 the Odds ratios (OR), with 95 % confidence interval, of BTB status with animal group and region in positive farms with at least 5 tested animals are presented. The category 'calf' (animals of <2 years age) was taken as a reference for the 'physiological status' groups. Compared to calves animals in all different 'physiological statuses' groups were significantly more at risk of being skin test positive, with pregnant lactating cows being most at risk (Table 3, Model 1). Animals in Maekel were most at risk to be SICTT positive reactors, followed by those in Debub when compared to the Anseba region.

The odds ratios of BTB statuses of 2,356 adult animals in 326 positive herds with 'physiological status' as potential risk factor in relation to positive skin test results, taking 'region' into account are presented in table 4 (Model 2). Compared to bulls, as reference, females tend to have a higher risk, with pregnant animals significantly so.

Discussion

The current study included more than half of the dairy cattle population of Eritrea and reports an overall prevalence of BTB at animal (11.3 %) as well as at herd (17.3 %) level. The individual animal BTB prevalence differed between the three regions included, ranging from 21.5 % in Maekel to 7.5 % in Debub and 0.2 % in Anseba. The BTB prevalence in skin test positive herds was not uniform; it was highest (37 %) in Maekel, followed by Debub (30 %), and Anseba (2 %).

The risk of having a BTB skin test positive animal was 13 times more likely in Maekel and ~9 times more likely in Debub than in Anseba (Table 3, Model 1). Differences in BTB prevalence may be due to risk factors varying across regions, as geographical location is in general an accepted risk factor. Herds and individual animal factors [17] may be attributed to the presence or absence of proper housing, adequate space for grazing and exercise, the type of breeds kept, and differences in climate.

Apart from differences in climatic circumstances between the regions the dairy housing system differs as well. In the highlands the farmers keep their animals indoors, in small houses with few or no windows [18], and as reported poor ventilation and housing may facilitate transmission of *M.bovis* [2, 19–21]. Close contact between animals is known to be a major risk factor as BTB is mainly transmitted via the respiratory route [22–24]. Similarly, close proximity of dairy farms and fence-line contact between animals, like in Maekel may be another important risk factor [25, 26]. These might have contributed over time to the persistence and high prevalence of BTB in the region, especially since no control program is in place in Eritrea. In contrast, in Anseba the dairy houses have adequate windows for ventilation and the animals spend more time outside in shaded areas and these factors might have contributed to the low number of SICTT reactors in the dairy farms of the region.

Asmara and its surroundings (Maekel region) was the first place in the country where exotic breeds were introduced for dairying. It was also in this area that BTB was reported for the first time in Eritrea, in 1929, and as already indicated in the previous study [13] as well as by the present study, the BTB prevalence is still on-going with an increasing trend. This might indicate that BTB has established in this region as it is likely for herd breakdowns to occur repeatedly in the same areas [27]. It may also be reasonable to argue that the subsequent development of an intensive dairy production system contributed to the establishment of *M.bovis* in this production environment as previously stated by various investigators [11, 28]. There is a good reason therefore to suspect that Maekel region might be acting as a source for *M.bovis* spread in the country, as it is a pioneering region for commercial dairying and a source of stocking and restocking of the dairy farms of the different regions in the country.

A major limitation of the present study was that the 'breed' of the animals tested was not sufficiently recorded by the operators during the skin testing period and thus could not be included in the analysis. However, we assume that the different breeds kept in the different regions of the country may have contributed to the relatively higher BTB prevalence observed in Maekel and Debub regions where the presence of the HF breed predominates at the farms as opposed to Anseba. Several studies have shown indigenous zebu cattle to be relatively tolerant to *M.bovis* infection as compared to exotic dairy breeds [3, 13, 20, 28–30]. To address this important aspect a relevant study is currently underway.

The present study indicated that pregnancy and lactation were significantly linked with the reactivity in SICTT. Lactating -pregnant cows were most at risk (OR ~ 11) to show positive reactions in the skin test compared to calves, and 2 times more compared to bulls. These results are in agreement with a similar study in Ethiopia [20] that reported that among 'physiological status' group, pregnant lactating animals had the highest prevalence of BTB. Kazwala et al., [31] also reported high reactivity in SICTT in lactating cows. Pregnant animals are reported to have two to threefold more severe inflammatory lesions with a rise to miliary tuberculosis, compared to non-pregnant young female adults [32]. It is difficult to explain the strongly increased odds ratio for SICTT positive responses in lactating-pregnant animals. Physiological alterations in immune responsiveness during pregnancy and lactation might influence skin test responsiveness and potentially lead to improved detectability of skin test positive females.

A case control study has shown that susceptibility of animals to BTB infection increases when they are fed deficient rations [33]. Currently, the available animal feeds in Eritrea are deficient quantitatively (inadequate in amount) as well as qualitatively (inadequate in their contents of carbohydrate, proteins, minerals, etc.) [34]. Feeding on such feeds might have contributed to the higher susceptibility to BTB in the pregnant cows as they are more demanding in their nutrient requirement.

Our risk factor analyses showed that 'physiological status' was strongly associated with SICTT positive reactivity (Table 4, Model 2). Since age and herd sizes have been indicated to be associated with the prevalence of BTB in several studies [13, 35–39], we included age and herd size as risk factors in the model. When the variable 'age' (3 levels) was added to the adult animal model on 'physiological status', age was not significant and did not influence the model estimates (results not shown). In the model with all age groups, calves clearly had the lowest risk of being SICTT positive as expected. Besides, including animals from all herd sizes, instead of only herds of 5 animals and above, did not have an important effect on the odds ratios of 'physiological status' vis-a-vis BTB prevalence (results not shown).

When compared to calves, bulls were at higher risk (OR ~ 5) to react to the skin test (Table 3, Model 1). The increased risk of bulls to be skin test positive as shown may be due to frequent contact of these animals with other herds during breeding, as sharing of bulls is a common practice in Eritrea [40], which is also considered to be a risk factor for BTB transmission between herds of animals [33]. In addition, in view of age, the potential cumulative exposure time for bulls is higher than for calves.

The overall results of the current study resemble those of similar studies in Ethiopia [1, 41–43]. The prevalences in Maekel were, however, higher than those reported by a study conducted previously by Omer et al. in the same region in Eritrea [10], which reported lower (14.5 %) animal, and comparable (41.7 %) herd BTB prevalences. As expected the animal prevalence was higher than in the previous study since no BTB control measures were in place. The slightly higher herd prevalence in the previous study [10] might be due to the fact that they only considered animal herd sizes of >9 whereas in our study herd sizes of ≥1 were taken into account. Smaller herds have a lower prevalence [37, 39, 40].

In this study bacterial culture was not conducted due to the fact that skin test positive dairy cattle were not slaughtered after the disclosure of the results due to a lack of a compensation scheme.

In order to get more insight in the importance of BTB and its epidemiology in Eritrea, as a potential basis for a control program, further investigations in dairy cattle as well as in indigenous cattle, goats and camels kept under traditional livestock production systems are underway.

Conclusion

The current study has brought to light the persisting prevalence of BTB as well as some animal characteristics, related to pregnancy and lactation, linked with SICTT positive skin test in the dairy sector in Eritrea. The BTB risk tended to vary by region and adult lactating-pregnant animals were the most likely to be test positive.

Additional file

Additional file 1: Frequency of skin thickness (mm) after 72 h post inoculation in both the avian and bovine sites and the differences within the two (Bovine 72 h - Avian 72 h). The first two columns show the skin thicknesses on the avian site of injection after 72 h of the PPD-A inoculation and their frequencies (frq). The following four columns show the skin thicknesses on the bovine site of injection after 72 h of PPD-B inoculation and their frequencies. The following six columns show the skin thickness differences on the bovine sites versus avian sites of injection after 72 h of inoculations of both PPD-A and PPD-B (PPD-B minus PPD-A) and their subsequent frequencies.

Abbreviations

BTB, bovine tuberculosis; HF, Holstein Friesian; MOA, Ministry of Agriculture; NUFFIC, Netherlands University Foundation for International Cooperation; OIE, World Organization for Animal Health; *M.bovis mycobacterium bovis;* PPD, purified protein derivative; PPD-A, purified protein derivative for *mycobacterium avium;* PPD-B, purified protein derivative for *mycobacterium bovis;* SICTT, single intradermal comparative tuberculin test.

Acknowledgements

We thank the Ministry of Agriculture, Eritrea, for implementing the project. The authors thank all animal health staff in the Ministry of Agriculture, Eritrea, who assisted in the tuberculin testing. We would like to thank Berhane Tecle (Maekel), Samuel Abraha (Debub), and Tekeste (National Animal and Plant Health Laboratory) for organizing the field survey. Efreme Ghebremeskel and Tsige Kifleyesus (National Animal and Plant Health Laboratory) are acknowledged for allowing us to use the laboratory and the office during the data collection and compilation. We thank NUFFIC for funding the first author's PhD fellowship (Grant award CF8608/2012). Many thanks to Hamelmalo Agricultural College for granting the first author a study leave. We thank also Institute Zooprofilatticco Abruzzo and Molise "G. Caporale", Teramo, Italy, for supplying the PPDs for the skin test.

Funding

The expenses for skin testing, including purchase of equipment, PPDs and daily subsistence allowances for those animal health staff who participated in the survey was funded by the Ministry of Agriculture, Eritrea. Expenses for the data collection and international conference attendance were funded by NUFFIC.

Authors' contributions

MKG involved in the design of the study, collected data, conducted descriptive data analysis and the write up of the manuscript. RVPMG involved in the design of the study and write-up of the manuscript. VJCM conducted analysis of the data and its interpretation and involved in the write up of the manuscript. UK participated in the implementation of the project, skin testing and data collection in the Debub region. TT participated in the data collection, compilation and write of the manuscript. BT involved in the implementation of the project, skin testing and data collection in Anseba region. MN involved in the design of the study, data analysis, data interpretation and write up of the manuscript. IAM involved in the implementation of the project and data collection in the Maekel region. MAL involved in the conceptualization of the study and write up of the manuscript. All authors read and approved the final manuscript.

Competing interests

The authors declare that they have no competing interest.

Consent for publication

Not applicable.

Author details

[1]Department of Infectious Diseases and Immunology, Faculty of Veterinary Medicine, Utrecht University, Utrecht, The Netherlands. [2]Department of Veterinary Sciences, Hamelmalo Agricultural College, Keren, Eritrea. [3]Department of Veterinary Tropical Diseases, Faculty of Veterinary Science, University of Pretoria, Pretoria, South Africa. [4]Department of Farm Animal Health, Faculty of Veterinary Medicine, Utrecht University, Utrecht, The Netherlands. [5]Veterinary Services, Ministry of Agriculture (MOA), Mendefera, Debub Region, Eritrea. [6]National Animal and Plant Health Laboratory, MOA, Asmara, Eritrea. [7]Veterinary Services, MOA, Keren, Anseba Region, Eritrea. [8]Veterinary Services, MOA, Asmara, Maekel Region, Eritrea.

References

1. Regassa A, Tassew A, Amenu K, Megersa B, Abuna F, Mekbib B, Macrotty T, Ameni G. Across-sectional study on bovine tuberculosis in Hawassa town and its surroundings, South Ethiopia. Trop Anim Health Prod. 2010;42:915–20.
2. Radostits OM, Gay CC, Hinchcliff KW, Constable PD. Diseases caused by mycobacterium species. In: Veterinary medicine- a text book of the diseases of cattle, sheep, pigs, goats and horses. 10th ed. London: Sounders; 2007. p. 1007–44.

3. Munyeme M, Muma JB, Samui KL, Skjerve E, Nambota AM, Phiri IGK, Rigouts L, Tryland M. Prevalence of bovine tuberculosis and animal level risk factors for indigenous cattle under different grazing strategies in the livestock/wildlife interface areas of Zambia. Trop Anim Health Prod. 2009;41:345–52.
4. Shirima GM, Kazwala RR, Kambarage DM. Prevalence of bovine tuberculosis in cattle in different farming systems in the eastern zone of Tanzania. Prev Vet Med. 2003;57:167–72.
5. Katale BZ, Mbugi EV, Karimuribo ED, Keyyu JD2, Kendall S, Kibiki GS5, Godfrey-332 Faussett P, Michel AL, Kazwala RR, van Helden P and Matee MI. Prevalence and risk factors for infection of bovine tuberculosis in indigenous cattle in the Serengeti ecosystem, Tanzania. BMC Vet Res. 2013;9:267.
6. World Organisation for Animal Health (OIE): Manual of Diagnostic Tests and Vaccines for Terrestrial Animals. OIE, 6th Ed. OIE, Paris 2008; p.683-697.
7. Monaghan ML, Doherty ML, Collins JD, Kazda JF, Quinn PJ. The tuberculin test. Vet Microbiol. 1994;40:111–24.
8. Goodchild AV, Downs SH, Upton P, Wood JLN, de la Rua-Domenech R. Specificity of the comparative skin test for bovine tuberculosis in Great Britain. Vet Rec. 2015. doi:10.1136/vr.102961.
9. Karolemeas K, de la Rua-Domenech R, Cooper R, Goodchild AV, Clifton-Hadley RS, Conlan AJK, Mitchell AP, Hewinson RG, Donnelly CA, Wood JLN, McKinley TJ. Estimation of the relative sensitivity of the comparative tuberculin skin test in tuberculous cattle herds subjected to depopulation. PLoS One. 2012;7(8), e43217. doi:10.1371/journal.pone.0043217.
10. Michel AL, Muller B, van Helden PD. *Mycobacterium bovis* at the animal–human interface: A problem, or not? Vet Microbiol. 2010;140:371–81.
11. Cosivi O, Grange JM, Daborn CJ, Raviglione MC, Fujikura T, Cousins D, Robinson RA, Huchzermeyer HFAK, de Kantor I, Meslin FX. Zoonotic Tuberculosis due to *Mycobacterium bovis* in developing countries. Emerg Infect Dis. 1998;4:59–70.
12. Amanfu W. The situation of tuberculosis and tuberculosis control in animals of economic interest. Tuberculosis. 2006;86:330–5.
13. Omer MK, Skjerve E, Woldehiwet Z, Holstad G. A cross sectional study of bovine tuberculosis in dairy farms in Asmara, Eritrea. Trop Anim Health Prod. 2001;33(4):295–303.
14. Ghebremariam WK, Ortmann GF, Nsahlai IV. A production function analysis of commercial dairy farms in the Highlands of Eritrea using ridge regression. Agrekon. 2006;45(2):25.
15. R Core Team: R: a language and environment for statistical computing. R Foundation for Statistical Computing; 2014, Vienna, Austria. URL: http://www.R-project.org/.
16. Bates D, Maechler M, Bolker B, Walker S. lme4: Linear mixed-effects models using Eigen and S4. R package version 1.1-7, 2014, URL: http://CRAN.R-project.org/package=lme4.
17. Skuce RA, Allen AR, McDowell SW. Herd-level risk factors for bovine tuberculosis: a literature review. Vet Med Int. 2012;621210.
18. Mogos A. Dairy Constraint Analysis in Eritrea, with special emphasis on Asmara and Surrounding Dairy Farms (Unpublished MSc thesis, University of Natal, South Africa), 2003
19. Costello E, Doherty ML, Monaghan ML, Quigley FC, O'Reilly PF. A study of cattle-to-cattle transmission of *Mycobacterium bovis* infection. Vet J. 1998;155:245–50.
20. Elias K, Hussein D, Asseged B, Wondwossen T, Gebeyehu M. Status of bovine tuberculosis in Addis Ababa dairy farms. Rev Sci Tech. 2008;27:915–23.
21. O'Reilly LM, Daborn CJ. The epidemiology of *Mycobacterium bovis* infections in animals and man: a review. Tubercl Lung Dis. 1995;1 suppl 76:1–46.
22. Ameni G, Aseffa A, Engers H, Young D, Hewinson G, Vordermeier M. Cattle husbandry in Ethiopia is a predominant factor affecting the pathology of bovine tuberculosis and gamma interferon responses to mycobacterial antigens. Clin Vaccine Immunol. 2006;13:1030–6.
23. Menzies FD, Neill SD. Cattle-to-cattle transmission of bovine tuberculosis. The Vet J. 2000;2000(160):92–106.
24. Johnston WT, Vial F, Gettinby G, Bourne FJ, Clifton-Hadley RS, Cox DR,Crea P, Donnelly CA, Mc Inerney JP, Mitchell AP, Morrison WI, Woodroffe R. Herd-level risk factors of bovine tuberculosis in England and Wales after the 2001foot-and-mouth disease epidemic. Int J Infect Dis. 2011;15:e833–40.
25. Kaneene JB, Bruning-Fann CS, Granger LM, Miller R, Porter-Spalding A. Environmental and farm management factors associated with tuberculosis on cattle farms in northeastern Michigan. JAVMA. 2002.
26. Munroe FA, Dohoo IR, Mc Nab WB, Spangler L. Risk factors for the between-herd spread of *Mycobacterium bovis* in Canadian cattle and cervids between 1985 and 1994. Prev Vet Med. 1999;41:119–33.
27. White PCL, Benhin JKA. Factors influencing the incidence and scale of bovine tuberculosis in cattle in southwest England. Prev Vet Med. 2004;63:1–7.
28. Vordermeier M, Ameni G, Berg S, Bishop R, Robertson BD, Aseffa A, Hewinson RG, Young DB. The influence of cattle breed on susceptibility to bovine tuberculosis in Ethiopia. Comp Immunol Infect Dis. 2012;35:227–32.
29. Ameni G, Erkihun A. Bovine tuberculosis on small scale dairy farms in Adama Town, central Ethiopia, and farmer awareness of the disease. Rev Sci tech Off Int Epiz. 2007;26:711–9.
30. Inangolet FO, Demelash B, Oloya J, Opuda- Asibo J, Skjerve E. A cross-sectional study of bovine tuberculosis in the transhumant and agro-pastoral cattle herds in the border areas of Katakwiand Moroto districts. Uganda. Trop Anim Health Prod. 2008;40:501–8.
31. Kazwala RR, Kambarage DM, Daborn CJ, Nyange J, Jiwa SFH, Sharp JM. Risk factors associated with the occurrence of bovine tuberculosis in cattle in the Southern Highlands of Tanzania. Vet Res Commun. 2001;25:609–14.
32. Weinberg ED. Pregnancy-associated depression of cell mediated immunity. Rev Inf Dis. 1984;6(6):814–31.
33. Griffin JM, Hahesy T, Lynch K, Salman MD, McCarthy J, Hurley T. The association of cattle husbandry practices, environmental factors and farmer characteristics with the occurrence of chronic bovine tuberculosis in dairy herds in the Republic of Ireland. Prev Vet Med. 1993;17(3–4):145–60.
34. Kayouli C, Tesfai T, Tewolde A. Country pasture/forage resource profiles, Eritrea FAO 2006
35. Cleaveland S, Shaw DJ, Mfinanga SG, Shirima G, Kazwala RR, Eblate E, Sharp M. *Mycobacterium bovis* in rural Tanzania: risk factors for infection in human and cattle populations. Tuberculosis. 2007;87:30–43.
36. Cook AJC, Tuchili LM, Buve A, Foster SD, Godfrey-Faussett P, Pandey GS, McAdam KPWJ, 1996. Human and bovine tuberculosis in the Monze district of Zambia – a cross-sectional study. Br Vet J. 1996;152:37–46.
37. Griffin JM, Martin SW, Thorburn MA, Eves JA, Hammond RF. A case–control study on the association of selected risk factors with the occurrence of bovine tuberculosis in the Republic of Ireland. Prev Vet Med. 1996;27:75–87.
38. Humblet MF, Gilbert M, Govaerts M, Fauville-Dufaux M, Walravens K, Saegerman C. New assessment of bovine tuberculosis risk factors in Belgium based on nationwide molecular epidemiology. JCM. 2010;48(8):2802–8.
39. Wright DM, Reid N, Montgomery WI, Allen AR, Skuce RA, Kao RR. Herd-level bovine tuberculosis risk factors: assessing the role of low-level badger population disturbance. Sci Rep. 2015;5:13062. doi:10.1038/srep13062.
40. Humblet MF, Boschiroli ML, Saegerman C. Classification of worldwide bovine tuberculosis risk factors in cattle: a stratified approach. Vet Res. 2009; 40:50. doi:10.1051/vetres/2009033.
41. Ameni G, Ameni K, Tibbo M. Bovine tuberculosis: Prevalence and risk factor assessment in cattle and cattle owners in Wuchale-Jida district central Ethiopia. J Appl Res Vet Med. 2003;1:1–4.
42. Shitaye JE, Getahun B, Alemayehu T, Skoric M, Treml F, Fictum P, Vrbas V, Pavlik I. A prevalence study of bovine tuberculosis by using abattoir meat inspection and tuberculin skin testing data, histopathological and *IS6110* PCR examination of tissues with tuberculous lesions in cattle in Ethiopia. Vet Med. 2006;51:512–22.
43. Gumi B, Schelling E, Firdessa R, Aseffa A, Tschopp R, Yamuah L, Young D, Zinsstag J. Prevalence of bovine tuberculosis in pastoral cattle herds in the Oromia region, southern Ethiopia. Trop Anim Health Prod. 2011;43:1081–7.

15

Evaluation of effects of *Mycoplasma* mastitis on milk composition in dairy cattle from South Australia

Abd Al-Bar Al-Farha[1,2*], Farhid Hemmatzadeh[1,3], Manouchehr Khazandi[1,3], Andrew Hoare[5] and Kiro Petrovski[1,3,4]

Abstract

Background: *Mycoplasma* mastitis is increasingly posing significant impact on dairy industry. Although the effects of major conventional mastitis pathogens on milk components has been widely addressed in the literature, limited data on the effects of different *Mycoplasma* and *Acholeplasma* spp. on milk quality and quantity is available. The aim of this study was to determine the casual relationship of *Mycoplasma* spp. and *A. laidlawii* to mastitis and compare them to subclinical mastitis caused by conventional mastitis pathogens from a single dairy herd in South Australia; *Mycoplasma* spp. and *A. laidlawii* were detected using PCR applied directly to milk samples. The herd had mastitis problem with high somatic cell count and low response rate to conventional antimicrobial therapy. A total of 288 cow-level milk samples were collected aseptically and used in this study.

Results: Conventional culture showed a predominance of coagulase-negative staphylococci, followed by coagulase-positive staphylococci, *Streptococcus* spp., *Enterococcus* spp., *E. coli*, and *Klebsiella* spp. PCR results showed a high prevalence of mycoplasmas (76.7%), including *A. laidlawii* (10.8%), *M. bovis* (6.2%), *M. bovirhinis* (5.6%), *M. arginini* (2%), and (52.1%) of cows were co-infected with two or more *Mycoplasma* and *Acholeplasma* species. *Mycoplasma* co-infection significantly increased somatic cell counts (SCC) similar to conventional mastitis pathogens and compared to non-infected cows with 389.3, 550.3 and 67.3 respectively; and decreased the milk yield with 29.0, 29.9 and 34.4 l, respectively. *Mycoplasma* co-infection caused significant increase in protein percentage, and significant decrease in fat percentage and total milk solids, similar to other conventional mastitis pathogens. In contrast, changes in milk composition and yield caused by various individual *Mycoplasma* species were non-significant.

Conclusions: *Mycoplasma* mastitis had on-farm economic consequences similar to common conventional mastitis pathogens. Results of our study indicate that co-infection *Mycoplasma* mastitis caused similar effect on milk composition to other mastitis pathogens and we hope these findings raise the awareness of the importance of their detection on routine diagnostic panels.

Keywords: *Mycoplasma*, Mastitis, Dairy cattle, Milk composition, Somatic cell count (SCC)

Background

The genus *Mycoplasma* belongs to the class *Mollicutes* and is responsible for many diseases in cattle, including respiratory disorders, arthritis, otitis media, and mastitis [12, 23, 41]. *Mycoplasma* mastitis is highly resilient to antimicrobial therapy and can be easily missed during laboratory culture and susceptibility testing diagnostic panels [25]. Among the 200 species of *Mycoplasma* discovered to date, several have been reported to be involved in bovine mastitis such as *M. bovis, M. bovigenitalium, M. californium, M. bovirhinis, M. arginini, M. dispar, M. canadense, M. bovoculi*, and *Mycoplasma* spp. bovine groups 7 and F-38 [13]. Some studies have also indicated that *Acholeplasma* spp. can be a milk contaminant or non-pathogenic saprophyte [5, 27]. However, others have reported isolation of *A. laidlawii* from clinical and subclinical cases of bovine mastitis, suggesting a causal relationship [31, 61, 63].

* Correspondence: abdal-barahmednoori.al-farha@adelaide.edu.au
[1]School of Animal and Veterinary Sciences, The University of Adelaide, Roseworthy, South Australia 5371, Australia
[2]Northern Technical University, Technical Foundation, Mosul, Iraq
Full list of author information is available at the end of the article

In dairy herds, mycoplasmas can cause clinical, subclinical or chronic mastitis [16]. *M. bovis* is considered the most common pathogen among mycoplasmas [14]. The possibility of isolating *Acholeplasma* spp. from mastitis cases is not excluded [8, 31, 61, 63]. The incubation period of *Mycoplasma* mastitis is 10–14 days [48], and shedding of the causative pathogen may occur during this period contributing to the spread of bacteria. Economic consequences of *Mycoplasma* mastitis in cattle are due to decreased milk production, cost of implementing control procedures, and cost of diagnosis and treatment [39]. For instance, the cost of *M. bovis* infection in cattle is more than US$140 million annually in the United States, and even higher losses have been reported in Europe [2].

Bacteriological culture of mycoplasmas from milk samples was once the most common method of detection. However, this method is relatively slow often taking one to 2 weeks with potential non-growth of these bacteria due to their fastidious culture requirements [45, 51], *Mycoplasma* mastitis is usually excluded from general mastitis screening tests due to its special growth requirements and time delay [30]. Similarly, serological detection method is time-consuming as antibody formation requires approximately 2 weeks [26]. Furthermore, there is a variation in the growth requirements of different species of *Mycoplasma* [20] which consequently affects disease detection, particularly when *Mycoplasma* co-infection occurs. However, most mastitis diagnostic investigations focus on the predominant species of mycoplasma associated with infection, *M. bovis*, and disregard other causative *Mollicutes* [52]. Studies of clinical *Mycoplasma* co-infection also deserve more attention, especially for epidemiological and treatment investigations. Therefore, development of a rapid and reliable screening diagnostic method is required which will be capable in distinguishing between different mycoplasma genera and species.

The association between *Mycoplasma* mastitis, individual somatic cell count (SCC) and milk yield also requires clarification. The association between conventional pathogens causing mastitis, such as *Streptococcus* and *Staphylococcus* spp., and elevated SCC has been previously reported [10, 19]. *Mycoplasma* mastitis can also affect SCC patterns [32, 47]. A marked decrease in milk production has been estimated particularly from mastitis caused by *S. agalactiae*, *Mycoplasma* spp. and *Pasteurella* spp. [62]. However, the effect of *Mycoplasma* mastitis compared to conventional bacterial pathogens on other milk composition has yet to be evaluated. Furthermore, the pathogenicity of each individual *Mycoplasma* spp., either as a single or co-infection, needs to be explored.

This study had two aims. The first aim of this study was to determine the effects of different *Mycoplasma* spp. and *A. laidlawii* compared to conventional mastitis pathogens on milk yield and other milk components in cattle with high SCC (subclinical mastitis) from a single dairy herd in South Australia with low response rate to conventional antimicrobial therapy. The second aim was to identify *Mycoplasma* spp. and *A. laidlawii* using novel PCR applied directly to milk samples.

Methods

Sample collection

Milk samples were collected aseptically once from each individual cow from a single commercial dairy farm near Mount Gambier, South Australia. This farm had a history of repeated failure of mastitis treatment with high SCC. Samples originated from cows aged 2–10 years in the hospital mob or main milking mob if they had a high SCC. Composite milk samples were collected aseptically in sterile 50 mL tubes from each functional quarter of individual cows ($n = 288$). Samples were kept on ice and were sent immediately to the PC2 laboratory at The University of Adelaide, Roseworthy Campus. In the laboratory, milk samples were subjected to conventional microbial culture using 10 µL aliquots according to National Mastitis Council guidelines [24]. All samples were then frozen at −20 °C for further analysis.

Milk analysis data

A database of individual cow yield production parameters (yield, total milk solids, fat and protein percentage) and SCC for sampled and non-sampled cows (the rest of the herd's cows that have also high and low SCC) was obtained from herd testing information for the 4 herd tests closest to the sampling points ($n = 7609$ cow data points). SCC was performed for each milk sample using a FOSS Fossomatic 5000 (Hillerød, Denmark), and instrumental milk components assay by a commercial laboratory (NHIS, Cohuna, VIC, Australia).

DNA extraction

After thawing milk at ambient temperature, 2 mL of each sample was centrifuged at 8000 X g for 20 min to remove supernatant fat and excess liquid. DNA extraction was performed directly from milk using the QIAmp DNA extraction kit (Qiagen, Germany) according to the manufacturer's instructions with minor modifications. In order to increase the DNA yield, a much larger milk sample (1 mL) was centrifuged at 5000 X g for 30 min and the lipid layer removed. Genomic DNA concentration was measured using Nanodrop 1000c (Thermofisher Scientific Inc., Waltham, MA, USA) and stored at −20 °C until further use.

PCR probes and protocol

A modified PCR protocol using four pairs of species-specific primers previously published elsewhere was developed (Table 1). In-vitro amplification of DNA to

Table 1 Oligonucleotides primers for detection of mycoplasmas and *A. laidlawii* identified by conventional PCR

Primers	Sequencing 5′-3′	Species	Targets	Amplicon size (bp)	Ref
Mb-F	CTTGGATCAGTGGCTTCATTAGC	*M.bovis*	*VspA* gene	400	[1]
Mb-R	GTCATCATGCGGAATTCTTGGGT				
Arg-F	AGAGTTTGATCCTGGCTCAGGA	*M. arginini*	16S rRNA gene	449	[11]
Arg-R	TCAACCAGGTGTTCTTTCCC				
Mbv-F	GCTGATAGAGAGGTCTATCG	*M. bovirhinis*	16S rRNA gene	316	[34]
Mbv-R	ATTACTCGGGCAGTCTCC				
Acho-F	AGCCGGACTGAGAGGTCTAC	*A. lailawii*	16S rRNA gene	505	[11]
Acho-R	TAATCCTGTTTGCTCCCCAC				

detect *Mycoplasma* and *Acholeplasma* spp. was carried out in 25 μL volumes consisting of 0.25 μL Taq DNA polymerase, 5 μL of 5× reaction buffer (Biolab, UK), 1 μL forward primer, 1 μL reverse primer, 1 μL of template DNA and 16.75 μL of DEPC-treated water. Amplifications were performed for 35 PCR cycles using the T100™ Thermal Cycler (Biorad Australia) and consisted of pre-heating activation for 5 min at 95 °C, denaturation at 95 °C for 30 s, and annealing at 60 °C for *M. bovis* and A. *laidlawii,* 55 °C for *M. arginini* and 64 °C for *M. bovirhinis.* The final extension step was performed at 72 °C for 10 min. The PCR products were subjected to electrophoresis in 1.5% agarose gels and visualised by staining with Gel Red.

Statistical analysis

For statistical analysis individual cows were grouped as follows:

a. Bacteriologically analysed milk samples (all with SCC of >250,000 cells/mL and assumed to have subclinical mastitis, $n = 288$): which were positive for pure *M. bovis* detection (GROUP 1, $n = 11$); positive for pure *A. laidlawi* detection (GROUP 2, n = 28); positive for pure *M. bovirhinis* detection (GROUP 3, $n = 13$); positive for pure *M. arginini* detection ($n = 6$); positive for *Mycoplasma* co-infection by two or more detected species (GROUP 4, $n = 119$); positive for conventional mastitis pathogens and *Mycoplasma* negative (GROUP 5, $n = 58$), and cows with positive *Mycoplasma/Acholeplasma* detection and positive on conventional mastitis culture (GROUP 6, $n = 53$).
b. Milk samples not bacteriologically analysed: cows with SCC of >250,000 cells/mL – assumed to have subclinical mastitis but not sampled (GROUP 7, $n = 1146$); and cows with SCC of ≤250,000 cells/mL assumed to be non-affected by subclinical mastitis 'healthy cows', (GROUP 8, $n = 6181$).

Statistical analysis software (SAS) version 9.4 (Cary Inc., USA) was used to analyse data.

For assessment of the effect of subclinical mastitis (1) or no subclinical mastitis (0) on SCC, herd test data were log transformed (somatic cell score; SCS). Due to the small number of positive detections (n = 6), the effect of *M. arginini* on milk composition and yield was not estimated.

The effect of subclinical mastitis on the test-day geometric mean of SCC was estimated using LOGISTIC REGRESSION in PROC GLIMMIX of SAS. The values obtained by modelling for the SCS were back-transformed to obtain the geometric mean of SCC. The effect of subclinical mastitis on test-day milk production parameters (yield, fat, protein and total milk solids) with pathogen as fixed effect and a cow as a random effect, was estimated using ANOVA in PROC MIXED of SAS. The effect of age and stage of production (i.e. days in milk) were not included in the final analysis as the preliminary model demonstrated no significant effect. A statistical difference of $P < 0.05$ was set as significant and $P < 0.001$ was highly significant.

Results

Results showed that 221 of the 288 milk samples (76.7%) collected from the dairy farm were positive for *Mycoplasma* spp. and *A. laidlawii.* Among these, *Mycoplasma* co-infection with two or more genera/species predominated (52.1%), followed by single infections with *A. laidlawii* (10.8%), *M. bovis* (6.2%), *M. bovirhinis* (5.6%), and *M. arginini* (2%) (Table 2). Agarose gel electrophoresis for PCR products revealed amplicon sizes stated in (Table 1). Conventional culture for all milk samples, independently of *Mollucites* isolation, showed a predominance of coagulase-negative staphylococci (CoNS) (12.2%), and relatively low occurrence of coagulase-positive staphylococci (CoPS) (2.4%), *Streptococcus* spp. (2.1%), *Enterococcus* spp. (1.7%), *E. coli* (1.4%), and *Klebsiella* spp. (0.4%) (Table 3).

Significant difference in test-day SCC was detected between groups (*Mycoplasma* co-infection; GROUP 4, conventional mastitis pathogens; GROUP 5, mycoplasmas and other pathogens; GROUP 6, and not tested high SCC cows; GROUP 7) and assumed healthy cows (low SCC;

Table 2 Distribution of *Mycoplasma* spp. detected by PCR for various conventional mastitis culture as detected directly from milk samples

Conventional mastitis culture	*Mycoplasma* detected spp.						Total
	A. laidlawii	*M. bovis*	*M. bovirhinis*	*M. arginini*	Mixed	Negative	
Negative (%)	28 (9.7)	11 (3.8)	13 (4.5)	3 (1)	119 (41.3)	56 (19.4)	230 (79.9)
Positive (%)	3 (1)	7 (2.4)	3 (1)	3 (1)	31 (10.76)	11 (3.8)	58 (20.1)
Total (%)	31 (10.8)	18 (6.2)	16 (5.6)	6 (2)	150 (52.1)	67 (23.2)	288 (100)

GROUP 8; $P < 0.001$) with 389.32, 550.26, 611, 960.7 and 67.33 SCCx10^3 cell/mL respectively. The *Mycoplasma* co-infected cows (GROUP 4) showed significant difference of SCC from assumed healthy cows (GROUP 8) at 376.15 SCCx10^3 cell/mL ($P \leq 0.05$). However, no significant difference was observed between cows infected with *M. bovis* (GROUP 1) and *A. laidlawii* (GROUP 2) in comparison with assumed healthy cows (GROUP 8) with 255.09 and 216.14, SCCx10^3 cell/mL respectively ($P > 0.05$). Correspondingly, test day milk yield was similarly affected significantly in *Mycoplasma* co-infection (GROUP 4), conventional mastitis pathogens (GROUP 5) and mycoplasmas and conventional pathogens (GROUP 6) in comparison with assumed healthy cows (GROUP 8) with 29.0, 29.9, 27.9 and 34.4 l, respectively ($P < 0.001$). In contrast, milk yield did not decline significantly in cows infected with *M. bovis* (GROUP 1), *A. laidlawii* (GROUP 2), or *M. bovirhinis* (GROUP 3) with 32.4, 32.4 and 30.0 l/day, respectively ($P > 0.05$).

Fat percentage was significantly decreased in cows co-infected with *Mycoplasma* (GROUP 4) with 3.1% ($P < 0.01$), while no differences were observed in all other tested groups with 3.4, 3.3, 3.3, 3.2 and 3.3% for groups *M. bovis* (GROUP 1), *A. laidlawii* (GROUP 2), *M. bovirhinis* (GROUP 3), conventional mastitis pathogens (GROUP 5) and mixed *Mycoplasma/Acholeplasma* and conventional mastitis culture (GROUP 6) respectively ($P > 0.05$). Protein percentage increased significantly in cows with *Mycoplasma* co-infection (GROUP 4), conventional mastitis pathogens (GROUP 5), mixed *Mycoplasma/Acholeplasma* and conventional mastitis culture (GROUP 6) and non-tested high SCC (GROUP 7) in comparison with cows assumed to be healthy (GROUP 8) with 3.4, 3.3, 3.4, 3.4 and 3.3%, respectively ($P < 0.05$), while other *Mycoplasma* groups showed no differences being 3.4, 3.4 and 3.3% for GROUPs s 1, 2 and 3 respectively ($P > 0.05$). In comparison, total milk solids decreased significantly in GROUPs 4, 5, 6 and 7 in comparison to GROUP 8 with 1.8, 1.9, 1.8, 2.0 and 2.2%, respectively ($P < 0.001$). However, there were no differences in total milk solids for GROUPs 1, 2 and 3 with 2.1, 2.1 and 2.0%, respectively ($P > 0.05$) (Table 4).

Table 3 Prevalence of non-*Mycoplasma* bacteria isolated by conventional mastitis culture from 288 milk samples collected purposively to identify *Mycoplasma* infections

Species.	Frequency (%)
CoNS	35 (12.2)
CoPS	7 (2.4)
Streptococcus	6 (2.1)
Enterococcus	5 (1.7)
E.coli	4 (1.4)
Klebsiella	1 (0.4)
ND	230 (79.86)
Total	288 (100%)

Discussion

Results showed a high prevalence of *Mycoplasma* and *Acoleplasma* species isolated from the purposive sampled cows. In addition, *Mycoplasma* co-infection significantly changed milk quality and quantity, similarly to other mastitis pathogens. Details on the results of culture and PCR, including test characteristics, were not subject of this study and are not further detailed.

The study farm had a history of mastitis treatment failure and high SCC. The long persistence of this problem may be due to the variety of transmission methods, such as via direct contact, milking machines and other fomites [29]. Intermittent shedding of the pathogen from cows suffering chronic mastitis may be another important reason for the relatively high prevalence of *Mycoplasma* mastitis in the studied herd. Chronic mastitis results from the capability of *Mycoplasma* spp. to form multiple micro-abscesses within the infected mammary gland [28]. Due to the high occurrence of mastitis, the studied farm is not representative of the South Australian dairy cattle population. As the primary objective of this study was not to carry out an epidemiological investigation, but to purposively sample and collect a significant number of isolates for research, the validity of the study is not decreased. In addition, study results may raise awareness of the importance of *Mycoplasma* mastitis to the dairy industry. In order to establish the prevalence of this disease in Australia, further attempts in affected herds using methodology similar to that described in this study are required. It is possible that the PCR technique used in

Table 4 The mean values (± SE) of SCC, milk yield, milk fat, protein and total solids for dairy cattle infected with *Mycoplasma* and other mastitis pathogens (in individual cow-testing points)

		Group	Frequency	Milk yield (litre/day)	Fat percentage	Protein percentage	Total milk solids	SCCx10^3 cell/mL
Tested	1	*Mycoplasma bovis*	11	32.4 ± 3.73	3.4 ± 0.26	3.4 ± 0.09	2.1 ± 0.20	255.1 ± 140.07
	2	*Acholeplasma laidlawii*	28	32.4 ± 2.33	3.3 ± 0.16	3.4 ± 0.06	2.1 ± 0.13	216.1 ± 87.79
	3	*M. bovirhinis*	13	30.0 ± 3.43^a	3.3 ± 0.24	3.3 ± 0.09	2.0 ± 0.19	376.2 ± 128.85^a
	4	*Mycoplasma* co-infection	119	29.0 ± 1.13^A	3.1 ± 0.08^a	3.4 ± 0.03^A	1.9 ± 0.06^A	389.3 ± 42.58^A
	5	Conventional mastitis pathogens	58	29.9 ± 1.57^a	3.2 ± 0.11	3.3 ± 0.04^a	1.9 ± 0.09^A	550.3 ± 59.00^A
	6	Mycoplasmas and conventional pathogens	53	27.9 ± 1.79^A	3.3 ± 0.12	3.4 ± 0.04^A	1.8 ± 0.11^A	611 ± 12.08^A
Total			282 (3.7%)					
Non tested	7	High SCC	1146	29.2 ± 0.36^A	3.6 ± 0.02^A	3.4 ± 0.01^A	2.0 ± 0.02^A	960.7 ± 13.72^A
	8	Low SCC	6181	34.4 ± 0.15	3.4 ± 0.01	3.3 ± 0.00	2.2 ± 0.01	67.3 ± 5.90
Total			7327 (96.3%)					

Lower case superscripts indicate statistical difference within column of $P < 0.05$ or capital superscripts indicate statistical difference within column of $P < 0.001$

this study detects foreign pieces of DNA. To ensure validity of our results we also carried out conventional culture for *Mycoplasma* (data not shown). Indeed, some culture-negative samples yielded positive PCR. To ensure that we were detecting *Mycoplasma* spp. and *A. laidlawii,* 16S rRNA sequencing was also carried out (data not shown).

Results of this study highlight the tendency of *Mycoplasma* co-infection to cause a significant alteration in milk composition in comparison to any individual *Mycoplasma* spp. group. This seems to be a result of developing evidence for pathogenicity of *Mycoplasma* co-infection [17], and can be a reflection of the advanced stage of the disease. It is often thought that the mechanism of milk alteration is attributed to the pathogen severity, proportion of involved mammary glands alveoli, interference with blood or hormonal nourishment to these alveoli, epithelial integrity disruption, and milk decomposition due to enzymatic activity.

Differences between the effects of each individual *Mycoplasma* spp. and *A. laidlawii* on milk composition in this study is noteworthy. It has been reported worldwide that *M. bovis* is the primary contagious pathogen in bovine *Mycoplasma* mastitis [14, 35, 39, 43]. In our study, cows infected only with *M. bovis* (GROUP 1) showed no significant impacts on milk composition apart from the SCC. Cows in this group may have been at an early stage of the disease. In addition, the limited sample size of this group ($n = 11$) needs to be considered. However, the effect of *M. bovis* was significant when it contributed to the *Mycoplasma* co-infection group. Multiple *Mycoplasma* spp. infection tends to be more common in *Mycoplasma*-associated diseases [59]. Although, some studies indicate the possibility of *A. laidlawii* being involves in mastitis cases [8, 31, 63], our data showed no effect on milk composition as an individual pathogen, being similar to common veterinary literature that establishes these bacteria as a milk contaminant [4, 27, 40]. We considered increased SCC and decreased milk production in affected cows as a crucial factor confirming the contribution of *Mycoplasma* in bovine mastitis.

M. arginini was another species detected in this study. However, we excluded it from the comparison as an individual group due to the limited number of detection in milk samples ($n = 6$). *M. arginini* has been reported elsewhere with no significant impact on milk composition, but can be considered a predisposing agent to *Str. dysagalactiae* leading to severe mammary gland inflammation [58]. It is important to consider that *M. arginini* is associated with some diseases in human [56, 64]. Evidence of transmission of *M. arginini* to human through bovine milk is available as it has been isolated from blood of a man with a history of milk product consumption who had eosinophilic ascites for 2 years [57]. *M. bovirhinis* (GROUP 3) showed remarkable effect on SCC and milk yield in our study. *M. bovirhinis* was first isolated in mastitis cows in England [21] and named by Leach [37]. These bacteria showed predominance to other mycoplasmas in some mastitis studies [22, 50]. SCC can be elevated in cows inoculated experimentally with four different strains of *M. bovirhinis* causing subclinical mastitis [6]. However, the study of Brownlie et al. [6] lacks information on other milk composition changes.

Somatic cells mainly include macrophages, lymphocytes, polymorphonuclear and epithelial cells [44]. The elevation of these cells in bovine quarters reflects the possibility of infection and is the standard method to discriminate between healthy and mastitic cows [7, 42, 46, 54]. The acceptable limit of individual cow SCC in raw milk in Australia has been established at 250,000 cell/mL [9]. Elevated SCC appears to be an immune response to mastitis caused by *Mycoplasma,* often followed by spontaneous recovery [15]. The negative correlation between high SCC and low milk production has been widely accepted for mastitis cases. In this study, low milk yield also appeared to be associated with positive *Mycoplasma* infection which in turn was consistently associated with higher SCC, confirming the economic impact of mastitis caused by *Mycoplasma* mastitis. This decline in milk yield can be attributed to reduction of synthesis ability of the secretory mammary tissue [53].

The role of *Mycoplasma* mastitis in declining milk fat percentage during mastitis has not been established as yet. For other mastitis causing pathogens, contradictory results have been stated for both elevation [46, 55] and depression of milk fat percentage [3, 33, 53]. The decline in fat percentage seems to be a result of leukocyte ingestion to fat globule [49] or due to a decrease in the synthesis and secretion activities of mammary glands [36]. It is also important to note that variation of fat percentage can also affected by stage of lactation, genetics, management, nutrition and hormonal changes [38]. This most likely explains the elevation of fat percentage in the non-tested high SCC (GROUP 7) in this study. Our results show that milk protein percentage can be elevated significantly during mastitis caused by *Mycoplasma* co-infection (GROUP 5) or other non-*Mycoplasma* pathogens (GROUP 6). It is generally thought that milk proteins increase during mastitis due to increases in blood albumins and immunoglobulins influx as an immune response [3, 18]. Protein percentage increases have been reported during *M. agalactiae* experimental mastitis in small ruminant [60].

Decreases in total milk solids were tested to allow farmers, practitioners and scientists who deal with mastitis to directly compare effects of *Mycoplasma* infection on milk yield to each other and to other mastitis pathogens. SCC are usually available only at herd testing level, while total milk solids can be detected daily by in-line milk meters on some farms. Hence, the effect of different mastitis pathogens on total milk solids tends to be more important than SCC records to the current dairy industry. We are aware that SCC are more sensitive to detecting early intramammary infections. With time, in-line SCC measures may be available on many farms, and the importance of total milk solids may not be as high as at this stage. Data on decreases of total milk solids will also allow for further research into economic effects of *Mycoplasma* mastitis to a particular dairy herd or to the entire dairy industry.

Effects on milk production and SCC detected in our study may vary between farms and this should be an area of future research.

Conclusions

In summary, we report bovine milk composition alteration during *Mycoplasma* mastitis. In addition, effects of *Mycoplasma* mastitis were compared with conventional mastitis pathogens. Results of our study indicate that co-infection *Mycoplasma* mastitis (GROUP 4) caused similar effect on milk composition to other conventional mastitis pathogens (GROUP 5), and we hope these findings raise the awareness of the importance of their detection on routine diagnostic panels. Ignoring timely detection may lead to developing *Mycoplasma* co-infection which may result in severe alterations in milk composition. However, roles of each individual *Mycoplasma* spp. in mastitis economics, and produced milk quantity and quality, needs further investigations, particularly when present as co-infections.

Abbreviations

CoNS: Coagulase-negative staphylococci; CoPS: Coagulase-positive staphylococci; PCR: Polymerase chain reaction; SAS: Statistical analysis software; SCC: Somatic cell counts; SCS: Somatic cell scour

Acknowledgements

The authors would like to express their special thanks to the farmer and farm staff for their cooperation and interest in the study. The technical support to this study by Ali Khazandi, Benjamin Linn, Budoor Ayoob, Chelsea Smith, Dana Thompson, Darren Trott, David Oswald, Gabe Laucke, Hannah Cohen, Ken Lee, Noor Haliza Hassan, Rhyanna Tedesco, and Taylor Urand is also appreciated. The guidance by Dr. Rick Tearle on statistical analysis is also appreciated.

Funding

The first author was partially funded by The Higher Committee for Education Development in Iraq in form of a scholarship. All field and laboratory procedures were financially supported by the University of Adelaide.

Authors' contributions

AA, FH and KP participated in the study design and coordination. AA, KP, MK & AH contributed to sample collection. AH contributed in herd test data collection and coordination with the farm. AA, MK and FH contributed in sample processing. AA and FH contributed in bacterial culture and PCR in the PC2 laboratory. KP performed the statistical analyses. All authors were involved in drafting the manuscript, read, corrected and approved the final manuscript.

Consent for publication

Not applicable as farm identity is not disclosed.

Competing interests

The authors declare that they have no competing interests.

Author details

[1]School of Animal and Veterinary Sciences, The University of Adelaide, Roseworthy, South Australia 5371, Australia. [2]Northern Technical University, Technical Foundation, Mosul, Iraq. [3]Australian Centre for Antimicrobial Resistance Ecology, The University of Adelaide, Adelaide, South Australia 5000, Australia. [4]Davies Centre, School of Animal and Veterinary Sciences, The University of Adelaide, Roseworthy, South Australia 5371, Australia. [5]South East Vets, Mt Gambier, South Australia 5290, Australia.

References

1. Alberti A, Addis MF, Chessa B, Cubeddu T, Profiti M, Rosati S, Ruiu A, Pittau M. Molecular and antigenic characterization of a Mycoplasma bovis strain causing an outbreak of infectious keratoconjunctivitis. J Vet Diagn Investig. 2006;18:41–51.

2. Amram E, Freed M, Khateb N, Mikula I, Blum S, Spergser J, Sharir B, Ozeri R, Harrus S, Lysnyansky I. Multiple locus variable number tandem repeat analysis of Mycoplasma bovis isolated from local and imported cattle. Vet J. 2013;197:286–90.
3. Auldist M, Hubble I. Effects of mastitis on raw milk and dairy products. Aust J Dairy Technol. 1998;53:28.
4. Boonyayatra S. Diagnosis of mycoplasma mastitis: validation and development. Washington: Washington State University; 2010.
5. Boonyayatra S, Fox LK, Gay JM, Sawant A, Besser TE. Discrimination between Mycoplasma and Acholeplasma species of bovine origin using digitonin disc diffusion assay, nisin disc diffusion assay, and conventional polymerase chain reaction. J Vet Diagn Investig. 2012;24:7–13.
6. Brownlie J, Howard C, Gourlay R. Pathogenicity of certain mycoplasma species in the bovine mammary gland. Res Vet Sci. 1976;20:261–6.
7. Carrillo-Casas EM, Miranda-Morales RE. Bovine mastitis pathogens: prevalence and effects on somatic cell count. Croatia: INTECH Open Access Publisher; 2012.
8. Counter D. A severe outbreak of bovine mastitis assoicated with Mycoplasma bovigenitalium and Acholeplasma laidlawii. Vet Rec. 1978;103:130–1.
9. DairyAustralia. Mastitis. Australia: Dairy Australia; 2014.
10. De Haas Y, Veerkamp R, Barkema H, Gröhn Y, Schukken Y. Associations between pathogen-specific cases of clinical mastitis and somatic cell count patterns. J Dairy Sci. 2004;87:95–105.
11. Dussurget O, Roulland-Dussoix D. Rapid, sensitive PCR-based detection of mycoplasmas in simulated samples of animal sera. Appl Environ Microbiol. 1994;60:953–9.
12. Foster AP, Naylor RD, Howie NM, Nicholas RA, Ayling RD. Mycoplasma bovis and otitis in dairy calves in the United Kingdom. Vet J. 2009;179:455–7.
13. Fox LK. Mycoplasma mastitis: causes, transmission, and control. Vet Clin North Am Food Anim Pract. 2012;28:225–37.
14. George LW, Divers TJ, Ducharme N, Welcome FL. Diseases of the Teats and Udder, Rebhun's diseases of dairy cattle, Second Edition ed. Missouri: Elsevier Health Sciences; 2007. p. 327–93.
15. Ghadersohi A, Hirst RG, Forbes-Faulkener J, Coelen RJ. Preliminary studies on the prevalence of Mycoplasma bovis mastitis in dairy in cattle in Australia. Vet Microbiol. 1999;65:185–94.
16. González RN, Wilson DJ. Mycoplasmal mastitis in dairy herds. Vet Clin N Am Food Anim Pract. 2003;19:199–221.
17. Gourlay RN, Thomas LH, Wyld SG. Increased severity of calf pneumonia associated with the appearance of Mycoplasma bovis in a rearing herd. Vet Rec. 1989;124:420–2.
18. Gráff M, Miko E. Analysis of mastitis in Holstein-Fresian cows and economic effects of mastitis. Agricult Manage/Lucrari Stiintifice Ser I Manage Agricol. 2015;17
19. Green M, Green L, Schukken Y, Bradley A, Peeler E, Barkema H, De Haas Y, Collis V, Medley G. Somatic cell count distributions during lactation predict clinical mastitis. J Dairy Sci. 2004;87:1256–64.
20. Hahn RG, Kenny GE. Differences in arginine requirement for growth among arginine-utilizing Mycoplasma species. J Bacteriol. 1974;117:611–8.
21. Harbourne J, Hunter D, Leach R. The isolation of Mycoplasma from bovine lungs and nasal swabs. Res Vet Sci. 1965;6:178–88.
22. Hirose K, Kawasaki Y, Kotani K, Tanaka A, Abiko K, Ogawa H. Detection of mycoplasma in mastitic milk by PCR analysis and culture method. J Vet Med Sci. 2001;63:691–3.
23. Hirose K, Kobayashi H, Ito N, Kawasaki Y, Zako M, Kotani K, Ogawa H, Sato H. Isolation of Mycoplasmas from nasal swabs of calves affected with respiratory diseases and antimicrobial susceptibility of their isolates. J Veterinary Med Ser B. 2003;50:347–51.
24. Hogan J, Gonzalez R, Harmon R, Nickerson S, Oliver S, Pankey J, Smith KL. Laboratory handbook on bovine mastitis. Madison: National Mastitis Council; 1999. p. 6–10.
25. Horwood PF, Schibrowski MI, Fowler EV, Gibson JS, Barnes TS, Mahony TJ. Is Mycoplasma bovis a missing component of the bovine respiratory disease complex in Australia? Aust Vet J. 2014;92:185–91.
26. Hotzel H, Frey J, Bashiruddin J, Sachse K. Detection and differentiation of ruminant mycoplasmas. PCR Detect Microbial Pathog. 2003;216:231–45.
27. Jasper D, Dellinger J, Rollins M, Hakanson H. Prevalence of mycoplasmal bovine mastitis in California. Am J Vet Res. 1979;40:1043–7.
28. Jasper DE. The role of Mycoplasma in bovine mastitis. J Am Vet Med Assoc. 1982;181:158–62.
29. Justice-Allen A, Trujillo J, Corbett R, Harding R, Goodell G, Wilson D. Survival and replication of Mycoplasma species in recycled bedding sand and association with mastitis on dairy farms in Utah. J Dairy Sci. 2010;93:192–202.
30. Kalogridou-Vassiliadou D. Mastitis-related pathogens in goat milk. Small Rumin Res. 1991;4:203–12.
31. Kirk JH, Glenn K, Ruiz L, Smith E. Epidemiologic analysis of Mycoplasma spp isolated from bulk-tank milk samples obtained from dairy herds that were members of a milk cooperative. J Am Vet Med Assoc. 1997;211:1036–8.
32. Kirk JH, Lauerman LH, Roberts C. Mycoplasma mastitis in dairy cows. Compend Contin Educ Pract Vet. 1994;16:541–52.
33. Kitchen BJ. Bovine mastitis: milk compositional changes and related diagnostic tests. J Dairy Res. 1981;48:167–88.
34. Kobayashi H, Hirose K, Worarach A, Paugtes P, Ito N, Morozumi T, Yamamoto K. In vitro amplification of the 16S rRNA genes from Mycoplasma bovirhinis, Mycoplasma alkalescens and Mycoplasma bovigenitalium by PCR. J Vet Med Sci. 1998;60:1299–303.
35. Kunkel JR. Isolation of Mycoplasma bovis from bulk milk. Cornell Vet. 1985; 75:398–400.
36. Le Maréchal C, Thiéry R, Vautor E, Le Loir Y. Mastitis impact on technological properties of milk and quality of milk products-a review. Dairy Sci Technol. 2011;91:247–82.
37. Leach R. Comparative studies of mycoplasma of bovine origin. Ann N Y Acad Sci. 1967;143:305–16.
38. Linn J. Factors affecting the composition of milk from dairy cows. Designing foods: animal product options in the marketplace. National Research Council (US) committee on technological options to improve the nutritional attributes of animal products, ed. Washington, DC: Natl. Acad. Press; 1988. p. 224.
39. Maunsell FP, Woolums AR, Francoz D, Rosenbusch RF, Step DL, Wilson DJ, Janzen ED. Mycoplasma bovis infections in cattle. J Vet Intern Med. 2011;25: 772–83.
40. Nicholas, R., Ayling, R., McAuliffe, L., 2008. Mycoplasma diseases of ruminants. CABI.
41. Nicholas RA, Ayling RD. Mycoplasma bovis: disease, diagnosis, and control. Res Vet Sci. 2003;74:105–12.
42. Peters M, Silveira I, Fischer V. Impact of subclinical and clinical mastitis on sensitivity to pain of dairy cows. Animal. 2015;9:2024–8.
43. Pfutzner H, Busch S, Hauke H, Landsiedel U, Schimmel D, Templin G, Wehnert C. Studies of Mycoplasma mastitis in cattle. 7. Mycoplasma mastitis in 3 dairy cattle herds. Arch Exp Vet. 1979;33:685–97.
44. Pillai S, Kunze E, Sordillo L, Jayarao B. Application of differential inflammatory cell count as a tool to monitor udder health. J Dairy Sci. 2001; 84:1413–20.
45. Pinnow CC, Butler JA, Sachse K, Hotzel H, Timms LL, Rosenbusch RF. Detection of Mycoplasma bovis in preservative-treated field milk samples. J Dairy Sci. 2001;84:1640–5.
46. Pyörälä S. Indicators of inflammation in the diagnosis of mastitis. Vet Res. 2003;34:565–78.
47. Radaelli E, Castiglioni V, Losa M, Benedetti V, Piccinini R, Nicholas RA, Scanziani E, Luini M. Outbreak of bovine clinical mastitis caused by Mycoplasma bovis in a north Italian herd. Res Vet Sci. 2011;91:251–3.
48. Razin S. Methods in Mycoplasmology V1: Mycoplasma characterization. New York: Elsevier; 2012.
49. Reiter B, Bramley A. Defence mechanisms of the udder and their relevance to mastitis control. Doc Int Dairy Fed. 1975;85:210–22.
50. Saad N, Hameed K. Detection of Mycoplasma species in raw milk of lactating animals in Assiut and Qena city of Egypt. Vet World. 2012;5:80-85.
51. Sachse K, Pfutzner H, Hotzel H, Demuth B, Heller M, Berthold E. Comparison of various diagnostic methods for the detection of Mycoplasma bovis. Rev Sci Tech. 1993;12:571–80.
52. Schnee C, Schulsse S, Hotzel H, Ayling RD, Nicholas RA, Schubert E, Heller M, Ehricht R, Sachse K. A novel rapid DNA microarray assay enables identification of 37 Mycoplasma species and highlights multiple Mycoplasma infections. PLoS One. 2012;7:e33237.
53. Schultz L. Somatic cells in milk-physiological aspects and relationship to amount and composition of milk. J Food Prot®. 1977;40:125–31.
54. Schwarz D, Diesterbeck U, König S, Brügemann K, Schlez K, Zschöck M, Wolter W, Czerny C-P. Flow cytometric differential cell counts in milk for the evaluation of inflammatory reactions in clinically healthy and subclinically infected bovine mammary glands. J Dairy Sci. 2011;94:5033–44.

55. Shuster D, Harmon R, Jackson J, Hemken R. Suppression of milk production during endotoxin-induced mastitis. J Dairy Sci. 1991;74:3763–74.
56. Sillis M. Mycoplasma arginini–a new human zoonosis? Clin Infect Dis. 1994; 18:488.
57. Sillo P, Pinter D, Ostorhazi E, Mazan M, Wikonkal N, Ponyai K, Volokhov DV, Chizhikov VE, Szathmary S, Stipkovits L, Karpati S. Eosinophilic fasciitis associated with Mycoplasma arginini infection. J Clin Microbiol. 2012;50:1113–7.
58. Stipkovits L, Somogyi M, Asvanyi B, Toth A, Szathmary S. Short communication: role of Mycoplasma arginini in mastitis caused by streptococcus dysgalactiae. J Dairy Sci. 2013;96:1661–7.
59. Szacawa E, Niemczuk K, Dudek K, Bednarek D, Rosales R, Ayling R. Mycoplasma bovis infections and co-infections with other Mycoplasma spp. with different clinical manifestations in affected cattle herds in eastern region of Poland. Bull Vet Inst Pulawy. 2015;59:331–8.
60. Todaro M, Puleio R, Sabelli C, Scatassa M, Console A, Loria G. Determination of milk production losses in Valle del Belice sheep following experimental infection of Mycoplasma agalactiae. Small Rumin Res. 2015;123:167–72.
61. Watts JL. Etiological agents of bovine mastitis. Vet Microbiol. 1988;16:41–66.
62. Wilson DJ, Gonzalez RN, Das HH. Bovine mastitis pathogens in New York and Pennsylvania: prevalence and effects on somatic cell count and milk production. J Dairy Sci. 1997;80:2592–8.
63. Windsor HM, Windsor GD, Noordergraaf J. The growth and long term survival of Acholeplasma laidlawii in media products used in biopharmaceutical manufacturing. Biologicals. 2010;38:204–10.
64. Yechouron A, Lefebvre J, Robson HG, Rose DL, Tully JG. Fatal septicemia due to Mycoplasma arginini: a new human zoonosis. Clin Infect Dis. 1992; 15:434–8.

16

Assessment of foot health and animal welfare: clinical findings in 229 dairy Mediterranean Buffaloes (Bubalus bubalis) affected by foot disorders

Jacopo Guccione[1], Christian Carcasole[2], Maher Alsaaod[3], Luigi D'Andrea[1], Antonio Di Loria[4], Angela De Rosa[1], Paolo Ciaramella[1*] and Adrian Steiner[3]

Abstract

Background: Lameness represents the third most important health-related cause of economic loss in the dairy industry after fertility and mastitis. Although, dairy Mediterranean Buffaloes (MB) and dairy cows share similar breeding systems predisposing to similar herd problems, published studies exploring its relevance and role in these ruminants are still rare and incomplete. The aims of this study were to describe the clinical findings of foot disorders (FDs) in dairy MB and their influence on animal welfare, determined by assessment of locomotion score (LS), body condition score (BCS) and cleanliness score (CS).

Results: Of 1297 multiparous MB submitted to routine trimming procedures, 229 buffaloes showed at least one FD. The prevalence of buffaloes affected by FDs was 17.7 %, while motility and lameness indexes were 84.1 % (1091/1297) and 15.9 % (206/1297), respectively. Overgrowth was present in 17.0 % (220/1297), corkscrew claw in 15.8 % (205/1297), interdigital phlegmon in 0.9 % (12/1297), white line abscess in 0.8 % (11/1297), digital dermatitis in 0.1 % (1/1297) and interdigital hyperplasia in 0.1 % (1/1297). Simultaneous presence of FDs was recorded in 17. 0 % of MB (221/1297): overgrowth and corkscrew claw occurred together in 15.8 % of cases (205/1297), overgrowth and interdigital phlegmon in 0.3 % (4/1297), overgrowth and white line abscess in 0.8 % (11/1297), digital dermatitis and interdigital hyperplasia in 0.1 % (1/1297). The presence of FDs was always associated with lameness ($LS > 2$), except from 23 MB with simultaneous overgrowth and interdigital phlegmon occurrence. The majority of MB within the under-conditioned group (95.5 %, 43/45) and all those with $CS > 2$ (122/122) had a locomotion score above the threshold of normality ($LS > 2$). Furthermore, foot diseases such as interdigital hyperplasia, white line abscess and digital dermatitis or interdigital hyperplasia seemed to occur more frequently associated with decreased BCS and increased CS scores.

Conclusions: This study describes for the first time the involvement of white line disease, interdigital phlegmona, digital dermatitis and interdigital hyperplasia in foot disorders of dairy Mediterranean buffalo and shows their association with an impairment of animal welfare.

Keywords: Foot disorders, Claw, Welfare, Lameness, Mediterranean Buffalo, Corkscrew claw, White line disease, Interdigital phlegmona, Digital dermatitis, Interdigital hyperplasia

* Correspondence: paociara@unina.it
[1]Department of Veterinary Medicine and Animal Productions, University of Napoli "Federico II", Via F. Delpino 1, 80137 Naples, Italy
Full list of author information is available at the end of the article

Background

Foot health has been classified as the most important welfare problem in dairy cows, and its surveillance is the most representative animal-based indicator of welfare in dairy cattle [1, 2]. Additionally, lameness due to foot disorders represents the third most important health-related cause of economic loss in dairy industry after reproduction disorders and mastitis [3]. The financial loss can be measured in terms of decreased milk yield [4], reduced reproductive performance [5], increased culling rate, and increased production costs [4].

In Europe, the number of Mediterranean Buffaloes (MB) bred is ~450.000 animals, of which ~84 % is from Italy [6]. Although, these ruminants and dairy cows share similar breeding systems predisposing to similar herd problems [7], only reproductive disorders [8] and mastitis [9, 10] have been recognized as cause of important economic loss in MB so far. Foot diseases were always considered virtually non-existent in MB herds [11]; as a consequence, only a few and incomplete studies exploring the role and economic relevance of foot disorders (FDs) in this dairy population have been performed [11, 12].

As widely known in cows, FDs impair animal welfare and some of the most important are associated with lameness [13], reduction in body condition score (BCS) [14], and impairment of cleanliness score (CS) [15]. Indeed in herd with high percentage of FDs, lameness and behavioral changes (e.g. reduced total eating time, meals per day, increased lie down time, etc.) of the affected animals are often observed [16, 17], as a consequence of discomfort and a painful status [18–20]. As in cows, a Welfare Quality® [21] system (www.welfarequality.net) was developed also for Mediterranean buffaloes; lameness, BCS and CS are some of the criteria and measures for welfare assessment introduced by this project [22, 23].

The most common FDs in dairy cows comprise: sole ulcers (SU), white line disease (WLD), digital dermatitis (DD) and interdigital necrobacillosis (interdigital phlegmon – IP) [24, 25], while in MB, claw overgrowth (OG) and formation of corkscrew claws (CC) are the only claw disorders described until now [26]; no data are available about other type of FDs causing lameness.

In the field, the diagnosis of FDs can sometimes be challenging, because not all lesions cause lameness, and animals often have more than one type of lesion per foot; SU and IP, when present, are likely to be the cause of lameness while WLD, DD and heel horn erosion can frequently be present in non-lame cows [13]. In these animals a remarkable clinical support could be provided by novel electronic sensor systems able to automatically detect lameness due to foot lesions, actually not available for MB [27–29]. Considering these premises, the aim of this study was to describe the clinical findings of foot disorders (FDs) observed during routine foot trimming in dairy MB and their influence on animal welfare.

Methods

Animals and farms

The survey was performed on 1297 MB from 4 free-stall dairy farms in the Latina District - Middle Italy. This samples size was calculated by using the formula proposed by Thrusfield [30] considering the following values: study population (378.000 MB head in 2015) [31], expected prevalence of FDs (39.5 % in multiparous cows) [32], confidence interval 95 % and desired absolute precision (5 %).

Farms were selected following convenient sampling by the local veterinary practicing claw-trimmer. All of them were characterized by barns with solid grooved concrete floors in the walking and feeding alleys. The lying area consisted either of elevated cubicles covered with rubber mattresses for milking MB or of a roofed deep straw yard area for dry MB. In the farms enrolled, similar cleaning procedures were employed. All the animals enrolled were fed with a total mixed ratio including hay, silage and a multi-vitamin integrator. In all farms, milking animals were fed and milked two times/day. All the enrolled farms were characterized by a spring-summer deseasonalized calving system.

Data acquisition

All the procedures performed in this study followed the common *good clinical practices* and received an institutional approval by Ethical Animal Care and Use Committee of University of Naples "Federico II"; moreover the farmers were previously informed and in agreement with purpose and methods used. Dairy farms were visited between April 2014 and March 2015. Routine foot trimming procedures were performed on all multiparous buffalo cows (≥2 lactations), whether lame or not, by the same investigator, one-time per year.

Foot disorders

Presence or absence of FDs, as well as their localization, were recorded at claw trimming. The lesions observed in buffalo cows were categorized according to the ICAR Claw Health Atlas for cows [33], because of the lack of previous studies describing FDs in buffaloes.

Similar to the definition in cows [34], claw overgrowth (hereafter referred to as "overgrowth" [OG]) in MB was considered as an excessive elongation of the claws. No studies describing claws' anatomical differences between MB and cows were published so far. According to the experience of the authors, the normal dorsal wall length (distance from coronet to tip of claw) in MB (after the foot-trimming) is usually considered 9 cm. As a consequence, lengths ranging from ≥ 9.1 to ≤ 10 cm were considered as moderate OG, but still within an acceptable range; a dorsal wall length exceeding 10 cm was instead considered affected by severe OG and categorized as FD by the authors in the current study.

Regarding CC, as for cows, it was considered as any torsion of either the outer or inner claw characterized by presence of the dorsal edge of the wall deviating from a straight line [33]. Corkscrew claw was diagnosed, if complete correction of the torsion at claw trimming was not possible.

Animal welfare

All the animals affected by FDs were submitted to a complete clinical examination and their welfare status was determined by assessment of locomotion score (LS), body condition score (BCS) and cleanliness score (CS).

Locomotion score

The LS of each animal was assessed and recorded before trimming by one and the same investigator to avoid inter-observer differences in scoring. Observations were performed from the side and the rear of the buffalo cow, while it was moved to the trimming chute on flat and non-slippery surfaces. Buffalo cows were evaluated for their lameness status using the 5-point LS system described by Sprecher et al. [35] based on observation of posture and gait. The score was also dichotomized in a group with lameness status "non-lame" (scores 1–2) and a group with lameness status "lame" (scores 3–5) for the statistical analysis. A mobility score system validated for cows was used in the present study due to the lack of a classification system specific for MB. A LS > 2 was considered indicative of impairment of welfare.

Body condition score

All the scorings were performed by one and the same investigator when the animals affected by FDs were confined in the trimming chute. The BCS was estimated by visual inspection of the animals, and manual palpation of four areas of the body where MB store fat: ribs, spine, hips and base of the tail (hereafter referred to as "tail") according to a 5-point scoring method introduced by [36] and slightly modified by the authors (Table 1). A score within the range of 1 (emaciated) and 5 (obese), with increments of 0.25 was given to each animal enrolled (score 3 = ideal BCS). For the analysis, BCS was grouped as under-conditioned (BCS ≤ 2.50), normally conditioned (BCS ≥ 2.75 and ≤ 3.50), and over-conditioned (BCS ≥ 3.75). Impairment of welfare was considered in animals that were under or over-conditioned (BCS ≤ 2.5 or ≥3.75, respectively).

Cleanliness score

Cleanliness score was assessed using a scoring method previously reported by Nielsen et al. [15] for cows. Briefly, hind feet and limbs were the anatomical area evaluated always by the same investigator before foot-trimming. The scoring system aimed to track the amount of manure present and the distance it extended proximally up the leg by means of an ordinal scale from 1 to 4, in which: 1 = little or no manure above the coronary band; 2 = minor splashing above the coronary band; 3 = distinct plaques of manure above the coronary band, but with hair coat visible; 4 = solid plaques of manure extending high up to the hock and stifle. To the knowledge of the authors, no studies describing the effects of CS on MB welfare were published so far; as a consequence, according to the experience of the authors, values of CS > 2 were considered indicative of impairment of welfare.

Table 1 Description of assessment criteria for individual components of the BCS in Mediterranean buffaloes (ribs, spine, hips, and tail) according to Ezenwa et al. [36] modified

Region of body				
Score	Ribs	Spine	Hips	Tail
5 (Obese)	Not visible; fatty layer on and between ribs	Spine bones not visible. Spine sits in slight depression between fatty bulges left and right of spine	Convex, smooth rear, hip bones not visually apparent	Tail base sits in depression surrounded by soft fatty tissue
4	Few ribs visible towards abdomen; ribs can be felt	Spine bones not visible. Spine feels flat; bone and surrounding tissue are on level	Hip bones can be seen, round smooth appearance and feel	Tail base on level with surrounding fatty tissue
3 (Ideal)	Some ribs visible in center of ribcage; abdominal ribs feel ridged	Spine palpable as a slightly elevated bone center-line	Points of hips distinctly visible; bone easy to feel but not protruding	Tail base protrudes slightly; obvious by touch, but not by sight
2	Ribs visible throughout; all have ridged feel	Individual spinal vertebrae clearly palpable	Points of hips protrude; flanks are concave	Tail base visibly sticks up from surrounding tissue
1 (Emaciated)	Ribs clearly visible with deep depressions between them; very ridged feel	Vertebrae distinguishable by sight and touch	Hip bones protrude beyond the hip point; emaciated rear	Tissue surrounding tail base forms round hollow defined by pelvis

Data regarding type of FDs, LS, BCS and CS were recorded on a spreadsheet and stored before the end of the foot trimming procedure, using the ear tag number of each buffalo cow to unequivocally identify individual animals. Information regarding farm management (type of feeding system, birth seasonality, type of barn, number of milking/day) as well as productive status of the animals with FD (milking or dried-off MB) were obtained from the farmers using a standardized questionnaire.

Statistical analysis

All parameters were analyzed by standard descriptive statistics, and data distribution was assessed using histograms, normal probability plots and Shapiro Wilk tests. Data were expressed as absolute numbers, percentages and averages. Prevalences (PR), motility and lameness indexes (MI and LI, respectively) were also calculated. PR = $E_+/(E_+ + E_-)$, where E_+ represents the number of animals showing the "*Event*" studied, while E_- represents the number of animals without the *Event* but that can express it (animals at risk); MI = number of cows not lame/number of multiparous cows in the herd; LI = number of cows lame/number of multiparous cows in herd [37]. Significant differences between expected and observed frequencies of categorical data (type of FD, LS, BCS, CS and productive status) were assessed by means of contingency tables, using χ^2-test. Probabilities of $P < 0.05$ were considered significant. Data were analyzed using dedicated software (SPSS, Version 17.0, Chicago, IL).

Results

Prevalence of foot disorders

The prevalence of animals affected by FDs was 17.7 % (229/1297): overgrowth (OG) was present in 17.0 % (220/1297) of the MB enrolled, corkscrew claw (CC) in 15.8 % (205/1297), interdigital phlegmon (IP) in 0.9 % (12/1297), white line abscess (WLA) in 0.8 % (11/1297), digital dermatitis (DD) in 0.1 % (1/1297) and interdigital hyperplasia (IH) in 0.1 % (1/1297). Considering the whole sampling population, simultaneous presence of FDs was recorded in 17.0 % of MB (221/1297): OG + CC occurred together in 15.8 % of cases (205/1297), OG + IP in 0.3 % (4/1297), OG + WLA in 0.8 % (11/1297) and DD + IH in 0.1 % (1/1297). Significant associations between OG and CC [$\chi^2 = 1191.967$, degree of freedom (df) = 1, $P < 0.0001$] and between OG and WLA ($\chi^2 = 54.311$, df = 1, $P < 0.0001$) were detected. No lameness caused by orthopedic pathology located outside the feet was detected during the investigation.

Prevalence, localizations and type of disorders in the population affected by FDs are indicated in Table 2. Briefly, considering the group of MB affected by a FD, 26.2 % (60/229) of FDs were located in the front limbs, while 73.8 % (169/229) were located in the hind limbs, and 7.8 % (16/205) both in front and hind limbs. Of all FDs identified, 216 (94.3 %) were non-infectious disorders and 13 (5.7 %) infectious disorders.

Table 2 Type, prevalence and localization of foot disorders recorded in Mediterranean buffaloes ($n = 229$)

FDs	Prevalence	Limb	Prevalence	Claw	Prevalence
OG	96.1 % (220)	FL + HL	100 % (220/220)	/	/
CC	89.5 % (205)	FL	27.8 % (57/205)	MC	100 % (57/57)
				LC	/
		HL	72.2 % (148/205)	MC	93.9 % (139/148)
				LC	6.1 % (9/148)
IP	5.2 % (12)	FL	8.3 % (1/12)	/	/
		HL	91.7 % (11/12)	/	/
WLA	4.8 % (11)	FL	27.3 % (3/11)	MC	100 % (3/3)
		HL	72,7 % (8/11)	LC	100 % (8/8)
DD	0.4 % (1)	HL	100 % (1/1)	/	/
IH	0.4 % (1)	HL	100 % (1/1)	/	/

(x) = number of buffaloes; *FD* foot disorder, *OG* overgrow, *CC* corkscrew, *IP* interdigital phlegmon, *WLA* white line abscess, *DD* digital dermatitis, *IH* interdigital hyperplasia, *FL* front limb, *HL* hind limb, *MC* medial claw, *LC* lateral claw

Animal welfare

Of all buffaloes affected by FDs, 90.0 % (206/229) showed LS score > 2, 19.6 % (45/229) a BCS ≤ 2.5 and 59.8 % (137/229) a CS > 2.

Lameness prevalence and locomotion score

Motility and lameness indexes recorded in the overall sample population were 84.1 % (1091/1297), and 15.9 % (206/1297), respectively; 57.3 % (118/206) of lame MB were milking while the 42.7 % (88/206) were dried-off. considering all the animals enrolled, lameness was always associated with the presence of at least one FD (LS > 2) ($\chi^2 = 1142.13$, df = 1, $P < 0.0001$), except for 23 MB with simultaneous occurrence of OG + CC. Significant differences between expected and observed frequencies of various LS on buffaloes affected by FDs were found for OG ($\chi^2 = 94.125$, df = 3, $P < 0.0001$), CC ($\chi^2 = 229.000$, df = 3, $P < 0.0001$), IP ($\chi^2 = 157.922$, df = 3, $P < 0.0001$) and WLA ($\chi^2 = 133.072$, df = 3, $P < 0.0001$). Prevalence, type of FDs and associated degree of lameness in the 229 buffaloes with FDs are described in detail in Table 3. Briefly, 10.0 % (23/229) of MB were scored as LS2 (non-lame group), while 79.5 % (182/229), 7.4 % (17/229) and 3.1 % (7/229) as LS3, LS4 and LS5, respectively (lame group). No statistically significant difference was found between MB with different productive status (milking versus dried-off) concerning their lameness status (lame versus non-lame).

Table 3 Type and prevalence of foot disorders in lame and non-lame Mediterranean buffaloes ($n = 229$) associated with different locomotion scores

LS	OG + CC	OG + IP	OG + WLA	DD + IH	IP
2	10.0 % (23)	-	-	-	-
3	79.5 % (182)	-	-	-	-
4	-	1.7 % (4)	2.2 % (5)	-	3.5 % (8)
5	-	-	2.6 % (6)	0.4 % (1)	-

(x) = number of buffaloes, *LS* locomotion score, *FD* foot disorder, *BCS* body condition score, *CS* cleanliness score, *OG* overgrow, *CC* corkscrew, *IP* interdigital phlegmon, *WLA* white line abscess, *DD* digital dermatitis, *IH* interdigital hyperplasia

Body condition score

Regarding the BCS, none of the MB with FDs were over-conditioned (BCS ≥ 3.75), while 80.3 % (184/229) were normal-conditioned (BCS ≥ 2.75 and ≤ 3.5) and 19.7 % (45/229) were under-conditioned (BCS ≤ 2.5). No statistically significant difference was found in MB with FDs between lameness status lame ($n = 206$) versus non-lame ($n = 23$) concerning BCS distribution. Of all MB with FDs, such with OG or CC were mainly associated with a BCS ranging between 2.50 and 3.00 while MB with WLA or IP mainly showed a BCS of 2.25 and 2.5; the only case of DD + IH detected showed a BCS of 2.75. A significant association between BCS and OG ($\chi^2 = 33.408$, df = 3, $P < 0.0001$) as well as between BCS and CC ($\chi^2 = 123.877$, df = 3, $P < 0.0001$), BCS and IP ($\chi^2 = 46.006$, df = 3, $P < 0.0001$), BCS and WLA ($\chi^2 = 206.966$, df = 3, $P < 0.0001$) were detected.

Prevalence and type of lesions in lame and non-lame MB enrolled affected by FDs, according to BCS observed, are reported in Table 4.

Cleanliness score

The CS 4 was recorded in 50.2 % (115/229) of MB affected with FDs, while it was 3 in 9.6 % (22/229), 2 in 6.6 % (15/229) and 1 in 33.6 % (77/229), respectively. In the MB with FD (lame and non-lame), the CS ranged between 1 and 4. Overgrowth and CC were detected at any scoring point, WLA and IP mainly at scores 3 and 4, and the only case of HH + ID showed score 4. Significant associations were found between CS and CC ($\chi^2 = 15.335$, df = 3, $P < 0.01$) as well as between CS and WLA ($\chi^2 = 7.824$, df = 3, $P < 0.05$). No associations were instead recorded for CS and OG ($\chi^2 = 7.135$, df = 3, $P = 0.068$), and CS and IP ($\chi^2 = 7.106$, df = 3, $P = 0.069$).

In Table 4, prevalence and types of lesions are given in lame and non-lame buffaloes enrolled, according to CS observed.

Table 4 Type and prevalence of foot disorders in lame and non-lame Mediterranean buffaloes associated with different body condition scores and cleanliness scores

Groups	BCS	FD	Prevalence	LS
Non-lame (23)	2.50	OG + CC	8.7 % (2)	2
	2.75	OG + CC	65.2 % (15)	
	3.00	OG + CC	26.1 % (6)	
Lame (206)	2.25	OG + IP	0.5 % (1)	4
		OG + WLA	2.4 % (5)	
		OG + WLA	1.9 % (4)	5
	2.50	OG + CC	10.2 % (21)	3
		IP	3.9 % (8)	4
		OG + IP	1.0 % (2)	
		OG + WLA	1.0 % (2)	5
	2.75	OG + CC	75.7 % (156)	3
		OG + IP	0.5 % (1)	4
		DD + IH	0.5 % (1)	5
	3.00	OG + CC	2.4 % (5)	3
Group	CS	FD	Prevalence	LS
Non-lame (23)	1	OG + CC	30.4 % (7)	2
	2	OG + CC	4.3 % (1)	
	3	OG + CC	8.7 % (2)	
	4	OG + CC	56.5 % (13)	
Lame (206)	1	OG + CC	34.0 % (70)	3
	2	OG + CC	6.3 % (13)	3
		IP	0.4 % (1)	4
	3	OG + CC	8.2 % (17)	3
		OG + IP	0.5 % (1)	4
		OG + WLA	0.5 % (1)	
		OG + WLA	0.5 % (1)	5
	4	OG + CC	40.0 % (82)	3
		OG + IP	1.5 % (3)	4
		IP	3.4 % (7)	
		OG + WLA	1.9 % (4)	
		OG + WLA	2.4 % (5)	5
		DD + IH	0.5 % (1)	

(x) = number of buffaloes, *LS* locomotion score, *FD* foot disorder, *BCS* body condition score, *CS* cleanliness score, *OG* overgrow, *CC* corkscrew, *IP* interdigital phlegmon, *WLA* white line abscess, *DD* digital dermatitis, *IH* interdigital hyperplasia

Discussion

In this study, the clinical findings of foot disorders (FDs) affecting animal welfare are outlined for the first time in Mediterranean Buffaloes. Some FDs reported were never observed before in these ruminants. As a consequence, a comparative analysis with cows was done; in this context, the sample size was reached using the prevalence data originating from similar studies in cows [32], and it was more than tripled to increase the data's reliability.

The overall lameness index detected in the 1297 MB enrolled was 15.9 % (206/1297); similar results were found in analogous studies performed in dairy cow farms in several European countries. Indeed in recent studies, the prevalence of lameness in dairy cows ranged from 14.8 to 19.0 % [24, 38]. Farmers, farm staff as well as veterinary practitioners traditionally tended to underestimate lameness in MB for a long time. This may be due to the low level of expression of the signs of pain by MB on one hand, and the lack of serious investigations performed in this field on the other hand. As described for cows, farmers may also do not feel the real necessity to investigate the problem, since they became tolerant to the levels of lameness in their herd, because the situation was similar to that of other herds they know [39]. Furthermore, unlike mastitis and infertility, from the farmers' perspective, lameness does not overtly impact on farm economics, as lame animals continue to produce milk. So, for busy farmers to care for a lame animal is of low priority [39]. In the current study, the analysis did not reveal a significant difference of lameness prevalence according to the productive status; however it is necessary to highlight that the lack of more accurate data regarding the exact phase of lactation (e.g.: fresh MB or early-, mid-, or late-lactation animals) or dry-off (e.g.: dry or close-up MB) and milk yield may represent one of the restrictions of this study.

Foot disorders observed were mainly localized on the hind feet (73.8 %) and less frequently on the front feet (26.2 %), confirming similar results recently described by Somers and O'Grady [25] for dairy cows, where 11.8 and 89.6 % of the cows showed lesions recorded on front and hind feet, respectively.

Claw overgrowth (OG) represented the most frequent FD found in this study (96.1 %–220/229). Size and shape of the claw at any one time is a balance between growth and wear rate; several factors such as age of the animals, high concentrate feeding, degree of exercise and surface characteristics of the paddock and the stall alleys significantly influence the horn growth [40]. As shown in Table 3, severe OG was always detected in association with foot diseases (CC, WLA and IP) and the most frequent combination observed was represented by OG + CC (205/229). As widely described for cows [34], claw OG predisposes for the occurrence of others FDs potentially causing lameness also in MB.

During the current investigation, claw corkscrew (CC) was the most common non-infectious foot disease recorded both in the lame (182/206) and non-lame (23/23) group of MB (Tables 2 and 3), and it was observed only associated with OG as reported also by Cammarano and Marino [26]. In MB such as in cows, CC has been mainly ascribed to genetic heritable disorders, transmitted by bulls carrying the defects which should basically be excluded from reproduction [12, 40].

Regarding the influences of these two FDs on lameness, our data suggest that both OG and CC, when not associated with other foot diseases, are responsible for a mild (L2) to moderate (L3) lameness ($P < 0.0001$) (Table 3); the result obtained is in contrast with Napolitano et al. [11], who indicated this clinical sign as virtually absent in the 3 buffalo farms examined. Nevertheless, the LS values observed did not significantly influence the other welfare parameters considered (BCS and CS, Table 4), showing that these two FDs may not substantially affect buffaloes' welfare.

Regarding the white line abscess (WLA), it represents a white line disorder mainly secondary to weakness of the white line, resulting, at the same time, in an abnormal horn production as a consequence of a laminitic insult (coriosis) [34]. White line disorders are considered some of the most important causes of severe lameness in cows [32]. While 4.8 % of the MB with FDs showed signs of WLA (Table 2), the values observed were lower than those recently described for lame cows, where the prevalence of WLA ranged from 17.2 to 23.5 % [25, 32].

In the present investigation, WLA was significantly associated with OG ($P < 0.0001$, Table 3); in agreement with Blowey [40], also in MB claw overgrowth could be considered as predisposing factor for white line disorders, increasing significantly the weakness of the hoof. Furthermore, as reported by Barker et al. [41] in cows, several factors can significantly increase the risk of WLA occurrence such as herd size, access to pasture and solid grooved concrete floors; in particular shallow grooves or loss of friction of solid grooved concrete over time lead to slippery floor surfaces associated with an increase in lameness, alterations in gait and increasing pressure on the sole and white line [42]. Solid grooved concrete floors were present in the alleys of all the farms enrolled and may have contributed to the appearance of the FDs, although this hypothesis lacks further scientific evidence. As proven by Whay et al. [19], white line disorders can also reduce cow's nociceptive threshold increasing the sensitivity to pain. As shown in Tables 3 and 4, WLA was always associated with severe clinical lameness, suggesting that the pathology is painful also in MB ($P < 0.0001$, LS $\geq$ 4). As a consequence, the stressful condition experienced by the MB affected by WLA seemed to significantly influence the other welfare parameters considered: all of the animals were indeed in the under-conditioned group (BCS $\leq$ 2.5–11/11; Table 4) and showed values of cleanliness score above the threshold (CS3: 2/11–CS4: 9/11; Table 4). Similarly to severely lame cows, the authors suppose that, also in MB, this impairment of welfare might mainly be the consequence of less active animals, reduced total eating time, numbers of meals per day or dry matter intake as well as a longer lying time [16, 43]; further studies are necessary to confirm this hypothesis.

Interdigital phlegmon (IP) represents the second most frequent foot infectious disease observed in the group of lame MB (5.2 %, Table 2); a similar prevalence (4.4 %) was described by DeFrain et al. [32] in cattle. Injuries of the interdigital skin are usually a prerequisite for the infection; stones, stubble, kale stems, hardened dung in the interdigital area, sticks, very dry pasture, or rough flooring can easily damage the area of the feet involved [34, 40]. A bacterial invasion of the skin of the interdigital cleft mainly due to *Fusobacterium necrophorum* is usually associated with IP in cattle [39]. Unfortunately, no information regarding the injuries promoting the infection and the type of microorganisms primarily involved in MB' IP are available, representing another restriction of the study.

As described by Whay et al. [18, 19], lesions of the interdigital skin are often related to hyperalgesia and consequently to lameness. This seems to hold true also in MB, because all the animals affected by IP had a severe lameness with gaits showing hesitating steps ($P < 0.0001$) (LS4, Table 3). Moreover, considering the effects on BCS and CS, 91.7 % (11/12) were in the underconditioned group ($BCS \leq 2.75$, $P < 0.0001$) and rated with $CS > 2$ (CS3: 1/11–CS4: 10/11, $P = 0.068$, Table 4). These results showed a lower effect of IP on impairment of behavior and welfare of affected animals than WLA.

Regarding digital dermatitis (DD), a single case of a DD-associated interdigital hyperplasia (IH) was detected in the present study. Digital dermatitis is known as bacterial skin infection primarily affecting the epidermis and seriously compromising the welfare and productivity of affected cattle [44]. It was described for the first time in Italy by Cheli and Mortellaro [45] and since then, it rapidly spread all over the world. The DD lesion occurred on the tip of IH and was diagnosed according to the clinical picture, showing close similarity to that in dairy cattle. Neither histological evaluation nor PCR for the relevant *Treponema* species were performed. In comparison to the situation in dairy cattle [24, 25, 32], the prevalence of DD was very low in MB, although management of both dairy species is very similar. The lack of a previous description in MB could be both related to the inability to recognize DD (if associated with a low LS) as well as to a lower sensitivity of this species to slurry exposure, commonly considered a major risk factor of DD in dairy cows [46]. This clinical findings merit further scientific attention.

As last, it is interesting to underline that other noninfectious foot diseases, such as sole ulcers (SU), were not found although they occur frequently in dairy cows. The pathogenesis of SU in dairy cow was recognized as the consequence of abnormal compression of the corium and nutritional imbalances, leading to ischemic necrosis, altered horn formation and formation of horn of poor quality, finally reaching the bearing surface of the sole [40]. The reason for the obvious difference in the prevalence of SU between MB and dairy cows is not known.

Making an overall judgement on the effect of FDs on welfare, as described for cows by Tadich et al. [13], a significant association between the presence of FDs and various degrees of lameness has been detected also in MB ($P < 0.0001$). The majority of MB within the underconditioned group (95.5 %, 43/45) and all those with $CS > 2$ (122/122) had a locomotion score above the threshold of normality ($LS > 2$), adversely affecting buffaloes' welfare. Furthermore, foot diseases such as IP, WLA and DD or IH but not CC and OG seemed to occur more frequently associated with decreased BCS and increased CS scores (Table 4).

Conclusions

This study represents the first clinical description of involvement of white line disease, interdigital phlegmona, digital dermatitis and interdigital hyperplasia in foot disorders of dairy Mediterranean buffalo, revealing impairment of animal welfare of affected animals. More complete studies including the assessment of the pathogenesis, risk factors and economic impact of FDs on MB dairy industry are warranted. By improving the scientific knowledge concerning FDs of MB, awareness of the problem and empathy with the welfare of these dairy ruminants can be increased.

Abbreviations

BCS, body condition score; CC, corkscrew; CS, cleanliness score; DD, digital dermatitis; FD, foot disorder; IH, interdigital hyperplasia; IP, interdigital phlegmon; LI, lameness indexes; LS, locomotion score; MB, Mediterranean buffaloes; MI, motility index; OG, overgrow; PR, prevalence; WLA, white line abscess

Acknowledgements

The Authors acknowledge the precious collaboration of all the farmers involved during the study.

Funding

No funding was obtained for this study.

Authors' contributions

JG, CC, PC & AS participated in the conceptual aspect and design of the study. JG, CC, LDA, ADL, ADR & AS involved in data collection. JG, MA, PC & AS statistical evaluation and data interpretation. All authors provided consultation and coordination. JG, PC & AS wrote the first draft of the manuscript, with all authors involved in reviewing. All authors read and approved the final version of the manuscript.

Competing interests

The authors declare that they have no competing interests.

Consent for publication

Not applicable.

Author details
[1]Department of Veterinary Medicine and Animal Productions, University of Napoli "Federico II", Via F. Delpino 1, 80137 Naples, Italy. [2]Veterinary practitioner, Latina District 04100, Italy. [3]Clinic for Ruminants, Department of Clinical Veterinary Medicine, Vetsuisse-Faculty, University of Berne, 3012 Bern, Switzerland. [4]Department of Health Science, University of Magna Graecia of Catanzaro, 88100 Catanzaro, Italy.

References

1. Whay HR, Main DC, Green LE, Webster AJ. Assessment of the welfare of dairy cattle using animal-based measurements: direct observations and investigation of farm records. Vet Rec. 2003;153:197–202.
2. European Food Safety Authority. Scientific report of EFSA prepared by the animal health and animal welfare unit on the effects of farming systems on dairy cow welfare and disease. Annex EFSA J. 2009;1143:1–38.
3. Booth CJ, Warnick LD, Grohn YT, Maizon DO, Guard CL, Janssen D. Effect of lameness on culling in dairy cows. J Dairy Sci. 2004;87:4115–22.
4. Green LE, Hedges VJ, Schukken YH, Blowey RW, Packington AJ. The impact of clinical lameness on the milk yield of dairy cows. J Dairy Sci. 2002;85: 2250–56.
5. Garbarino EJ, Hernandez JA, Shearer JK, Risco CA, Thatcher WW. Effect of lameness on ovarian activity in postpartum Holstein cows. J Dairy Sci. 2004; 87:4123–31.
6. Vidu L, Diaconescu C, Udroiu A, Bacila V, Popa D, Popa R, Stanciu M. The buffalo -part of animal biodiversity in Romania and the importance for the bioeconomy. In 14th international multidisciplinary scientific geoconference surveying geology and mining ecology management. SGEM. 2014;5:665–72.
7. Guccione J, Cosandey A, Pesce A, Di Loria A, Pascale M, Piantedosi D, et al. Clinical outcomes and molecular genotyping of *Staphylococcus aureus* isolated from milk samples of dairyprimiparous Mediterranean buffaloes (Bubalus bubalis). J Dairy Sci. 2014;97:7606–13.
8. Vecchio D, Neglia G, Di Palo R, Prandi A, Gasparrini B, Balestrieri A, et al. Is a delayed treatment with GnRH, hCG or progesterone beneficial for reducing embryonic mortality in buffaloes? Reprod Domest Anim. 2008;45:614–8.
9. Guccione J, Pesce A, Pascale M, Tommasini N, Garofalo F, Di Loria A, et al. Short communication: effects of systemic treatment with penethamate hydriodide on udder health and milk yields in dry primiparous Mediterranean buffaloes (Bubalus bubalis). J Dairy Sci. 2014;97:2219–25.
10. Guccione J, Perreten V, Steiner A, Thomann A, Pesce A, Ciaramella P, et al. Short communication: role of *Streptococcus pluranimalium* in Mediterranean buffaloes (Bubalus bubalis) with different udder health statuses. J Dairy Sci. 2016;99:2945–9.
11. Napolitano F, Grasso F, Bordi A, Tripaldi C, Saltalamacchia F, Pacelli C, et al. On-farm welfare assessment in dairy cattle and buffaloes: evaluation of some animal-based parameters. Ital J Anim Sci. 2005;4:223–31.
12. De Rosa G, Napolitano F, Grasso F, Pacelli C, Bordi A. On the development of a monitoring scheme of buffalo welfare at farm level. Ital J Anim Sci. 2005;4:115–25.
13. Tadich N, Flor E, Green LE. Associations between hoof lesions and locomotion score in 1098 unsound dairy cows. Vet J. 2010;184:60–5.
14. Bicalho RC, Machado VS, Caixeta LS. Lameness in dairy cattle: a debilitating disease or a disease of debilitated cattle? A cross-sectional study of lameness prevalence and thickness of the digital cushion. J Dairy Sci. 2009; 92:3175–84.
15. Nielsen BH, Thomsen PT, Sørensen JT. Identifying risk factors for poor hind limb cleanliness in Danish loose-housed dairy cows. Anim. 2011;5:1613–19.
16. O'Callaghan KA, Cripps PJ, Downham DY, Murray RD. Subjective and objective assessment of pain and discomfort due to lameness in dairy cattle. Anim Welf. 2003;12:605–10.
17. Flower FC, Sedbauer M, Carter E, Von Keyserlingk MAG, Sanderson DJ, Weary DM. Analgesics improve the gait of lame dairy cows. J Dairy Sci. 2008;91:3010–14.
18. Whay HR, Waterman AE, Webster AJF. Associations between locomotion, claw lesion and nociceptive threshold in dairy heifers during the peri-partum period. Vet J. 1997;154:155–61.
19. Whay HR, Waterman AE, Webster AJF, O'Brien JK. The influence of lesion type on the duration of hyperalgesia associated with hind limb lameness in dairy cattle. Vet J. 1998;156:23–9.
20. Belge F, Bildik A, Belge A, Kikcalp D, Atasoy N. Possible association between chronic laminitis and some biochemical parameters in dairy cattle. Aust Vet J. 2004;82:556–57.
21. Welfare Quality®. Welfare quality assessment protocol for cattle. Lelystad: Welfare Quality Consortium; 2009.
22. De Rosa G, Napolitano F, Grasso F, Bilancione A, Spadetta M, Pacelli C, et al. Welfare quality®: a pan-European integrated project including buffalo. Ital J Anim Sci. 2007;2(6):1360–63.
23. De Rosa G, Grasso F, Winckler C, Bilancione A, Pacelli C, Masucci F, et al. Application of the welfare quality protocol to dairy buffalo farms: prevalence and reliability of selected measures. J Dairy Sci. 2015;98: 6886–96.
24. Becker J, Steiner A, Kohler S, Koller-Bähler A, Wüthrich M, et al. Lameness and foot lesions in swiss dairy cows: I prevalence. Schweiz Arch Tierheilkd. 2014;156:71–8.
25. Somers J, O'Grady L. Foot lesions in lame cows on 10 dairy farms in Ireland. Irish Vet J. 2015;68:10.
26. Cammarano A, Marino AL. Theorical-practical course of podology. Bubalus Bubalis. 2003;9:13–7.
27. Miekley B, Stamer E, Traulsen I, Krieter J. Implementation of multivariate cumulative sum control charts in mastitis and lameness monitoring. J Dairy Sci. 2013;96:5723–33.
28. Alsaaod M, Syring C, Dietrich J, Doherr MG, Gujan T, Steiner A. A field trial of infrared thermography as a noninvasive diagnostic tool for early detection of digital dermatitis in dairy cows. Vet J. 2003;199:281–85.
29. Nechanitzky K, Starke A, Vidondo B, Müller H, Reckardt M, Friedli K, et al. Analysis of behavioral changes in dairy cows associated with claw horn lesions. J Dairy Sci. 2016;99:2904–14.
30. Thrusfield M. Surveys. In: Thrusfield M, editor. Veterinary epidemiology. Blackwell science Ltd. London: UK; 2007. p. 228–42.
31. Italian Ministry of Health. Banca Dati Nazionale dell'Anagrafe Zootecnica. 2015. http://statistiche.izs.it/portal/page?_pageid=73,12918&_dad=portal&_schema=PORTAL&op=view_rep&p_liv=N&p_sigla_liv=&p_report=plet_rep_r1&p_anno=2015. Accessed Mar. 2015.
32. DeFrain JM, Socha MT, Tomlinson DJ. Analysis of foot health records from 17 confinement dairies. J Dairy Sci. 2013;96:7329–39.
33. Egger-Danner C, Nielsen P, Fiedler A, Müller K, Fjeldaas T, Döpfer D, et al. ICAR Claw Health Atlas. 2015. http://www.icar.org/wp-content/uploads/2016/02/ICAR-Claw-Health-Atlas.pdf.
34. Weaver AD, Jean GS, Steiner A. Lameness. In: Weaver AD, Jean GS, Steiner A, editors. Bovine Surgery and Lameness. 2nd ed. Oxford: Blackwell Science Ltd; 2005. p. 198–257.
35. Sprecher DJ, Hostetler DE, Kaneene JB. A lameness scoring system that uses posture and gait to predict dairy cattle reproductive performance. Theriogenology. 1997;47(6):1179–87.
36. Ezenwa VO, Jolles AE, Brien POM. A reliable body condition scoring technique for estimating condition in African buffalo. Afr J Ecol. 2009;47: 476–81.
37. Bell NJ, Huxley JN. Locomotion, lameness and mobility in dairy cows. Vet Rec. 2009;164:726–6.
38. Rutherford KMD, Langford FM, Jack MC, Sherwood L, Lawrence AB, Haskell MJ. Lameness prevalence and risk factors in organic and non-organic dairy herds in the United Kingdom. Vet J. 2009;180:95–105.
39. Huxley J, Archer S, Bell N, Burnell M, Green L, Potterton S, et al. Control of lameness. In: Green M, editor. Dairy herd health. Oxford: CABI International; 2012. p. 169–204.
40. Blowey RW. Lameness in the foot. In: Andrews AH, Blowey RW, Boyd H, Eddy RG, editors. Bovine medicine disease and husbandry of cattle. 2nd ed. Oxford: Blackwell Science Ltd; 2004. p. 409–31.
41. Barker ZE, Amory JR, Wright JL, Mason SA, Blowey RW, Green LE. Risk factors for increased rates of sole ulcers, white line disease, and digital dermatitis in dairy cattle from twenty-seven farms in England and Wales. J Dairy Sci. 2009;92:1971–78.
42. Telezhenko E, Bergsten C. Influence of floor type on the locomotion of dairy cows. Appl Anim Behav Sci. 2005;93:183–97.
43. Cook NB, Bennett TB, Nordlund KV. Effect of free stall surface on daily activity patterns in dairy cows with relevance to lameness prevalence. J Dairy Sci. 2004;87:2912–22.
44. Gomez A, Cook NB, Bernardoni ND, Rieman J, Dusick AF, Hartshorn R, et al. An experimental infection model to induce digital dermatitis infection in cattle. J Dairy Sci. 2012;95:1821–30.

17

Serum hepatokines in dairy cows: periparturient variation and changes in energy-related metabolic disorders

Jianguo Wang[1*†], Xiaoyan Zhu[1,2†], Guanghui She[1], Yezi Kong[1], Yazhou Guo[1], Zhe Wang[3], Guowen Liu[3] and Baoyu Zhao[1*]

Abstract

Background: During peripartum period, dairy cows are highly susceptible to energy metabolism disorders such as fatty liver and ketosis. Angiopoietin-like protein 4 (ANGPTL4) and fibroblast growth factor 21 (FGF21), known as hepatokines, play important roles in lipid metabolism. The purposes of our study were to evaluate variations of serum ANGPTL4 and FGF21 concentrations in periparturient dairy cows and changes in these serum analyte concentrations of energy-related metabolic disorders in early lactation dairy cows.
This study was divided into two experiments. *Experiment I:* Blood parameters were measured in healthy periparturient Holstein cows from 4 wk antepartum to 4 wk postpartum ($n = 219$). In this experiment, weekly blood samples were obtained from 4 wk before the expected calving date through 4 wk after calving. *Experiment II:* Blood parameters were measured in healthy cows ($n = 30$) and cows with clinical ketosis ($n = 29$) and fatty liver ($n = 25$) within the first 4 wk of lactation. In the present study, all blood samples were collected from the coccygeal vein in the early morning before feeding.

Results: Serum ANGPTL4 and FGF21 concentrations peaked at parturition, and declined rapidly over the following 2 wk Serum ANGPTL4 and FGF21 concentrations were positively correlated with serum non-esterified fatty acids (NEFA) concentration ($r = 0.856$, $P = 003$; $r = 0.848$, $P = 0.004$, respectively). Cows with clinical ketosis and fatty liver had significantly higher serum ANGPTL4 and FGF21 concentrations than healthy cows ($P < 0.01$).

Conclusion: Serum ANGPTL4 and FGF21 concentrations were elevated during peripartum period, suggesting that energy balance changes that were associated with parturition contributed significantly to these effects. Although FGF21 and ANGPTL4 could play important roles in the adaptation of energy metabolism, they may be involved in the pathological processes of energy metabolism disorders of dairy cows in the peripartum period.

Keywords: Dairy cow, Peripartum period, Angiopoietin-like protein 4, Fibroblast growth factor 21, Energy metabolism disorder

Background

The periparturient period, from 4 wk antepartum to 4 wk postpartum, is vitally important to the health status and the reproductive performance of dairy cows. During this period, dairy cows experience a dramatic transition from the pregnant, non-lactating state to the non-pregnant, lactating state [1]. Postpartum milk production leads to a large increase in energy requirements that cannot be met solely through dietary intake. As a result, dairy cows enter a negative energy balance (NEB) that is induced by nutrient prioritization towards the mammary gland [2, 3]. Under NEB conditions, cows start to mobilize their body reserves from adipose tissue, resulting in elevated plasma concentrations of non-esterified fatty acids (NEFA). Although NEFA can be used directly as a fuel source by various tissues such as muscle, and used for milk fat synthesis by the mammary gland.

* Correspondence: jgwang0625@163.com; zhaobaoyu12005@163.com
†Jianguo Wang and Xiaoyan Zhu contributed equally to this work.
[1]College of Veterinary Medicine, Northwest A&F University, Yangling 712100, Shaanxi, China
Full list of author information is available at the end of the article

Liver removes 15 to 20% of NEFA from the bloodstream, in which NEFA can undergo mitochondrial β-oxidation to produce energy [4]. The liver may receive up to 38% of cardiac output, therefore it is susceptible to an oversupply of NEFA when arterial concentrations are elevated [5]. Excessive NEFA will be incompletely oxidized by the liver to generate ketone bodies or re-esterified to form triacylglycerols (TAG), thereby potentially resulting in the development of metabolic disorders associated with NEB, such as ketosis and fatty liver [6]. These disorders are associated with increased veterinary costs, longer calving intervals, decreased milk production, and decreased average lifetime of cows, causing huge economic losses for the dairy industry [6, 7].

Angiopoietin-like protein 4 (ANGPTL4), also known as the fasting-induced adipose factor (FIAF), and fibroblast growth factor 21 (FGF21) have been identified as hepatokines, which are predominantly expressed in the liver and adipose tissue [8, 9]. In dairy cows, most circulating ANGPTL4 and FGF21 originate from the liver [10, 11]. ANGPTL4 is a secreted glycoprotein that can inhibit lipoprotein lipase (LPL) activity and stimulate lipolysis, thereby regulating systemic energy and lipid metabolism. FGF21 has also been identified as a hormonal factor involved in the regulation of metabolic adaptations, such as promotion of hepatic lipid oxidation and ketogenesis, during energy deprivation [12]. Nutritional status can drive expression and secretion of ANGPTL4 and FGF21 in human, mouse and bovine subjects [13]. Under fasting conditions, expression levels of ANGPTL4 and FGF21 are strongly upregulated in the liver and white adipose tissue [9, 14]. Moreover, our previous report demonstrated that NEFA can promote ANGPTL4 and FGF21 expression and secretion in bovine hepatocytes cultured in vitro [15, 16]. However, the in vivo kinetics of ANGPTL4 and FGF21 are not entirely clear in dairy cows. In addition to our interest in whether ANGPTL4 and FGF21 participate in the pathological process of energy metabolic disorders in cows, we hypothesized that the serum concentration of these two hepatokines in dairy cows with ketosis and fatty liver are different from the contemporary healthy cows. Therefore, the purposes of this study were to evaluate variations in serum concentrations of ANGPTL4 and FGF21 in periparturient dairy cows, to examine correlations between serum NEFA and β-hydroxybutyrate (BHBA) concentrations and ANGPTL4 and FGF21 concentrations in the serum of the experimental cows, and to compare the concentrations of hepatokines in serum samples from cows with spontaneous ketosis and fatty liver, and healthy cows.

Results

Changes of serum hepatokines in periparturient dairy cows

In the first part of the study, 310 nonlactating multiparous Chinese Holstein dairy cows in late gestation were housed in a free-stall barn and included in the study from 4 wk before the anticipated calving date through 4 wk after parturition. The cows were 3–6 years old with 1–4 parities and had similar initial body condition scores (BCS, range 3.0–3.5, on a scale of 1–5). During the experiment, 65 cows had subclinical or clinical hypocalcemia, one cow had twins, and 18 cows calved much earlier and 7 cows calved much later than their anticipated calving dates. Ninety-one animals were eventually excluded from the experiment. Finally, 219 cows were chosen.

As shown in Fig. 1, similar to the NEFA concentration, serum ANGPTL4 and FGF21 concentrations peaked on the day of parturition and were significantly higher during parturition than in late pregnancy (2 to 4 wk antepartum) and early lactation (2 to 4 wk postpartum) ($P < 0.01$). Serum BHBA concentration was significantly higher at 1 wk postpartum than during late pregnancy (1 to 4 wk antepartum) and early lactation (2 to 4 wk postpartum) ($P < 0.05$).

Serum NEFA concentration was positively correlated with serum concentrations of ANGPTL4 ($r = 0.856$, $P = 0.003$) and FGF21 ($r = 0.848$, $P = 0.004$) in experimental cows during the periparturient period. However, serum BHBA concentration did not appear to be associated with serum concentrations of ANGPTL4 ($r = 0.066$, $P = 0.866$) and FGF21 ($r = 0.295$, $P = 0.442$) in experimental cows.

Serum analytes in cows with energy-related metabolic disorders

Compared with healthy cows, cows with fatty liver and clinical ketosis exhibited significant higher serum concentrations of NEFA, ANGPTL4 and FGF21 ($P < 0.01$, Fig. 2). In addition, BHBA concentrations were significantly higher in clinically ketotic cows ($P < 0.01$) and cows with fatty liver ($P < 0.05$) than in the control group (Fig. 2).

The activities of enzymes aspartate aminotransferase (AST), alanine aminotransferase (ALT), lactic dehydrogenase (LDH) and alkaline phosphatase (ALP), and the concentrations of two proinflammatory factors, tumor necrosis factor-α (TNF-α) and interleukin-6 (IL-6), in the serum of cows with normal and energy-related metabolic disorders are shown in Table 1. There were significant increases in the activities of AST, ALT and LDH, and the concentrations of TNF-α and IL-6 in the serum of cows with fatty liver ($P < 0.01$) than in the control group. The enzymes AST and LDH were significantly greater in the serum of ketotic cows ($P < 0.01$) than in the control group. ALT activity and IL-6 level were raised, while TNF-α was decreased in the serum of cows with ketosis although not statistically significant. There were no significant differences in the activity of serum

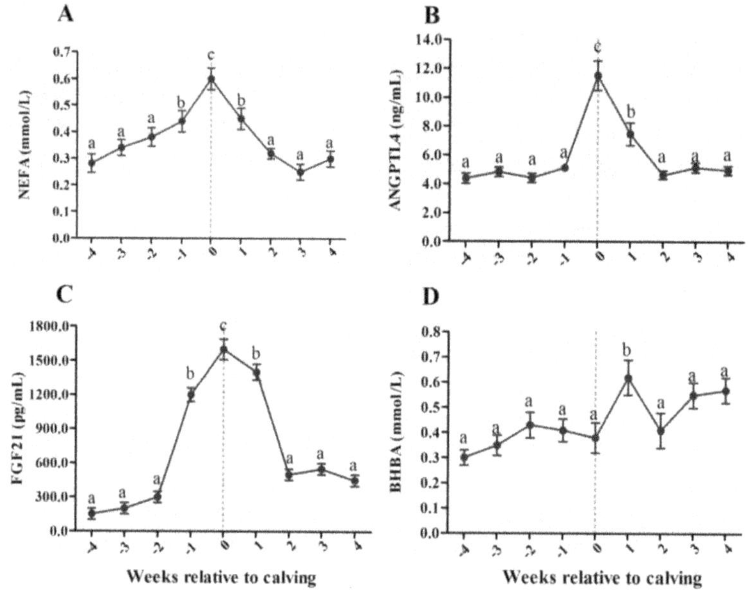

Fig. 1 The variation of serum ANGPTL4, FGF21, NEFA and BHBA concentrations in perparturient dairy cows ($n = 219$). Serum NEFA (**a**), ANGPTL4 (**b**) and FGF21 (**c**) concentrations peaked on the day of parturition and were significantly higher during parturition than in late pregnancy (2 to 4 wk antepartum) and early lactation (2 to 4 wk postpartum). Serum BHBA (**d**) concentration was significantly higher at 1 wk postpartum than during late pregnancy (1 to 4 wk antepartum) and early lactation (2 to 4 wk postpartum). Results are presented as means ± SE and were analyzed using repeated-measures analysis of variance (ANOVA) followed by Bonferroni's multiple comparison test. The neighboring lowercase letters show significant differences at $P < 0.05$, while the divided lowercase letters show very significant differences at $P < 0.01$. The same lowercase letters show no significant differences at $P > 0.05$

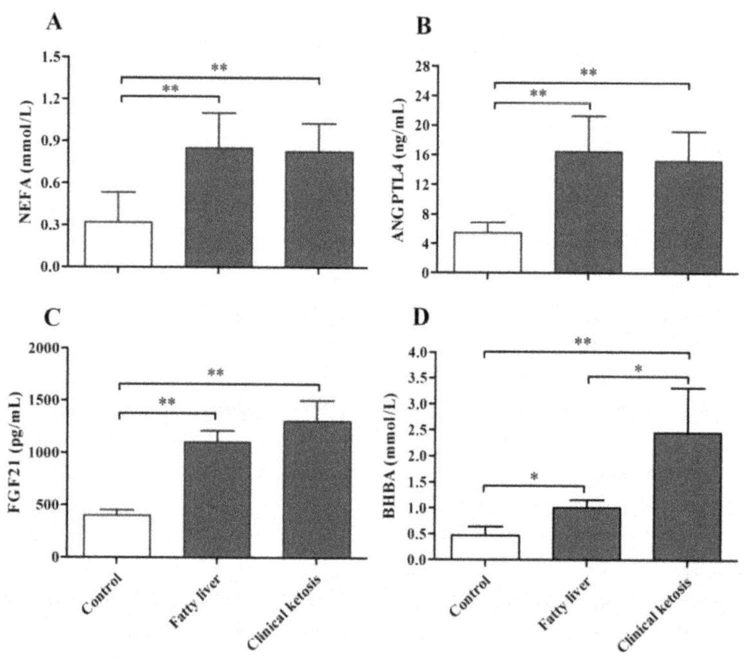

Fig. 2 Levels of ANGPTL4, FGF21, NEFA and BHBA were analyzed in the serum of healthy control cows ($n = 30$) and those affected by clinical ketosis ($n = 29$) and fatty liver ($n = 25$). In the serum, concentrations of NEFA (**a**), ANGPTL4 (**b**), FGF21 (**c**) and BHBA (**d**) were higher in the cows with fatty liver and clinical ketosis than in the healthy control cows. Results are presented as means ± SE and were analyzed by one-way ANOVA followed by Dunnett's multiple comparisons test. $^{*}P < 0.05$, $^{**}P < 0.01$ when compared to the healthy control cows. The comparisons between cows with clinical ketosis and fatty liver were analyzed by unpaired *t*-test. No significant differences in the serum concentrations of NEFA, ANGPTL4 and FGF21 were measured between cows with clinical ketosis and fatty liver ($P > 0.05$). As the basis of ketosis diagnosis, the serum BHBA levels were higher ($P < 0.05$) in the cows with clinical ketosis (2.45 ± 0.71) than in the cows with fatty liver (1.01 ± 0.15)

Table 1 Selected metabolic parameters in the experimental and healthy control groups (means ± SE)

Parameter	Healthy ($n = 30$)	Clinical Ketosis ($n = 29$)	Fatty liver ($n = 25$)
AST (U/L)	42.15 ± 3.12	160.28 ± 15.79**	348.13 ± 25.5**
ALT (U/L)	19.21 ± 0.45	22.29 ± 0.93	74.30 ± 2.15**
LDH (U/L)	862.31 ± 22.21	1012.73 ± 55.15**	1048.60 ± 63.67**
ALP (U/L)	51.33 ± 3.85	53.10 ± 2.18	49.96 ± 3.39
TNF-α (ng/L)	206.97 ± 35.39	170.65 ± 15.10	1551.12 ± 102.37**
IL-6 (ng/L)	6.42 ± 0.60	7.13 ± 1.95	59.23 ± 10.34**

**$P < 0.01$ versus healthy cows

ALP of cows with clinical ketosis and fatty liver when compared with healthy cows ($P > 0.05$).

Discussion

The periparturient period is challenging for both cows and producers due to the rapid increase in energy demands required to support milk production after parturition [17]. During the early lactation period, dietary intake alone cannot meet the enormous energy demands associated with milk production, thereby resulting in NEB, which then leads to the extensive mobilization of fatty acid stored in adipose tissue and causes marked elevations in blood NEFA concentrations. Therefore, serum concentrations of NEFA reflect the magnitude of the mobilization of fat from storage and mirror levels of dry matter intake (DMI) [18]. For exactly these reasons, serum NEFA is considered to be one of the most important indicators of energy balance status [19]. The present study showed that serum NEFA peaked on the day of parturition and declined rapidly over the following 1 wk This result is consistent with previous findings [20, 21].

Peroxisome proliferator-activated receptor α (PPARα) is one of the key transcription factors for regulating lipid metabolism [22]. Research has shown that the NEFA ligand can lead to the activation of PPARα in the liver during early lactation, thus promoting expression of its target genes [23, 24]. FGF21 and ANGPTL4 have been shown to be downstream targets of PPARα. Schlegel and colleagues observed that FGF21 expression was strongly upregulated during the transition period [12]. ANGPTL4 expression and secretion can be driven by the nutritional status of humans and rodents [25, 26], with the plasma ANGPTL4 concentration increasing with fasting and decreasing with high-fat feeding [9]. In the present study, we showed that serum FGF21 and ANGPTL4 concentrations were significantly increased during parturition. The serum NEFA concentration was significantly and positively correlated with concentrations of FGF21 and ANGPTL4 in experimental cows during the periparturient period. Some researchers have identified FGF21 and ANGPTL4 as hepatokines that are markedly responsive to energy balance changes in lactating dairy cows [27, 28]. Moreover, there is evidence that NEFA can trigger hepatic FGF21 and ANGPTL4 upregulation and production through the activation of PPARα in nonruminants. The effects of these fatty acids may explain the trends in serum FGF21 and ANGPTL4 concentrations that varied consistently with NEFA concentration trends in dairy cattle during the periparturient period.

It is worth noting that hepatokines may also influence lipid mobilization in a manner that is dependent on the levels of other hormones, such as growth hormone (GH), adiponectin and insulin. Khan and coworkers indicated that the activation of FGF21 could inhibit or block local GH signaling, and when in circulation, FGF21 can reduce adipose tissue insulin sensitivity and contribute to lipolysis in dairy cows [29]. The reduced insulin sensitivity of adipose tissue during early lactation might be associated with the decreased adiponectin sensitivity in adipose tissue of dairy cows after calving [30]. Adiponectin, a hormone secreted by adipocytes, plays crucial roles in regulating glucose and lipid metabolism and insulin sensitivity. Lin et al. [31] and Holland et al. [32] showed that in mice, FGF21 is a potent regulator of adiponectin secretion and critically depends on adiponectin to exert its energy glycemic and insulin sensitizing effects. In the present study, we found that serum FGF21 concentrations were significantly increased during parturition. In cattle, adiponectin mRNA is highly expressed in adipose tissues. However, adiponectin mRNA expression in adipose tissue is lower in lactating cows than in non-lactating cows [33, 34]. Although these results may be specific in ruminants, the details of these mechanisms are still unclear. There are several possible mechanisms that could account for the paradoxical dissociation between serum FGF21 levels and adiponectin expression in dairy cows during the periparturient period. First, FGF21 is not the only regulator of adiponectin production. Many factors that negatively regulate adiponectin expression and secretion, such as endoplasmic reticulum stress (ERS), and oxidative stress, are progressively increased during the transition period [35, 36]. Therefore, elevated FGF21 is not sufficient to counteract the effects of these negative regulators on adiponectin production under this circumstance. Second, it has been reported the decreases of adiponectin mRNA levels in adipose tissues and serum adiponectin concentrations induced by the NEB during lactation [37]. Therefore, it is likely that the ability of FGF21 to induce adipocyte production of adiponectin is diminished due to severe NEB during parturition. Some studies have shown that in non-ruminant animals, ANGPTL4 exerts distinct effects on lipid and glucose metabolism [25]. Activation of ANGPTL4 contributes to lipolysis by inhibiting LPL activity in adipose tissue, whereas the genetic and viral-mediated overexpression of ANGPTL4 potently decreases the blood glucose concentration by suppressing basal glucose output and inhibiting hepatic glucose production in mice [38]. The inhibitory effect of ANGPTL4 on hepatic glucose production is similar to adiponectin, although they might act through distinct pathways [25]. In addition, very recently, Kim et al. [39] found that ANGPTL4 may play a critical role in the regulation of insulin secretion. A limitation of our study is that we were, for technical reasons, unable to estimate insulin sensitivity in the dairy cows. The detailed metabolic mechanisms that underlie the action of ANGPTL4 on adipose tissue insulin sensitivity in cattle remain to be clarified.

Some studies have reported that hepatokines, especially FGF21, are involved in the regulation of hepatic lipid oxidation and ketogenesis during energy deprivation [40, 41]. However, the results of the present study showed that serum BHBA concentration did not appear to be associated with serum FGF21 and ANGPTL4 concentrations in dairy cows during the periparturient period. Serum BHBA concentration peaked at 1 wk postpartum, the timing of this peak was different from the peak in FGF21 and ANGPTL4 concentrations, which occurred on the day of parturition. Circulating levels of ketone bodies are determined by their rates of production and utilization. When the production of ketone bodies exceeds the extrahepatic tissue's capacity to use them, they accumulate in the bloodstream, resulting in hyperketonemia [18]. Hence, we believe that the results of this study are not contradictory to the theory that FGF21 and ANGPTL4 are involved in the regulation of ketogenesis. Experimental studies have observed a positive linear correlation between hepatic FGF21 mRNA abundance and abundance in the hepatic mRNA for the gene encoding the key enzyme of ketogenesis in postpartum dairy cows [12]. That finding suggested that FGF21 could play a role in the regulation of ketogenesis in dairy cows during lactation, although there was no correlation between hepatic FGF21 mRNA abundance and concentration of BHBA in plasma. However, an earlier study conducted by Hotta et al. showed that blood BHBA levels were markedly increased in fasting FGF21 knock-out mice, indicating that FGF21 was not physiologically required for hepatic ketogenesis [42]. Further studies are needed to investigate the roles and precise mechanisms of hepatokines in ketogenesis in dairy cows.

NEB is considered as the etiological basis of energy metabolic disorders in periparturient dairy cows. Under NEB conditions, excessive fat mobilization results in a marked influx of NEFA in the blood; this influx is crucial to the etiology of energy metabolic disorders, such as fatty liver and ketosis. Fatty liver and ketosis are metabolic disorders that usually develop in dairy cows, particularly in high yielding dairy cows, during the periparturient period [43, 44]. Our results demonstrated that serum FGF21 and ANGPTL4 concentrations were significantly higher in cows with fatty liver and clinically ketotic cows compared with the healthy control cows. Studies have shown that energy insufficiency during early lactation in dairy cows was associated with increased plasma FGF21 [11]. In mice, hepatic expression and circulating levels of FGF21 were induced by both ketogenic diets and fasting [14]. Coincidentally, Loor et al. reported that expression of ANGPTL4 mRNA in the livers of feed-restricted and ketotic dairy cows was significantly higher than that observed in the healthy cows [13]. Moreover, our preliminary study showed that high concentrations of NEFA significantly promoted in vitro ANGPTL4 expression and secretion in bovine hepatocytes [15]. However, as far as we know, no relevant studies have assessed changes in serum ANGPTL4 levels in dairy cows with energy metabolism disorders. In addition, the exact mechanisms of FGF21 and ANGPTL4 on energy metabolism in dairy cows are still unclear. There is evidence that impairment of the insulin regulation of energy metabolism is considered to be an etiologic key factor for metabolic diseases [45]. According to our results, we speculate that upregulation of FGF21 and ANGPTL4 due to NEB could diminish adipose tissue insulin sensitivity and further aggravate lipolysis during early lactation in dairy cows, thus playing a role in the pathological processes of energy metabolism disorders.

In the present study, cows with fatty liver did not show any clinical symptoms of ketosis, and their serum BHBA concentrations were lower than in clinically ketotic cows. However, serum BHBA concentrations still were significantly higher in cows with fatty liver than in the control group. Thus, we could not rule out the possibility of the effects of BHBA on the expressions of FGF21 and ANGPTL4. In addition, this study was implemented in a commercial dairy farm. To avoid the stress and subsequent diseases (such as peritonitis and hepatapostema), liver biopsies were not performed on control group cattle. This is a potential flaw in our study. Although the choice of control group cows was based on the combination of veterinary clinical examination and blood biochemical analysis, the subjectivity of veterinarian diagnosis and the difference of metabolic state may still affect the results. Further studies are needed to clarify the effects of liver TAG content to the expressions of two hepatokines.

Here we showed that the serum activity of the enzymes AST, ALT and LDH were increased in cows with energy-related metabolic disorders, particularly in cows with fatty liver, which are consistent with the previous studies [46, 47]. Elevations in serum activity of these hepatic enzymes have been associated with the liver damage, although they are not ideal indicators for the diagnosis of liver disease in cattle because of they are not liver specific. TNF-α and IL-6 are major proinflammatory cytokines which play important roles in inflammatory response and lipid metabolism. TNF-α can promote IL-6 production and release by various pathways [48, 49]. There is strong evidence that TNF-α plays a role in fatty liver in rodent models. Moreover, serum TNF-α activity was increased in cows with fatty liver, and TNF-α can induce lipolysis by affecting insulin sensitivity [50, 51]. The result of this study showed that the serum concentrations of TNF-α and IL-6 also increased in cows with fatty liver, which further support inflammation as a component in the etiology of fatty liver.

Conclusions

In summary, the present study showed that the serum FGF21 and ANGPTL4 concentrations change dramatically during the periparturient period in dairy cows, particularly during parturition and early lactation. Serum ANGPTL4 and FGF21 concentrations were positively correlated with serum concentration of NEFA in experimental cows during the periparturient period. Compared with healthy cows, cows with fatty liver and clinical ketosis exhibited significant higher serum concentrations of ANGPTL4 and FGF21. Although FGF21 and ANGPTL4 have important roles in the adaptation of energy metabolism, these two hepatokines may be involved in the pathological processes of energy metabolism disorders during the peripartum period in dairy cows.

Methods

A commercial dairy farm located in Heilongjiang Province of China was selected to participate in the study.

Selection of cows

The present study was divided into two experiments.

Experiment I: Two hundred and nineteen periparturient Holstein cows were housed together in a free-stall barn and had ad libitum access to water. All cows were fed the same basal diet as a total mixed ration (TMR), which was formulated to meet the nutrient requirements recommended by the Dairy Cows Breeding Standard of China (2000, Table 2). The TMR was fed twice daily. TMR samples were collected weekly and composited monthly to measure the crude protein (Association of Official Analytical Chemists (AOAC), 2000; method 976.05), neutral detergent fiber (NDF), acid detergent fiber (ADF) [52], Ca and P (AOAC, 2000; method 935.13). In this experiment, the cows with subclinical or clinical hypocalcemia were excluded from the study. Based on the criteria from Roche and Berry, serum Ca levels of 2.0 and 1.4 mmol/L at calving day were proposed as thresholds of subclinical and clinical hypocalcemia, respectively [53].

Experment II: Eighty-four cows of the same breed within 4 weeks after calving were provided by the commercial dairy farm. All cows were housed together in a free-stalls, fed the TMR twice a day, had ad libitum access to water, and milked twice daily. Individual cow milk yield were recorded by an electronic monitoring system daily throughout the experiment. Parity, body

Table 2 Ingredient and nutrient composition of the antepartum and postpartum diets on a dry matter (DM) basis

Item	Antepartum	Postpartum
Ingredient		
Corn silage, %	38.1	30.9
Hay crop silage, %	12.0	15.5
Grass hay, %	12.6	–
Alfalfa hay, %	–	8.6
Ground shelled corn, %	21.1	24.6
Soybean meal (47.5% CP), %	4.0	3.7
Expeller soybean meal, %	5.6	7.7
Whole cottonseeds, %	3.6	5.6
Sodium bicarbonate, %	–	0.5
Mineral and vitamin mix, %	2.9	2.9
Energy and nutrients		
NE_L, Mcal/kg	1.54	1.74
Crude protein (CP), %	15.6	16.4
MP supply, g/d	1433	2099
Neutral detergent fiber (NDF), %	35.7	34.3
Acid detergent fiber (ADF), %	23.4	22.0
Calcium (Ca), %	0.91	1.07
Phosphorus (P), %	0.45	0.49
Magnesium (Mg), %	0.46	0.32
Sodium (Na), %	0.14	0.13
Potassium (K), %	1.48	1.40
Chloride (Cl), %	0.76	0.39
Sulfur (S), %	0.20	0.22

weight (BW), and milk production of the experimental cows are shown in Table 3. Currently, fatty liver can be diagnosed reliably only by performing biochemical analysis to determine the TAG content of a liver puncture biopsy sample (% wet weight). Cows with TAG contents > 10% in their livers were considered to suffer from fatty liver [6]. Cows with typical clinical signs of ketosis (such as anorexia, lethargy, constipation, loss of body weight, and decreased milk production) and serum BHBA concentrations > 1.2 mmol/L were considered to have clinical ketosis [51, 54]. A reference value for liver TAG contents of < 1% was used to exclude a diagnosis of fatty liver in the selected clinically ketotic cows. Cows with normal blood parameters (BHBA< 0.6 mmol/L, NEFA< 0.4 mmol/L) and no clinical signs of metabolic disorders were considered to be healthy controls. In the present study, 54 cows were liver biopsied in order to generate the group of cows with fatty liver. To narrow the screening process, general symptoms can be assessed to make a preliminary diagnosis, such as anorexia, depression, rapid loss of weight, and decreased milk yield. Meanwhile, serum BHBA concentrations < 1.2 mmol/L and no clinical signs of ketosis (including lethargy, constipation and nervous signs) were used to exclude a diagnosis of ketosis in the selected cows with fatty liver. Finally, 25 cows with fatty liver were selected according to the TAG contents of their livers. Twenty-nine cows with ketosis were chosen according to the serum BHBA concentration and TAG contents of their livers. Thirty cows were used to be healthy controls.

Sampling and analyses

Blood

In the first part of the study, blood samples collected from the coccygeal vein in the early morning before feeding were obtained weekly from 4 wk antepartum to 4 wk postpartum. Within 1 h of collection, samples were carried to the laboratory, placed at room temperature for 30 min and centrifuged at 4000×*g* for 10 min to obtain serum. Sera were stored in sterile tubes at − 80 °C until further analysis.

Serum NEFA, BHBA, AST, ALT, LDH and ALP concentrations were assayed using corresponding kits (enzymatic method, Randox Laboratories Ltd., Ibach, Switzerland), respectively. All assays were analyzed using a Hitachi 7170 auto-analyzer (Hitachi Co., Tokyo, Japan). Serum ANGPTL4, FGF21, TNF-α and IL-6 levels were measured using commercial bovine ELISA kits (Cusabio Biotech Co. Ltd., Wuhan, China; ANGPTL4, Catalog No. CSB-EL001712BO; FGF21, Catalog No. CSB-EL008627BO; TNF-α, Catalog No. CSB-E12020B; IL-6, Catalog No. CSB-E12899B). All kits were used according to the manufacturer's instructions.

Liver

In the second part of the study, liver tissue samples were collected by an experienced veterinarian using a liver biopsy needle (Shanghai Surgical Equipment Factory, Shanghai, China). Immediately after collection, liver tissue biopsies were rinsed with saline, frozen in liquid nitrogen, and stored at − 80 °C for determination of liver TAG content. Liver TAG content was measured after extraction according to Schlegen et al. [24]. In brief, lipids from liver biopsy specimens were extracted with a mixture of n-hexane and isopropanol (3:2, vol/vol). An aliquot of the extracts containing 25 to 50 nmol of TAG was pipetted into a glass vial (1.5 mL). After evaporation of the solvent by vacuum, the lipids were resolved in a 20 μL portion of a 1:1-mixture of chloroform and Triton X-100, and again, the solvent was evaporated. TAG content was measured by colorimetry using a commercially available enzymatic TAG kit (Applygen Technologies Inc., Beijing, China) and an automatic biochemical analyzer (Shenyang EKSV Medical Equipment Co., Ltd., China).

Data analysis

Statistical analysis was performed using GraphPad Prism 5, version 5.01 (GraphPad Software Inc., USA). The normality distribution was tested by using the Kolmogorov-Smirnov (KS). Data about serum concentrations of ANGPTL4, FGF21, NEFA and BHBA in periparturient dairy cows, that were normally distributed, were analyzed using repeated-measures analysis of variance (ANOVA) followed by Bonferroni's multiple comparison test. The estimated effects of sampling time on analyte levels were expressed as means ± standard errors (SE). The correlations of serum NEFA and BHBA concentrations and ANGPTL4 and FGF21 concentrations in experimental cows were analyzed by Pearson's correlation analysis. Mean serum

Table 3 Parity, BW, and milk production of cows with clinical ketosis and fatty liver and healthy cows (means ± SE)

Item	Healthy (*n* = 30)	Clinical Ketosis (*n* = 29)	Fatty liver (*n* = 25)
Parity	3.73 ± 0.94	3.65 ± 0.71	3.58 ± 0.69
BW (kg)	661.89 ± 52.15	599.87 ± 71.54*	572.43 ± 66.23*
Milk production (kg/d)	31.13 ± 2.96	28.35 ± 3.23	26.44 ± 4.45

*$P < 0.05$ versus healthy cows

concentrations of ANGPTL4, FGF21, NEFA, BHBA, AST, ALT, LDH, ALP, TNFα and IL-6 in cows with clinical ketosis and fatty liver, and healthy cows (control group) were compared using one-way ANOVA followed by Dunnett's multiple comparisons test. The results were expressed as the means ± SE. $P < 0.05$ was considered statistically significant, and $P < 0.01$ was considered highly statistically significant.

Abbreviations

ALP: Alkaline phosphatase; ALT: Alanine aminotransferase; ANGPTL4: Angiopoietin-like protein 4; AST: Aspartate aminotransferase; BCS: Body condition scores; BHBA: β-hydroxybutyrate; FGF21: Fibroblast growth factor 21; IL-6: Interleukin-6; LDH: Lactic dehydrogenase; LPL: Lipoprotein lipase; NEB: Negative energy balance; NEFA: Non-esterified fatty acids; PPARα: Peroxisome proliferator-activated receptor α; TAG: Triacylglycerols; TNF-α: Tumor necrosis factor-α

Acknowledgments

We thank Yanbin Li and his colleagues for expert technical assistance in the commercial dairy farm located in Heilongjiang Province of China.

Funding

The research was granted by the National Natural Science Foundation of China (31502129), the China Postdoctoral Science Foundation funded project (2014 M560811), the Scientific and Technological Research and Development Program of Shaanxi Province (2016NY-100) and the Fundamental Research Funds for the Central Universities (2014YB014).

Authors' contributions

WJG and ZXY designed and performed experiments and prepared the manuscript; ZXY, SGH and KYZ revised the manuscript; KYZ and GYZ were involved in the statistical analysis; WZ, ZBY and LGW participated in the study design, coordinated all activities of this project and revised the manuscript. All authors read and approved the final manuscript.

Consent for publication

Not applicable.

Competing interests

The authors declare that they have no competing interests.

Author details

[1]College of Veterinary Medicine, Northwest A&F University, Yangling 712100, Shaanxi, China. [2]UMR 1195 Inserm and University Paris-Saclay, Kremlin-Bicêtre, France. [3]College of Veterinary Medicine, Jilin University, Changchun 130062, Jilin, China.

References

1. DeGaris PJ, Lean IJ. Milk fever in dairy cows: a review of pathophysiology and control principles. Vet J. 2008;176(1):58–69.
2. Contreras GA, Sordillo LM. Lipid mobilization and inflammatory responses during the transition period of dairy cows. Comp Immunol Microb. 2011; 34(3):281–9.
3. Leroy JLMR, Vanholder T, Van Knegsel ATM, Garcia-Ispierto I, Bols PEJ. Nutrient prioritization in dairy cows early postpartum: mismatch between metabolism and fertility? Reprod Domest Anim. 2008;43:96–103.
4. Drackley JK, Andersen JB. Splanchnic metabolism of long-chain fatty acids in ruminants. In: Ruminant Physiology: Digestion, Metabolism and Impact of Nutrition on Gene Expression, Immunology and Stress; 2006. https://doi.org/10.3920/978-90-8686-566-6.
5. Reynolds CK, Aikman PC, Lupoli B, Humphries DJ, Beever DE. Splanchnic metabolism of dairy cows during the transition from late gestation through early lactation. J Dairy Sci. 2003;86(4):1202–17.
6. Bobe G, Young JW, Beitz DC. Invited review: pathology, etiology, prevention, and treatment of fatty liver in dairy cows. J Dairy Sci. 2004; 87(10):3105–24.
7. Guard CL. The costs of common diseases of dairy cattle. In: Abstract: proceedings of the central veterinary conference West, San Diego, California, USA; 2008. p. 695–700.
8. Yoon JC, Chickering TW, Rosen ED, Dussault B, Qin YB, Soukas A, et al. Peroxisome proliferator-activated receptor gamma target gene encoding a novel angiopoietin-related protein associated with adipose differentiation. Mol Cell Biol. 2000;20(14):5343–9.
9. Kersten S, Mandard S, Tan NS, Escher P, Metzger D, Chambon P, et al. Characterization of the fasting-induced adipose factor FIAF, a novel peroxisome proliferator-activated receptor target gene. J Biol Chem. 2000;275(37):28488–93.
10. Mamedova LK, Robbins K, Johnson BJ, Bradford BJ. Tissue expression of angiopoietin-like protein 4 in cattle. J Anim Sci. 2010;88(1):124–30.
11. Schoenberg KM, Giesy SL, Harvatine KJ, Waldron MR, Cheng C, Kharitonenkov A, et al. Plasma FGF21 is elevated by the intense lipid mobilization of lactation. Endocrinology. 2011;152(12):4652–61.
12. Schlegel G, Ringseis R, Keller J, Schwarz FJ, Windisch W, Eder K. Expression of fibroblast growth factor 21 in the liver of dairy cows in the transition period and during lactation. J Anim Physiol Anim Nutr. 2013;97(5):820–9.
13. Loor JJ, Everts RE, Bionaz M, Dann HM, Morin DE, Oliveira R, et al. Nutrition-induced ketosis alters metabolic and signaling gene networks in liver of periparturient dairy cows. Physiol Genomics. 2007;32(1):105–16.
14. Badman MK, Pissios P, Kennedy AR, Koukos G, Flier JS, Maratos-Flier E. Hepatic fibroblast growth factor 21 is regulated by PPAR alpha and is a key mediator of hepatic lipid metabolism in ketotic states. Cell Metab. 2007;5(6): 426–37.
15. Wang J, Zhu X, Wang Z, Yao J, Zhao B, Liu G. Non-esterified fatty acids promote expression and secretion of angiopoietin-like protein 4 in calf hepatocytes cultured in vitro. Mol Cell Biochem. 2015;401(1-2):141–6.
16. Wang JG, Guo YZ, Kong YZ, Dai S, Zhao BY. High non-esterified fatty acid concentrations promote expression and secretion of fibroblast growth factor 21 in calf hepatocytes cultured in vitro. J Anim Physiol Anim Nutr. 2018;102(1):e476–81.
17. Caixeta LS, Ospina PA, Capel MB, Nydam DV. The association of subclinical hypocalcemia, negative energy balance and disease with bodyweight change during the first 30 days post-partum in dairy cows milked with automatic milking systems. Vet J. 2015;204(2):150–6.
18. Adewuyi AA, Gruys E, van Eerdenburg F. Non esterified fatty acids (NEFA) in dairy cattle. A review Vet Quart. 2005;27(3):117–26.
19. Petrera F, Napolitano F, Dal Pra A, Abeni F. Plasma parameters related to energy and lipid metabolism in periparturient Modenese and Italian Friesian cows. J Anim Physiol Anim Nutr. 2015;99(5):962–73.
20. Kerestes M, Faigl V, Kulcsar A, Balogh O, Foeldi J, Febel H, et al. Periparturient insulin secretion and whole-body insulin responsiveness in dairy cows showing various forms of ketone pattern with or without puerperal metritis. Domest Anim Endocrin. 2009;37(4):250–61.
21. Schwalm JW, Schultz LH. Relationship of insulin concentration to blood metabolites in the dairy cow. J Dairy Sci. 1976;59(2):255–61.
22. Rakhshandehroo M, Sanderson LM, Matilainen M, Stienstra R, Carlberg C, de Groot PJ, et al. Comprehensive analysis of PPARalpha-dependent regulation of hepatic lipid metabolism by expression profiling. PPAR Res. 2007;2007:26839.
23. White HM, Koser SL, Donkin SS. Differential regulation of bovine pyruvate carboxylase promoters by fatty acids and peroxisome proliferator-activated receptor-alpha agonist. J Dairy Sci. 2011;94(7):3428–36.
24. Schlegel G, Keller J, Hirche F, Geissler S, Schwarz FJ, Ringseis R, et al. Expression of genes involved in hepatic carnitine synthesis and uptake in dairy cows in the transition period and at different stages of lactation. BMC Vet Res. 2012;8(1):28.

25. Xu AM, Lam MC, Chan KW, Wang Y, Zhang JL, Hoo RLC, et al. Angiopoietin-like protein 4 decreases blood glucose and improves glucose tolerance but induces hyperlipidemia and hepatic steatosis in mice. P Nat Acad Sci USA. 2005;102(17):6086–91.
26. Robciuc MR, Tahvanainen E, Jauhiainen M, Ehnholm C. Quantitation of serum angiopoietin-like proteins 3 and 4 in a Finnish population sample. J Lipid Res. 2010;51(4):824–31.
27. Koltes DA, Spurlock DM. Adipose tissue angiopoietin-like protein 4 messenger RNA changes with altered energy balance in lactating Holstein cows. Domest Anim Endocrin. 2012;43(4):307–16.
28. Akbar H, Batistel F, Drackley JK, Loor JJ. Alterations in hepatic FGF21, co-regulated genes, and upstream metabolic genes in response to nutrition, ketosis and inflammation in Peripartal Holstein cows. PLoS One. 2015;10(10):e0139963.
29. Khan MJ, Jacometo CB, Graugnard DE, Corrêa MN, Schmitt E, Cardoso F, et al. Overfeeding dairy cattle during late-pregnancy alters hepatic PPARα-regulated pathways including hepatokines: impact on metabolism and peripheral insulin sensitivity. Gene Regul Syst Biol. 2014;8:97–111.
30. De Koster J, Urh C, Hostens M, Van Den Broeck W, Sauerwein H, Opsomer G. Relationship between serum adiponectin concentration, body condition score, and peripheral tissue insulin response of dairy cows during the dry period. Domes Anim Endocrin. 2017;59:100–4.
31. Lin Z, Tian H, Lam KSL, Lin S, Hoo RCL, Konishi M, et al. Adiponectin mediates the metabolic effects of FGF21 on glucose homeostasis and insulin sensitivity in mice. Cell Metab. 2013;17(5):779–89.
32. Holland WL, Adams AC, Brozinick JT, Bui HH, Miyauchi Y, Kusminski CM, et al. An FGF21-Adiponectin-Ceramide Axis controls energy expenditure and insulin action in mice. Cell Metab. 2013;17(5):790–7.
33. Ohtani Y, Yonezawa T, Song S-H, Takahashi T, Ardiyanti A, Sato K, et al. Gene expression and hormonal regulation of adiponectin and its receptors in bovine mammary gland and mammary epithelial cells. Anim Sci J. 2011; 82(1):99–106.
34. Komatsu T, Itoh F, Sakumoto R, Hodate K, Obara Y, Kushibiki S. Changes in the gene expression of adiponectin and glucose transporter 12 (GLUT12) in lactating and non-lactating cows. Anim Sci J. 2007;78(1):98–102.
35. Gessner DK, Schlegel G, Ringseis R, Schwarz FJ, Eder K. Up-regulation of endoplasmic reticulum stress induced genes of the unfolded protein response in the liver of periparturient dairy cows. BMC Vet Res. 2014;10(1):46.
36. Sordillo LM, Aitken SL. Impact of oxidative stress on the health and immune function of dairy cattle. Vet Immunol Immunop. 2009;128(1–3):104–9.
37. Ohtani Y, Takahashi T, Sato K, Ardiyanti A, Song S-H, Sato R, et al. Changes in circulating adiponectin and metabolic hormone concentrations during periparturient and lactation periods in Holstein dairy cows. Anim Sci J. 2012; 83(12):788–95.
38. McCabe M, Waters S, Morris D, Kenny D, Lynn D, Creevey C. RNA-seq analysis of differential gene expression in liver from lactating dairy cows divergent in negative energy balance. BMC Genomics. 2012; 13(1):193.
39. Kim H-K, Kwon O, Park K-H, Lee KJ, Youn B-S, Kim S-W, et al. Angiopoietin-like peptide 4 regulates insulin secretion and islet morphology. Biochem Bioph Res Co. 2017;485(1):113–8.
40. Douris N, Melman T, Pecherer JM, Pissios P, Flier JS, Cantley LC, et al. Adaptive changes in amino acid metabolism permit normal longevity in mice consuming a low-carbohydrate ketogenic diet. B BA-Mol Basis Dis. 2015;1852(10):2056–65.
41. Akbar H, Grala TM, Riboni MV, Cardoso FC, Verkerk G, McGowan J, et al. Body condition score at calving affects systemic and hepatic transcriptome indicators of inflammation and nutrient metabolism in grazing dairy cows. J Dairy Sci. 2015;98(2):1019–32.
42. Hotta Y, Nakamura H, Konishi M, Murata Y, Takagi H, Matsumura S, et al. Fibroblast growth factor 21 regulates lipolysis in white adipose tissue but is not required for ketogenesis and triglyceride clearance in liver. Endocrinology. 2009;150(10):4625–33.
43. Goff JP, Horst RL. Physiological changes at parturition and their relationship to metabolic disorders. J Dairy Sci. 1997;80(7):1260–8.
44. Bell AW. Regulation of organic nutrient metabolism during transition form late pregnancy to early lactation. J Anim Sci. 1995;73(9):2804–19.
45. Holtenius P, Holtenius K. A model to estimate insulin sensitivity in dairy cows. Acta Vet Scand. 2007;49:29.
46. Kalaitzakis E, Roubies N, Panousis N, Pourliotis K, Kaldrymidou E, Karatzias H. Clinicopathologic evaluation of hepatic lipidosis in periparturient dairy cattle. J Vet Intern Med. 2007;21(4):835–45.
47. Pechova A, Illek J, Halouzka R. Diagnosis and control of the development of hepatic steatosis in dairy cows in the periparturient period. Acta Vet Brno. 1997;66(4):235–43.
48. Milagres D, Rueff-Barroso C, Bolognese A, Costa A, Porto L. Immunohistochemical localization of tumor necrosis factor-alpha and interleukin-6 during orthodontic movement in rats. Braz J Morphol Sci. 2009;26(1):42–8.
49. Zhang Y, Lin J-X, Vilcek J. Interleukin-6 induction by tumor necrosis factor and interleukin-1 in human fibroblasts involves activation of a nuclear factor binding to a kappa B-like sequence. Mol Cell Biol. 1990;10(7):3818–23.
50. Ohtsuka H, Koiwa M, Hatsugaya A, Kudo K, Hoshi F, Itoh N, et al. Relationship between serum TNF activity and insulin resistance in dairy cows affected with naturally occurring fatty liver. J Vet Med Sci. 2001;63(9):1021–5.
51. Sun LW, Zhang HY, Wu L, Shu S, Xia C, Xu C, et al. H-1-nuclear magnetic resonance-based plasma metabolic profiling of dairy cows with clinical and subclinical ketosis. J Dairy Sci. 2014;97(3):1552–62.
52. Van Soest PJ, Robertson JB, Lewis BA. Methods for dietary fiber, neutral detergent fiber, and nonstarch polysaccharides in relation to animal nutrition. J Dairy Sci. 1991;74(10):3583–97.
53. Roche JR, Berry DP. Periparturient climatic, animal, and management factors influencing the incidence of milk fever in grazing systems. J Dairy Sci. 2006; 89(7):2775–83.
54. Li P, Li XB, Fu SX, Wu CC, Wang XX, Yu GJ, et al. Alterations of fatty acid beta-oxidation capability in the liver of ketotic cows. J Dairy Sci. 2012;95(4): 1759–66.

18

Field application of an indirect gE ELISA on pooled milk samples for the control of IBR in free and marker vaccinated dairy herds

Barbara Colitti[1], Elvira Muratore[1], Maria Elena Careddu[2], Luigi Bertolotti[1*], Bryan Iotti[1], Mario Giacobini[1], Margherita Profiti[1], Chiara Nogarol[1], Jens Böttcher[3], Andreino Ponzo[4], Roberto Facelli[5] and Sergio Rosati[1]

Abstract

Background: The aim of the present study was to assess the reliability of a new strategy for monitoring the serological response against Bovine Herpesvirus 1 (BoHV1), the causative agent of infectious bovine rhinotracheitis (IBR). Bulk milk samples have already been identified as cost effective diagnostic matrices for monitoring purposes. Nevertheless, most eradication programs are still based on individual standard assays. In a region of northwestern Italy (Piedmont), the voluntary eradication program for IBR has become economically unsustainable. Being the prevalence of infection still high, glycoprotein E-deleted marker vaccines are commonly used but gE blocking ELISAs are less sensitive on bulk milk samples compared to blood serum.

Results: A recently developed indirect gE ELISA showed high versatility when applied to a wide range of matrices. In this study, we applied a faster, cost effective system for the concentration of IgG from pooled milk samples. The IgG enriched fractions were tested using a gE indirect ELISA for monitoring purposes in IBR-positive and IBR-marker-vaccinated herds. Official diagnostic tests were used as gold standard. During a 3 years study, a total 250 herds were involved, including more than 34,500 lactating cows. The proposed method showed a very good agreement with official diagnostic protocols and very good diagnostic performances: only 37 positive animals were not detected across the entire study.

Conclusions: The results highlighted the ability of the proposed method to support the surveillance of IBR in the Piedmont region, reducing the costs without affecting the diagnostic performances.

Keywords: BoHV1, gE ELISA, Pooled milk, IBR control

Background

Infectious bovine rhinotracheitis (IBR) is the most severe disease caused by Bovine Herpesvirus 1 (BoHV1), an etiological agent that is also responsible of other clinical forms such as infectious pustular vulvovaginitis (IPV), abortions and infectious pustular balanophostitis (IBV) [1, 2]. The reactivation from the ganglia of latently infected cattle and re-excretion of BoHV1 results in infection of susceptible or even vaccinated hosts. Control measures for IBR are justified by the substantial economic losses due to the within-farm persistence and to the trade restrictions imposed by IBR-free countries [3].

Several conventional (whole-virus strains) and glycoprotein E-deleted (gE) marker-vaccines are available to contain the clinical forms and reduce the spread of BoHV1. In parallel gE companion diagnostic tests were developed. The gE blocking ELISAs detect antibodies against the missing antigen, differentiating infected from vaccinated animals (DIVA). The use of gE blocking ELISA on milk is discouraged because the level of IgG is about 1/15th of that found in serum and sensitivity issues are not overcome even if un-diluted milk is tested [4–6]. However, milk can easily be collected, pooled and tested, thereby reducing the total costs of surveillance as well as animal stress. According to EU Regulation (2004/

* Correspondence: luigi.bertolotti@unito.it
[1]Department of Veterinary Science, University of Turin, Largo Paolo Braccini, 2 10095 Grugliasco, Turin, Italy
Full list of author information is available at the end of the article

558/EC), some prerequisites of BoHV1-bulk-milk testing have to be met: pools must include no more than 50 individual milks (PM) and at least the 30% of cattle on premises must be lactating; finally, the pool size is calibrated according to the limit of detection (LOD) of the method. Whole-virus ELISAs were demonstrated fit for monitoring purposes in IBR-free areas, where vaccination is forbidden. Nevertheless, if adopting a PM approach in marker-vaccinated (MV) farms only DIVA tests could be employed, and blocking ELISAs could lack in sensitivity and specificity [6, 7]. Concentration and purification of IgG from milk by affinity chromatography (AC) partially improved the LOD of blocking gE assays [8]. This approach is however discouraged by the costs and the complexity of the AC protocol. An alternative to AC could consist in testing the IgG-enriched fraction, using cheaper chemical reagents with no adverse effects on ELISA testing.

An indirect ELISA for the detection of IgG against the gE of the BoHV1 was developed in previous studies [9, 10] and showed good performance on serum, individual milk, and purified/concentrated IgG from bulk milk (BM) samples. Since the LOD of the method in concentrated BM was estimated to be equal to a prevalence of 2.5% infected animals per pool, it could be reliable also in MV farms monitoring.

In Piedmont, a voluntary eradication program is currently highly demanding in terms of human and material costs and a further step towards a compulsory eradication program would be unsustainable. This surveillance system is currently operating only in IBR-MV and IBR-free herds, but by changing from blood serum to PM based monitoring, substantial resources could become available for a larger number of farms. Moreover, animals belonging to IBR positive farms are tested only for trade purposes. This aspect leaded to an inaccurate definition on their IBR status at the begin of the study. This lack of information was overcame during the study, updating the official status of all the included farms.

In the present study, individual milk samples, collected for functional feature analysis, were used for PM preparation and an integrated system of partners involved in a cost effective IBR surveillance study was established. Samples were treated with an ammonium-sulphate based concentration protocol, replacing the more expensive AC method. The concentrated PM samples, collected from IBR-positive and IBR marker-vaccinated-farms were tested in gE indirect ELISA. The aim of this study was to evaluate the field diagnostic performance of the method in comparison with the standard individual tests employed in IBR surveillance.

Results

In the present study, approximately 65,000 individual milks were used for the preparation of isovolumetric pooled milk samples. The mean number of cows per pool was 35 (each pool size ranged from 20 to 40 animals). Approximately 34,500 lactating cows were involved in at least one pooled milk investigation, while 1500 animals (approximately 4%) were excluded for dry period or for treatment with antibiotics. In accordance with EU Decision 2004/558/EC, the aforementioned 1500 animals were individually tested by standard assays. All the individual tests confirmed the previous status of each animal.

Twenty-seven farms were excluded from data analysis since the serological status of the lactating cows was not up-to-date and the estimation of within-pool prevalence could not be performed. Conversely, the IBR status of the remaining 253 recruited farms was updated, permitting the assessment of the diagnostic performance of the method. The updating of the official status allows the number of expected gE-negative farms to increase from 120 to 156, while the number of expected IBR-positive herds was reduced to 97. Description of the farms IBR status is reported in Table 1. In Fig. 1 each farm is depicted according to the highest percentage of OD detected in pooled samples across the entire study (y axis) and the maximum number of expected positive animal per pool (x axis).

The estimation of the diagnostic specificity (DSp) was conducted on the results on all the pools including only gE negative animals belonging to farms with both IBR negative and positive farms status. Among all expected negative pools ($n = 1224$), only 30 resulted in ELISA positive outcome (diagnostic specificity at pool level dSp = 97.55, 95%CI: 96.52–98.43%; at farm level dSp = 95.91, 95%CI: 96.15–100.00%). All of them belonged to four gE negative farms. Further investigations revealed a recent BoHV-1 wild type circulations, promptly detected by pools testing (Fig. 2). The outbreaks were confirmed by the gE blocking ELISA applied to serum samples of the lactating cows. Considering the aforementioned outbreaks, the final evaluation of the DSp at pool level was calculated as 100% (95%CI: 99.69–100.00%). Considering

Table 1 Description of IBR status of the investigated farms according with the presence or absence of antibodies against the glycoprotein E (gE). Thirty involved farms (unknown status) have been never investigated for BoHV1 (Bovine herpesvirus 1) antibodies prior the present study

Farms	May 2015[1]	November 2016[2]
gE-positive	152	97
gE-negative MV	98	156
Unknown status	30	0
Total	280	253

[a]Marker-vaccinated farms. The number of farms belonging to each group at the beginning (1) and at the end (2) of the study is reported

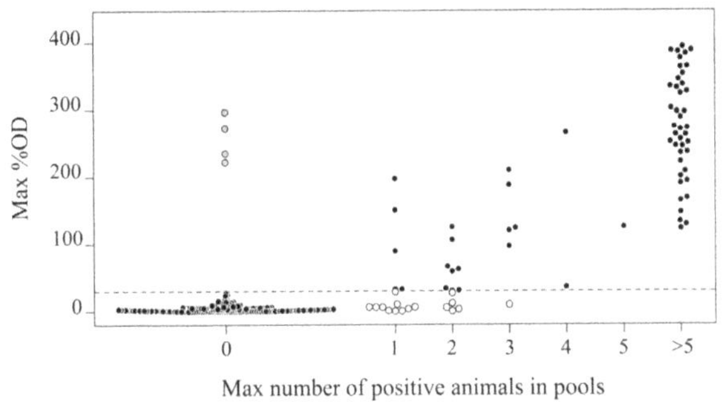

Fig. 1 Summary of ELISA Results. All the farms are showed: the position in the plot is given by the maximum number of gE (glycoprotein E) positive animals identified into the pools (x axis) and on the maximum ELISA results recorded among all tested pools. Black dots represent the results in agreement with Official IBR status; grey dots represent the 4 recorded IBR re-infection in expected gE-negative farms; white dots represents the 16 expected gE-positive farms which always resulted negative to gE indirect ELISA. The dotted lines indicate the threshold values of the gE indirect ELISA

only gE negative farms, the dSp was 100% (95% CI: 96.15–100.00%).

Out of the 97 expected gE-positive farms, 81 confirmed their status in gE indirect ELISA performed on IgG enriched fractions while 16 resulted negative across the entire study, suggesting a raw estimation of the diagnostic sensitivity equal to 83.5% (95% CI: 74.6–90.3%). All the 16 false negative farms were sampled 2 or 3 times and the maximum number of expected positive animals was recorded for each pool. The maximum number of within-pool gE-positive animals ranged from 1 to 3 cows (Fig. 1). A deeper investigation on those farms revealed that 37 gE-positive animals were included in those false negative pools. All the 37 animals have updated IBR statuses, confirming their gE positive status and were 3 to 12 years old (median = 5). Analyses conducted at farm level indicated a very good agreement between the official status and the ELISA results, showing a Cohen's Kappa equal to 0.8619 (95% CI: 0.7971–0.9268).

Discussion

Bulk milk or pooled milk testing is a valuable tool for eradication and/or monitoring programs, where a large number of cattle must be tested without inducing stress on the animals and incurring in high costs [11, 12]. Nevertheless, the size of pooled milk samples, in accordance with Decision 558/2004/CE, must be tailored to the LOD of the serological method, ensuring the detection of a single weak positive milk diluted in a modulated pooled milk.

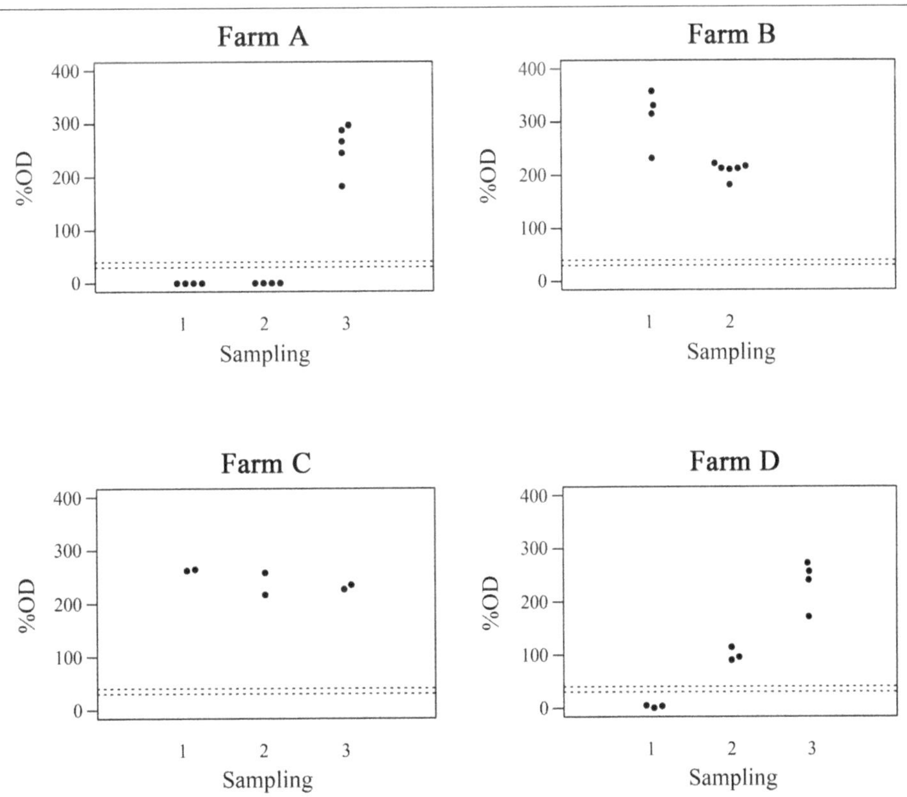

Fig. 2 Distribution of IgG enriched fractions ELISA optical density (%OD) related to two/three sampling in the four farms (**a**, **b**, **c**, **d**) with new BoHV1(Bovine Herpesvirus 1) outbreaks detected across the 18 months of the present study. The dotted lines indicate the threshold values of the gE (glycoprotein E) indirect ELISA

The whole-virus ELISAs have been demonstrated the most sensitive tool for the detection of antibodies against BoHV1 in bulk milk samples [6]. However, in farms where marker vaccination is adopted, whole virus ELISAs fail to distinguish infected from vaccinated cows, and only gE assays are suitable to monitor MV-herds. In the last two decades, the only available tests for IBR in MV-herds were gE blocking ELISAs, which were shown to be less sensitive when applied to PM samples than on individual blood samples (OIE Terrestrial Manual 2010, [6]). Indeed, previous studies demonstrated the importance of the within-pool prevalence and of the stage of lactation (variable levels of IgG) in the sensitivity of gE conventional assays [13]. Nevertheless, new strategies can be adopted to increase sensitivity in BM or PM serological investigations [14] as well as in the surveillance plan strategies [15]. In a recent study [10] we described a new approach for IBR surveillance, based on the use of a novel gE indirect ELISA applied to concentrated/purified BM samples. The concentration protocol consisted in the use of an affinity matrix to bind the IgG that are present in the whey and that can then be purified with a spin-column system.

In the present study, we used the individual milk samples collected for routine functional feature evaluation and identified by a barcode for the preparation of isovolumetric PM samples (1 ml/head). Therefore, this approach bypasses the differences in daily milk production that are often observed when using bulk tank milk. Moreover, the planning of 2–3 sample collections across 18 months reduced the number of animals that were never included in PM investigation to 1500 cows (dried cows and bulls, namely 4.35% of the total), and these had to be tested by individual standard assays.

The diagnostic specificity evaluation involved the 156 gE-negative farms, tested in two independent sample collections and using the gE indirect ELISA applied to IgG enriched fractions. Across the large number of PM samples tested, four cases of BoHV1-seroconversion were promptly detected in the group of gE-negative herds. Indeed, gE blocking ELISA applied to serum samples collected from the four farms confirmed the presence of gE specific antibodies. Even if marker vaccination was in place in 3 out 4 of the new infected herds, the method was demonstrated fit for the purpose of surveillance.

Regarding the 97 expected IBR-positive farms, 81 confirmed their actual IBR status in the outcomes of gE indirect ELISA, while 16 resulted negative. In the current regional intervention program, recognized IBR-positive farms are excluded from yearly monitoring and their official status can stay unchanged for years, despite the progressive elimination of infected heads. Therefore, the number of expected positive farms changed from the beginning of the study and a substantial number of herds were reclassified as gE-negative. Only 16 farms scored negative for the entire study even though recent individual assays have identified 37 gE positive animals included in the respective pooled samples. Those animals represent around the 2% of all the positive animals ($n = 1805$) belonging to gE-positive farms enrolled in the present study.

The apparently false negative results can be partially justified by reasons other than a possible lack in sensitivity of the proposed diagnostic approach. However this condition has been recorded in very few occasions, at least in the area of study and seems correlated to the use of old conventional vaccines (whole virus vaccine, forbidden since 2003) rather than to true viral circulation. Moreover we can not exclude that those animals showed an antibodies titer in the milk samples lower than the limit of detection of the proposed method. On the other hand, new cases of infection were promptly detected and tended to show very high reactivity in all the tested pools regardless of vaccination status. Therefore, the probability of a very small number of truly positive animals being present within the herd without the emergence of an outbreak can be considered quite low. Moreover, the absence of a true gold standard should be taken into account as well as the diagnostic specificity (DSp) value of the commercial gE blocking ELISA applied to serum samples of tested animals. Indeed, the test DSp was measured at 92% [6] and this aspect could, at least in part, justify our results, if we consider these 37 animals (less than one one-thousandth of the total number of heads) as possible false-positive results in gE blocking ELISAs.

Conclusions

Our findings provide compelling evidence in favor of the reliability a new strategy for IBR surveillance, which includes a traceable and standardized system of PM preparation.

Investigation for IBR surveillance using PM samples could represent a cost-effective solution for monitoring herd-status. Moreover the pooled milk approach can increase the number of serological investigations per year from 1 to 2–3 (according to the herd-qualification), allowing an earlier detection of new BoHV1 outbreaks. We previously demonstrated that the concentration/purification of IgG associated with a gE indirect ELISA represent an effective strategy to overcome the sensitivity issue in BM surveillance. When compared to previous work, in this case we adopted a cheaper, easier and faster protocol for the concentration of IgG from milk, and evaluated its fitness for monitoring purposes. A PM-based surveillance program seems reliable if the pool size is determined accounting for the LOD of the method. Moreover the humoral response against BoHV-1 wild-type strains was found to be high in the target population, regardless the presence of marker-vaccination. The present work was carried out using the preexisting infrastructure for quality assurance programs, which was further improved with a highly automated system of data collection and analysis. The proposed method was proven fit for IBR surveillance in gE-negative farms (DSp of 100%) but also suggested its field applicability for eradication purposes. Indeed, bulk milk samples could be collected from gE-positive farms, processed, and tested until the relative IgG enriched fractions became negative in gE ELISA. This finding will suggest that the within-herd prevalence is low and would justify the adoption of a pooled-milk-based approach. When all pooled milk of those farms test negative, none or only few gE-positive heads are expected to be still in lactation. Therefore, the final step in IBR eradication would consist in the employment of the standard individual assays on blood serum to identify and slaughter the last eventual gE-positive animals. Finally, another advantage of the present study was the active participation of different partners involved in sanitary procedures (ASL and IZSPLV) and functional feature evaluations (ARAP), ensuring the creation of an integrated system with future applicability in different fields of health monitoring in dairy herds.

Methods

Farms involved and planning of surveillance

Currently, the regional eradication program in Piedmont includes different levels of IBR status qualifications, based on the herd-immunity against BoHV1. IBR-free farms do not adopt any vaccination program and the entire herd must test negative to whole-virus assays. IBR-marker-vaccinated farms include gE marker vaccinated and IBR-free animals, and new infections can be detected using gE blocking ELISA through individual testing. IBR-positive farms are defined by the presence of antibodies against gE in at least one animal in diagnostic age (> 24 months).

At the beginning of 2015, a group of 280 farms including all IBR official statuses were voluntarily recruited for the present study. All the farms included animals belonging to dairy breeds, in most cases Friesian breed, with a mean size of 80 lactating cows per farm. Milk was collected during the ordinary lactating procedure: this aspect allowed to select only the animals older than 24 months, and, consequently, to exclude the presence of maternal antibodies and colostrum in the analyzed pools.

A total amount of 98 gE-negative IBR-marker-vaccinated farms and 152 gE-positive farms (IBR-positive) were included into two or three PM collection sessions. In addition, 30 farms with unknown IBR status were enrolled in the study and all animals older than 24 months of age were serologically tested using standard assays (SVANOVIR® IBR whole virus ELISA and IDEXX IBR gE Ab blocking ELISA). At the end of the project 254 farms out of 280 were serologically tested using official IBR-ELISAs, updating their IBR status and calculating the within-PM prevalence.

Sample collection was performed across the full 18 months of the study and planned according to the guidelines of Decision 2004/558/EC for IBR eradication programs approved by the European Community: two samples collections were performed in IBR-free and in IBR-MV farms, while three PM preparations, four months apart, were scheduled for the IBR-positive farms.

Pooled samples: Preparation, identification system and data analysis

Individual milk samples are routinely collected once a month by the Provincial Breeder Association (ARAP) of Cuneo to monitor milk quality through functional feature evaluation. This existing source was used for the preparation of PM samples. From each individual milk sample, 1 ml was collected and used for the preparation of pooled milk that could be processed with the downstream concentration protocol. Each PM sample was identified using a barcode that uniquely identified it and allowed retrieval of the individual ear tags of included animals. Data was stored into a SQL relational database and an ad hoc developed web application was used to interact with the Regional Veterinary database, where diagnostic information is stored by the Official Veterinary Service. This interaction allows retrieval of IBR status (i.e negative, vaccinated or positive) and diagnostic history of each animal included in the pool. This web application also guides the operator in the identification of the optimal pool size (20–40 animals) to optimize LOD based on farm composition.

Ammonium-sulphate based IgG concentration

Ten ml from each PM were treated with 100 µl of a rennet based solution to precipitate the caseins during an incubation time of 20 min at 37 °C. The obtained curd was broken by manual agitation and incubated on ice for 10 min. The samples were centrifuged at 3600 g for 10 min at 4 °C to obtain three different fractions: a fat layer on top, the whey in the middle and a casein pellet at the bottom of the tube. The whey was transferred to a new 14 ml tube, mixed with an equal volume of a saturated solution of ammonium sulphate (4.1 M at 25 °C) and incubated on a platform shaker for 1 h. After additional centrifugation (3600 g for 10 min at 18 °C), an enriched IgG fraction was visible at the bottom of the tube [16–18]. The supernatant was discarded and the pellet was briefly drained upside down on a paper towel before being resuspended in 400 µl of PBS 1.25% casein. The described protocol allowed a concentration of the milk IgG of about 15X (data not showed). This result is in agreement with previous published data [10]. Even if a different purification method was used, the ammonium sulphate based precipitation method showed a very comparable efficiency. The enriched fractions of IgG were used immediately in the downstream ELISA protocol or alternatively stored at – 20 °C.

Serological assays

The gE indirect ELISA (ERADIKIT ™ Bulk Milk surveillance Kit, In3diagnostic, Italy), described in two previous papers [10], originally developed for the detection of antibodies in serum or concentrated/purified BM samples, was adapted for application on IgG enriched fractions obtained from PM samples collected in IBR-MV and in IBR-positive farms. The ELISA protocol was slightly modified based on the different diagnostic matrix to be investigated. Briefly, the enriched fractions were placed into two adjacent wells (200 μl/well), the former coated with the recombinant BoHV1 gE, while the latter with a negative antigen. The starting incubation was performed at room temperature (RT) for 120′ minutes. The plate was washed four times and a peroxidase-labeled secondary antibody diluted at 10 ng/ml in PBS 1.25% casein was added to each well. After 45 min of incubation at room temperature, four new washing cycles were performed before the addition of the substrate solution (3,3′,5,5′-tetramethylbenzidine, TMB). The reaction was stopped after 15 min using 100 μl/well of stop solution and read in a microplate spectrophotometer at 450 nm. The reactivity of each sample was calculated as the ratio, expressed as percentage, between the net OD value of the sample (gE-OD value – negative antigen OD value) and the net OD value of the positive control (gE-OD value – negative antigen OD value).

Samples with a percentage of reactivity equal to or greater than 40% were classified as positive, those with reactivity between 30 and 40% as doubtful, and those with a reactivity lower than 30% were considered negative.

Data analyses

To evaluate the diagnostic performances of the proposed protocol, the ELISA results were compared to the number of positive animals within each milk pool. The goal was to evaluate the ability of the diagnostic protocol to show positive results if at least one gE positive animal was included in the pool. Results of official serological diagnostic procedures were used as gold standard. Diagnostic sensitivity and specificity were evaluated at pool level, as well as at farm level (i.e. all pools collected from a farm during the 18 months surveillance period, fitting the surveillance purpose of the method). The concordance between the results obtained during the study and the official serological investigations was calculated at farm level by Kappa Cohen coefficient. In more details, a farm from which at least one tested milk pool included at least one gE positive animal was considered as expected positive farm. Confidence interval of diagnostic sensitivity and specificity was calculated using Exact Binomial test (R Statistical software ver. 3.2.0).

Abbreviations

AC: Affinity chromatography; BM: Bulk milk; BoHV1: Bovine Herpesvirus 1; DIVA: Differentiating infected from vaccinated animals; Dsp: Diagnostic specificity; IBR: Infectious bovine rhinotracheitis; IBV˙: Infectious pustular balanophostitis; IgG: Immunoglobulin G; IPV: Infectious pustular vulvovaginitis; LOD: Limit of detection; MV: Marker vaccinated; PM: Pooled milk

Acknowledgements

The authors acknowledge all the farmers and Breeders Association technicians participating in the "Pooled milk-based IBR monitoring project". The authors would like to remember Dr. Andreino Ponzo, who passed away prematurely: without him, this work would not have been possible.

Funding

The present work was funded by the Fondazione Cassa di Risparmio di Cuneo (CRC) (BANDO RICERCA SCIENTIFICA 2013). The funding body have no role in the design of the study, collection, analysis, and interpretation of data nor in writing the manuscript.

Authors' contributions

SR, RF, AP conceived the study; BC and EM conducted the lab work on milk, supported by MEC, MP, CN and JB; LB, BI and CN analyzed the data, supported by MG. All authors were involved in drafting the manuscript; all authors read and approved the final manuscript.

Consent for publication

Not applicable

Competing interests

Concerning the gE indirect ELISA, a patent application has been registered by In3diagnostic on the 8th May 2015 and published on the 11th November of the same year (EP 2942628 A1 Kit and in vitro method for identifying Bovine Herpes Virus 1 infections). The patent is based on the ELISA test and the protocol developed in a previous study. The authors have received no reimbursements, fees, funding, or salary linked to this patent application. In3diagnostic, manufacturer of the gE indirect ELISA, had no role in the study design, the collection, analysis, or interpretation of the data, the writing of the manuscript, or the decision to submit it for publication.

Author details

[1]Department of Veterinary Science, University of Turin, Largo Paolo Braccini, 2 10095 Grugliasco, Turin, Italy. [2]Istituto Zooprofilattico Sperimentale del Piemonte della Liguria e della Valle D'Aosta (IZSPLV), 12100 Cuneo, Italy. [3]Tiergesundheitsdienst bayern e.V, 85586 Poing, Germany. [4]Azienda sanitaria locale Cuneo (ASL CN1), 12100 Cuneo, Italy. [5]Associazione Regionale Allevatori Piemonte (ARAP), 12020 Madonna dell'Olmo, Cuneo, Italy.

References

1. Pastoret PP, Thiry E, Brochier B, Derboven G. Bovid herpesvirus 1 infection of cattle: pathogenesis, latency, consequences of latency. Ann Rech Vet. 1982; 13:221–35.
2. Letellier C, Kerkhofs P, Wellemans G, et al. Outbreak of bovine virus diarrhea on Dutch dairy farms induced by a bovine herpesvirus 1 marker vaccine contaminated with bovine virus diarrhea virus type 2. Vet Microbiol. 2014; 13:15. https://doi.org/10.1016/j.rvsc.2012.11.016.
3. Muylkens B, Thiry J, Kirten P, et al. Bovine herpesvirus 1 infection and infectious bovine rhinotracheitis. Vet Res. 2007;38:181–209. https://doi.org/10.1051/vetres:2006059.
4. van Oirschot JT. Diva vaccines that reduce virus transmission. J Biotechnol. 1999;73:195–205.

5. Marnila P, Korhonen H. Immunoglobulins. In: Encyclopedia of Dairy Science. London: Academic Press; 2002. p. 1950–6.
6. Kramps JA, Banks M, Beer M, et al. Evaluation of tests for antibodies against bovine herpesvirus 1 performed in national reference laboratories in Europe. Vet Microbiol. 2004;102:169–81. https://doi.org/10.1016/j.vetmic.2004.07.003.
7. Wellenberg GJ, Verstraten ER, Mars MH, Van Oirschot JT. Detection of bovine herpesvirus 1 glycoprotein E antibodies in individual milk samples by enzyme-linked immunosorbent assays. J Clin Microbiol. 1998;36:409–13.
8. Schroeder C, Horner S, Bürger N, et al. Improving the sensitivity of the IBR-gE ELISA for testing IBR marker vaccinated cows from bulk milk. Berl Munch Tierarztl Wochenschr. 2012;125:290–6.
9. Bertolotti L, Muratore E, Nogarol C, et al. Development and validation of an indirect ELISA as a confirmatory test for surveillance of infectious bovine rhinotracheitis in vaccinated herds. BMC Vet Res. 2015;11:300. https://doi.org/10.1186/s12917-015-0612-5.
10. Muratore E, Bertolotti L, Nogarol C, et al. Surveillance of infectious bovine Rhinotracheitis in marker-vaccinated dairy herds: application of a recombinant gE ELISA on bulk milk samples. Vet Immunol Immunopathol. 2017;185. https://doi.org/10.1016/j.vetimm.2017.01.003.
11. Hartman A, van Wuijckhuise L, Frankena K, et al. Within-herd BHV-1 prevalence prediction from an ELISA on bulk milk. Vet Rec. 1997;140:484–5.
12. Reber A, Reist M, Schwermer H. Cost-effectiveness of bulk-tank milk testing for surveys to demonstrate freedom from infectious bovine rhinotracheitis and bovine enzootic leucosis in Switzerland. Schweiz Arch Tierheilkd. 2012; 154:189–97. https://doi.org/10.1024/0036-7281/a000329.
13. Nylin B, Strøger U, Rønsholt L. A retrospective evaluation of a Bovine Herpesvirus-1 (BHV-1) antibody ELISA on bulk-tank milk samples for classification of the BHV-1 status of Danish dairy herds. Prev Vet Med. 1999; 47:91–105.
14. Casarin E, Lucchese L, Grazioli S, et al. A new ELISA using the ANANAS technology showing high sensitivity to diagnose the bovine rhinotracheitis from individual sera to pooled milk. PLoS One. 2016;11:e0145912. https://doi.org/10.1371/journal.pone.0145912.
15. Veldhuis A, Santman-Berends I, Schauer B, et al. Epidemiological performance and subsequent costs of different surveillance strategies to control bovine herpesvirus type 1 in dairy farms. Prev Vet Med. 2017;139: 105–14. https://doi.org/10.1016/j.prevetmed.2016.12.003.
16. Wingfield P. Protein precipitation using ammonium sulfate. Curr Protoc protein Sci Appendix 3:Appendix 3F. 2001;https://doi.org/10.1002/0471140864.psa03fs13.
17. Forschner E, Bünger I, Krause HP, Küttler D. Control measures in officially acknowledged brucellosis-free and leukosis unsuspected dairy herds on the basis of bulk milk samples in combination with ELISA tests. Dtsch Tierarztl Wochenschr. 1989;96:475–86.
18. Böttcher J, Meier N, Stengel K-H, et al. Feasibility of serological bulk milk testing as a method for BVD surveillance. Berl Munch Tierarztl Wochenschr. 2003;116:244–51.

19

Quantification of *Clostridioides (Clostridium) difficile* in feces of calves of different age and determination of predominant *Clostridioides difficile* ribotype 033 relatedness and transmission between family dairy farms using multilocus variable-number tandem-repeat analysis

Petra Bandelj[1*], Céline Harmanus[2], Rok Blagus[3], Marko Cotman[1], Ed J. Kuijper[2], Matjaz Ocepek[1] and Modest Vengust[1]

Abstract

Background: Community acquired *Clostridioides (Clostridium) difficile* infection (CA-CDI) is a significant health problem in human and veterinary medicine. Animals are often considered as potential reservoirs for CA-CDI. In Europe, family farming is the most predominant farming operation, with a complex interaction between animals and the community. Therefore, it is pertinent to evaluate transmission patterns of *C. difficile* on such prominent European farming model. Fecal samples from calves ($n = 2442$) were collected biweekly over a period of one year on 20 mid-size family dairy farms. Environmental samples ($n = 475$) were collected in a three month interval. *Clostridioides difficile* was detected using qPCR in 243 fecal samples (243/2442); positive samples were then quantified. Association between prevalence/load of *C. difficile* and age of the calves was estimated with logistic regression model. Most common *C. difficile* isolate from calves ($n = 76$) and the environment ($n = 14$) was *C. difficile* ribotype 033, which was further analyzed using multilocus variable-number tandem-repeat analysis (MLVA) to assess intra- and between-farm relatedness.

Results: *Clostridioides difficile* was detected in feces of calves less than 24 h old. Results showed a non-linear statistically significant decrease in shedding load of *C. difficile* with age ($P < 0.0001$). A nonlinear relationship was also established between the number of calves and the farm *C. difficile* prevalence, whereas the prevalence of *C. difficile* ribotype 033 increased linearly with the number of calves. MLVA revealed close intra-farm relatedness among *C. difficile* ribotypes 033. It also revealed that the between-farms close relatedness of *C. difficile* ribotypes 033 can be a direct result of farm to farm trade of calves.

Conclusions: Implementation of better hygiene and management measures on farms may help decrease the risk of spreading CA-CDI between animals and the community. Trading calves older than 3 weeks would decrease the possibility *C. difficile* dissemination in the community because of lower prevalence and lower load of *C. difficile* in feces.

Keywords: *Clostridioides (Clostridium) difficile*, Ribotype 033, community-acquired infection, Dairy cattle, Epidemiology

* Correspondence: petra.bandelj@vf.uni-lj.si
[1]Veterinary faculty, University of Ljubljana, Cesta v Mestni log 47, SI-1115 Ljubljana, Slovenia
Full list of author information is available at the end of the article

Background

The increasing number of community acquired *Clostridioides difficile* infection (CA-CDI) in the past decade has prompted investigations into animal source of CDI [1, 2]. Several studies have shown that *Clostridioides difficile* can cause gastrointestinal disease in dogs, horses, piglets and possibly calves [3–9]. It remains unclear, however, if *C. difficile* strains found in animals can cause CDI in humans [9].

Family dairy farms were included in this study. Family farms are the most common operating farming system in the European Union. It represents the sustainable agriculture/agribusiness, and is beneficial for the local community [10]. To estimate the significance of a possible infection source, shedding numbers/load of *C. difficile* had to be quantified and *C. difficile* strains phylogenetically assessed. One of the major risk factor for increased *C. difficile* shedding in the environment is the age of the animal [11–14]. Regardless of the farming management, the prevalence of *C. difficile* has been shown to decrease dramatically with age from 10 to 56% at and near birth, to 0–3.8% at the time of slaughter [11, 14–16].

Several studies evaluated the impact of *C. difficile* shedding with calves feces by reporting *C. difficile* prevalence [8, 17–20], using longitudinal models [11, 14–16], or more recently, enumeration with viable plate counts from feces and carcasses of newborn calves [21]. None of them, however, quantified *C. difficile* in calves in relation to their age, which would indicate the age related risk for *C. difficile* dissemination in the community.

Calves on family farms have been shown to harbor several *C. difficile* ribotypes, with *C. difficile* ribotype 033 being the most prevalent [22]. Most previous reports suggest that *C. difficile* ribotype 033 is of less clinical importance compared to ribotypes 078, 027, 014 and 012, which are frequently isolated from feces of calves raised in big veal raising operations [6, 11, 14, 15, 18, 23, 24]. In humans, due to different genotype and phenotype, ribotype 033induced CDIs might be underdiagnosed [25]. However, one report has associated *C. difficile* ribotype 033 with diarrhea and eventually death in an elderly hospitalized patient in Italy [26], which necessitates further epidemiological assessment of all *C. difficile* ribotypes. Detailed epidemiological investigation can be achieved employing multilocus variable-number tandem-repeat analysis (MLVA), which is the method of choice to identify routes of transmission between patients and the environment [27–29]. This method shows a high level of discrimination and was proven useful for geographical tracking of several outbreak strains of bacteria [30–32].

The aim of this study was to quantify *C. difficile* in calves' feces from birth to six months of age, to determine how much calves contribute to the shedding of *C. difficile* into environment, and to evaluate the relatedness of the most predominant *C. difficile* ribotype 033 between family dairy farms.

Results

Overall, 243 fecal samples positive for *C. difficile* were collected from 155 calves. A hundred and seven calves ($n = 107/155$, 69%) were positive only once (Additional file 1 and Additional file 2). Forty eight calves ($n = 48/155$, 31%) were positive multiple times (2-6×) (Table 1). *Clostridioides difficile* prevalence in calves feces decreased significantly with the increasing age of the calves ($P < 0.0001$) (Fig. 1).

Association between *C. difficile* prevalence and the number of calves

A nonlinear association between the number of calves and the farm prevalence was established for *C. difficile* ($P < 0.0001$, Table 2). A steady increase in *C. difficile* prevalence is observed when the number of calves on the farm is up to eight (8), whereas for larger number of animals the *C. difficile* prevalence seems to be only mildly affected by the number of calves (Fig. 2). The prevalence of ribotype 033, however, increases linearly with the number of calves on a farm ($P = 0.032$, Table 2).

Quantification

The load of *C. difficile*/g of calves feces decreased with age ($P < 0.0001$) (Fig. 3). One day old calves had the highest load of *C. difficile* in feces (mean 3.4×10^6 cells/g feces; 453–13.7×10^6 *C. difficile* cells/g feces), followed by calves that were 8, 11 and 12 days old with the mean of 1.8–1.9×10^6 cells/g feces (50–10.8×10^6 *C. difficile* cells/g feces). Some calves (less than 24 h old) were also positive for *C. difficile* and had a mean of 6.2×10^3 *C. difficile* cells/g feces (126–2.2×10^4 *C. difficile* cells/g feces).

MLVA

The minimum spanning tree (Fig. 4) revealed close relatedness among most *C. difficile* ribotype 033. Isolates from calves from the same farm (farm 4, 17 and 18) or local community (farms 5 and 6; farms 14, 15 and 16) were mostly clonal (STRD = 0) or in one clonal complex (STRD≤2). Most calves that were positive for *C. difficile* ribotype 033 several times during the sampling period, harbored the same *C. difficile* clone (T2, T10, T11, T15, T16, T20, T34, T39, T44), had *C. difficile* isolates in the same clonal complex (T40, T41, T42), or were genetically related with STRD≤10 (T27, T28). Two male calves, T20 and T39, from farms 15 and 16 were sold to farm 14. Both were positive with the same clone of *C. difficile* ribotype 033 before and after their relocation. The *C. difficile* clone from calf (T20) was found subsequently in calves born on farm 14 (T21 and T23). Same clone was not present on farm 14 before introduction of calf T20. *Clostridioides*

Table 1 Quantification results for *C. difficile* in feces of calves that were positive multiple times

Calf										
1	Age (days)	10	24	38	52	66	80	94	108	
	No. CD	102	0	0	0	0	0	88	0	
2	Age (days)	11	25	39	53	67	81	95		
	No. CD	0	3716	0	0	0	LOQ	0		
3	Age (days)	12	27	40	54	68	82			
	No. CD	147	578	0	LOQ	194	0			
4	Age (days)	10	24	38	52	66				
	No. CD	0	0	318	174	0				
5	Age (days)	6	20	34						
	No. CD	2758	732	0						
6	Age (days)	5	19	33	47					
	No. CD	0	3300	110	0					
7	Age (days)	2	16	30	44					
	No. CD	684	0	LOQ	0					
8	Age (days)	3	17	31	45	59				
	No. CD	295	0	0	LOQ	0				
9	Age (days)	2	16	30						
	No. CD	1530	441	0						
10	Age (days)	5	19	33	47	61	75			
	No. CD	135,486	92	0	183	LOQ	0			
11	Age (days)	5	19	33	47	61	75	89	103	117
	No. CD	2676	572	241	0	209	0	0	LOQ	0
12	Age (days)	9	23	37	51	65				
	No. CD	805	0	942	129	0				
13	Age (days)	9	22	36	50	92	106	120		
	No. CD	662	408	1189	0	0	610	0		
14	Age (days)	1	15	29	43	57	71			
	No. CD	96,398	37,640	0	0	LOQ	0			
15	Age (days)	11	25	39	53	67				
	No. CD	3569	2683	1901	LOQ	0				
16	Age (days)	1	15	29	43	57				
	No. CD	13,705,988	32,219	1623	368	0				
17	Age (days)	0	14	28	42	56	70	84		
	No. CD	1519	0	LOQ	0	41	198	0		
18	Age (days)	0	14	29	42	56	70	0		
	No. CD	0	0	333	0	274	0	0		
19	Age (days)	12	26	40	54	67	82			
	No. CD	0	0	890	65,507	2867	0			
20	Age (days)	59	73	87	101	115	129	143	157	171
	No. CD	0	242	0	0	0	0	0	LOQ	0
21	Age (days)	3	17	31						
	No. CD	3937	4115	0						
22	Age (days)	1	15	29	43	57	71	85		
	No. CD	114	0	287	0	0	356	0		

Table 1 Quantification results for *C. difficile* in feces of calves that were positive multiple times *(Continued)*

Calf										
23	Age (days)	13	27	41						
	No. CD	23,470	322	0						
24	Age (days)	3	17	31						
	No. CD	54,900	4,278,134	0						
25	Age (days)	1	15	29	43	57	71	85	99	113
	No. CD	103,235	1177	0	0	0	0	0	LOQ	0
26	Age (days)	11	25	39	53	67	81	95	109	123
	No. CD	274,108	0	0	0	0	0	0	631	0
27	Age (days)	12	25	40						
	No. CD	10,811,054	938	0						
28	Age (days)	2	16	30	44	58	72			
	No. CD	0	105,727	0	0	LOQ	0			
29	Age (days)	9	23	37	51					
	No. CD	2919	0	LOQ	0					
30	Age (days)	7	21	35						
	No. CD	2,589,994	1152	0						
31	Age (days)	4	18	32	46					
	No. CD	0	806	38,702	LOQ					
32	Age (days)	7	21	35						
	No. CD	9630	1827	0						
33	Age (days)	6	20	34						
	No. CD	1,587,504	5278	0						
34	Age (days)	9	23	37	51	65	79	93		
	No. CD	3,906,984	1316	1055	0	0	783	0		
35	Age (days)	13	27	42	55	69	83	97		
	No. CD	53,735	66,424	LOQ	20,396	0	LOQ	0		
36	Age (days)	12	26	40	54	68				
	No. CD	842,730	1084	0	304	0				
37	Age (days)	3	17	31	45	59				
	No. CD	7,260,792	1346	632	LOQ	0				
38	Age (days)	5	19	33	47	61	75			
	No. CD	11,629	137	3715	0	723	0			
39	Age (days)	34	48	62	76	90				
	No. CD	294	0	0	LOQ	0				
40	Age (days)	9	23	37	51					
	No. CD	207	LOQ	LOQ	0					
41	Age (days)	10	24	38						
	No. CD	428	401	0						
42	Age (days)	6	20	34						
	No. CD	28,022	779	0						
43	Age (days)	20	34	48	62	76	90	104	118	132
	No. CD	0	30,971	0	0	0	0	0	0	LOQ
44	Age (days)	9	22	37						
	No. CD	3,593,368	4064	0						

Table 1 Quantification results for *C. difficile* in feces of calves that were positive multiple times *(Continued)*

Calf							
45	Age (days)	8	21	36			
	No. CD	678,040	1627	0			
46	Age (days)	11	25	39	53	67	81
	No. CD	129	13,724	0	0	537	0
47	Age (days)	0	14	28	42	56	
	No. CD	0	3,200,151	7897	2160	0	
48	Age (days)	12	26	40			
	No. CD	48,901	8297	0			

LOQ under the limit of quantification; *CD Clostridioides difficile*

difficile ribotype 033 clone recovered from calf T39 was introduced from farm 15 to farm 14. However, the same clone wasn't found in any other calves born on farm 14.

The same MLVA profile ($n = 6$) that was identified from the environmental samples, could be linked to the *C. difficile* ribotype 033 isolated form calves of the same farm or from the same geographical area (Fig. 4). Several ($n = 7$) were more genetically related to isolates from other unrelated farms. Interestingly, *C. difficile* ribotype 033 recovered from a barn fly on farm 5, had the same MLVA profile as the *C. difficile* recovered from calves (T19, T26) on an epidemiologically unrelated farm (farm 16).

Discussion

The aim of the study was to quantify *C. difficile* in feces of calves and to evaluate the relatedness of the most common *C. difficile* ribotype 033 between economically related and unrelated family dairy farms.

We also established that the *C. difficile* prevalence is non-linearly related to the number of calves on the farm. While a steady increase in *C. difficile* prevalence is observed up to a certain number of animals, it then remains roughly at the same level, when the number of animals increases further. The prevalence of *C. difficile* ribotype 033, however, is linear to the number of calves on the farm.

Newborn calves had the highest prevalence and load of *C. difficile* /g feces, which decreased over time. The decrease in prevalence over the increasing age of animals has previously been established [11, 14]. However, this is the first report associating the fecal load of *C. difficile* with age of calves. Repeated sampling of the same calves over their first 6 months of life has also shown that calves can

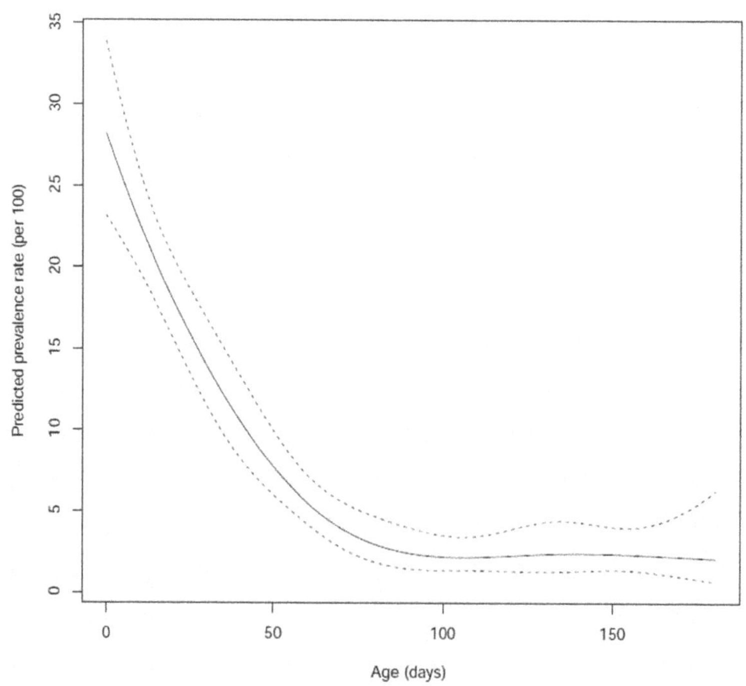

Fig. 1 *C. difficile* prevalence rates in calves from 0 to 180 days old per 100 and 95% confidence intervals

Table 2 The association between the prevalence of *C. difficile* and ribotype 033 and the number of calves

Outcome	Variable	Estimate	SE	*p*	(Exp)estimate	95% CI
C. difficile	Number of Calves			< 0.0001		
	linear	0.33	0.08	< 0.0001	1.39	1.18–1.66
	non-linear	−0.39	0.16	0.0174	0.68	0.48–0.93
	Time of sampling	0.02	0.01	0.0251	1.02	1.01–1.04
Ribotype 033	Number of Calves	0.13	0.06	0.0323	1.13	1.01–1.27
	Time of sampling	0.01	0.01	0.4932	1.01	0.98–1.04

The time of sampling is included only to control for the effect of the meteorolical season. The linear/non-linear effect of *C. difficile* prevalence should be estimated from Fig. 2
SE Standard error

be positive for *C. difficile* more than once with the same or different ribotype or MLVA profile.

In this study calves aged 1 day and around 12 days had the highest load of *C. difficile* cells/g feces. Interestingly, some calves sampled in the first 24 h after birth, were also shedding *C. difficile* with feces (meconium). This is in concordance with Hopman et al. [33], who demonstrated CD shedding in piglets after their first hour of life and the prevalence at day one increased from 8.3 to 62%.

The predominance of *C. difficile* ribotype 033 on sampled farms gave us the opportunity to investigate the epidemiology of *C. difficile* between economically related and unrelated family dairy farms. Most studies to date were performed on big veal farms which congregate large numbers of calves of different geographical origins [11, 34]. Our study included only family farms located in a driving distance between each other (≤ 2 h from farm to farm), where some are economically connected with trading of calves or products. As expected the *C. difficile* isolates 033 from the same farm sampled on different dates or different calves were clonal or in the same clonal complex as was recently found also in piglets [35]. The similarity was also shown in *C. difficile* isolates from farm sharing the same private farm road (farm 5 and 6) or trading of the male calves (farm 14 and 15; farm 14 and 16). We even suggested a possible transmission of *C. difficile* from one farm (farm 16) to another (farm 14) through a colonized calf, where the calf remained colonized for several weeks, shedding *C. difficile* in the environment. Two other calves, born on the farm (farm 14), and later throughout the study placed in the same individual box as the

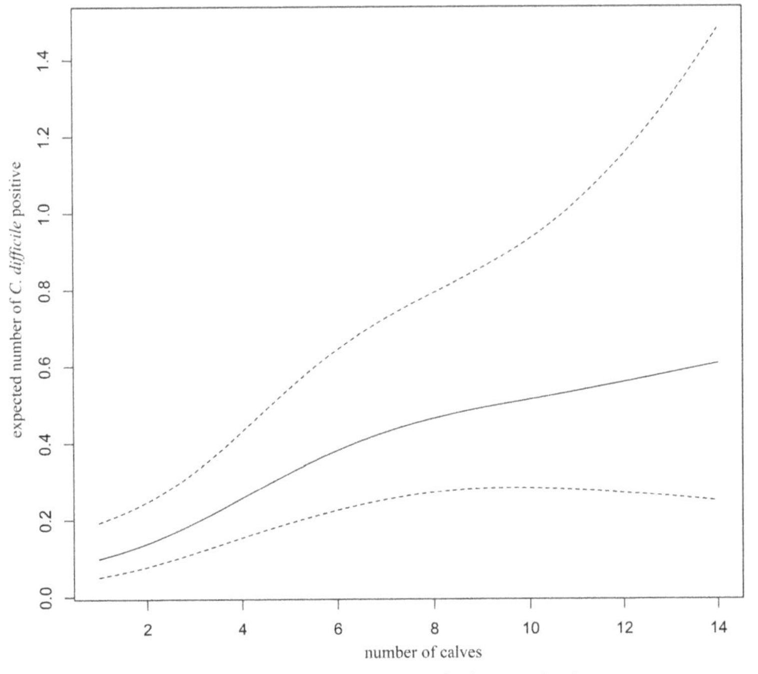

Fig. 2 Expected prevalence of *C. difficile* positive calves versus the total numer of calves on the farms

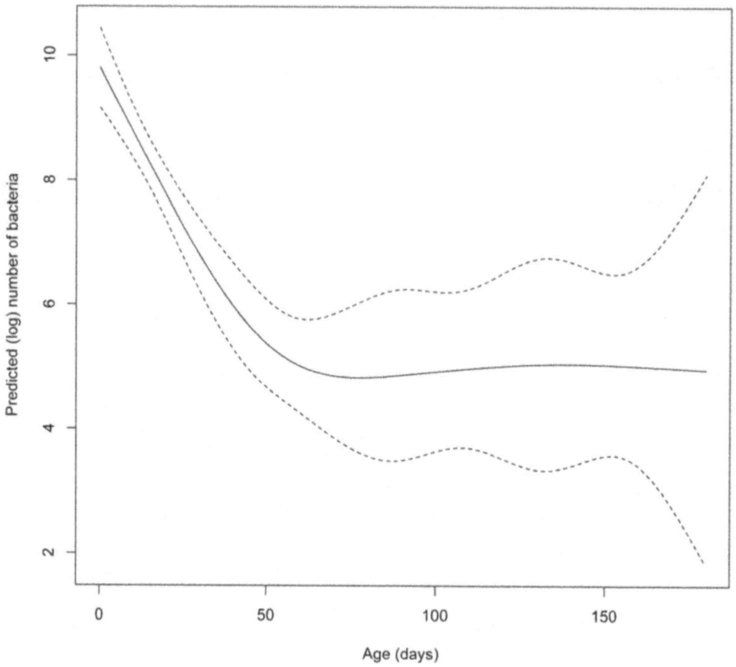

Fig. 3 Predicted log transformed load of *C. difficile* and age (in days) of calves and 95% confidence intervals

colonized calf, were found colonized with the same clonal *C. difficile* isolate. This is the first suggestion of a possible calf to calf and farm to farm transmission of *C. difficile*. However, due to lower sensitivity of the culture method compared to qPCR [36] and the history of calf trading between these two farms, we could also assume that transmission of this clonal *C. difficile* isolate could have happened before the start of our study.

Interestingly, one calf (T19 from farm 16) was positive for three consecutive times with *C. difficile* ribotype 033, which were not related (STRD≤10) or clonal (STRD ≤2). Calves have been shown to harbor different *C. difficile* ribotypes during different life stages [14]. Results of this study also indicate that they can be colonized with different MLVA types of the same *C. difficile* ribotype. Some environmental *C. difficile* strains (farm 5 and farms 2, 13, 14, 15) were clonal to strains from calves found on seemingly unrelated farms (farms 2, 4, 16 and farms 5, 6). There might be some epidemiological connection that we are unaware of, since farms have fields scattered across the area and sometimes cross paths with each other. Or simply, the high relatedness of all the samples tested could be the consequence of less natural variability in ribotype 033 than in other ribotypes as stated in the article from Bakker et al. [28]. Another possibility that could contribute to the spreading of different MLVA types of *C. difficile* in the community is flying insects. In our study, we found a fly on farm 5 to harbor the same clone of *C. difficile* as calves on farm 16 (distance between farms approximately 5 km).

Conclusions

In conclusion, we demonstrated that calves can shed high loads of *C. difficile* from birth and that there is a non-linear statistically significant decrease of *C. difficile* prevalence and load with age. The superior prevalence of ribotype 033 compared to other ribotypes gained from a previous study [22] gave us the opportunity to assess the epidemiology of *C. difficile* between farms. We have suggested a farm to farm transmission through trading of a colonized calf. However, environmental and calves *C. difficile* strains from the same farm weren't always related. Nevertheless, implementing better hygiene and management measures may help decrease the risk of spreading CA-CDI between animals and the community. Trading calves older than 3 weeks would decrease the possibility for *C. difficile* dissemination in the community not only because of lower prevalence, but also because of lower load of *C. difficile* in feces.

Methods

Material

Twenty mid-size family dairy farms with 9 to 40 cows located in Slovenian Prealps were included in this study. Farms run by family member only were selected based on several factors; location, accessibility, farmers compliance and number of dairy cows in production. All data with regards to the farms characteristics were described before [22]. Feces from all calves on the farm were collected at the time of sampling. All calves had mandatory ear tags used for individual identification. Fecal samples

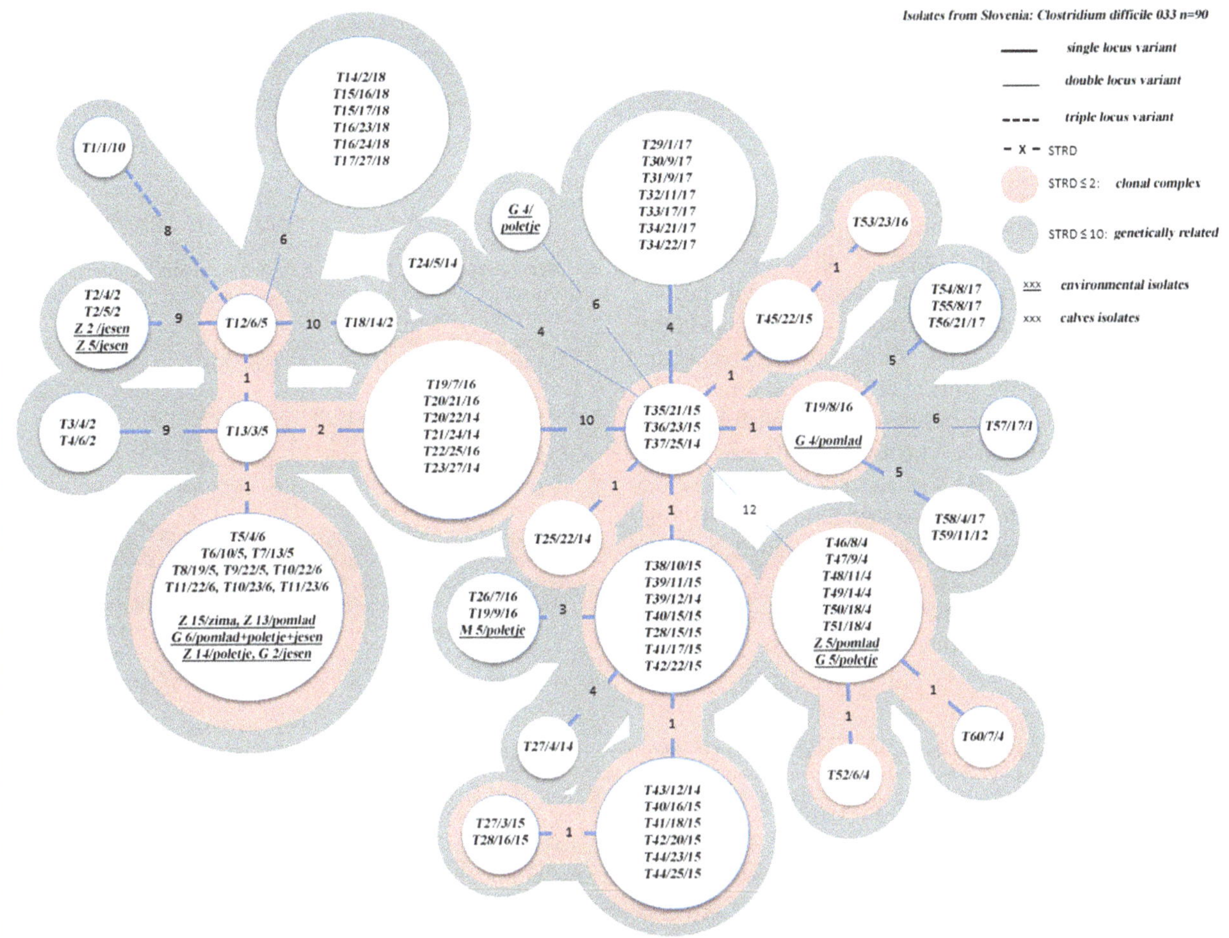

Fig. 4 Minimum spaning tree for *C. difficile* ribotype 033 isolated from calves feces and environmental samples: Clonal (STRD = 0), clonal complex (STRD ≤2), genetically related (STRD ≤10). Legend: Tn/n/n – individual calf number/sampling time/farm; Z n – soil sample on farm n; G n – manure sample from farm n; M n – flies from farm n

from calves ($n = 2442$) were collected individually from the rectum with clean latex gloves in two weeks intervals over a period of one year [22]. From these samples, 243 were positive for CD with qPCR and subsequently, 76 were identified as CD ribotype 033 with selective culture and ribotyping [22].

Environmental samples collected on every farm during each meteorological season were soil, manure, water, feed, other animals present on the farm, barn fly (*Stomoxys calcitrans*) and droppings from Barn Swallow (*Hirundo rustica*) [22]. Fourteen environment samples (manure = 7, soil = 6, barn fly (*Stomoxys calcitrans*) = 1) that were used in this study, were identified as CD ribotype 033 by selective culture and ribotyping [22].

Methods

Quantification of C. difficile in calves fecal samples

Calves fecal samples were tested for *C. difficile* specific fragment of 16S gene using a quantitative real-time PCR (qPCR) reported by Bandelj et al. [36] with a LOD and LOQ of 7.72 CD cells/g feces and 77.2 CD cells/g feces, respectively. Samples ($n = 243$) were retested in duplicates and in 1:10 dilutions to evaluate for possible inhibitory effects of the matrix.

Multilocus variable-number tandem-repeat analysis (MLVA) of C. difficile ribotype 033

For MLVA, we used 90 *C. difficile* ribotype 033 isolates belonging to 76 calves and 14 environment samples. The *C. difficile* isolates were tested for relatedness with a modified MLVA. All MLVA PCRs for six loci were performed in singleplex format as described previously [27, 28]. To determine the genetic distance between isolates the minimum spanning tree was constructed. The number of differing loci and the summed tandem repeat difference (STRD) was used as coefficients for the genetic distance in BioNumerics, version 7.0 (Applied Maths) as previously described [37]. Genetically related were isolates with a STRD ≤10, whereas clonal complexes were defined by a STRD ≤2.

Statistical analysis

The association between the prevalence and load of bacteria of *C. difficile* and age was estimated by the logistic regression model; the unit of analysis was an animal. Restricted cubic splines (using 5 knots) were used to account for a highly non-linear association between the prevalence/load of bacteria and age.

The association between the number of *C. difficile* and ribotype 033 positives and number of calves was estimated with Poisson generalized linear model with log link, including farm ID as a random effect (random intercept) and time of sampling as a fixed effect to account for repeated measurements; here the unit of analysis was a farm. For *C. difficile* a non-linear association was modeled by using restricted cubic splines with 3 knots. A possible non-linear association for ribotype 033 was also considered, however since the non-linear effect was not significant ($P > 0.05$), the results assuming a linear association are presented.

A *p*-value of less than 0.05 was considered as statistically significant. The analysis was performed with R language for statistical computing (R version 3.0.3) [38].

Abbreviations

CA-CDI: Community acquired *Clostridioides (Clostridium) difficile* infection; CDI: *Clostridioides (Clostridium) difficile* infection; MLVA: Multilocus variable-number tandem-repeat analysis; qPCR: quantitative real-time PCR; STRD: Summed tandem repeat difference

Acknowledgments

The authors thank Alenka Magdalena Usenik, Milojka Setina, dr. Katarina Logar, dr. Tina Pirs, dr. Darja Kusar, mag. Maja Kavalic, dr. Maja Rupnik and all participating farmers for their valuable assistance.

Funding

This work was financed with the support of Slovenian National Research Agency (grant nos. J4–2236, J3–4298, P4–0092, P4–0053 and grant for young researchers – P. Bandelj).

Authors' contributions

The study was designed by PB, MV and MO. The qPCR and MLVA analysis were performed by PB, MC and CH. Interpretation of MLVA data was performed by EJK and CH. The statistical analysis was performed by RB and assessed by RB and MV. The manuscript was written by PB, MV and RB. It was critically reviewed by EJK, MO, CH and MC. All authors read and approved the final manuscript.

Competing interest

The authors declare that they have no competing interests.

Consent for publication

Not applicable.

Author details

[1]Veterinary faculty, University of Ljubljana, Cesta v Mestni log 47, SI-1115 Ljubljana, Slovenia. [2]Department of Medical Microbiology, Center of Infectious Diseases, Leiden University Medical Center, Leiden, Netherlands. [3]Institute for biostatistics and Medical informatics, University of Ljubljana, Vrazov trg 2, SI-1104 Ljubljana, Slovenia.

References

1. Wilcox MH, Mooney L, Bendall R, Settle CD, Fawley WN. A case-control study of community-associated *Clostridium difficile* infection. J Antimicrob Chemother. 2008;62:388–96.
2. Khanna S, Pardi DS, Aronson SL, Kammer PP, Orenstein R, St Sauver JL, Harmsen WS, Zinsmeister AR. The epidemiology of community-acquired *Clostridium difficile* infection: a population-based study. Am J Gastroenterol. 2012;107:89–95.
3. Weese JS, Staempfli HR, Prescott JF, Kruth SA, Greenwood SJ, Weese HE. The roles of *Clostridium difficile* and enterotoxigenic *Clostridium perfringens* in diarrhea in dogs. J Vet Intern Med. 2001;15:374–8.
4. Weese JS, Staempfli HR, Prescott JF. A prospective study of the roles of *Clostridium difficile* and enterotoxigenic *Clostridium perfringens* in equine diarrhoea. Equine Vet J. 2001;33:403–9.
5. Yaeger MJ, Kinyon JM, Songer JG. A prospective, case control study evaluating the association between *Clostridium difficile* toxins in the colon of neonatal swine and gross and microscopic lesions. J Vet Diagn Investig. 2007;19:52–9.
6. Rodriguez-Palacios A, Staempfli HR, Duffield T, Peregrine AS, Trotz-Williams LA, Arroyo LG, Brazier JS, Weese JS. *Clostridium difficile* PCR ribotypes in calves, Canada. Emerg Infect Dis. 2006;12:1730–6.
7. Songer JG, Anderson MA. *Clostridium difficile*: an important pathogen of food animals. Anaerobe. 2006;12:1–4.
8. Hammitt MC, Bueschel DM, Keel MK, Glock RD, Cuneo P, DeYoung DW, Reggiardo C, Trinh HT, Songer JG. A possible role for *Clostridium difficile* in the etiology of calf enteritis. Vet Microbiol. 2008;127:343–52.
9. Keessen EC, Gaastra W, Lipman LJA. *Clostridium difficile* infection in humans and animals, differences and similarities. Vet Microbiol. 2011;153:205–17.
10. Agriculture and rural development – Family farming. European Commission. 2015. http://ec.europa.eu/agriculture/family-farming/index_en.htm. Accessed 20 May 2015.
11. Costa MC, Stämpfli HR, Arroyo LG, Pearl DL, Weese JS. Epidemiology of *Clostridium difficile* on a veal farm: prevalence, molecular characterization and tetracycline resistance. Vet Microbiol. 2011;152:379–84.
12. Hawken P, Weese JS, Friendship R, Warriner K. Longitudinal study of *Clostridium difficile* and methicillin-resistant *Staphylococcus aureus* associated with pigs from weaning through to the end of processing. J Food Prot. 2013;76:624–30.
13. Weese JS, Wakeford T, Reid-Smith R, Rousseau J, Friendship R. Longitudinal investigation of *Clostridium difficile* shedding in piglets. Anaerobe. 2010;16:501–4.
14. Zidaric V, Pardon B, dos Vultos T, Deprez P, Brouwer MSM, Roberts AP, Henriques AO, Rupnik M. Different antibiotic resistance and sporulation properties within multiclonal *Clostridium difficile* PCR ribotypes 078, 126, and 033 in a single calf farm. Appl Environ Microbiol. 2012;78:8515–22.
15. Rodriguez-Palacios A, Pickworth C, Loerch S, LeJeune JT. Transient fecal shedding and limited animal-to-animal transmission of *Clostridium difficile* by naturally infected finishing feedlot cattle. Appl Environ Microbiol. 2011; 77:3391–7.
16. Knight DR, Thean S, Putsathit P, Fenwick S, Riley TV. Cross-sectional study reveals high prevalence of *Clostridium difficile* non-PCR ribotype 078 strains in Australian veal calves at slaughter. Appl Environ Microbiol. 2013;79:2630–5.
17. Pirs T, Ocepek M, Rupnik M. Isolation of *Clostridium difficile* from food animals in Slovenia. J Med Microbiol. 2008;57:790–2.
18. Koene MGJ, Mevius D, Wagenaar JA, Harmanus C, Hensgens MPM, Meetsma AM, Putirulan FF, Van Bergen MAP, Kuijper EJ. *Clostridium difficile* in Ducth animals: their presence, characteristics and similarities with human isolates. Clin Microbiol Infect. 2012;18:778–4.
19. Romano V, Albanese F, Dumontet S, Krovacek K, Petrini O, Pasquale V. Prevalence and genotypic charaterization of *Clostridium difficile* from ruminants in Switzerland. Zoonoses Public Health. 2012;59:545–8.
20. Rodriguez C, Taminiau B, Van Broeck J, Avesani V, Delmée M, Daube G. *Clostridium difficile* in young farm animals and slaughter animals in Belgium. Anaerobe. 2012;18:621–5.
21. Knight DR, Putsathis P, Elliott B, Riley TV. Contamination of Australian newborn calf carcasses at slaughter with *Clostridium difficile*. Clin Microbiol Infect. 2016;22:266.e1–7.
22. Bandelj P, Blagus R, Briski F, Frlic O, Vergles Rataj A, Rupnik M, Ocepek M, Vengust M. Identification of risk factors influencing *Clostridium difficile* prevalence in middle-size dairy farms. Vet Res. 2016;47:41.
23. Keel K, Brazier JS, Post KW, Weese JS, Songer JG. Prevalence of PCR Ribotypes among *Clostridium difficile* isolates from pigs, calves, and other species. J Clin Microbiol. 2007;45:1963–4.

24. Schneeberg A, Neubauer H, Schmoock G, Grossmann E, Seyboldt C. Presence of *Clostridium difficile* PCR ribotype clusters related to 033, 078 and 045 in diarrheic calves in Germany. J Med Microbiol. 2013;62:1190–8.
25. Krutova M, Wilcox MH, Kuijper EJ. The pitfalls of laboratory diagnostics of *Clostridium difficile* infection. Clin Microbiol Infect. 2018;24:682–3.
26. Grandesso S, Arena F, Eseme FE, Panese S, Henrici De Angelis L, Spigaglia P, Barbanti F, Rossolini GM. *Clostridium difficile* ribotype 033 colitis in a patient following broad-spectrum antibiotic treatment for KPC-producing *Klebsiella pneumoniae* infection, Italy. New Microbiol. 2016;39:235–6.
27. van den Berg RJ, Schaap I, Templeton KE, Klaassen CH, Kuijper EJ. Typing and subtyping of *Clostridium difficile* isolates by using multiple-locus variable-number tandem-repeat analysis. J Clin Microbiol. 2007;45:1024–8.
28. Bakker D, Corver J, Harmanus C, Goorhuis A, Keessen EC, Fawley WN, Wilcox MH, Kuijper EJ. Relatedness of human and animal *Clostridium difficile* PCR ribotype 078 isolates determined on the basis of multilocus variable-number tandem-repeat analysis and tetracycline resistance. J Clin Microbiol. 2010;48:3744–9.
29. Rodriguez C, Avesani V, Taminiau B, Van Broeck J, Brévers B, Delmée M, Daube G. Investigation of *Clostridium difficile* interspecies relatedness using multilocus sequence typing, multilocus variable-number tandem-repeat analysis and antimicrobial susceptibility testing. Vet J. 2015;206:349–55.
30. Killgore G, Thompson A, Johnson S, Brazier J, Kuijper E, Pepin J, Frost EH, Savelkoul P, Nicholson B, van den Berg RJ, Kato H, Sambol SP, Zukowski W, Woods C, Limbago B, Gerding DN, McDonald LC. Comparison of seven techniques for typing international epidemic strains of *Clostridium difficile*: restriction endonuclease analysis, pulsed-field gel electrophoresis, PCR-ribotyping, multilocus sequence typing, multilocus variable-number tandem-repeat analysis, amplified fragment length polymorphism, and surface layer protein a gene sequence typing. J Clin Microbiol. 2008;46:431–7.
31. Manzoor SE, Tanner HE, Marriott CL, Brazier JS, Hardy KJ, Platt S, Hawkey PM. Extended multilocus variable-number tandem-repeat analysis of *Clostridium difficile* correlates exactly with ribotyping and enables identification of hospital transmission. J Clin Microbiol. 2011;49:3523–30.
32. Eyre DW, Fawley WN, Best EL, Griffiths D, Stoesser NE, Crook DW, Peto TE, Walker AS, Wilcox MH. Comparison of multilocus variable-number tandem-repeat analysis and whole-genome sequencing for investigation of *Clostridium difficile* transmission. J Clin Microbiol. 2013;51:4141–9.
33. Hopman NE, Keessen EC, Harmanus C, Sandersm IM, van Leengoed LA, Kuijper EJ, Lipman LJ. Acquisition of *Clostridium difficile* by piglets. Vet Microbiol. 2011;149:186–92.
34. Magistrali CF, Maresca C, Cucco L, Bano L, Drigo I, Filippini G, Dettori A, Broccatelli S, Pezzotti G. Prevalence and risk factors associated with *Clostridium difficile* shedding in veal calves in Italy. Anaerobe. 2015;33:42–7.
35. Krutova M, Zouharova M, Matejkova J, Tkadlec J, Krejčí J, Faldyna M, Nyc O, Bernardy J. The emergence of *Clostridium difficile* PCR ribotype 078 in piglets in the Czech Republic clusters with *Clostridium difficile* PCR ribotype 078 isolates from Germany, Japan and Taiwan. Int J Med Microbiol. 2018.
36. Bandelj P, Logar K, Usenik AM, Vengust M, Ocepek M. An improved qPCR protocol for rapid detection and quantification of *Clostridium difficile* in cattle feces. FEMS Microbiol Lett. 2013;341:115–21.
37. Goorhuis A, Bakker D, Corver J, Debast SB, Harmanus C, Notermans DW, Bergwerff AA, Dekker FW, Kuijper EJ. Emergence of *Clostridium difficile* infection due to a new hypervirulent strain, polymerase chain reaction ribotype 078. Clin Infect Dis. 2008;47:1162–70.
38. R Core Team. R: a language and environment for statistical computing. R Foundation for statistical computing, Vienna. 2014. https://doi.org/10.1016/j.ijmm.2018.05.006. http://www.R-project.org/. Accessed 20 Feb 2017.

20

Risk factors for the presence of *Fasciola hepatica* antibodies in bulk-milk samples and their association with milk production decreases, in Cuban dairy cattle

Amilcar Arenal[1*†], Yipsi García[1†], Lídice Quesada[1], Dayamis Velázquez[1], Diamela Sánchez[1], Mayelin Peña[1], Asnaldo Suárez[1], Arnielis Díaz[1], Yuliet Sánchez[1], Stijn Casaert[2], Jan van Dijk[3], Jozef Vercruysse[2] and Johannes Charlier[4]

Abstract

Background: Worldwide, *Fasciola hepatica* infection causes high production losses in the livestock industry. Recently, studies have analyzed the association between measurements of *F. hepatica* infection intensity and herd management practices. The aim of the present study, the first of its kind in a subtropical region, was to evaluate associations between *F. hepatica* bulk-tank milk ELISA results with herd management factors and milk yield in dairy herds, in Camagüey, Cuba. The SVANOVIR® *F. hepatica*-AB ELISA was used to measure *F. hepatica* antibody levels in a random sample of 516 dairy herds during the period of May–July of 2014. Farm management practice data were collected using a questionnaire.

Results: With 82% of the herds testing positive, the results indicate that *F. hepatica* is very widespread in this area. Reductions in milk production of 18 and 32% were observed in herds with Optical Density Ratios (ODR) of 0.3–0.6 and > 0.6, respectively, when compared to herds with ODR < 0.3. Overall, the longer the milking cows were put out to pasture, the higher the levels of anti-parasite antibodies. Co-grazing with sheep and goats also significantly increased the risk of high ODR.

Conclusions: Our data show a widespread occurrence of the parasite as well as a major potential impact of the infection on the Cuban development goal of becoming self-sufficient in milk production. Our risk factor analysis suggests that the prevention of infection around water sources, and the separation of cattle from small ruminants could be useful control measures. This is the first epidemiological survey of *F. hepatica* abundance, and associated reductions in milk yield, in dairy herds in Cuba.

Keywords: Bulk-tank, Milk production, Risk factors, Liver fluke, Helminth, Ruminants, ELISA

Background

During the 1980s, Cuban dairy cattle production levels enjoyed the highest growth in Latin America. In 1989, production peaked at 1134 million liters of milk. Since then, the Cuban dairy industry has faced momentous changes and challenges. During the Cuban economic crisis of the Nineties, milk production dropped back and was recorded at 353 million liters in 2005 [1]. At the same time, pure-bred Holstein herds, which had made up 72% of all herds, were reduced to 12%, with cross-breeding of Holstein dairy cows and Zebu cattle becoming the norm [2]. The percentage of dairy cows kept on privately-owned, as opposed to state-owned, farms increased from 20 to 80%. In recent years, milk production rebounded to approximately 600 million liters. However, this is estimated to be only 50% of the current Cuban milk demand [2]. To raise self-sufficiency levels in

* Correspondence: amilcar.arenal@reduc.edu.cu
†Amilcar Arenal and Yipsi García contributed equally to this work.
[1]Department of Morpho-Physiology, University of Camagüey Ignacio Agramonte Loynaz, Km 5 ½, 74650 Camagüey, Cuba
Full list of author information is available at the end of the article

subtropical countries like Cuba, it is clearly important to evaluate existing milk production limitations.

In Cuba, milk production is based on the utilization of pastures in the rainy season and green and preserved forages, supplemented by sugar-industry by-products, in the dry season. The most important limitation on milk production in Cuba is thought to be that these nutritional resources contain less than desirable energy density [3]. This lack manifests itself especially in the dry season, by halving of milk production. Another likely factor reducing yield milk is infections, especially with helminths. However, the prevalence of economically important helminths on dairy farms, and their impact on milk production, has not been quantified in Cuba. Existing impact studies were all carried out in different climatic zones, and for very different farming systems [4, 5], and therefore it is unlikely that the results of such studies can be applied to milk production systems in subtropical regions.

Helminth infections are recognized as a major limitation for livestock production throughout the tropics and elsewhere [6]. Among these, infections with *Fasciola hepatica* are responsible for significant economic losses in the cattle industry, due to mortality, reduced production of meat and milk and costs of deworming.

Various diagnostic methods based on detecting antibodies specific for *F. hepatica* in feces, serum, meat juice and milk have been described previously [7–9]. The wide availability and simplicity of these tests have facilitated large epidemiological studies [10] and evaluation of the association between fluke infection status and milk production parameters [11]. The magnitude of such effects has been shown to depend on the production system [4, 12, 13], lending argument to the need to study such losses in disparate epidemiological and production settings.

To date, the only epidemiological data available in Cuba are prevalence data from routine inspections in slaughterhouses in the central provinces showing prevalences of 20–50% for *F. hepatica* [14, 15]. To define the potential constraint of helminth infections on dairy productivity and initiate the development of *F. hepatica* herd management recommendations, we conducted a targeted survey in the major milk producing province of Camagüey and deployed a bulk-tank milk (BTM) ELISA test as a tool for diagnosis of fasciolosis in Cuban dairy cattle.

Results

Fasciola hepatica antibodies

The mean, SD and range of the *F. hepatica* ODR were 0.510, 0.201 and 0.049 to 1.192, respectively. According to the manufacturer's interpretation criteria 82.2% of the herds tested positive for *F. hepatica* (> 0.3 ODR, 95% confidence interval: 0.561–0.591), while 35.7% of herds were likely to suffer significant production decreases (> 0.6 ODR, 95% confidence interval: 0.705–0.736).

Associations of *Fasciola hepatica* antibodies and management factors with milk yield

Complete data (consisting of BTM ELISA results combined with complete questionnaire and milk production information) was obtained from 516 out of the 650 selected farms. The observed average milk yield per dairy cow per year was 1024 kg (95% confidence interval: 996–1051 kg). The average milk yield per dairy cow per year of the negative herds (< 0.3 ODR) was 1266 kg (95% confidence interval: 1200–1333 kg). There was a significant negative correlation between ODR and milk yield ($R = -0.44$; $P < 0.01$). In the one-way ANOVA, the differences in average milk yield per cow per year between *F. hepatica* > 0.6 ODR and 0.3–0.6 ODR comparing to negative herds (< 0.3 ODR) were 401 kg (32%) and 226 kg (18%), respectively (Fig. 1).

The management factors that were significantly associated through univariable regression with milk yield are listed in Table 1. Grass proportion in the dry season (higher milk yield with higher grass proportion), watering place (lower milk yield for pool/pond/brook vs. well), farm total area (lower milk yield in smaller farms), municipality, nutritional supplement in dry (lower milk yield in case of supplement) and rainy season (higher milk yield in case of supplement), and ODR (lower milk yield with higher ODR) were all associated with milk yield. The multivariable regression model to investigate the association between ODR and milk yield (Table 2) retained ODR and municipality as the two significant predictors. According to this model, an increase in *F. hepatica* ODR over the interquartile range (0.33–0.64) is associated with a reduction in average milk yield of 183 kg/cow per year (14%).

Associations of *Fasciola hepatica* antibodies with management factors

The frequencies at which different categories of management variables were measured, and their corresponding ODR are listed in Table 3. The UEB farms (state sector) presented significantly higher ODR values compared to the private sector (CPA and CCS) and also UBPC (state sector). Farms with more than 30 ha had a significantly higher ODR compared to the smaller ones. When the cows were grazed with sheep and goats the ODR was significantly higher compared to the farms where cows are grazed without other species or only with horses. Herds with access to pool/pond/brookhad significantly higher ODR compared to herds with wells as water source. It was also observed that farms with a lower grass proportion in the cow's diet in the dry season presented a significantly higher ODR. In the dry season, a lower ODR value was also found when the grazing time was less than 6 h. The multivariate regression model identified grass proportion

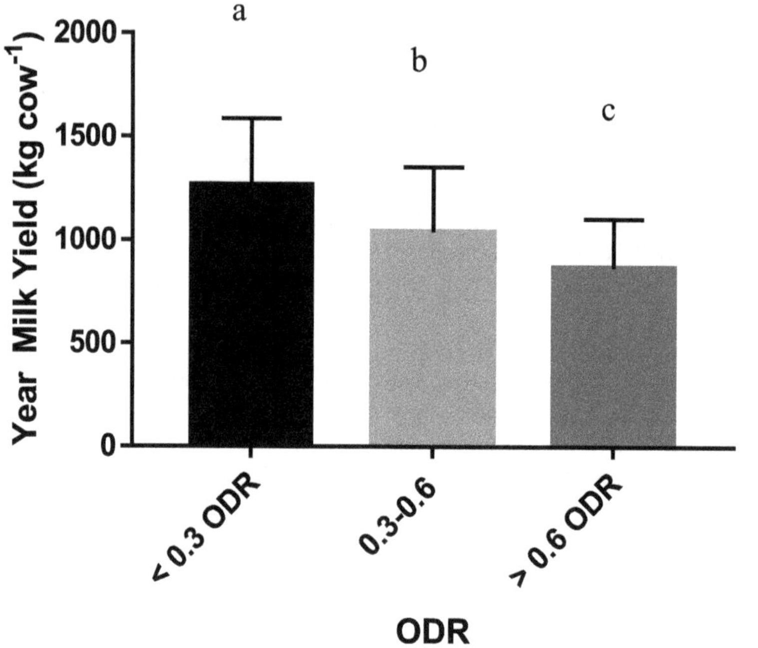

Fig. 1 Milk production per cow per year according to the ELISA cut-off of the relative optical density ratio (ODR). Data are presented as mean ± SE (< 0.300, $N = 92$; 0.300–0.600, $N = 257$; > 0.600, $N = 167$). Bar indicates the SE. Different letters indicate significant differences among groups ($P < 0.001$)

in the dry season and municipality as the two most significant associated factors with *F. hepatica* ODR (Table 4).

Discussion

Our study found evidence of *F. hepatica* infection on 4 out of 5 farms in the major milk producing province in Cuba. Moreover, this infection was not only highly prevalent, but also significantly associated with decreases in milk yield.

In previous abattoir-based studies in Cuba, *F. hepatica* parasites were observed to be present between 20 and 50% of the animals. However, the latter studies were conducted at the individual cow level [14, 15] and in a different geographical region of Cuba. Moreover, it is known that meat inspection at the slaughterhouse has a lower sensitivity than serology-based methods [16].

The evidence of widespread *F. hepatica* infection in Cuban dairy herds, together with the known deleterious

Table 1 Variables significantly ($P < 0.05$) associated through univariable regression with annual milk yield (kg/cow/year) in Camagüey province, Cuba

Variable	Parameter	B	Std. Error	R^2	Sig.
ODR	Intercept	1387.5	34.8	0.197	
	ODR	− 712.3	63.5		< 0.001
Municipality[a]	Intercept	978.1	34.9	0.153	
	Municipality				< 0.001
Grass proportion[a]	Intercept	940.7	25.8	0.051	
	Grass proportion				< 0.001
Watering place [a]	Intercept	1044.0	15.9	0.015	
	Watering place				0.006
Farm total area[a]	Intercept	1050.6	24.7	0.012	
	Farm total area				0.048
Nutritional supplement in dry season[a]	Intercept	1054.7	16.8	0.022	
	Nutritional supplement dry season				0.001
Nutritional supplement in rainy season[a]	Intercept	945.3	21.9	0.040	
	Nutritional supplement in rainy season				< 0.001

[a]Different levels not shown

Table 2 Association between *F. hepatica* antibody level and milk yield (kg/cow per year) in Camagüey province, Cuba[a]

Parameter	B	Std. Error	Sig.
Intercept	1300.3	49.4	< 0.001
ODR	– 590.9	68.1	< 0.001
Municipality			< 0.001
Camagüey	– 107.0	50.3	0.034
Céspedes	– 138.1	81.9	0.093
Esmeralda	−102.2	72.2	0.157
Florida	−100.2	66.9	0.135
Guaimaro	129.7	46.5	0.005
Jimaguayú	16.8	50.5	0.739
Minas	−35.1	66.9	0.600
Najasa	86.5	54.5	0.113
Nuevitas	13.3	67.0	0.842
S.Cubitas	64.9	62.7	0.302
Santa Cruz del Sur	144.8	54.6	0.008
Sibanicú	70.2	55.8	0.210
Vertientes	Base	.	.

[a]Multivariable linear regression model ($R^2 = 0.264$, $N = 516$ dairy herds)

effects of *F. hepatica* on animal welfare and productivity suggest than these infections should be considered of major importance in Cuban dairy farms. However, on high-input, intensive, Holstein-pedigree farms, milk production per cow per year was more than 6 times that of the Cuban mixed breed cattle studied here. Cuban cattle are not genetically capable of achieving such high levels of milk production, they eat a less nutritious diet and, in subtropical climates, they often face higher parasite burdens. Because Cuban cattle are likely under lower metabolic stress than their European counterparts in intensive production systems, it could be proposed that parasite-ascribed decreases in milk production in Cuba should be lower than, for example, in Europe. However, in the UK, in high yielding herds, *F. hepatica* – associated decreases were estimated at 15% [4], compared to estimated decreases of 18% to 32% in the present study. The Cuban estimate is substantially higher than the 3% reduction described in Belgium in herds with high ODR [11] and the 6% in Spanish herds with high infection levels [13]. Partly, these differences may be ascribed to the fact that we did not control for some confounding factors, such as lactation stage, age composition or somatic cell count data in our analysis. This was not possible, as these data are not routinely collected in Cuban dairy farms. Therefore, further elucidation of the true and recoverable production impact would require an intervention trial using anthelmintic treatment under field conditions [17].

The impact of parasite infections on food security may be more keenly felt in countries where demand is already outstripping supply. At the same time, options for control are likely to be more limited in subtropical systems. For example, with very few water sources available, options for pasture rotation are limited. In Cuba, anthelmintic treatments for *F. hepatica* are not used routinely either because of a lack of availability in the Cuban market and/or a lack of diagnostic routine. This study made a start with the identification of risk factors, which should aid in the development of control recommendations for the Cuban dairy sector. Different farming systems had different ODR levels. UEB farms, which are normally the larger farms, with more extensive access to suitable habitats for lymnaeid snails, had higher ODRs. Similarly, farms with a higher number of hectares available had higher ODRs. In Denmark, larger dairy herds were also more prone to *F. hepatica* infection [18]. In Turkey and Tanzania, large-scale and traditional (stationary herds without effective disease control) dairy farms presented higher prevalence of *Fasciola* sp. than small-scale farms [19, 20]. This may be related to intensively grazed pastures and to an increased likelihood of cattle encountering fluke-contaminated snail habitats on larger farms.

Access to suitable habitats for lymnaeid snails, usually man-made ponds of stagnant water used to water cattle, indeed appears to be an important factor contributing to higher levels of infection. In this study, farms with less grass as a proportion of the total diet, available during the dry season, had significantly higher antibody titers. On these farms, cattle will normally be congregated around these habitats for lymnaeid snails and receive supplementation with other food sources, such as sugar cane byproducts. They will therefore have increased contact time with metacercaria-contaminated snail-infested areas. Access to, and type of, water sources could be the key overriding factor in fluke transmission in Cuba. This may be an important area to focus on in terms of limiting losses to the parasite.

Grazing alongside horses was not a significant risk factor whereas, in agreement with other studies [21], co-grazing with small ruminants clearly increased the risk of higher ODR levels.

There were significant differences in ODR levels between municipalities. The reason for this could include local environmental differences as well as differences in local farm management practices [22]. In the UK, McCann, Baylis and Williams [23] detected rainfall as the main responsible factor of variation (23%) in *F. hepatica* BTM antibody levels, whilst farm management explained about 21% of variation. Bennema et al. [24] found that in regions with relatively homogenous climatic and environmental conditions, management

Table 3 Frequency and percentages of the herd characteristics of dairy herds sampled in Cuba ($N = 516$) in a cross-sectional questionnaire survey conducted in March–July, 2014

Variable			%	N	Mean ODR	SD	Range
Type of production	State[a]	UEB	10.1	52	0.600[a]	0.188	0.05-0.99
		UBPC	25.6	132	0.522[b]	0.187	0.05–0.98
	Private[a]	CPA	15.1	78	0.496[b]	0.177	0.07–0.89
		CCS	49.2	254	0.490[b]	0.212	0.05–1.19
Type of Herd	Dairy only		54.8	283	0.511 [a]	0.208	0.05–1.19
	Both beef and dairy		45.2	233	0.509 [a]	0.192	0.05–1.03
Farm total area	<=13.42 ha		32.8	169	0.474[b]	0.194	0.12–1.03
	> 13.42 and < =30 ha		14.0	72	0.472[b]	0.186	0.05–0.88
	> 30 ha		53.3	275	0.543[a]	0.204	0.05-1.19
Herd size (adult cows: lactating + dry):	< 30		55.4	286	0.499[a]	0.217	0.05-1.03
	30–60		30.6	158	0.539[a]	0.200	0.05-1.19
	> 60		14.0	72	0.534[a]	0.215	0.07-0.89
Deworming of cows			3.1	16	0.539[a]	0.187	0.28-0.98
Not dewormed							
Dewormed when worm problems			22.9	118	0.480[a]	0.197	0.05-0.89
Preventive treatment			74.0	382	0.518[a]	0.202	0.05-1.19
Cows grazed together with other species							
Sheep and goats			19.2	99	0.552[a]	0.178	0.05-1.03
Horse			21.5	111	0.501[b]	0.196	0.12–0.99
None			59.3	306	0.500[b]	0.208	0.05–1.19
Stocking rate: average number of cows per hectare on a parcel?							
< 1			40.3	208	0.496[a]	0.185	0.07-1.19
1–2			56.2	290	0.521[a]	0.210	0.05-0.99
> 2			3.5	18	0.501[a]	0.232	0.05-0.85
Watering place							
Pool/pond/brook			20.9	108	0.561[a]	0.206	0.07-1.19
Pump on pasture			79.1	408	0.497[b]	0.198	0.05–0.99
Rotational grazing of cows							
Yes			5.4	28	0.511[a]	0.190	0.15-0.82
No			94.6	488	0.502[a]	0.201	0.05-1.19
What was the cow's grazing time per day during the dry season?							
Day and night			14.0	72	0.542[a]	0.165	0.05-0.76
< 6 h per day			62.6	322	0.498[b]	0.217	0.05–1.19
> 6 h per day			23.4	122	0.541[a]	0.167	0.05-0.98
What was the cow's grazing time per day during the rainy season?							
Day and night			11.0	57	0.513[a]	0.168	0.05-0.80
< 6 h per day			42.4	219	0.504[a]	0.205	0.05-1.19
> 6 h per day			46.5	240	0.515[a]	0.205	0.05-0.99
Grass proportion dry season							
81–100			20.5	106	0.406[c]	0.201	0.05–0.88
51–80			50.6	261	0.511[b]	0.198	0.07–0.99
< =50			28.9	149	0.583[a]	0.172	0.05-1.19

Table 3 Frequency and percentages of the herd characteristics of dairy herds sampled in Cuba ($N = 516$) in a cross-sectional questionnaire survey conducted in March–July, 2014 *(Continued)*

Variable		%	N	Mean ODR	SD	Range
Grass proportion rainy season						
81–100		30.4	157	0.501[a]	0.188	90.05-0.91
51–80		52.3	270	0.519[a]	0.205	0.05-1.19
< =50		17.2	89	0.501[a]	0.215	0.12-1.03
Grass Mowing						
51–100%		27.9	144	0.496[a]	0.187	0.05-0.88
< 50%		43.0	222	0.512[a]	0.210	0.05-1.03
Never		29.1	150	0.522[a]	0.200	0.07-1.19
Municipality	Camagüey	10.7%	55	0.630[a]	0.114	0.37-.87
	Céspedes	2.7%	14	0.634 [a]	0.303	0.05–1.19
	Esmeralda	3.7%	19	0.493 [a,b,c]	0.196	.15–.82
	Florida	4.5%	23	0.548 [a,b]	0.171	0.25–.86
	Guaimaro	15.7%	81	0.381 [c,d]	0.179	0.05–.88
	Jimaguayú	10.3%	53	0.567 [a,b]	0.179	0.07–.99
	Minas	4.5%	23	0.511 [a,b]	0.254	0.09–.88
	Najasa	8.1%	42	0.455 [b,c,d]	0.142	0.17–.77
	Nuevitas	4.5%	23	0.598[a,b]	0.263	0.07–.99
	S.Cubitas	5.6%	29	0.360 [d]	0.192	0.11–.91
	Santa Cruz del Sur	7.9%	41	0.572 [a,b]	0.177	0.15–.91
	Sibanicú	7.6%	39	0.464 [b,c,d]	0.171	0.17.84
	Vertientes	14.3%	74	0.545 [a,b]	0.179	0.05.88

Different letters indicate significant differences between groups ($P < 0.05$)
[a]UEB (Basic unit from the state), UBPC (Cooperative Unit Basic of Production), CCS (Credit and Service Cooperative) and CPA (Agropecuary Production Cooperative)

factors are the primary factors determining *F. hepatica* infection risk. Further research is recommended to determine the importance of water source as well as of infection and the local environmental (soil type, local pasture, infection with other parasite, bacterial interactions, landscape features) and climatic conditions affecting the infection risk. Moreover, it is necessary to evaluate the impact of host factors such as age and genetic make-up. Ultimately, this could result in local risk maps and evidence-based and practical management recommendations such as sanitation of pastures and water sources and targeted anthelmintic treatment during periods of highest infection pressure [24, 25].

Conclusions

We have provided baseline *F. hepatica* exposure data for the major milk production area of Cuba. Our data show a widespread occurrence of the parasite, as well as a major potential impact of this infection on the Cuban development goal to become self-sufficient in milk production. Our risk factor analysis suggests that the prevention of infection around habitats suitable for lymnaeid snails, and that the separation of cattle and small ruminants could be useful control recommendations. However, further research to confirm the importance of these risk factors as well as to understand the basic *F. hepatica* epidemiology in relation to temporal and regional changes in climate and landscape in Cuba is needed.

Methods

Study area

The study was conducted in Camagüey province, eastern Cuba. Camagüey has a surface of 15,615 km^2 and a tropical climate with an average annual temperature and rainfall of 24.7 °C and 1200 mm (www.one.cu), respectively. Elevation varies slightly, from sea level at the coast to 100 m in the center. According to the milk industry department of the Ministry of Agriculture, in Camagüey, approximately 10,000 dairy farms provide milk to a dairy cooperative during the rainy season (March–July); however, during the dry season (August–February) the number of dairy farms providing milk dropped to below 6000 (Reynaldo González, personal communication).

Table 4 Multivariable linear regression model of management factors associated with *F. hepatica* ODR measured in bulk-tank milk samples in Camagüey province, Cuba[a]

Parameter	B	Std. Error	Sig.
Intercept	0.579	0.025	< 0.001
Grass proportion dry season (%)			0.004
81–100	− 0.110	0.033	0.001
50–80	−0.022	0.022	0.317
< 50	Base		
Municipality			< 0.001
Camagüey	0.070	0.033	0.033
Céspedes	0.060	0.054	0.268
Esmeralda	−0.062	0.048	0.197
Florida	−0.010	0.044	0.827
Guaimaro	−0.099	0.035	0.005
Jimaguayú	0.009	0.033	0.791
Minas	−0.046	0.046	0.308
Najasa	−0.112	0.036	0.002
Nuevitas	0.040	0.046	0.374
S.Cubitas	−0.198	0.042	0.000
Santa Cruz del Sur	0.006	0.036	0.874
Sibanicú	−0.093	0.038	0.015
Vertientes	Base	.	.

[a]($R^2 = 0.204$, $N = 516$ dairy herds)

Sampling and laboratory procedure

The farms were selected based on the following criteria: (a) storage of farm production data in the milk industry department of the Ministry of Agriculture, in Camagüey; (b) providing milk during the whole year; (c) proportionally according to the total farm per municipality and (d) farmers agree to participate. Using the RANDBETWEEN function in Microsoft® Excel, 650 BTM samples were randomly chosen out of all available regional samples (N ≈ 6000). We collected the samples during the period of May–July 2014. The 650 dairy farms were located across the 13 municipalities in Camagüey. We transported the samples to the laboratory within 4 h after collection. The samples were kept at 4 °C between collections on the farms. After arrival at the laboratory, the milk samples were centrifuged (16,000 × *g*, 5 min), fat was skimmed off and the supernatant was collected and frozen at − 20 °C (for a maximum period of three months) until further analysis.

Samples were analyzed using a commercially available ELISA test (SVANOVIR® *F. hepatica*-Ab, Svanova Biotech, Uppsala) according the manufacturer's instructions. The ELISA results are expressed as optical density ratios (ODR). ODR = (OD - NC) / (PC - NC), where OD is the optical density at 405 nm of the sample and NC and PC are the OD at 405 nm of the negative and positive controls, respectively.

Questionnaire

We collected the management data by interviewing the farmers in person. Information was collected on location, herd size, type of production (private or state), watering place [pool/pond/brook or well], pasture management and other husbandry practices, and anthelmintic control measures in adult cows. A complete list of the collected variables is provided in Table 1. Milk production data were collected from the milk industry department of the Ministry of Agriculture.

Associations between *F. hepatica* antibody levels and milk yield

The association between *F. hepatica* BTM antibody level (ODR) and the average milk yield per cow per year (referred to as "milk yield") was first investigated by the Pearson correlation coefficient. Next, milk yield was compared by a one-way ANOVA with a Student–Newman–Keuls multiple comparisons post-hoc test between the quartiles of the *F. hepatica* ODR. In addition, one-way ANOVA, was also used to evaluate the association of farm management factors with milk yield, for each management factor independently. Finally, a multivariable model was built to assess the association of *F. hepatica* ODR with farm management factors (=independent variables) and milk yield (= outcome variable).

Associations between *F. hepatica* antibody levels and management variables

First, a one-way ANOVA was used to test for significant differences in ODR between the different categories of each farm management factor. The differences were further analyzed using the Student–Newman–Keuls multiple comparisons post-hoc test. For comparisons of factors with only two categories, the Student t-test was used.

Factors that were significant in this first screening ($\alpha = 0.05$) were evaluated in a multivariable linear regression model that was constructed by forward stepwise selection of variables with a nominal significance level of $\alpha = 0.05$ and 0.10 for the entry and removal of a variable, respectively. Two-way interactions between the variables included in the final model were evaluated for significance. In all the above models, a level of $\alpha = 0.05$ was used to declare a variable to be statistically significant. The analysis was conducted in SPSS v21.0 (SPSS Inc., Chicago, USA).

Abbreviations

BTM: Bulk tank milk; CCS: (Cooperativa de Crédito y Servicio) Credit and Service Cooperative; CPA: (Cooperativa de Producción Agropecuaria) Agricultural Production Cooperative; ODR: Optical Density Ratios; UBPC: (Unidades Básicas de Producción Cooperativa) Basic Units of Cooperative Production (The new farms which now make up the largest sector in Cuban agriculture); UEB: (Unidades Empresariales de Base) State-owned Basic Business Units

Acknowledgments
We express our gratitude to the farmers that participated in the survey and the ministry of Agriculture for their support. We thank Dr. John van der Meer (Pan-American Marine Biotechnology Association) for advice on English improvement of the manuscript.

Funding
This research has been supported by the Belgian Development Cooperation through VLIR-UOS. VLIR-UOS supports partnerships between universities and university colleges in Flanders (Belgium) and the South looking for innovative responses to global and local challenges. Visit www.vliruos.be for more information.

Authors' contributions
Conception and study design: AA, JV, JC; Collection of milk samples and milk processing by ELISA: YG, LQ, DV, DS, MP, AS, AD, YS, SC; questionnaire interviews: YG, LQ, DV, DS, MP, AS, AD, YS; Statistical analysis: AA, JC; Drafting the manuscript: AA, JC, and JvD; Interpretation of data, revising manuscript critically AA, JV, JC; JvD. All authors read and approved the final manuscript.

Consent for publication
Not applicable

Competing interests
The SVANOVIR®*F. hepatica*-Ab ELISA is commercialized by Boehringer Ingelheim Svanova under a license agreement with Ghent University.

Author details
[1]Department of Morpho-Physiology, University of Camagüey Ignacio Agramonte Loynaz, Km 5 ½, 74650 Camagüey, Cuba. [2]Department of Virology, Parasitology and Immunology, Faculty of Veterinary Medicine, Ghent University, Salisburylaan 133, 9820 Merelbeke, Belgium. [3]Centre for Preventive Medicine, Animal Health Trust, Lanwades Park, Kentford, Newmarket, Suffolk CB8 7UU, UK. [4]Kreavet, Hendrik Mertensstraat 17, 9150, Kruibeke, Belgium.

References

1. AHDB. World Milk Production. 2015. (dairy.ahdb.org.uk)
2. Ponce P. Un enfoque crítico de la lechería internacional y cubana. Rev Sal Anim. 2009;31:77–85.
3. Milera MC. López O, Alonso O. Principios generados a partir de la evolución del manejo en pastoreo para la producción de leche bovina en Cuba. Pastos y Forrajes. 2014;37:382–91.
4. Howell A, Baylis M, Smith R, Pinchbeck G, Williams D. Epidemiology and impact of *Fasciola hepatica* exposure in high-yielding dairy herds. Prev Vet Med. 2015;121:41–8.
5. Charlier J, van der Voort M, Kenyon F, Skuce P, Vercruysse J. Chasing helminths and their economic impact on farmed ruminants. Trends Parasitol. 2014;30:361–7.
6. Githiori J, Höglund J, Waller P, Baker R. Evaluation of anthelmintic properties of some plants used as livestock dewormers against *Haemonchus contortus* infections in sheep. Parasitology. 2004;129:245–53.
7. Mezo M, González-Warleta M, Carro C, Ubeira FM. An ultrasensitive capture ELISA for detection of Fasciola hepatica coproantigens in sheep and cattle using a new monoclonal antibody (MM3). J Parasitol. 2004;90:845–52.
8. Reichel MP, Vanhoff K, Baxter B. Performance characteristics of an enzyme-linked immunosorbent assay performed in milk for the detection of liver fluke (*Fasciola hepatica*) infection in cattle. Vet Parasitol. 2005;129:61–6.
9. Charlier J, De Cat A, Forbes A, Vercruysse J. Measurement of antibodies to gastrointestinal nematodes and liver fluke in meat juice of beef cattle and associations with carcass parameters. Vet Parasitol. 2009;166:235–40.
10. Bennema S, Vercruysse J, Claerebout E, Schnieder T, Strube C, Ducheyne E, et al. The use of bulk-tank milk ELISAs to assess the spatial distribution of *Fasciola hepatica*, *Ostertagia ostertagi* and *Dictyocaulus viviparus* in dairy cattle in Flanders (Belgium). Vet Parasitol. 2009;165:51–7.
11. Charlier J, Duchateau L, Claerebout E, Williams D, Vercruysse J. Associations between anti-*Fasciola hepatica* antibody levels in bulk-tank milk samples and production parameters in dairy herds. Prev Vet Med. 2007;78:57–66.
12. Bennema SC, Vercruysse J, Morgan E, Stafford K, Höglund J, Demeler J, et al. Epidemiology and risk factors for exposure to gastrointestinal nematodes in dairy herds in northwestern Europe. Vet Parasitol. 2010;173:247–54.
13. Mezo M, González-Warleta M, Castro-Hermida JA, Muiño L, Ubeira FM. Association between anti-*F. hepatica* antibody levels in milk and production losses in dairy cows. Vet Parasitol. 2011;180:237–42.
14. Brito AE, Hernández MA, Fe P, Silveira EA. Prevalence, liver confiscation and economic losses by *Fasciola hepatica* in bovine slaughterhouses of three provinces of the central region of Cuba. Rev Electron Vet. 2010;11:1–7.
15. González R, Pérez Ruano M, Brito S. Fasciolosis bovina. Evaluación de las principales pérdidas provocadas en una empresa ganadera. Rev Sal Anim. 2007;29:167–75.
16. Mazeri S, Sargison N, Kelly RF, Barend M, Handel I. Evaluation of the performance of five diagnostic tests for *Fasciola hepatica* infection in naturally infected cattle using a Bayesian no gold standard approach. PLoS One. 2016;11:e0161621.
17. Charlier J, Hostens M, Jacobs J, Van Ranst B, Duchateau L, Vercruysse J. Integrating fasciolosis control in the dry cow management: the effect of closantel treatment on milk production. PLoS One. 2012;7:e43216.
18. Takeuchi-Storm N, Denwood M, Hansen TVA, Halasa T, Rattenborg E, Boes J, et al. Farm-level risk factors for *Fasciola hepatica* infection in Danish dairy cattle as evaluated by two diagnostic methods. Parasit Vectors. 2017;10:555.
19. Keyyu J, Kassuku A, Msalilwa L, Monrad J, Kyvsgaard NC. Cross-sectional prevalence of helminth infections in cattle on traditional, small-scale and large-scale dairy farms in Iringa district, Tanzania. Vet Res Comm. 2006;30: 45–55.
20. Yildirim A, Ica A, Duzlu O, Inci A. Prevalence and risk factors associated with Fasciola hepatica in cattle from Kayseri province. Turkey Rev Med Vet (Toulouse). 2007;158:613.
21. Brockwell YM, Elliott TP, Anderson GR, Stanton R, Spithill TW, Sangster NC. Confirmation of *Fasciola hepatica* resistant to triclabendazole in naturally infected Australian beef and dairy cattle. Int J Parasitol Drugs Drug Resist. 2014;4:48–54.
22. Bennema S, Ducheyne E, Vercruysse J, Claerebout E, Hendrickx G, Charlier J. Relative importance of management, meteorological and environmental factors in the spatial distribution of *Fasciola hepatica* in dairy cattle in a temperate climate zone. Int J Parasitol. 2011;41:225–33.
23. McCann CM, Baylis M, Williams DJL. The development of linear regression models using environmental variables to explain the spatial distribution of *Fasciola hepatica* infection in dairy herds in England and Wales. Int J Parasitol. 2010;40:1021–8.
24. Charlier J, Vercruysse J, Morgan E, Van Dijk J, Williams D. Recent advances in the diagnosis, impact on production and prediction of *Fasciola hepatica* in cattle. Parasitology. 2014;141:326–35.
25. Knubben-Schweizer G, Torgerson PR. Bovine fasciolosis: control strategies based on the location of Galba truncatula habitats on farms. Vet Parasitol. 2015;208:77–83.

Chi-miR-3031 regulates beta-casein via the PI3K/AKT-mTOR signaling pathway in goat mammary epithelial cells (GMECs)

Kaiwen Chen[1†], Jinxing Hou[2†], Yuxuan Song[1], Xiaochuan Zhang[1], Yuhan Liu[1,2], Gonghai Zhang[1], Kai Wen[1], Haidong Ma[1], Guang Li[1], Binyun Cao[1] and Xiaopeng An[1*]

Abstract

Background: MicroRNAs can regulate gene expression at the posttranscriptional level through translational repression or target degradation. Our previous investigations examined the differential expression levels of chi-miR-3031 in caprine mammary gland tissues in colostrum and common milk stages.

Results: The present study detected the role of chi-miR-3031 in the lactation mechanisms of GMECs. High-throughput sequencing was used to analyze transcriptomic landscapes of GMECs transfected with chi-miR-3031 mimics (MC) and a mimic negative control (NC). In the MC and NC groups, we acquired 39,793,503 and 36,531,517 uniquely mapped reads, respectively, accounting for 85.85 and 81.66% of total reads. In the MC group, 180 differentially expressed unigenes were downregulated, whereas 157 unigenes were upregulated. KEGG pathway analyses showed that the prolactin, TNF and ErbB signaling pathways, including *TGFα*, *PIK3R3*, *IGF2*, *ELF5*, *IGFBP5* and *LHβ* genes, played important roles in mammary development and milk secretion. Results from transcriptome sequencing, real-time PCR and western blotting showed that chi-miR-3031 suppressed the expression of IGFBP5 mRNA and protein. The expression levels of β-casein significantly increased in the MC and siRNA-IGFBP5 groups. We observed that the down-regulation of IGFBP5 activated mTOR at the Ser2448 site in GMECs transfected with MC and siRNA-IGFBP5. Previous findings and our results showed that chi-miR-3031 activated the PI3K-AKT-mTOR pathway and increased β-casein expression by down-regulating IGFBP5.

Conclusions: These findings will afford valuable information for improving milk quality and contribute the development of potential methods for amending lactation performance.

Keywords: Dairy goats, miRNAs, Gene expression, Transcriptome, Milk components

Background

Previous studies have shown that mammary epithelial cells (MECs) play key roles in milk protein and lipid biosynthesis [1–3]. During lactation, MECs secrete large quantities of milk proteins, notably caseins. Caseins (αS1-, αS2-, β-, and κ-CN in most species) are the main milk-specific proteins [4]. Transcriptomic regulations affect the changes of milk composition and yield in MECs. Thus, in this study, MECs were harvested from the mammary gland.

* Correspondence: anxiaopengdky@163.com
†Kaiwen Chen and Jinxing Hou contributed equally to this work.
1College of Animal Science and Technology, Northwest A&F University, No. 22 Xinong Road, Yangling, Shaanxi 712100, People's Republic of China
Full list of author information is available at the end of the article

RNA sequencing (RNA-seq) provides a holistic view of transcriptomes, allows the refinement of gene structures and also uncovers novel splice isoforms, SNPs and transcribed regions [5–7], in addition, it also produces complete gene expression measurements instead of relative measurements, thereby providing more significant insights and accuracy than microarrays [8, 9]. Using a transcriptome analysis, Suarez-Vega et al. (2017) found that RNA transport, and fatty acid elongation Jak-STAT signaling pathways could affect milk yield, protein percentage and fat percentage [10]. Via an RNA-seq analysis of mammary epithelial cells, Wanyonyi et al. (2017) suggested that seven pathways, including lysosome, galactose metabolism, p53 signaling, cell adhesion molecule (CAM), complement, coagulation and Nod-like receptor signaling pathways, are only significantly responsive to extracellular matrices in the presence of lactogenic hormones [11].

MiRNAs are noncoding small RNA molecules (about 22 nucleotides) that regulate gene expression post-transcriptionally by binding to the 3′-UTR of their target mRNAs [12, 13], which refer to proliferation [14], apoptosis [15], reproduction [16] and milk performance [17, 18]. The overexpression of miR-130a notably decreases the levels of cellular triacylglycerol (TAG) and suppresses lipid droplet formation, but the inhibition of miR-130a results in increased TAG accumulation and lipid droplet formation in bovine mammary epithelial cells [19]. The miRNA target network shows that bta-miR-574 can affect lactation and mammary gland development by leptin receptor, which is involved in the adipocytokine signaling pathway. MiR-3880 regulates the development of mammary gland and lactation by serine/threonine-protein phosphatase, which participates in the oxytocin signaling pathway [20]. The miR-15b expression decreases during lactation and oppositely correlates with lipid synthesis proteins, suggesting that miR-15b may refer to milk production and lipid synthesis [21]. The overexpression of miR-145 notably increases TAG accumulation, fat droplet formation, and proportions of unsaturated fatty acids. Oppositely, silencing miR-145 impairs fatty acid synthesis [22].

The insulin-like growth factor (IGF) system mainly includes IGF-I and IGF-II ligands, IGF-IR and IGF-IIR receptors and a family of IGF-binding proteins (IGFBPs) in vertebrates [23]. This system performs indispensable roles in adjusting growth, development and reproduction [24]. The functions of IGFs are regulated by their receptors and IGF-IGFBPs that serve as major regulators of IGF activity through acting as carriers [25]. These molecules also dominate the bioactivity and bioavailability of IGFs that interact with IGF receptors to modulate downstream signal transduction networks [26]. IGFBP5 is more conserved than other IGFBPs in mammals [27, 28]. A further study showed that IGFs act as important upstream activators of PI3K/AKT-mTOR signaling pathways to regulate milk protein synthesis [29–31]. The effect of milking frequency on milk yield is locally controlled within mammary glands and can be a function of changes in either the number or activity of secretory mammary epithelial cells. IGF-I signaling can mediate these effects because it can be locally controlled within mammary glands [32].

Our previous study examined the differential expression levels of chi-miR-3031 in caprine mammary gland tissues in colostrum (2 days postpartum) and common (90 days postpartum) milk stages [20]. Bioinformatics analysis suggested chi-miR-3031 and IGFBP5 are key signaling factors that regulate protein synthesis and cell proliferation in GMECs. IGFBP5 can regulate IGFs downstream of signal transduction networks by interacting with IGF receptors [26–29]. Until now, there have been no reports on the effects of chi-miR-3031 and IGFBP5 on milk protein synthesis. Therefore, in the present study, we used chi-miR-3031 mimics and mimic negative controls to transfect GMECs. We performed RNA-Seq to identify DEGs and the relative signal pathways affecting protein synthesis in GMECs. Among the identified DEGs, due to their involvement in lactation mechanisms, we selected IGFBP5 based on bioinformatics for further studies. This study aimed to determine the regulatory mechanisms of chi-miR-3031 on casein synthesis in GMECs.

Results

Summary of transcriptome sequencing

RNA-Seq was used to analyze the transcriptomic landscapes of GMECs transfected with MC and NC. Total RNAs from GMECs were extracted to construct RNA libraries for Illumina sequencing. Reads with adaptors and low-quality reads were removed prior to assembly. In total, we acquired 46,351,307 and 44,748,620 clean reads from each library (Table 1). In the MC and NC groups, uniquely mapped reads totaled 39,793,503 and 36,531,517, respectively, accounting for 85.85 and 81.66% of total reads. Only the uniquely mapped reads were considered in this analysis. Distribution of reads in different regions of the reference genome was evaluated using the following standard metrics: exon, intron and intergenic reads. Table 2 summarizes these results.

Identification of DEGs

Although a number of genes were differentially expressed between the two libraries, the present study focused on genes meeting the designated criteria of p values < 0.05 and $|\log_2 \text{FoldChange}| > 1$ (Additional file 1: Table S1). According to changes in relative gene abundance between the two libraries, 180 differentially expressed unigenes were down-regulated, and 157 unigenes were up-regulated in the MC group compared with the NC group of GMECs (Additional file 1: Table S1). Differential expression of *transglutaminase 3, thromboxane synthase, ATPase, H+ transporting, lysosomal accessory protein 1-like, fragile histidine triad* and seven other genes was observed at three more times in the MC and NC groups. Compared with the NC group, *IGFBP5, SRY-box containing gene 11, phosphatidylethanolamine-binding protein 4, keratin 79, parkin 2* and *fructosamine-3-kinase* genes were upregulated in the MC group. *FGF5, sprouty RTK signaling antagonist 4, mesenchymal stem cells, PTGS2, neuregulin 1* and *TGFα* genes were downregulated in the MC group compared with their expression in the NC group. Other differentially expressed genes between the two libraries included *POU domain, class 2, transcription factor 3; IGFBP3, TIMP3,*

Table 1 Summary of sequence read alignments to the reference genome

Category	Chi-miR-3031		Negative control	
	Reads number	Percentage	Reads number	Percentage
Total reads	46,351,307	100%	44,748,620	100%
Total mapped reads	43,734,071	94.35%	40,333,973	90.16%
Multiple mapped reads	3,940,568	8.50%	3,802,457	8.50%
Uniquely mapped reads	39,793,503	85.85%	36,531,517	81.66%
Reads mapped in proper pairs	37,712,533	81.36%	33,479,622	74.82%

Note: Total reads: total number of sequencing reads. Total mapped reads: the reads that aligned to reference sequence and its ratio. Multiple mapped reads: in total mapped reads, reads aligned to two or more places. Uniquely mapped reads: in total mapped reads, reads aligned to only one position. Reads mapped in proper pairs: paired-end sequencing sequence aligned to the chromosome reasonable direction and position

VEGFC, *G protein-coupled receptor 151* and *SOCS3* (Additional file 1: Table S1).

GO and KEGG pathway analysis of DEGs

To further investigate the biological function of DEGs, GO analysis was performed through running queries for each DEG against the GO database. Additional file 2: Figure S1 presents these GO functional annotations. Overall, GO terms were assigned to DEGs and classified into three independent ontology categories: (1) 334 terms in biological processes, (2) 209 terms in molecular functions and (3) 284 terms in cellular components. The most important enriched terms are bound up with biological regulation, single-organism processes and cellular processes. These terms were followed by developmental and metabolic processes. The *IGFBP5* and *IGF2* genes in the GO:0005576 term are known to affect mammary development and milk secretion [33, 34], which were downregulated in the MC group. In the cellular component, GO terms with high levels of significance comprised cell parts, organelles and membrane parts (Additional file 2: Figure S1). This organelle is linked with lipid secretor mechanisms of mammary epithelial cells [35]. Additionally, cell and membrane parts have effects on mammary gland involution [36].

Different genes typically cooperate with each other to exercise their biological functions. Pathway-based analyses provide additional insight into the biological functions of genes. Pathway enrichment analysis identifies significantly enriched metabolic pathways or signal transduction pathways related to DEGs. Overall, 22 pathways were significantly enriched (Q value < 0.05) for DEGs (Additional file 3: Figure S2, Additional file 4: Table S2). The prolactin signaling pathways included *FOS, ELF5, LHβ, SOCS3* and *PIK3R3* genes. *IGF2, FOS, TGFα, PIK3R3,* and *EREG* genes regulate mammary cell growth, apoptosis, proliferation and survival through the TNF and ErbB signaling pathways and proteoglycans [37–39]. KEGG pathway analyses and previous studies [40–44] showed that *TGFα, IGFBP5, PIK3R3, IGF2, ELF5* and *LHβ* play key roles in mammary development and milk secretion.

Table 2 The distribution of reads in the goat genome in different areas

Category	Chi-miR-3031	Negative control
Exon reads	85.53%	85.93%
Intron reads	7.36%	6.95%
Intergenic reads	7.11%	7.12%

Effect of chi-miR-3031 on IGFBP5 expression level

To examine whether chi-miR-3031 can reduce endogenous IGFBP5 expression, GMECs were transfected with MC, siRNA and NC. After 36 h of culture, total RNAs were extracted from the MC, siRNA and NC groups to determine the transfection efficiency of chi-miR-3031 and IGFBP5 mRNA expression. RT-qPCR results showed that the expression level of chi-miR-3031 in the MC group were 500 times higher than that of the NC group (Fig. 1; $P < 0.01$). Additionally, the expression level of *IGFBP5* mRNA in GMECs markedly decreased in the MC group (Fig. 2a; $P < 0.01$), confirming transcriptome sequencing results. Western blot results showed that the expression level of IGFBP5 protein was significantly reduced in the MC group when compared with the NC group (Fig. 3; $P <$ 0.05). SiRNA-IGFBP5 significantly reduced the expression levels of IGFBP5 mRNA and protein (Fig. 2b and Fig. 3; P < 0.01). These results suggest that chi-miR-3031 plays a negative regulatory role in IGFBP5 expression.

Chi-miR-3031 and IGFBP5 promote β-casein protein expression

To detect the concentrations of β-casein and κ-casein, GMECs were transfected with NC or MC by direct lysis after 36 h of culture. ELISA results showed that the expression levels of β-casein protein significantly increased in the MC and siRNA-IGFBP5 groups when compared with the NC group (Fig. 4b; $P < 0.01$ or $P < 0.05$). However, the expression levels of κ-casein protein showed no significant differences among the NC, MC and siRNA -IGFBP5 groups (Fig. 4a).

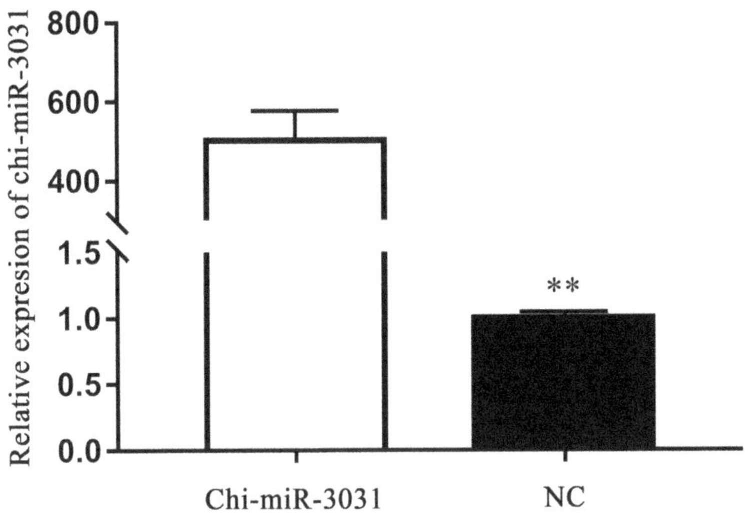

Fig. 1 Expression levels of chi-miR-3031 in GMECs transfected with MC and NC. $^{**}P < 0.01$. NC: Negative control

Chi-miR-3031 induces p-mTOR expression via downregulating IGFBP5

Previous studies have suggested that PI3K/AKT-mTOR signaling pathways regulate milk protein synthesis [45]. Thus, this study investigated the expression levels of p-mTOR in GMECs transfected with NC, siRNA or MC. Western blot results showed that chi-miR-3031 and siRNA-IGFBP5 induced the expression of p-mTOR (Fig. 5), suggesting that chi-miRNA-3031 can promote the expression of p-mTOR by downregulating IGFBP5.

Discussion

MiRNAs play key roles in the development of milk components [46, 47] and mammary glands [48]. However, only a limited number of genes are known to be directly or indirectly regulated by miRNAs in GMECs. Thus, the detection and identification of relevant genes are necessary to determine their function. Our previous studies investigated the differential expression of miRNAs in caprine mammary gland tissues during colostrum and common milk stages using high throughput sequencing and bioinformatics. Herein, we focused on chi-miR-3031 due to its prominent roles in influencing mammary gland development and lactation performance [20]. In MC and NC with transfected GMECs, we selected DEGs that were directly or indirectly regulated by chi-miR-3031. A total of 157 up-regulated and 187 downregulated genes were detected in the MC group compared with the NC group. Singh et al. (2016) showed that SOCS1 and SOCS3 mRNA levels increased by 72 h post-milking compared with levels at 6 h post-milking, additionally, IGF1 mRNA levels increased by 192 h post-milking, whereas IGF2 mRNA decreased by 18 h post-milking compared with 6 h post-milking [49]. In this study, *Fos* and *SOCS3* genes in the prolactin signaling pathway were up-regulated, whereas *ELF5*, *LHβ* and *PIK3R3* genes were down-regulated in the MC group. In cultured mammary gland adipocytes, Prokesch et al. (2014) showed that

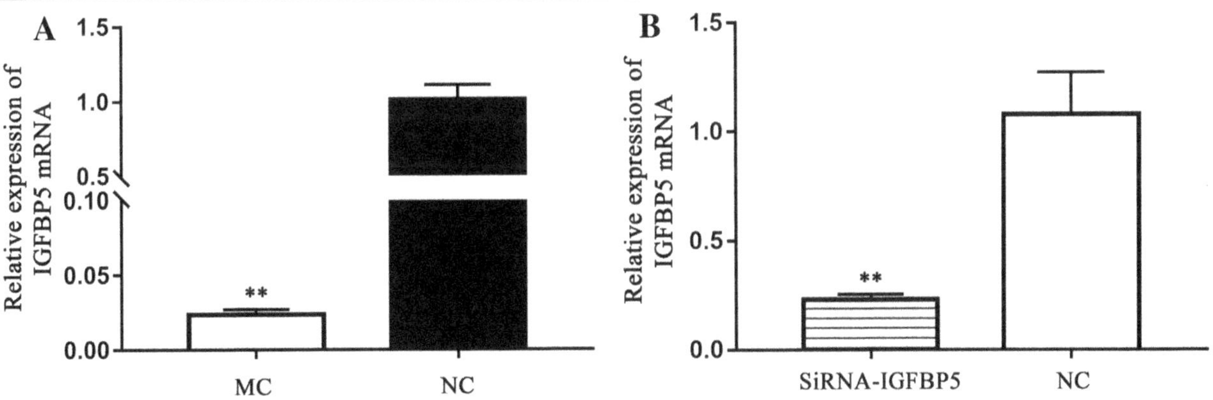

Fig. 2 Increasing chi-miR-3031 and siRNA-IGFBP5 levels notably decreased IGFBP5 mRNA expression. **a**: Effect of chi-miR-3031 on IGFBP5 mRNA. **b**: Effect of siRNA-IGFBP5 on IGFBP5 mRNA. $P < 0.01$ is shown as **. MC: Chi-miR-3031 mimic. NC: Negative control

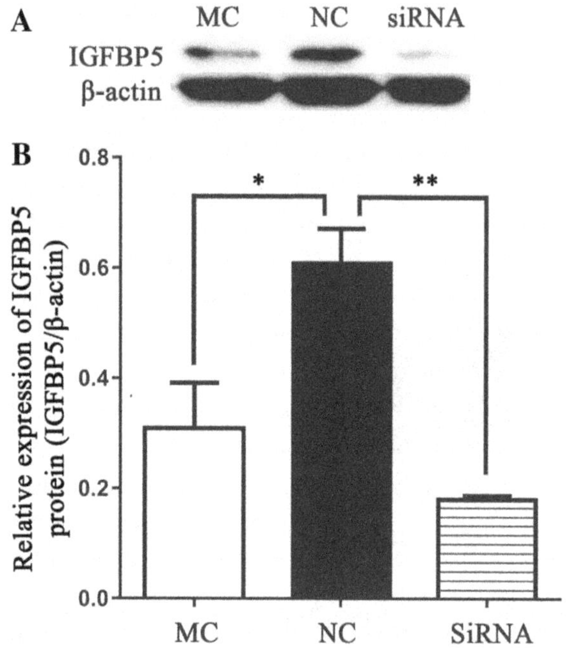

Fig. 3 Effects of chi-miR-3031 and siRNA-IGFBP5 on IGFBP5 protein levels. **a**: Western blot analysis results. **b**: Densitometric quantification of western blot results. Protein levels were normalized to β-actin. $P < 0.05$ is shown as *, and $P < 0.01$ is shown as **. MC: Chi-miR-3031 mimic. NC: Negative control

transient and stable ectopic expression of ELF5 induces the expression of whey acidic protein, a milk component; however, the general adipocyte phenotype remained unaffected, suggesting that additional pioneering factors are necessary [50]. Fifteen members of cancer pathways, including *FOS, TGFα, FZD8, MMP1, laminin subunit gamma 2, PTGS2, laminin subunit beta 3, ITGA2,* laminin subunit alpha 3, *VEGFC, WNT2B, heat shock protein 90 beta family member 1, WNT5B, FGF5* and *WNT3A,* were up-regulated in the MC group. *WNT10A, PIK3R3,* and *VEGFD* were down-regulated. *FZD8, plasminogen activator, urokinase receptor, ITGA2, WNT2B, TIMP3, WNT5B, FGF5,* heparanase, *HBEGF* and *WNT3A* were up-regulated in the proteoglycan pathways in cancer, whereas *WNT10A, PIK3R3, IGF2* were down-regulated. Nine genes, including *Fos, FGF5, ITGA2, FZD8, PIK3R3, WNT2B, WNT3A, WNT5B* and *WNT10A,* are involved in two or three pathways. Thus, the differential expression of these genes could play roles in regulating lactation or mammary gland development.

Several authors have showed the impact of miRNAs on mammary gland development or lactation [21, 46, 51]. Recently, Do et al. (2017) suggested that 19 miRNAs are differentially expressed across lactation in remarkable and time-dependent manners [52]. Additionally, miR-221/222 cluster, miR-18a and transcription factors homeobox A7, and NOTCH 3 and 4 have important roles in regulating lactose. MiR-186, miR-148a and miR-200a are remarkably associated with milk yield and are likely the most important miRNAs for component traits [53]. MiR-130a overexpression decreased *FABP3, PPARG, FATP1* and *PLIN2* mRNA expression levels, whereas miR-130a downregulation increased *PPARG, C/EBP beta, C/EBP alpha, PLIN2, FABP3* and *FATP1* mRNA expression levels [19]. In this study, results from transcriptome sequencing, real-time PCR and western blot showed that chi-miR-3031 can suppress IGFBP5 mRNA and protein expression levels. However, β-casein expression levels significantly increased in the MC and siRNA-IGFBP5 groups. IGFBP5 is an important member of the IGFBP family and has high affinity for IGF [54]. IGFs act as key upstream activators of PI3K/AKT-mTOR signaling pathways that regulate milk protein synthesis [29, 31]. We observed that the down-regulation of IGFBP5 can activate the phosphorylation of mTOR at the Ser2448 site in GMECs transfected with MC and siRNA-IGFBP5. Previous findings have suggested that

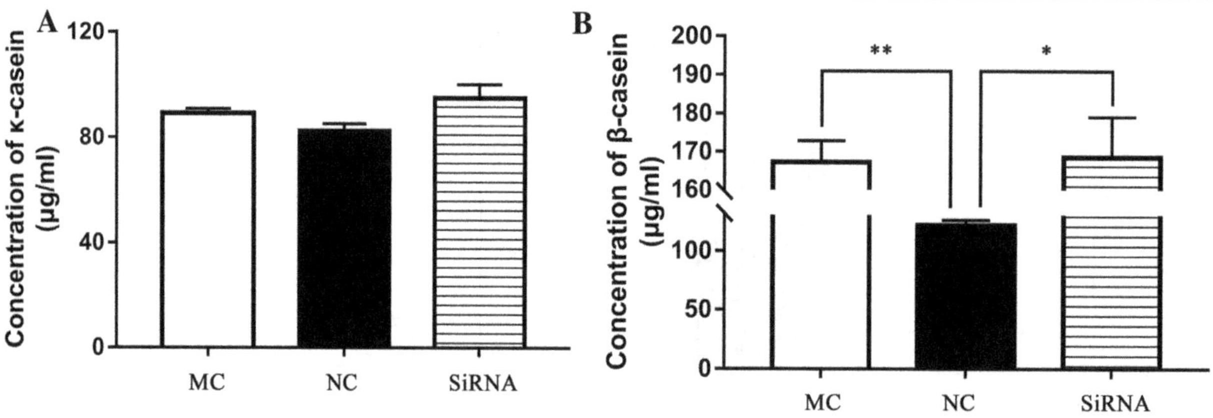

Fig. 4 Protein expression levels of κ-casein (**a**) and β-casein (**b**) in GMECs transfected with MC, NC and siRNA-IGFBP5. $P < 0.05$ is shown as *, and $P < 0.01$ is shown as **. MC: Chi-miR-3031 mimic. NC: Negative control

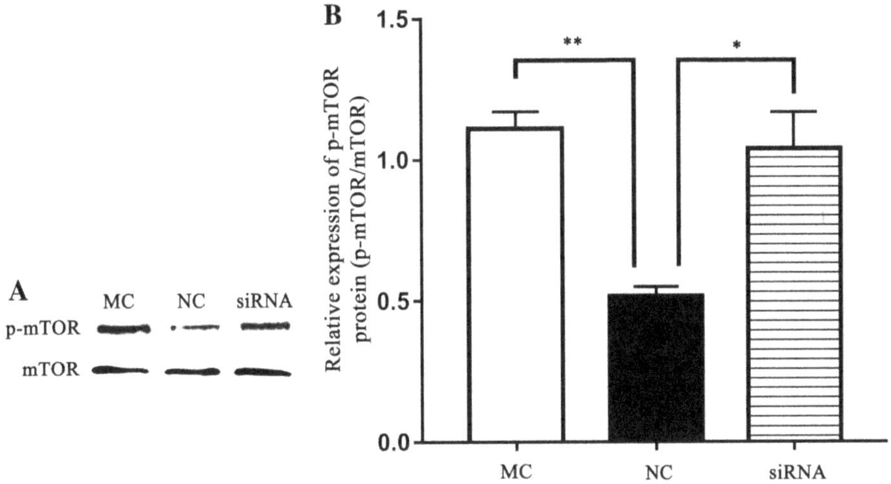

Fig. 5 Effects of chi-miR-3031 and siRNA-IGFBP5 on p-mTOR protein levels. **a**: Western blot analysis results. **b**: Densitometric quantification of western blot results. $P < 0.05$ is shown as *, and $P < 0.01$ is shown as **. MC: Chi-miR-3031 mimic. NC: Negative control

the activation of p-mTOR implies activation of the PI3K pathway [55]. We speculated that chi-miR-3031 can activate the PI3K/AKT-mTOR pathway and increase β-casein expression by down-regulating IGFBP5. However, κ-casein was not affected by chi-miR-3031.

Conclusion

In this study, 157 upregulated and 187 downregulated genes were detected by transcriptome sequencing in the MC group. KEGG pathway analyses and previous studies showed that *TGFα, PIK3R3, IGF2, IGFBP5, ELF5* and *LHβ* play key roles in mammary development and milk secretion. Thus, the differential expression of these genes can take part in the regulation of lactation or mammary gland development. Additionally, chi-miR -3031 can activate the PI3K/AKT-mTOR pathway and improve the expression of β-casein by down-regulating IGFBP5. Therefore, we conclude that chi-miR-3031 plays key roles in regulating β-casein synthesis in GMECs.

Methods

Animals and cell culture

Three Guanzhong dairy goats (3 years old, female) at peak lactation (90 days postpartum) periods were selected from a local farm in the Northwest A&F University of China. These Guanzhong dairy goats are research animals. Half-hour before the operation, these animals were anesthetized by intramuscular injection of 150 mg phenobarbital sodium. In the middle part of mammary gland, open the wound of 1 cm with a scalpel, remove about 1 cm^3 of mammary gland tissue, then sew and sterilize the wound. The special person kept the experimental animals, and after 1 week, the surgical line was removed, and all the animals recovered. Mammary gland tissues for culturing GMECs were stored in PBS containing streptomycin (100 mg/mL) and penicillin (100 U/mL). GMECs were cultured on the basis of a previous study [56].

Transfection and total RNA extraction

In accordance with the Lipofectamine RNAiMAX Reagent (Invitrogen, Carlsbad, CA, USA) protocol, GMECs were transfected with MC, NC and siRNA-IGFBP5 at final concentrations of 100 nmol/L. The MC, siRNA and NC were chemically synthesized and purified by Shanghai GenePharma (Shanghai, China). GMECs were cultured for another 36 h after transfection, and TRIzol reagent was used to extract total RNA. The quality of total RNA was detected using an Agilent 2100 Bioanalyzer (Agilent Technologies, Palo Alto, CA, USA). The RIN values of the MC group were 9.2, 9.0 and 8.8, and those of the NC group were 9.3, 9.1 and 8.9. Additional file 5: Table S3 provides the sequences of MC, NC and siRNA.

Transcriptome sequencing

Library construction and sequencing

For each independent experiment, three wells of a 6-well plate were mixed to extract the total RNAs to construct the MC group. The NC group was constructed in the same way. The total RNAs for each group were homogenized to maintain the same concentrations. Sequencing libraries were constructed with NEBNext Ultra RNA Library Prep Kit for Illumina (NEB, Beijing, China). Two MC groups were independently sequenced, respectively. The NC group was sequenced in the same way. Then, in accordance with instructions for use, libraries with different indices were loaded on an Illumina HiSeq instrument (Illumina, San Diego, CA, USA). Sequencing data is processed and analyzed by Personalbio (Shanghai, China).

Mapping reads on goat reference genome and analysis of gene expression

After removal of the 3´ adaptor sequence, low-quality and redundant reads were removed using Cutadapt (version 1.9.1) and clean data were aligned to caprine genome via Hisat2 software (v2.0.1). Caprine genome sequences and gene model annotation files were downloaded from the NCBI genome website (GCF_001704415.1_ASM170441v1).

The level of gene expression was measured by read density, and gene expression calculation was performed with Cuffdiff (v2.2.1), which calculates fragments per kilo bases per million (FPKM) reads as FPKM = total exon reads/mapped reads in millions × exon length in kb. The results from the cuffdiff analysis were further analyzed to identify genes with significant differential expression according to the criteria of fold-changes greater than 1 and p values less than 0.05.

Gene ontology (GO) and KEGG pathway analysis of DEGs

The DEGs were mapped to GO terms in the Gene Ontology database (http://www.geneontology.org/). Remarkably enriched GO terms in differentially methylated genes were defined by hypergeometric test [57–59]. Pathway-based analyses aid in further understanding gene function [60]. The calculation formula of KEGG (http://en.wikipedia.org/wiki/KEGG) [61] was the same with that of GO analysis. Pathway analysis identified remarkably enriched signal transduction pathways for DEGs [62].

Reverse transcription and real-time PCR

Table 3 shows the names of miRNA and genes for reverse transcription and real-time PCR. The reverse transcription system contained the following: (1) 800 ng of total RNA, 1 μL of gDNA Eraser, 2 μL of 5 × gDNA Eraser Buffer and RNase-Free dH_2O mixed to a final volume of 10 μL and incubated at 42 °C for 2 min. (2) 4 μL of 5 × Prime Script Buffer 2, 1 μL of RT Primer Mix, 1 μL of PrimeScript RT Enzyme Mix 1 and RNase-Free dH_2O mixed to a final volume of 20 μL and incubated at 42 °C for 15 min followed by 85 °C for 5 s using a PrimeScript RT reagent Kit (TaKaRa, Dalian, China). cDNA products were stored at – 20 °C.

Real-time PCR was performed with a 25 μL volume containing 2 μL of template cDNA, 12.5 μL of SYBR Premix Ex Taq II (TaKaRa, Dalian, China) and 1 μM of primers using the CFX Connect Real-Time PCR Detection System (Bio-Rad, CA, USA). Thermal cycling conditions were 95 °C for 10 min, followed by 40 cycles at 94 °C for 15 s, 60 °C for 30 s and 72 °C for 30 s. Primers are shown in Table 3. The *β-actin* or U_6 mRNA levels were used for normalization. Each experiment was independently repeated at least three times, and fold changes were analyzed via the $2^{-\Delta\Delta Ct}$ method.

Enzyme-linked immunosorbent assay

ELISA was performed to detect the concentration of caseins in cell lysates using goat β-casein and κ-casein kit (Tongwei, Shanghai, China). Briefly, GMECs were transfected with NC or MC after 36 h of culture in six-well plates and then lysed with 200 μL RIPA Lysis Buffer (Bioteke, Beijing, China). According to manufacturer instructions, 50 μL cell lysate or standards were used, and absorbance at 450 nm was measured through an Epoch microplate reader (Biotek, Winooski, USA). Concentrations were calculated after subtracting the standard curve. Kit sensitivity was 1.0 μg/mL, and mean intra- and inter-assay variable coefficient values were less than 15 and 10%, respectively.

Western blotting

GMECs were lysed in denaturing lysis buffer containing protease inhibitors (RIPA, Bioteke, Beijing, China) for 30 min on ice and centrifuged (12,000×g) for 15 min at 4 °C. Protein concentrations in the lysates were determined by

Table 3 Primer information for RT-qPCR

Name	Primer	Primer sequence (5′→3′)	Size (bp)	Tm (°C)
Chi-miR-3031	RT-Primer	GTCGTATCCAGTGCAGGGTCCGAGG TATTCGCACTGGATACGACGGGGCCGA		
	Forward primer	TTATGGGCTGGCTCCCTC	42	60
	Reverse primer	CAGTGCAGGGTCCGAGGT		
U6	Forward primer	CTCGCTTCGGCAGCACA	89	60
	Reverse primer	AACGCTTCACGAATTTGCGT		
IGFBP5	Forward primer	GTACCTGCCCAACTGTGACC	116	60
	Reverse primer	CTGGCAGCTTCATCCCATAC		
β-casein	Forward primer	ACAGCCTCCCACAAAACATCC	297	60
	Reverse primer	TGAGAAAGGGACAGCACGGA		
β-actin	Forward primer	TGACCCAGATCATGTTTGAGA	186	60
	Reverse primer	CAAGGTCCAGACGCAGGAT		

a BCA protein assay kit (Vazyme Biotech, Nanjing, China). Exactly 30 μg of protein were separated on a 12% SDS-PAGE and transferred to a polyvinylidene difluoride membrane that was blocked with 5% nonfat dried milk in Tris-buffered saline containing 0.1% Tween 20 (pH 7.6) for 1 h at 25 °C. Subsequently, the membranes were incubated overnight at 4 °C with a rabbit anti-IGFBP5 polyclonal antibody (1:300, Boster, USA), a rabbit anti-mTOR polyclonal antibody (1:100, Boster, USA), a rabbit anti -p-mTOR polyclonal antibody (1:100, Boster, USA) or a mouse β-actin monoclonal primary antibody (1:1000, Beyotime, Jiangsu, China). Then, the blot was incubated with a horseradish peroxidase-conjugated secondary antibody for 1 h at 25 °C. The quantification of protein was detected through the Quantity One program (Bio-Rad, California, USA).

Statistical analysis

The one-way ANOVA of SPSS 16.0 software package was used to analyze data. Means of multiple comparisons was analyzed by least significant difference (LSD). Data were presented as means ± standard deviation of three independent experiments. $P < 0.05$ or $P < 0.01$ was considered significant.

Additional files

Additional file 1: Table S1. The up and down-regulated genes in the chi-miR-3031 mimics VS. negative control groups.

Additional file 2: Figure S1. GO analysis results based on biological process, cellular component and molecular function.

Additional file 3: Figure S2. KEGG pathway analysis for DEGs.

Additional file 4: Table S2. Pathway annotations of differentially expressed genes.

Additional file 5: Table S3. Sequence information of chi-miR-3031 mimics, mimic negative control and siRNA-IGFBP5.

Abbreviations

AKT: Protein kinase B; C/EBP: CCAAT/enhancer binding protein; CAM: Cell adhesion molecule; DEGs: Differentially expressed genes; ELF5: E74-like factor 5; ELISA: Enzyme linked immunosorbent assay; EREG: Epiregulin; FABP3: Fatty acid binding protein 3; FATP1: Fatty acid transport protein; FGF5: Fibroblast growth factor 5; FOS: FBJ osteosarcoma oncogene; FPKM: Fragments Per Kilo base per Million reads; FZD8: Frizzled class receptor 8; GMECs: Goat mammary epithelial cells; GO: Gene ontology; HB -EGF: Heparin-binding EGF-like growth factor; IGF: Insulin-like growth factor; IGFBPs: IGF-binding proteins; ITGA2: Integrin alpha 2; KEGG: Kyoto encyclopedia of genes and genomes; LHβ: Luteinizing hormone beta subunit; miRNAs: MicroRNAs; MMP1: Metalloproteinase-1; mTOR: Phosphorylation of mechanistic target of rapamycin; PBS: Phosphate-buffered saline; PI3K: Phosphoinositide 3-kinase; PIK3R3: Phosphatidylinositol 3-kinase regulatory subunit gamma; PLIN2: Perilipin 2; p-mTOR: Phosphorylated mTOR; PPARG: Peroxisome proliferator activated receptor gamma; PTGS2: Prostaglandin-endoperoxide-synthase-1 gene; RNA-Seq: High-throughput sequencing of RNA; SOCS3: Suppressor of cytokine signaling 3; STAT: Signal transducer and activator of transcription; TAG: Triacylglycerol; TGFα: Transforming growth factor alpha; TIMP3: Tissue inhibitor of metalloproteinases-3; VEGFC: Vascular endothelial growth factor C; WNT2B: Wnt family member 2B; β-casein: Beta-casein

Acknowledgments

Not applicable.

Funding

This study was supported by the National Natural Science Foundation of China (31601925), China Postdoctoral Science Foundation (2016 T90954 and 2014 M552498), Shaanxi Science and Technology Innovation Project Plan (2017ZDXM-NY-081 and 2016KTZDNY02-04).

Authors' contributions

BYC and XPA designed the experiments and wrote the manuscript. KWC, JXH and YXS performed most of the experiments and analyzed the results. XCZ, YHL and GHZ conducted cell culture. KW and HDM prepared the Figs. GL revised the manuscript. All authors read and approved the final manuscript.

Consent for publication

Not applicable.

Competing interests

The authors declare that they have no competing interests.

Author details

[1]College of Animal Science and Technology, Northwest A&F University, No. 22 Xinong Road, Yangling, Shaanxi 712100, People's Republic of China. [2]Animal Engineering Branch, Yangling Vocational & Technical College, No. 10 Xinong Road, Yangling, Shaanxi 712100, People's Republic of China.

References

1. Zhang R, Ma HM, Gao Y, Wu YJ, Qiao YM, Geng A, Cai CG, Han YY, Zeng YA, Liu XL, Ge GX. Th-POK regulates mammary gland lactation through mTOR-SREBP pathway. PLoS Genet. 2018;14:e1007211.
2. Saipin N, Noophun J, Chumyim P, Rungsiwiwut R. Goat milk: non-invasive source for mammary epithelial cell isolation and in vitro culture. Anatomia Histologia Embryologia. 2018;47:187–94.
3. Yang Y, Fang X, Yang R, Yu H, Jiang P, Sun B, Zhao Z. MiR-152 regulates apoptosis and triglyceride production in MECs via targeting ACAA2 and HSD17B12 genes. Sci Rep. 2018;8:417.
4. Huppertz, T. 2013. Chemistry of the caseins. Pages 135–160 in Advanced dairy chemistry. Vol. 1A: Proteins: Basic Aspects. 4th ed. P. L. H. McSweeney and P. F. Fox, ed. Springer Science+Business Media, New York, NY.
5. Li H, Ruan J, Durbin R. Mapping short DNA sequencing reads and calling variants using mapping quality scores. Genome Res. 2008;18(11):1851.
6. Cloonan N, Grimmond SM. Transcriptome content and dynamics at single-nucleotide resolution. Genome Biol. 2008;9(9):1–4.
7. Lister R, O'Malley RC, Tonti-Filippini J, Gregory BD, Berry CC, Millar AH, Ecker JR. Highly integrated single-base resolution maps of the epigenome in Arabidopsis. Cell. 2008;133(3):523.
8. Marioni JC, Mason CE, Mane SM, Stephens M, Gilad Y. RNA-seq: an assessment of technical reproducibility and comparison with gene expression arrays. Genome Res. 2008;18(9):1509.
9. Grabherr MG, Haas BJ, Yassour M, Levin JZ, Thompson DA, Amit I, Adiconis X, Fan L, Raychowdhury R, Zeng Q. Full-length transcriptome assembly from RNA-Seq data without a reference genome. Nat Biotechnol. 2011;29(7):644.
10. Suarez-Vega A, Gutierrez-Gil B, Klopp C, Tosser-Klopp G, Arranz JJ. Variant discovery in the sheep milk transcriptome using RNA sequencing. BMC Genomics. 2017;18:170.

11. Wanyonyi SS, Kumar A, Du Preez R, Lefevre C, Nicholas KR. Transcriptome analysis of mammary epithelial cell gene expression reveals novel roles of the extracellular matrix. Biochem Biophys Rep. 2017;12:120.
12. Croce CM, Calin GA. miRNAs, cancer, and stem cell division. Cell. 2005;122(1):6–7.
13. Bartel DP. MicroRNA target recognition and regulatory functions. Cell. 2009; 136(2):215–33.
14. Shiah SG, Hsiao JR, Chang WM, Chen YW, Jin YT, Wong TY, Huang JS, Tsai ST, Hsu YM, Chou ST. Downregulated miR-329 and miR-410 promote the proliferation and invasion of oral squamous cell carcinoma by targeting Wnt-7b. Cancer Res. 2014;74(24):7560–72.
15. Ambros V. MicroRNA pathways in flies and worms: growth, death, fat, stress, and timing. Cell. 2003;113(6):673–6.
16. An X, Song Y, Hou J, Li G, Zhao H, Wang J, Cao B. Identification and profiling of microRNAs in the ovaries of polytocous and monotocous goats during estrus. Theriogenology. 2016;85(4):769–80.
17. Li HM, Wang CM, Li QZ, Gao XJ. MiR-15a decreases bovine mammary epithelial cell viability and lactation and regulates growth hormone receptor expression. Molecules. 2012;17(10):12037–48.
18. Cui YJ, Sun X, Jin LF, Yu GP, Li QZ, Gao XJ, Ao JX, Wang CM. MiR-139 suppresses beta-casein synthesis and proliferation in bovine mammary epithelial cells by targeting the GHR and IGF1R signaling pathways. BMC Vet Res. 2017;13:350.
19. Yang WC, Guo WL, Zan LS, Wang YN, Tang KQ. Bta-miR-130a regulates the biosynthesis of bovine milk fat by targeting peroxisome proliferator-activated receptor gamma. J Anim Sci. 2017;95(7):2898–906.
20. Hou JX, An XP, Song YX, Cao BY, Yang HP, Zhang Z, Shen WZ, Li YP. Detection and comparison of microRNAs in the caprine mammary gland tissues of colostrum and common milk stages. BMC Genet. 2017;18:38.
21. Chu M, Zhao Y, Yu S, Hao Y, Zhang P, Feng Y, Zhang H, Ma D, Liu J, Cheng M, et al. MiR-15b negatively correlates with lipid metabolism in mammary epithelial cells. Am J Physiol Cell Physiol. 2017;314:C43–52.
22. Wang H, Shi H, Luo J, Yi Y, Yao D, Zhang X, Ma G, Loor JJ. MiR-145 regulates lipogenesis in goat mammary cells via targeting INSIG1 and epigenetic regulation of lipid-related genes. J Cell Physiol. 2017;232(5):1030.
23. Hwa V, Oh Y, Rosenfeld RG. Insulin-like growth factor binding proteins: a proposed superfamily. Acta Paediatr. 1999;88:37.
24. Zhang H, Shi Y, He M. Molecular identification of an insulin growth factor binding protein (IGFBP) and its potential role in an insulin-like peptide system of the pearl oyster, Pinctada fucata. Comp Biochem Physiol B Biochem Mol Biol. 2017;214:27–35.
25. Murphy M, Pykett MJ, Harnish P, Zang KD, George DL. Identification and characterization of genes differentially expressed in meningiomas. Cell Growth Differ. 1993;4:715.
26. Denley A, Cosgrove LJ, Booker GW, Wallace JC, Forbes BE. Molecular interactions of the IGF system. Cytokine Growth Factor Rev. 2005;16:421.
27. Pera EM, Wessely O, Li SY, De Robertis EM. Neural and head induction by insulin-like growth factor signals. Dev Cell. 2001;1:655.
28. Salih DAM, Tripathi G, Holding C, Szestak TAM, Gonzalez MI, Carter EJ, Cobb LJ, Eisemann JE, Pell JM. Insulin-like growth factor-binding protein 5 (Igfbp5) compromises survival, growth, muscle development, and fertility in mice. Proc Natl Acad Sci U S A. 2004;101:4314.
29. Ma JC, Sawai H, Matsuo Y, Ochi N, Yasuda A, Takahashi H, Wakasugi T, Funahashi H, Sato M, Takeyama H. IGF-1 mediates PTEN suppression and enhances cell invasion and proliferation via activation of the IGF-1/PI3K/Akt signaling pathway in pancreatic Cancer cells. J Surg Res. 2010;160(1):90–101.
30. Bionaz M, Loor JJ. Gene networks driving bovine milk fat synthesis during the lactation cycle. BMC Genomics. 2008;9:21.
31. Sobolewska A, Gajewska M, Zarzyńska J, Gajkowska B, Motyl T. IGF-I, EGF, and sex steroids regulate autophagy in bovine mammary epithelial cells via the mTOR pathway. Eur J Cell Biol. 2009;88:117–30.
32. Murney R, Stelwagen K, Wheeler TT, Margerison JK, Singh K. The effects of milking frequency on insulin-like growth factor I signaling within the mammary gland of dairy cows. J Dairy Sci. 2015;98(8):5422–8.
33. Bomfim GF, Merighe GKF, de Oliveira SA, Negrao JA. Effect of acute stressors, adrenocorticotropic hormone administration, and cortisol release on milk yield, the expression of key genes, proliferation, and apoptosis in goat mammary epithelial cells. J Dairy Sci. 2018;101:6486–96.
34. De Silva D, Kunasegaran K, Ghosh S, Pietersen AM. Transcriptome analysis of the hormone-sensing cells in mammary epithelial reveals dynamic changes in early pregnancy. BMC Dev Biol. 2015;15:7.
35. Ghosal D, Shappell NW, Keenan TW. Endoplasmic reticulum lumenal proteins of rat mammary gland-potential involvement in lipid droplet assembly during lactation. Biochim Biophys Acta. 1994;1200:175–81.
36. Suárez-Vega A, Gutiérrez-Gil B, Klopp C, Robert-Granie C, Tosser-Klopp G, Arranz JJ. Characterization and comparative analysis of the milk transcriptome in two dairy sheep breeds using RNA sequencing. Sci Rep. 2015;5:18399.
37. Kosciuczuk EM, Lisowski P, Jarczak J, Majewska A, Rzewuska M, Zwierzchowski L, Bagnicka E. Transcriptome profiling of staphylococci -infected cow mammary gland parenchyma. BMC Vet Res. 2017;13:161.
38. Bade LK, Goldberg JE, Dehut HA, Hall MK, Schwertfeger KL. Mammary tumorigenesis induced by fibroblast growth factor receptor 1 requires activation of the epidermal growth factor receptor. Journal of Cell Sci. 2011;124:3106–17.
39. Williams MM, Vaught DB, Morrison-Joly M, Hicks DJ, Sanchez V, Owens P, Rahman B, Elion DL, Balko JM, Cook RS. ErbB3 drives mammary epithelial survival and differentiation during pregnancy and lactation. Breast Cancer Res. 2017;19:105.
40. O'Leary KA, Jallow F, Rugowski DE, Sullivan R, Sinkevicius KW, Greene GL, Schuler LA. Prolactin activates ER alpha in the absence of ligand in female mammary development and carcinogenesis in vivo. Endocrinology. 2013;154:4483–92.
41. Stahel P, Kim JJ, Cieslar SRL, Warrington JM, Xiao CT, Cant JP. Supranutritional selenium intake from enriched milk casein impairs hepatic insulin sensitivity via attenuated IRS/PI3K/AKT signaling and decreased PGC-1 alpha expression in male Sprague-Dawley rats. J Nutr Biochem. 2017;41:142–50.
42. Chean J, Chen CJ, Shively JE. ETS transcription factor ELF5 induces lumen formation in a 3D model of mammary morphogenesis and its expression is inhibited by Jak2 inhibitor TG101348. Exp Cell Res. 2017;359:62–75.
43. Gao YH, Jiang JP, Yang SH, Hou YL, Liu GE, Zhang SG, Zhang Q, Sun DX. CNV discovery for milk composition traits in dairy cattle using whole genome resequencing. BMC Genomics. 2017;18:265.
44. Brozos CN, Kiossis E, Fthenakis GC, Tsousis G, Boscos C. Supplementation of lactating ewes with a glucogenic preparation or beta-carotene in mid- to late lactation on subsequent milk yield and luteinizing hormone secretion. Can J Anim Sci. 2007;87:185–90.
45. Burgos SA, Cant JP. IGF-1 stimulates protein synthesis by enhanced signaling through mTORC1 in bovine mammary epithelial cells. Domest Anim Endoc. 2010;38:211–21.
46. Wang H, Luo J, He QY, Yao DW, Wu J, Loor JJ. miR-26b promoter analysis reveals regulatory mechanisms by lipid-related transcription factors in goat mammary epithelial cells. J Dairy Sci. 2017;100(7):5837–49.
47. Tang KQ, Wang YN, Zan LS, Yang WC. miR-27a controls triacylglycerol synthesis in bovine mammary epithelial cells by targeting peroxisome proliferator-activated receptor gamma. J Dairy Sci. 2017;100(5):4102–12.
48. Sandhu GK, Milevskiy MJG, Wesley W, Shewan AM, Brown MA. Non-coding RNAs in mammary gland development and disease. Adv Exp Med Biol. 2016;886:121.
49. Singh K, Vetharaniam I, Dobson JM, Prewitz M, Oden K, Murney R, Swanson KM, Mcdonald R, Henderson HV, Stelwagen K. Cell survival signaling in the bovine mammary gland during the transition from lactation to involution. J Dairy Sci. 2016;99(9):7523–43.
50. Prokesch A, Smorlesi A, Perugini J, Manieri M, Ciarmela P, Mondini E, Trajanoski Z, Kristiansen K, Giordano A, Bogner-Strauss JG, et al. Molecular aspects of adipoepithelial transdifferentiation in mouse mammary gland. Stem Cells. 2014;32(10):2756–66.
51. Zhen L, Liu H, Jin X, Lijan L, Liu J. Expression profiles of microRNAs from lactating and non-lactating bovine mammary glands and identification of miRNA related to lactation. BMC Genomics. 2012;13(1):731.
52. Do DN, Li R, Dudemaine PL, Ibeaghaawemu EM. MicroRNA roles in signalling during lactation: an insight from differential expression, time course and pathway analyses of deep sequence data. Sci Rep. 2017;7:44605.
53. Do DN, Dudemaine PL, Li R, Ibeagha-Awemu EM. Co-expression network and pathway analyses reveal important modules of miRNAs regulating milk yield and component traits. Int J Mol Sci. 2017;18:E1560.
54. Mcqueeney K, Dealy CN. Roles of insulin-like growth factor-I (IGF-I) and IGF-I binding protein-2 (IGFBP2) and −5 (IGFBP5) in developing chick limbs. Growth Hormon IGF Res. 2001;11:346.
55. Altomare DA, Wang HQ, Skele KL, De Rienzo A, Klein-Szanto AJ, Godwin AK, Testa JR. AKT and mTOR phosphorylation is frequently detected in ovarian cancer and can be targeted to disrupt ovarian tumor cell growth. Oncogene. 2004;23(34):5853–7.
56. Chen Z, Luo J, Sun S, Cao D, Shi H, Loor JJ. miR-148a and miR-17-5p synergistically regulate milk TAG synthesis via PPARGC1A and PPARA in goat mammary epithelial cells. RNA Biol. 2017;14(3):326.
57. Harris MA. The gene ontology (GO) database and informatics resource: WCB/McGraw-Hill; 2004.

58. Drabkin HJ, Blake JA, Database MGI. Manual gene ontology annotation workflow at the mouse genome informatics database. Database (Oxford). 2012;29:45.
59. Benjamini Y, Yekutieli D. The control of the false discovery rate in multiple testing under dependency. Ann Stat. 2001;29:1165–88.
60. Kanehisa M, Araki M, Goto S, Hattori M, Hirakawa M, Itoh M, Katayama T, Kawashima S, Okuda S. Tokimatsu T. KEGG for linking genomes to life and the environment. Nucleic Acids Res. 2008;36:D480–4.
61. Ogata H, Goto S, Sato K, Fujibuchi W, Bono H, Kanehisa MKEGG. Kyoto encyclopedia of genes and genomes. Nucleic Acids Res. 2000;27:29–34.
62. Kanehisa M, Goto S, Sato Y, Furumichi M, Mao T. KEGG for integration and interpretation of large-scale molecular data sets. Nucleic Acids Res. 2012;40:D109–14.

22

Sodium butyrate improves antioxidant stability in sub-acute ruminal acidosis in dairy goats

Nana Ma[†], Juma Ahamed Abaker[†], Muhammad Shahid Bilal, Hongyu Dai and Xiangzhen Shen[*]

Abstract

Background: Currently, little is known about the effect of sodium butyrate (NaB) on oxidative stress following grain-induced sub-acute ruminal acidosis in dairy goats. In the present study, 18 lactating dairy goats implanted with a ruminal cannula and permanent indwelling catheters in the portal and hepatic veins were randomly allocated into 3 treatment groups over 20 weeks: low grain (**LG**, 40% grain; $n = 6$), high grain (**HG**, 60% grain; $n = 6$) and high grain with sodium butyrate (**HG + NaB**, 60% grain + NaB; $n = 6$).

Results: When added to the HG diet, NaB increased the mean ruminal pH and reduced the levels of ruminal, portal and hepatic LPS; Additionally, we observed an increase in SOD1, SOD2, SOD3, GPX1 and CAT mRNA expression, increased levels of TSOD and CAT enzyme activity as well as increased total antioxidant capacity (T-AOC) and decreased malondialdehyde (MDA) in both the liver and plasma, while GPx activity increased in the liver of goats fed the HG + NaB diet. The mRNA expression of UGT1A1, NQO1, MGST3, and Nrf2, as well as total Nrf2 protein levels were increased in goats fed the HG + NaB diet.

Conclusions: Our study indicates that sodium butyrate could improve the oxidative status in sub-acute ruminal acidosis through the partial activation of Nrf2-dependent genes.

Keywords: Oxidative stress, High-grain diet, Sodium butyrate, Dairy goats

Background

Oxidative stress is an active research area in veterinary medicine, and it is implicated in the pathogenesis of many pathological conditions, including sepsis, mastitis, enteritis, pneumonia, respiratory and joint diseases [1] and hepatic insufficiency [2]. Moreover, oxidative stress is believed to play a significant role in regulating the metabolic activity of some organs and in productivity in farm animals [3]. In response to oxidative stress, organisms can redirect their metabolic flux from glycolysis to pentose phosphate pathway which provides the reducing power for the major cellular redox systems [4]. Pregnancy and lactation are physiological stages considered to induce some metabolic stress [5]. Recent studies have focused on metabolic diseases that affect dairy ruminants during the peripartum period [6]. Subacute ruminal acidosis (SARA) is a common metabolic disorder in high-producing dairy farms that leads to great economic losses [7]. In ruminants, SARA is a metabolic disease induced by feeding excessive amounts of highly fermentable carbohydrate and insufficient dietary coarse fibre [8]. Lactating goats were reported to have experienced a certain stress during subacute ruminal acidosis [9]. As a response to the demand for increased milk production and food scarcity, grain is commonly used to combat negative energy balance [10]. However, increasing grain-to-forage ratios leads to depression of ruminal pH [8, 11]. The accumulation of short chain fatty acids (SCFA) leads to SARA and affects the integrity of the rumen epithelium [11]. When SARA compromises epithelial integrity, it results in health problems such as liver abscesses and even laminitis [12].

Lipopolysaccharide (LPS) is a bacterial product reported to be translocated to the blood circulation when the gastrointestinal wall and epithelial tight junctions are perturbed [13]. Translocation of LPS into blood circulation in SARA was found to trigger an inflammatory response and leads to oxidative stress [13–15], due to an increase of reactive

* Correspondence: xzshen@njau.edu.cn
[†]Nana Ma and Juma Ahamed Abaker contributed equally to this work.
College of Veterinary Medicine, Nanjing Agricultural University, Nanjing 210095, People's Republic of China

oxygen species (ROS), associated with unbalanced dietary anti-oxidants [16]. The alteration of oxidative balance could induce oxidative stress and, consequently, cellular damage if not balanced with antioxidants [17].

Butyrate, a short-chain fatty acid, is a major microbial fermentation product of dietary fibre in the forestomach (rumen) of ruminants that contributes to approximately 70% of the daily metabolizable energy requirement of ruminants [18, 19]. Butyrate is known to maintain a barrier function that regulates several cellular processes and to modulate gene expression through histone deacetylase complex (HDAC) inhibition [20]. Butyrate shows a protective effect against oxidative DNA damage induced by H2O2 in isolated human colonocytes and in HT29 tumour cells [21]. Previous studies in ruminants showed that ingested butyrate could down-regulate the expression of virulence genes, increase ration digestibility, increase growth rate, and decrease oxidative stress and cytokine synthesis, even at very low doses [22, 23], resulting in control of the pathogen. Feeding sodium butyrate to lipopolysaccharide-challenged chickens increases serum superoxide dismutase (SOD) activity and decreases malondialdehyde (MDA) content in raised chickens [24]. LPS is believed to be a key point in the pathogenesis of SARA [25], whose diagnosis is laborious, but proper management and feeding can prevent SARA [25]. However, no information is currently available about the effects of sodium butyrate on stress parameters of ruminants fed a high-grain diet (HG). Therefore, the objective of this study is to examine the effect of adding sodium butyrate to a high-grain diet formula (HG + NaB) on antioxidant function and to determine whether NaB could prevent translocation of LPS into the blood and alleviate oxidative stress during SARA. We hypothesized that compared with the low-grain diet (LG), the high-grain diet (HG) would increase endotoxin concentrations and induce oxidative stress responses in lactating dairy goats as a result of decreased ruminal pH and that sodium butyrate (NaB) could stimulate fibre-digesting bacteria and consequently stabilize rumen pH and rumen LPS and prevent oxidative stress.

Methods

Goats and diets

Eighteen lactating Saanen dairy goats (average bodyweight, 38.9 ± 2.06 kg) in parity 1 or 2 (milk yield, 1.13 ± 0.0200 kg/day) were selected based on lactation time and were randomly assigned to three groups. All goats were ruminally cannulated, hepatic catheters were fitted surgically in the hepatic vein and portal vein, then exteriorized and fixed on the para-lumbar area beneath the transverse process. The first group (*n.* = 6; LG) was given a low-grain diet with a 4:6 grain-to-forage ratio, the second group (*n.* = 6; HG) received a high-grain diet with a 6:4 grain-to-forage ratio, and the third group (*n.* = 6; HG+ NaB) was given the high-grain diet mixed with sodium butyrate. Goats were housed in separate boxes and given food twice/day, at 8:00 am and at 18:00 pm. Dry matter intake was 1.9 ± 0.15 kg/d/head, with free access to water throughout the experiment. The goats were observed daily regarding feed intake, rectal temperature and respiratory rate. The ingredients and nutritional composition of the diets are presented in Table 1.

Blood samples and rumen fluid were collected from the three groups on the last three consecutive days of the 20th week. Blood was sampled from the jugular vein in 5 mL vacuum tubes containing sodium heparin as an anticoagulant. Plasma was separated by centrifugation at 3000×g at 4 °C for 15 min, and aliquots were frozen at – 20 °C for the analysis of LPS, thiobarbituric acid reactive substances (MDA), total antioxidant capacity (T-AOC) and antioxidant enzyme activity.

Rumen fluid was collected through a rumen fistula every 2 h for 7 h after feeding. The samples were centrifuged at 10,000×g for 45 min, and the supernatant was aspirated and filtered through a disposable 0.22-μm filter. The filtrate was collected in a sterile depyrogenated glass tube (previously heated at 250 °C for 2 h) and heated at 100 °C for 30 min. Samples were cooled at room temperature for 15 min and stored at – 20 °C for LPS measurements. The pH of the ruminal fluid was immediately detected using a pH meter (Sartorius, Basic pH Meter PB-10, PB-21, Goettingen, Germany). The goats were slaughtered after the 20th week after fasting overnight and liver samples were collected, immediately frozen in liquid nitrogen and stored at – 70 °C.

Volatile fatty acids analysis

Chromatography with FFAP 123–3233 30 m × 0.32 mm × 0.5 μm capillary column (Agilent Technologies, Stevens Creek Blvd, Santa Clara, CA, United States) in an Agilent 7890A system (Agilent Technologies) was used to assay the VFA concentration as previously described by Wood et al., (2010) [26].

Milk analysis

Goats were milked twice a day at 8:00 am and at 18:00 pm and milk yield were recorded daily. The milk sample was taken to assay the milk fat and milk protein concentrations using MilkoScan (FT1, Foss, Hillerod, Denmark).

Measurement of LPS

The concentration of LPS in the rumen fluid, portal vein, hepatic vein plasma was determined using a chromogenic endpoint limulus amoebocyte lysate assay kit (CE64406 & CE80545, Chinese Horseshoe Crab Reagent Manufactory Co., Ltd., Xiamen, China) as described by Dong et al. [27] to detect LPS concentrations in the ruminal fluid and plasma with a minimum detection limit of 0.01 EU/mL (EU, Endotoxin Unit).

Table 1 The Ingredients of the diets and nutritional composition

Ingredients, % of DM	High grain (HG)[a]	High grain with NaB (HG + NaB) [a]	Low grain (LG) [a]
Corn (g/100 g)	23.07	23.07	29.2
Bran (g/100 g)	28.03	28.03	0
Soybean meal (g/100 g)	2	2	8.43
Rape seed meal (g/100 g)	3.7	3.7	0
Stone powder (g/100 g)	1.43	1.43	0.57
Calcium hydrophosphate (g/100 g)	0.6	0.6	0.9
Premix[b] (g/100 g)	0.4	0.4	0.5
Salt (g/100 g)	0.5	0.5	0.4
Sodium butyrate (g/100 g)	–	1	–
Coarse material			0.00
Oats grass (g/100 g)	32	32	48
Alfalfa Oats grass (g/100 g)	8	8	12
Nutritional index			
NE (MJ/kg)	17.56	17.56	17.17
Fat (%)	3.6	3.6	2.39
CP (%)	16.76	16.76	17.79
Starch	19.914	19.914	17.295
NDF (%)	42.64	42.64	40.79
ADF (%)	7.22	7.22	2.83
Ca (%)	1.55	1.55	1.73
P (%)	0.76	0.76	0.79

NE Net Energy, *CP* Crude Protein, *NDF* Neutral Detergent Fibre, *ADF* Acid Detergent Fibre, *Ca* Calcium, *P* Phosphorus
[a]Treatment LG, 40% grain on DM basis; HG, 60% grain on DM basis, HG, 60% grain on DM basis + sodium butyrate
[b]Premix consists of the following ingredients per kg of diet: vitamin A, 6000 IU; vitamin D, 2500 IU; vitamin E, 53.6 IU; Cu, 6.25 IU/kg; Fe, 62.5 mg; Zn, 62.5 mg; Mn, 50 mg; mg I, 0.125 mg; Co, 0.125 mg; Mo, 0.125 mg

Radioimmunoassay

Radioimmunoassay was applied to determine the concentrations of master cytokines, including IL-1β, IL-6 and TNF-α in circulating blood. The concentrations of IL-1β, IL-6 and TNF-α were determined with commercially available human radioimmunoassay kits purchased from Beijing North Institute of Biological Technology. The detected range of radioimmunoassay kits for IL-1β (cat. C09DJB), IL-6 (cat. C12DJB) and TNF-α (cat. C06PJB) were 0.1–8.1 ng/mL, 50–4000 pg/ml and 9–590 fmol/mL, respectively.

Quantitative real time-PCR (qRT-PCR)

Total RNA was prepared from 50 mg liver tissue using TRIZOL (Invitrogen, USA) as described by the manufacturer's protocol. RNA quality was assessed by both agarose gel (1%) electrophoresis and a NanoDrop ND-1000 Spectrophotometer (Thermo Scientific, USA). Only samples with an A260 to A280 ratio between 1.8 and 2.1 were used in subsequent experiments. Reverse transcription (RT) was performed using 250 ng/μL RNA to synthesize cDNA with PrimeScript RT Master Mix Perfect Real Time (Takara Co., Otsu, Japan) according to the manufacturer's instructions.

Primers for the CAT, GPX1, GPx3, SOD1, SOD2, and SOD3 genes and for GAPDH as a reference gene were designed using the NCBI primer blast online software. Each cDNA sample was amplified using SYBR Green (Takara Co., Otsu, Japan) and an ABI 7300 Fast Real-time PCR System (Applied Bio system, USA). Briefly, the reaction mixture consisted of 2 μL of cDNA and 0.4 μM primers, 10 μL of SYBR green, 0.4 μL of ROX, and 6.8 mL of ddH_2O in a final volume of 20 μL. The qRT-PCR conditions were as follows: denaturation at 95 °C for 15 s, followed by 40 cycles of annealing at 60 °C for 31 s and extension at 72 °C. All the reactions were run in triplicate, and the specific PCR products were confirmed by gel electrophoresis on a 3.0% agarose gel. The primer sequences and the accession numbers of targeted genes are provided in Table 2.

The data were normalized to the mean of the housekeeping gene *GAPDH* to control for variability in expression levels and were analysed using the $2^{-\Delta\Delta Ct}$ method (($Ct_{target\ gene}$ - Ct_{GAPDH}) treatment - ($Ct_{target\ gene}$ - Ct_{GAPDH}) control) as previously described [28].

Table 2 Primer sequences used for real-time PCR amplification of selected genes in the livers of lactating goats

Gene[a]	Primers 5′ to 3′[b]	Size(bp)	Accession [c]
SOD1	ACACAAGGCTGTACCAGTGC TTCACATTGCCCAGGTCTCC	104	NM_001285550.1
SOD2	TCAATAAGGAGCAGGGACGC AGCAGGGGGATAAGACCTGT	85	XM_005684984.1
SOD3	ATCGACCCGAACGGTAACGC AGGACATAGAAGGGGTCTGCG	199	XM_005681678.1
CAT	ACAATGTCACTCAGGTGCGG TCTCACACAGGCGTTTCCTC	70	XM_005690077.1
GPx1	GTAGCATCGCTCTGAGGCAC TTGGCATTTTCCTGATGCCC	131	XM_005695962.1
GPx3	TTCGGTCTGGTCATTCTGGG CGAACATACTTGAGGGTGGC	95	XM_005683183.1
HMOX2	AGCTAGACAAAGGTGCCCTG CAGGATGCTGACAGGCAAAG	113	XM_005697930.1
MGST3	AGCTCACCAGAACACGTTGG GCCCAAACCAGAAGCTATGC	94	NM_001285615.1
NQO1	CATGGCTGTCAGAAAAGCACTG GACACAGTGACCTCCCATCC	120	XM_005692193.1
UGT1A1	TCCGGAGCAGAAAGCTATGG CCGAGTCTTTGGGTGACCAAG	160	XM_005678746.1
TXNRD1	TTTCGCTCAGTTTGCTCCAG TCCTGAGAAGCCTTCAGGGTC	200	XM_005680564.1
NF-kB	AGGTGGCGATCGTTGTTCTA TTGCCTTTGTTCTTCCTGCC	226	XM_00569893.1
NrF2	AGCCAGGTGAGATGGAACTG CCAGACTCCCTGTTTCGCTG	120	XM_005679848.2
MT2A	TCCTGCAAGAAGAGCTGCTG AACTGCACTTGTCCGAGGC	94	XM_005691999.1
SRXN1	GACACGATCCGGGAGAATCC GGTCTGAGAGGGTGGATTGG	175	XM_013968785.1
MT1A	ATAATAGCGCTCGGCTCCTG TTGGAGGAAAAGCGAGGTCC	98	XM_005692001.2
GAPDH	CCTGCCCGTTCGACAGATAG CCGTTCTCTGCCTTGACTGT	249	XM_005680968.2

[a]*SOD1* superoxide dismutase 1, *SOD2* superoxide dismutase 2, *SOD3* superoxide dismutase3, *GPX1* glutathione peroxidase1, *GPX3* glutathione peroxidase 3, *CAT* catalase, *HMOX2* haem oxygenase 2, *MGST3* microsomal glutathione S-transferase 3, *MT1A* metallothionein 1A, *NQO1* NAD(P)H, quinone 1, *UGT1A1* UDP glucuronosyltransferase 1 family, polypeptide A1, *MT2A* metallothionein 2A, *TXNRD1* thioredoxin reductase 1, *SRXN1* sulfiredoxin 1, *GAPDH* glyceraldehyde 3-phosphate dehydrogenase
[b]The first primer listed for each gene is the forward primer, and the second primer is the reverse primer
[c]The reference sequence numbers are given for primers whose source is the National Center for Biotechnology Information (NCBI) GenBank database (http://www.ncbi.nlm.nih.gov/genbank/)

Oxidative and antioxidative biomarker analysis in liver and plasma

The antioxidant assays were conducted using assay kits purchased from the Nanjing Jiancheng Institute of Bioengineering (Nanjing, Jiangsu, China), and the assays were conducted according to the method described by Zheng et al. [29]. Briefly, one gram of frozen liver tissue in 10 ml of homogenization buffer (0.9% cool physiological saline) was homogenized on ice for 30 s at 12500 rpm using a Polytron PT 1200 E (POLYTRON® PT 1200 E Manual Disperser, Luzern - Switzerland). The homogenate was centrifuged at 2500 rpm for 10 min at 4 °C, and the supernatant (10% concentration) was aliquoted. Physiological saline (0.9%) was used to dilute the supernatant to different concentrations, which was stored at − 20 °C for the following analyses:

Glutathione peroxidase

A Nanjing Jiancheng Bioengineering Institute commercial kit (Nanjing, Jiangsu, China) was used to determine GPx spectrophotometrically (UV3600, Daojin Corp., Japan) at 412 nm as previously described by Wu et al. [30] the oxidizing speed of GSH was expressed as the GSH reduction within a certain time. One unit of GPx activity was defined as 1 μmol/L GSH oxidized to glutathione disulfide (GSSG)/mg protein/ min.

Superoxide dismutase

Plasma and liver homogenate were used to determine SOD activity based on a reaction system that contained xanthine and xanthine oxidase using Nanjing Jiancheng Bioengineering Institute commercial kits (Nanjing, Jiangsu, China, Cat. No. A001−1 SOD). The superoxide anion (O-·2) oxidizes hydroxylamine to form nitrite and was assayed spectrophotometrically (UV3600, Daojin Corp., Japan) at 550 nm. One unit of SOD is defined as the amount of sample resulting in 50% inhibition of nitroblue tetrazolium reduction.

Catalase

Nanjing Jiancheng Bioengineering Institute commercial kits (Nanjing, Jiangsu, China) were used to determine catalase activity based on the decrease in H_2O_2 concentration in 15 s as previously described by Wang et al. [31]. Catalase was assayed spectrophotometrically (UV3600, Daojin Corp., Japan) at 405 nm. One unit of CAT activity was defined as the amount of enzyme decomposing 1 mol of H_2O_2 in 1 s.

Malondialdehyde (MDA) measurement

MDA levels were measured spectrophotometrically at 532 nm using the thiobarbituric acid reaction method as previously described by Chen et al. [32] using commercially available kits (Nanjing Jiancheng Bioengineering Institute). Lipid peroxidation was expressed as nmol MDA/mg protein or nmol MDA/mL plasma.

Determination of total antioxidants (T-AOC)

T-AOC was measured spectrophotometrically based on the reduction of Fe3+ to Fe2+, which combines with phenanthrene and forms a coloured compound that can be detected at 520 nm. One unit of T-AOC was defined as the extent to which the optical density is increased by 0.01 /mg protein or/mL plasma/min.

Western blotting analysis

Proteins were extracted from 100 mg liver tissue with RIPA buffer (Beyotime Biotechnology Co., Ltd., Shanghai, China). Protein concentrations were assayed using a bicinchoninic acid (BCA) protein assay kit (Pierce, Rockford, IL, USA). The concentration was normalized to 4 μg/μL, and 10 μL protein in sodium dodecyl sulfate (SDS) loading buffer was subjected to 10% SDS-polyacrylamide gel electrophoresis (PAGE) and transferred to nitrocellulose (NC) membranes (Millipore, Danvers, MA) at 4 °C for 90 min; 10% skim milk in Tris-buffered saline was used to block the membrane. Monoclonal goat antibodies against Nrf2 and GAPDH (Santa Cruz Biotechnology, CA, USA, 1:200 dilution) (Bio-world, CA, USA, 1:10,000 dilution) were used at 4 °C overnight, and then the membrane was washed and treated with horseradish peroxidase-conjugated donkey anti-goat secondary antibody. Immunoreactive proteins were detected by chemiluminescence using an ECL reagent (Super Signal West Pico Trial Kit, Pierce, USA) and subsequently analysed by autoradiography. Quantitation of the results was performed using a Bio-Rad Gel Doc 2000™ System (Bio-Rad, USA) and analysis with Bio-Rad TDS Quantity One software (Bio-Rad). The relative quantities of proteins were determined by a densitometer and expressed as absorbance units (AU).

Statistical analysis

All the data were analysed using SAS software (SAS version 9.2, SAS Institute Inc., USA); the normality of the distribution of variables was tested by the Shapiro–Wilk test. The mixed procedure of SAS was used to analyse pH and milk data with a repeated measures design. The effects of diet and time were considered fixed factors. The effects of goats were considered random. The time within treatments and goats were considered repeated measurements, and compound symmetry (CS) was used as the type of covariance. Differences between groups regarding ruminal volatile fatty acids and proinflammatory cytokines, gene expression, enzyme activity and other biomarkers were analyzed by analysis of variance (ANOVA) followed by post hoc Dunnett's multiple comparison tests. Differences were considered significant when p values ≤0.05. Graphs were made using Graph Pad Prism version 5.00 (Graph Pad Software).

Results

Ruminal pH

As shown in Fig. 1, significant differences were observed in ruminal pH among the HG diet, HG + NaB and LG diet groups ($p < 0.05$); NaB added to the HG diet significantly improved rumen pH compared to the HG diet alone; however, rumen pH still remained significantly lower than in the LG diet.

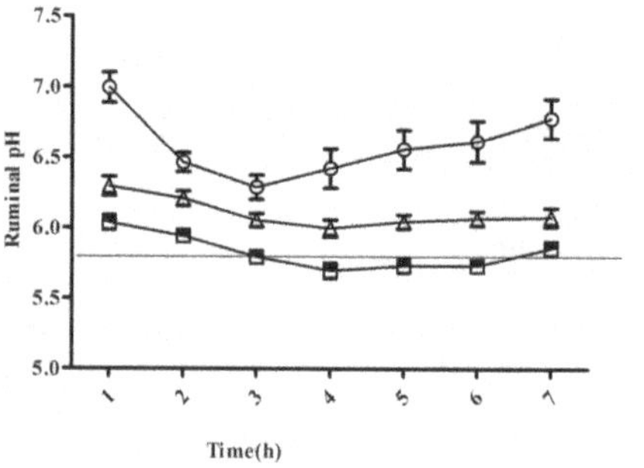

Fig. 1 Comparison of pH values in ruminal fluid among the low-grain (LG), high-grain (HG) and high-grain with sodium butyrate (HG + NaB) groups. Ruminal fluid samples were collected at times ranging from 0 to 7 h for three consecutive days during the 20th week. Data were analysed by mixed models in SAS (SAS version 9.2, SAS Institute Inc., USA) and are expressed as the mean ± S.E.M. Significant differences were observed at all sampling times ($p < 0.05$). The error bars indicate the standard error of the mean

Milk yield and composition

The results revealed that the milk yield, milk fat %, the milk fat yield, milk protein % and protein yield were significantly decreased in goats fed HG compared with HG + NaB and goats fed LG ($p = 0.01$, $p = 0.012$, $p = 0.001$, $p = 0.024$, and $p = 0.004$ respectively, Table 3).

LPS content in ruminal fluid, hepatic vein plasma and portal vein plasma

The concentration of LPS in each of the 3 groups is presented in Fig. 2. The highest LPS concentration in the rumen fluid was observed in goats fed the HG diet ($p \leq 0.01$), compared to those fed the LG and HG diets with sodium butyrate. Consistent with the ruminal LPS concentration, goats fed the HG diet had a higher level of LPS in their portal vein plasma than those in the other groups ($p \leq 0.01$), whereas no differences were observed between the LG diet and the HG diet with NaB. Moreover, the LPS concentration in hepatic vein plasma showed a significant difference among the three groups ($p \leq 0.01$).

Levels of pro -inflammatory cytokines in the peripheral blood

The results are shown in Table 4. The concentrations of IL-1β, IL-6 and TNF-α were remarkably increased ($p = 0.024$, $p = 0.003$ and $p = 0.01$ respectively) in goat fed high grain diet. However, sodium butyrate significantly decreases these cytokines when added to HG diet.

Rumen volatiles fatty acids

The results of ruminal VFA concentrations were shown in Fig. 3. The levels of butyrate, propionate, and total

Table 3 The effect of Sodium butyrate on milk yields and milk composition of goats fed HG diet versus those fed HG alone and LG diet

	Treatment[d]				*p* value		
Item	LG	HG	HG + NaB	SEM	Diet	week	Diet*week
Milk yield	0.962[a]	0.941[b]	1.032[c]	0.07	0.01	0.01	0.01
Fats %	3.365[a]	2.852[b]	3.275[a]	0.128	0.012	0.35	0.3
Fats yield, kg	0.034[a]	0.025[b]	0.035[a]	0.001	0.001	0.02	0.56
Protein %	4.192 [a]	3.801[b]	4.106 [a]	0.064	0.024	0.01	0.134
Protein yield, kg	0.043[a]	0.035[b]	0.044[a]	0.001	0.004	0.002	0.09

The milk yield and milk compositions were analyzed by the mixed model of SAS with repeated measures, $p \leq 0.05$ was considered significant
[a, b and c] The means with different superscript were considered significantly different
[d]*LG* low gain, *HG* high grain, *HG + NaB* high grain + sodium butyrate

VFA were significantly higher in the HG + NaB group than the HG group, meanwhile LG diet showed the lowest values of ruminal VFA ($p = 0.01$, $p = 0.001$, $p = 0.001$).

Antioxidant gene expression in goat liver tissue

Among the antioxidant genes that were evaluated, including SOD1, SOD2, SOD3, GPX1, GPX3 and CAT, significant increases in the expression of SOD1, SOD2, SOD3, GPX1 and CAT in the HG + NaB group ($p = 0.008$; $p = 0.001$; $p = 0.001$; $p = 0.001$ and $p = 0.001$, respectively) were observed (Fig. 4a), whereas in the HG group, lower levels of these genes were expressed than in the LG diet group. No significant differences were noticed between goats fed the HG + NaB and LG diets except for CAT, GPX1 and SOD3.

The mRNA expression levels of hepatic UGT1A1, NQO1 and MGST3 were significantly increased with the HG + NaB diet ($p = 0.008$; $p = 0.001$ and $p = 0.002$, respectively). Conversely, these genes were significantly decreased in the HG diet group compared to the LG and HG + NaB diets. The expression levels of hepatic TNXRD1, HMOX2, MT1E, MT1A, MT2A, and SRXN1 showed no difference among the three groups (Fig. 4b). NF-κB and Nrf2 mRNA were significantly different among the three groups ($p \leq 0.01$) (Fig. 4c).

Total antioxidant capacity and lipid peroxidation

The values of T-AOC and MDA in both liver and plasma are shown in Figs. 5a, b and 6a, b. Feeding the HG + NaB diet significantly decreased MDA in the liver and plasma of lactating goats ($p = 0.01$ and $p = 0.04$, respectively) compared to feeding the HG diet. However, T-AOC was not reduced in a HG + NaB group ($p = 0.03$ and 0.02) compared to the HG group, whereas no differences were observed between the goats fed the LG and HG + NaB diets.

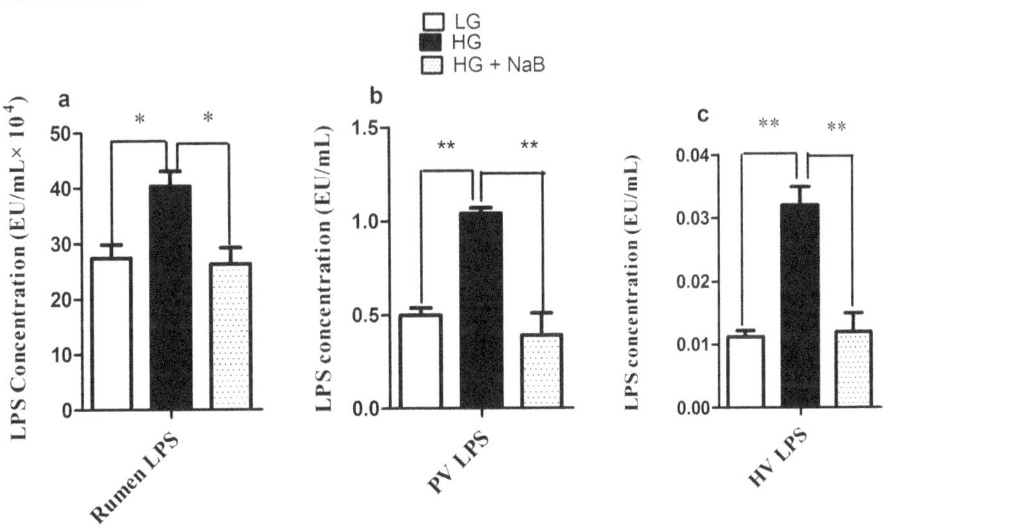

Fig. 2 Effect of sodium butyrate on LPS concentration in the rumen, portal vein, hepatic vein and jugular vein plasma of dairy goats fed three diets (LG, HG and HG + NaB). **a** Ruminal LPS, **b** Portal vein LPS, **c** Hepatic vein LPS. The concentration of lipopolysaccharide (LPS) in plasma was determined by a chromogenic endpoint limulus amoebocyte lysate assay kit (CE64406 &CE80545, Chinese Horseshoe Crab Reagent Manufactory Co., Ltd., Xiamen, China). Data were analysed by mixed model in SAS (SAS version 9.2, SAS Institute Inc., USA) and are expressed as the mean ± S.E.M. $^{**}P < 0.01$, $^{*}P < 0.05$. LG = low grain; HG = high grain; HG + NaB = high grain + sodium butyrate

Table 4 The effect of Sodium butyrate on proinflammatory cytokines in goats fed HG diet versus those fed HG alone and LG diet

Item	Treatment			SEM	*p* value
	LG	HG	HG + NaB		
IL-1β, ng/mL	0.1568^{a}	0.2500^{b}	0.1670^{a}	0.024	0.003
IL-6, pg/mL	289.01^{a}	612.5^{b}	404.8^{c}	25.40	0.01
TNF-α fmol/mL	4.008^{a}	7.236^{b}	2.728^{c}	0.234	0.007

The milk yield and milk compositions were analyzed by the mixed model of SAS
LG low gain, *HG* high grain, *HG + NaB* high grain + sodium butyrate
a, b and cThe means with different superscript were considered significant different

Antioxidant enzyme activity

Variables related to antioxidant enzyme activity in the liver, including the concentrations of TSOD, GPx and CAT, showed significant differences ($p = 0.005$, $p = 0.013$, $p = 0.001$, respectively) among the goats fed the HG + NaB, HG and LG diets; however, no differences were observed between goats fed the LG and HG + NaB diets (Fig. 5c, d, e). Additionally, the results showed significant increases in TSOD and CAT activity in the plasma of goats fed the HG diet with NaB compared with the HG diet alone ($p = 0.001$ and $p = 0.05$, respectively). These findings were consistent with the liver antioxidant enzyme activity; however, no significant differences were observed in the plasma GPx activity among the groups (Fig. 6c, d, and e).

Nrf2 protein expression

As shown in Fig. 7, we observed significant differences in Nrf2 protein expression among the three groups ($p = 0.01$). Although the HG + NaB diet increased the levels of Nrf2 protein, the Nrf2 level still remained lower than in goats fed the LG diet.

Discussion

In the present study, we determined the ability of sodium butyrate to mitigate the oxidative stress induced by long-term feeding of a high-grain diet in lactating dairy goats. In this study, we used three different feed formulas, low grain as a control, high grain to induce SARA, and high grain with sodium butyrate, to investigate the effect of NaB on SARA, and our results confirmed the presence of SARA in the group of goats fed a high-grain diet [33], in which ruminal pH fell below 5.8 for more than 180 min/d [34]. Moreover, our results showed lower levels of milk yield (kg/d), milk fat (%), and milk fat yield (kg/d) in goats fed HG alone, these have been previously reported by Dong et al., (2011)

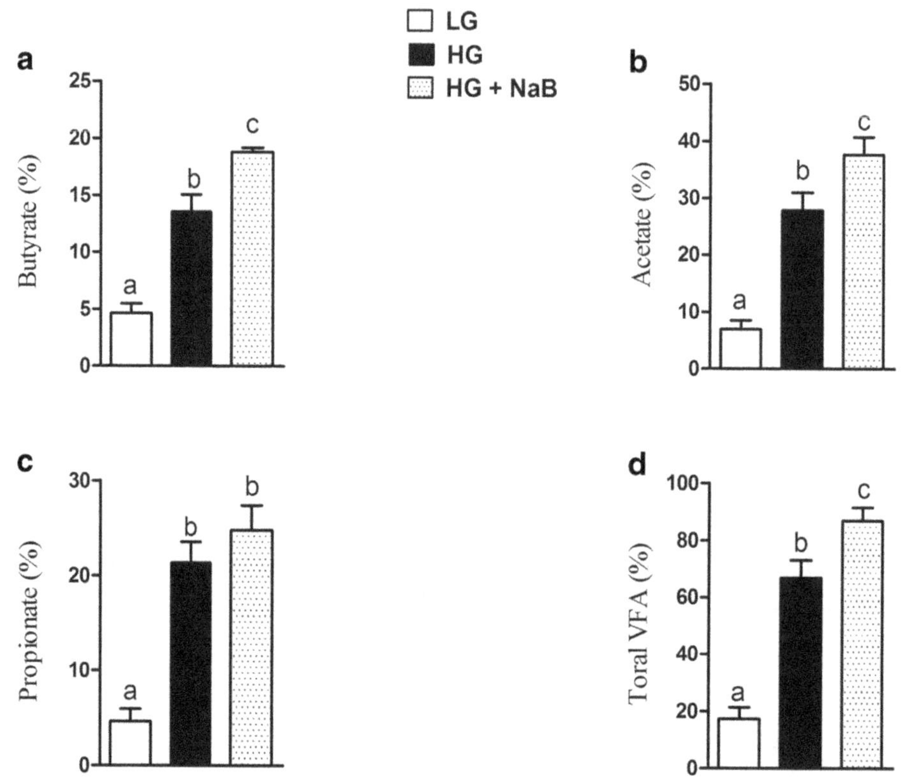

Fig. 3 Effect of sodium butyrate on volatile fatty acids on ruminal fluid of goats fed HG with sodium butyrate compared with high grain alone and low grain. **a** Butyrate. **b** Acetate. **c** Propionate. **d** Total volatile fatty acids. Data were analyzed by the mixed model of SAS (SAS version 9.2, SAS Institute Inc), and expressed as the mean ± S.E.M. The means with different superscript were considered significantly different

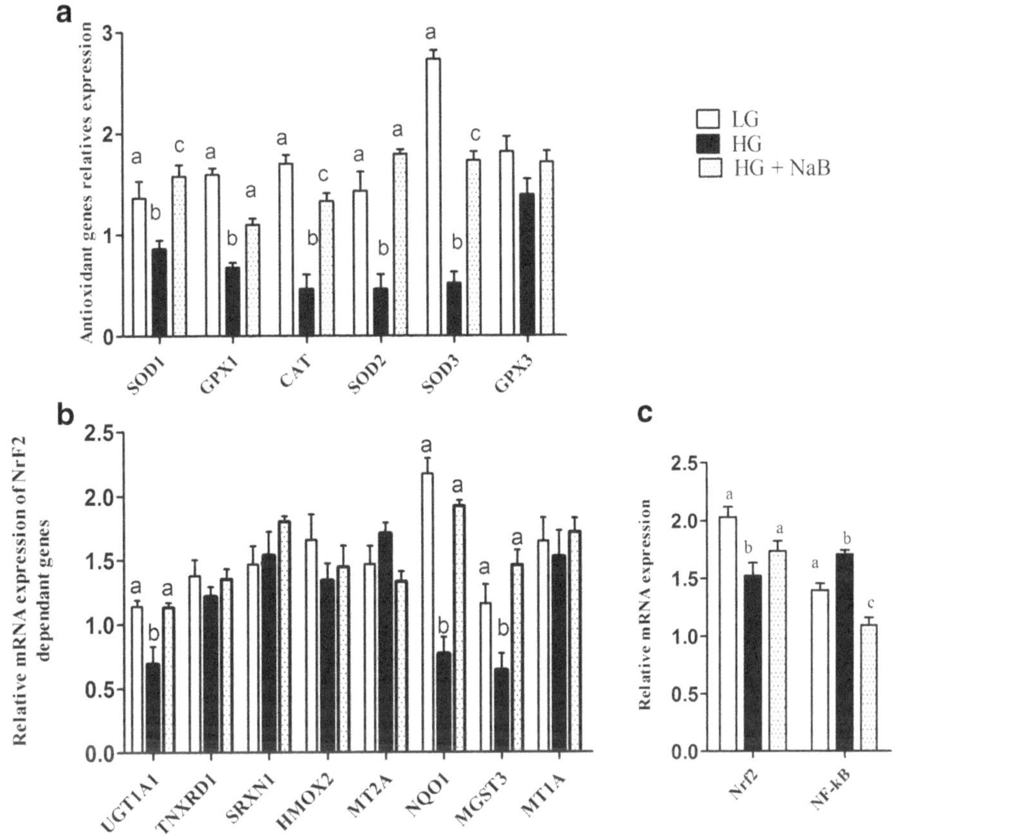

Fig. 4 Effect of sodium butyrate on relative mRNA expression in the liver of goats fed HG vs. HG alone and LG. Gene expression was analysed quantitative real time-PCR (qRT-PCR). **a** mRNA abundance of genes encoding antioxidant enzymes in the liver (**b**) mRNA abundance of Nrf2 target genes in the liver (**c**) NrfF2 and NF-kB mRNA expression. *GAPDH* was used as the reference gene for gene expression. Data were analysed by mixed model in SAS (SAS version 9.2, SAS Institute Inc., USA) and are expressed as the mean ± S.E.M. $^{**}P < 0.01$, $^{*}P < 0.05$. LG = low grain; HG = high grain; HG + NaB = high grain + sodium butyrate; SOD1 = superoxide dismutase 1; SOD2 = superoxide dismutase 2; SOD3 = superoxide dismutase 3; GPX1 = glutathione peroxidase1; GPX3 = glutathione peroxidase 3; CAT = catalase; HMOX2 = haem oxygenase 2; MGST3 = microsomal glutathione S-transferase 3; MT1A = metallothionein 1A; NQO1 = NAD(P)H dehydrogenase, quinone 1; UGT1A1 = UDP glucuronosyltransferase 1 family, polypeptide A1; MT1E = metallothionein 1E; MT2A = metallothionein 2A; TXNRD1 = thioredoxin reductase 1; SRXN1 = sulfiredoxin 1; NrfF2 = nuclear factor E2-related factor 2; NF-kB = nuclear factor kappa B; *GAPDH* = glyceraldehyde 3-phosphate dehydrogenase

[35]. These milk composition reduced because the free circulating LPS may decrease the activity of key enzymes in milk component syntheses, such as acetyl-CoA carboxylase, fatty acid synthase, and lipoprotein lipase [15]. Consistent with previous SARA studies we observed increased levels of proinflammatory cytokines in goats fed HG alone [36] and conversely these proinflammatory cytokines were prominently decreased in goat fed HG with NaB. These further confirmed the presence of SARA in a group of goats fed high grain diets. Meanwhile, sodium butyrate successfully improves the rumen pH when added to the HG diets [37, 38]; this is contradictory to the results of ruminal volatile fatty acids which is significantly higher in goats fed HG with sodium butyrate. However it's previously reported by Norton et al., that sodium butyrate infusion significantly increased rumen fluid VFA and pH [39], it is possible that butyrate altered the metabolic flux of the rumen, favoring the neutralization of H+ through the efflux of bicarbonate [40]. There are several conflicting results have been observed in previous studies about the effect of NaB on rumen pH; e, g., Laarman et al... demonstrated the presence of SARA in both groups of cows fed a total mixed ration and concentrate with or without NaB [41], whereas Herrick et al. reported a significant increase in rumen pH in lactating dairy cows ruminally dosed with butyrate [42]. Moreover, Schlau et al. observed a marked increase in the molar proportion of butyrate in acidosis-resistant steers compared to that in acidosis-susceptible steers and stated that this increase may have at least partially contributed to the higher pH observed in acidosis-resistant steers [37]. This discrepancy may be because of using different concentrations of infused butyrate. Consistent with these findings, lower levels of LPS in rumen fluid, portal vein plasma and hepatic vein plasma were observed in the current study. The present results are in agreement with those previously reported by Li et al [19].

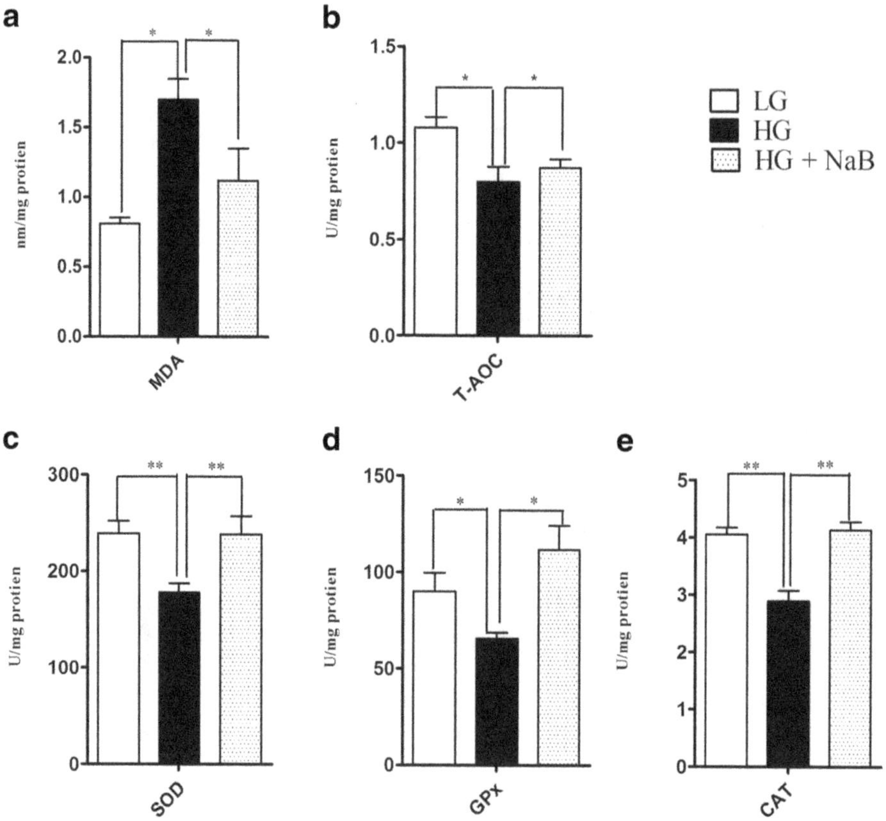

Fig. 5 Effect sodium butyrate on lipid peroxidation, total antioxidant capacity and antioxidant enzyme activity in the liver of goats fed HG vs. LG and HG alone. **a** MDA. **b** T- AOC. **c** Total SOD activity. **d** GPx activity. **e** Catalase activity. The results are expressed as the mean ± SEM. Data were analysed by mixed model in SAS (SAS version 9.2, SAS Institute Inc., USA) and are expressed as the mean ± S.E.M. $^{**}P < 0.01$, $^{*}P < 0.05$ vs. LG. HG = high grain; LG = low grain; HG + NaB = high grain + sodium butyrate; MDA = malondialdehyde; T-AOC = total antioxidant capacity; TSOD = total superoxide dismutase; GPx = glutathione peroxidase; CAT = catalase

Those results showed that exogenous butyrate had a stimulating effect on the native butyrate-producing bacterial population. This increase in butyrate-producing bacteria could maintain the stability of the fibre-degrading microbiota during SARA. Thus, the release of LPS in the rumen was reduced and gastrointestinal homeostasis was maintained. Furthermore, other studies demonstrated that butyrate maintains barrier function through the genomic regulation of several cellular processes, and it modulates gene expression through histone deacetylase complex (HDAC) inhibition [20], which in turn can reduce LPS production and prevent LPS translocation into the bloodstream, potentially reducing the systemic effect of LPS. Butyrate has been implicated in the down-regulation of bacterial virulence, both by direct effects on virulence gene expression and by acting on the proliferation of host cells [22]. Conversely consuming a HC diet leads to remarkable increases in LPS produced in the digestive tract and delivered into the liver via the portal vein [43]. The increased levels of LPS trigger a liver inflammatory response through the TLR4-signaling pathway, lead to an increase of pro- and anti-inflammatory cytokines and acute phase proteins (APPs) including LPS-binding protein (LBP), haptoglobin (Hp) and serum amyloid A (SAA) in the peripheral blood [13, 44]. This study demonstrated that supplementation of a high-grain diet (HG) with NaB promotes antioxidant activity and alleviates oxidative stress during SARA. The higher concentration of plasma and liver SOD, CAT and GPx in the goats fed the HG + NaB diet than in the goats fed the HG diet could be related to the higher ruminal pH and reduced levels of LPS in associated with the HG + NaB diet, and these factors may reduce the inflammatory reaction and reactive oxygen species (ROS) production. Moreover, among the possible other mechanisms by which butyrate may boost enzyme activity, antioxidant activity has also been recently suggested. Ma et al. reported significantly increased levels of antioxidant enzymes in piglet intestinal cells following butyrate incubation, and they concluded that the sodium butyrate-mediated alteration in antioxidant indices, including MDA, GSH, and antioxidative enzymes, suggests an improvement in the level of oxidative stress in the intestinal mucosa cells, which may result in improved wound healing, tight junctions, epithelial integrity and defence systems [45]. Furthermore, Dionissopoulos et al.

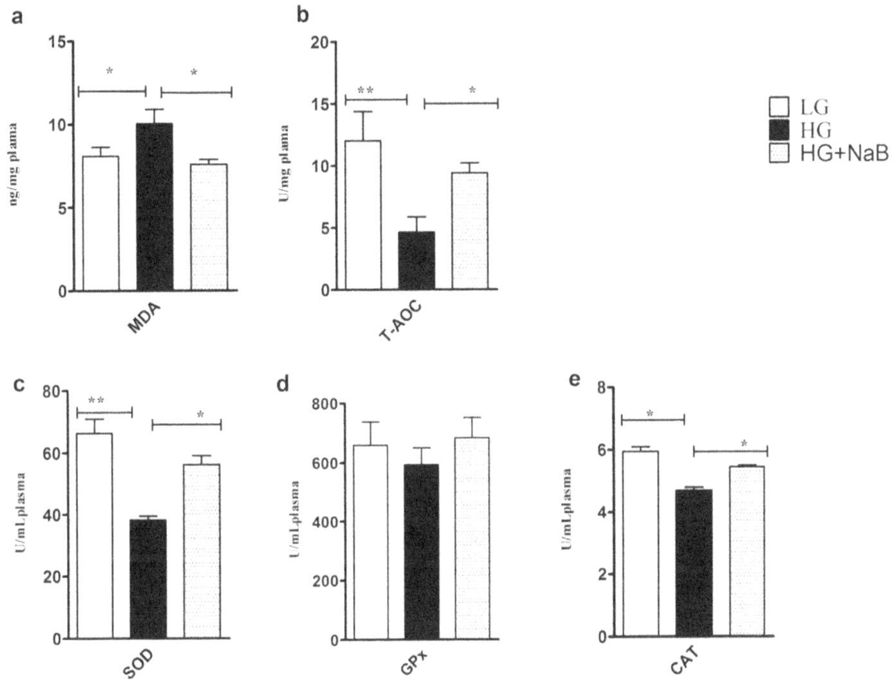

Fig. 6 Effect of sodium butyrate on lipid peroxidation, total antioxidant capacity and antioxidant enzyme activity in the plasma of goats fed the HG compared with HG alone and LG . **a** MDA. **b** T-AOC. **c** Total SOD activity. **d** GPx activity. **e** Catalase activity. All enzyme activities and biomarkers were evaluated spectrophotometrically, and the results are expressed as the mean ± SEM. Data were analysed by mixed model in SAS (SAS version 9.2, SAS Institute Inc., USA) and are expressed as the mean ± S.E.M. $^{**}P < 0.01$, $^{*}P < 0.05$. HG = high grain; LG = low grain; MDA = malondialdehyde; T-AOC = total antioxidant capacity; TSOD = total superoxide dismutase; GPx = glutathione peroxidase; CAT = catalase

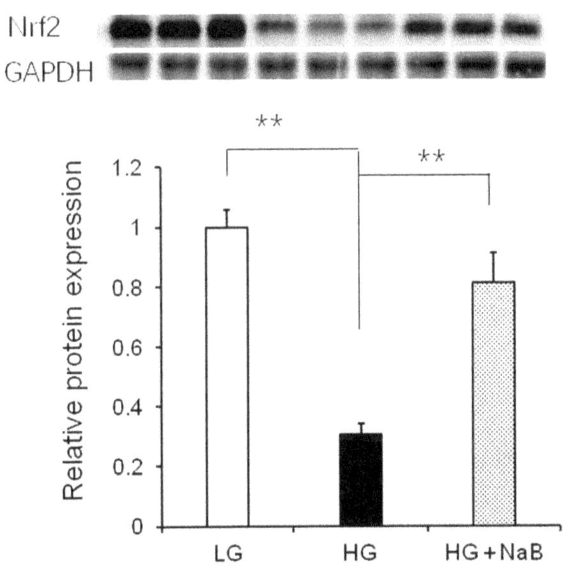

Fig. 7 Effect of sodium butyrate on the relative expression of Nrf2 protein in the livers of goats fed the HG compared with the HG alone or LG. GAPDH was used as the reference protein. Nrf2 protein was significantly higher than in goats fed the high-grain diet. Data were analysed by mixed model in SAS (SAS version 9.2, SAS Institute Inc., USA) and are expressed as the mean ± S.E.M. $^{**}P < 0.01$, $^{*}P < 0.05$

reported high expression of GPX2 in cow rumen papillae after butyrate supplementation; GPX2 is known to protect the rumen from ROS produced by ruminal bacteria [40]. Moreover, several clinical studies have demonstrated that butyric acid or its sodium salt mediate immune responses and antioxidant capacity in vitro and in vivo [46, 47]. Butyrate can decrease oxidative stress and cytokine synthesis, even at very low doses [22]. Conversely, in goats fed the HG diet, we observed lower activity of antioxidant enzymes, including SOD and CAT, in both liver and plasma and lower activity of GPx in the liver, which corresponded to lower levels of mRNA. These results are in accordance with evidence previously reported by Watson et al., who studied the activities of conjugating and antioxidant enzymes following endotoxin exposure in male Sprague Dawley rats and concluded that total SOD activity and CAT and GPx activity were decreased at different time points following endotoxin administration [48].

In our results, the high-grain diet significantly increased liver and plasma MDA, which is in agreement with the results of Guo et al., who observed higher concentrations of plasma MDA in cows fed a diet containing 20% finely ground wheat (W20) than in cows fed 0% finely ground wheat (W0) [49]. On the other hand, the liver and serum MDA were significantly lower in the sodium butyrate group than in the high-grain diet group;

this result was noted previously by Zhang et al. in stressed broiler chickens supplemented with microencapsulated NaB, and they concluded that NaB had partially elevated the oxidative stress and reduced the MDA levels [50]. A reduced level of MDA was also seen in intestinal cells treated with sodium butyrate accompanied by increased antioxidant enzyme indices [45]. Furthermore, Guo et al. reported decreased levels of MDA and increased levels of T-AOC and beta hydroxybutyric acid in cows with sub-acute ruminal acidosis regulated by pelleted beet pulp, which is a known butyrigenic carbohydrate source that may lead to increased levels of beta hydroxybutyric acid and alleviation of oxidative stress [49]. The increased levels of MDA and the decreased levels of T-AOC indicate that the pro-oxidants overwhelmed the antioxidants and exerted lipid peroxidation [51]. MDA is a highly toxic by-product generated by lipid peroxidation [52], and its toxicity depends on its rapid reaction with proteins and DNA [52]. If accumulated, these lipid peroxy radicals act on nearby fatty acids in the plasma membranes of the cells and induce radical formation by positive the feedback loop.

Nuclear factor E2-related factor-2 (Nrf2) is a transcription factor that orchestrates the expression of a battery of antioxidant and detoxification genes under both basal and stress conditions [53]. In our results, we observed up-regulation of UGT1A1, NQO1 and MGST3. These genes are known to be involved in butyrate-mediated defence against oxidative stress, providing evidence of favourable modulation of stress [54]. Butyrate was reported to have immune suppression [55] and anti-inflammatory properties, in part by suppressing NF-κB activity [56]. NF-κB is known to be antagonized by Nrf2. Recent studies indicated that sodium butyrate could attenuate inflammatory reactions by inhibiting the expression of inflammatory mediators such as NF-κB, tumour necrosis factor alpha (TNF-α) and interferon-gamma (IFN-γ) through modulation of antioxidant defence mechanisms [57]. These findings suggest that sodium butyrate could control the reactive oxygen species (ROS)-mediated activation of NF-κB. Furthermore, Yaku et al., (2012) [58] noted that sodium butyrate also causes enhancement of Nrf2 mRNA levels and suppression of p53 mRNA levels in normal intestinal epithelial cells, enhancing the activities of phase 2 enzymes via an increase in Nrf2 protein levels in the nucleus and a decrease in the mRNA and protein levels of p53.

Conclusions

In conclusion, our study indicates that feeding a high-grain diet to lactating dairy goats reduces the mean ruminal pH. When added to the HG diet, sodium butyrate (NaB) increased the mean ruminal pH and reduced the levels of ruminal, portal and hepatic LPS, indicating that sodium butyrate reduced rumen fermentation and improved rumen tight junctions. Data from this study also demonstrate that sodium butyrate improved antioxidant capabilities to some extent due to increased mRNA expression of antioxidant genes, including the nuclear factor E2 related factor2 (*Nrf2*)-dependent genes SOD1, SOD2, SOD3, GPX1, CAT, UGT1A1, NQO1 and MGST3. Additionally, in goats fed the HG + NaB diet, TSOD and CAT enzyme activity and total antioxidant capacity (T-AOC) increased, and malondialdehyde (MDA) levels in both the liver and plasma decreased compared to goats fed the high-grain diet alone. However, GPx activity was increased only in the livers of goats in the HG + NaB group. Western blotting revealed increased total Nrf2 protein associated with the HG + NaB diet compared to the HG diet; however, Nrf2 levels remained lower than those in the LG diet. Taking all the above results together, this study indicates that sodium butyrate could improve the oxidative status in sub-acute ruminal acidosis through partial activation of a set of Nrf2-dependent genes.

Abbreviations

CAT: Catalase; GAPDH: Glyceraldehyde 3-phosphate dehydrogenase; GPX: Glutathione peroxidase1; HG: High grain; HMOX2: Haem oxygenase 2; LG: Low gain; MGST3: Microsomal glutathione S-transferase 3; MT1A: Metallothionein 1A; MT2A: Metallothionein 2A; NaB : Sodium butyrate; NQO1: NAD(P)H, quinone 1;; Nrf 2: Nuclear factor E2-related factor 2; SOD: Superoxide dismutase; SRXN1: Sulfiredoxin 1; TXNRD1: Thioredoxin reductase 1; UGT1A1: UDP glucuronosyltransferase 1 family, polypeptide A1

Funding

The National Natural Science Foundation of China (No. 31672618), and the Priority Academic Program Development of Jiangsu Higher Education Institutions (PAPD) funded this work. The funders had no role in the design of the study and collection, analysis, and interpretation of data and in writing the manuscript.

Authors' contributions

XS conceived and designed the experiments. NM and JAA performed the experiments. MSB and HD contributed to sample collection and animal feeding. NM, JAA and XS analysed the data. NM and JAA performed statistical analysis. NM and JAA drafted the manuscript. XS obtained the funding. All authors read and approved the final manuscript.

Consent for publication

Not applicable.

Competing interests

The authors declare that they have no competing interests that would have predisposed the content of the paper or intervened with their objective consideration of the manuscript.

References

1. Lykkesfeldt J, Svendsen O. Oxidants and antioxidants in disease: oxidative stress in farm animals. Vet J. 2007;173:502–11.
2. Mudroň P, Rehage J, Qualmann K, Sallmann HP, Scholz H. A study of lipid peroxidation and vitamin E in dairy cows with hepatic insufficiency. Zentralbl Veterinarmed A. 1999;46:219–24.

3. Celi P, Di Trana A, Claps S. Effects of plane of nutrition on oxidative stress in goats during the peripartum period. Vet J. 2010;184:95–9.
4. Grant CM. Metabolic reconfiguration is a regulated response to oxidative stress. J Biol. 2008;7:1.
5. Drackley JK. Biology of dairy cows during the transition period: the final frontier? J Dairy Sci. 1999;82:2259–73.
6. Celi P. The role of oxidative stress in small ruminants' health and production. R Bras Zootec. 2010;39:348–63.
7. Mutsvangwa T, Wright TC. Sub-acute ruminal acidosis (SARA) in dairy cows. Ontario: Ministry of Agriculture and Food; 2003.
8. Gozho G, Plaizier J, Krause D, Kennedy A, Wittenberg K. Subacute ruminal acidosis induces ruminal lipopolysaccharide endotoxin release and triggers an inflammatory response. J Dairy Sci. 2005;88:1399–403.
9. Jia YY, Wang SQ, Ni YD, Zhang YS, Zhuang S, Shen XZ. High concentrate-induced subacute ruminal acidosis (SARA) increases plasma acute phase proteins (APPs) and cortisol in goats. Animal. 2014;8:1433–8.
10. Kennelly J, Robinson B, Khorasani G. Influence of carbohydrate source and buffer on rumen fermentation characteristics, milk yield, and milk composition in early-lactation Holstein cows. J Dairy Sci. 1999;82:2486–96.
11. Zebeli Q, Metzler-Zebeli B. Interplay between rumen digestive disorders and diet-induced inflammation in dairy cattle. Res Vet Sci. 2012;93:1099–108.
12. Krause KM, Oetzel GR. Understanding and preventing subacute ruminal acidosis in dairy herds: a review. Anim Feed Sci Technol. 2006;126:215–36.
13. Khafipour E, Krause D, Plaizier J. A grain-based subacute ruminal acidosis challenge causes translocation of lipopolysaccharide and triggers inflammation. J Dairy Sci. 2009;92:1060–70.
14. Asehnoune K, Strassheim D, Mitra S, Kim JY, Abraham E. Involvement of reactive oxygen species in toll-like receptor 4-dependent activation of NF-κB. J Immunol. 2004;172:2522–9.
15. Abaker JA, Xu TL, Jin D, Chang GJ, Zhang K, Shen XZ. Lipopolysaccharide derived from the digestive tract provokes oxidative stress in the liver of dairy cows fed a high-grain diet. J Dairy Sci. 2017;100:666–78. https://doi.org/10.3168/jds.2016-10871.
16. Miller J, Brzezinska-Slebodzinska E, Madsen F. Oxidative stress, antioxidants, and animal function. J Dairy Sci. 1993;76:2812–23.
17. Trevisan M, Browne R, Ram M, Muti P, Freudenheim J, Carosella AM, et al. Correlates of markers of oxidative status in the general population. Am J Epidemiol. 2001;154:348–56.
18. Bergman E. Energy contributions of volatile fatty acids from the gastrointestinal tract in various species. Physiol Rev. 1990;70:567–90.
19. Li R, Wu S, Vi RL, Li W, Li C. Perturbation dynamics of the rumen microbiota in response to exogenous butyrate. PLoS One. 2012;7:e29392.
20. Plöger S, Stumpff F, Penner GB, Schulzke JD, Gäbel G, Martens H, et al. Microbial butyrate and its role for barrier function in the gastrointestinal tract. Ann N Y Acad Sci. 2012;1258:52–9.
21. Rosignoli P, Fabiani R, De Bartolomeo A, Spinozzi F, Aqea E, Pelli MA. Protective activity of butyrate on hydrogen peroxide-induced DNA damage in isolated human colonocytes and HT29 tumour cells. Carcinogenesis. 2001;22:1675–80.
22. Guilloteau P, Martin L, Eeckhaut V, Ducatelle R, Zabielski R, Van Immerseel F. From the gut to the peripheral tissues: the multiple effects of butyrate. Nutr Res Rev. 2010;23:366–84.
23. Leonel AJ, Alvarez-Leite JI. Butyrate: implications for intestinal function. Curr Opin Clin Nutr Metab Care. 2012;15:474–9.
24. Zhang WH, Jiang Y, Zhu QF, Gao F, Dai SF, Chen J, et al. Sodium butyrate maintains growth performance by regulating the immune response in broiler chickens. Br Poult Sci. 2011;52:292–301.
25. Mathew MK, Ajithkumar S. Sub acute ruminal acidosis and its effects on production. J Agric Vet Sci. 2014;7:63–5.
26. Wood T, Ramos-Morales E, McKain N, Shen X, Atasoglu C, Wallace R. Chrysanthemum coronarium as a modulator of fatty acid biohydrogenation in the rumen. Anim Feed Sci Tech. 2010;161:28–37.
27. Dong H, Wang S, Jia Y, Ni Y, Zhang Y, Zhuang S, et al. Long-term effects of subacute ruminal acidosis (SARA) on milk quality and hepatic gene expression in lactating goats fed a high-concentrate diet. PLoS One. 2013;8:e82850.
28. Livak KJ, Schmittgen TD. Analysis of relative gene expression data using real-time quantitative PCR and the 2-ΔΔCT method. Methods. 2001;25:402–8.
29. Zheng W, Hou Y, Yao W. Lactulose increases equol production and improves liver antioxidant status in barrows treated with Daidzein. PLoS One. 2014;9:e93163.
30. Wu X, Huang K, Wei C, Chen F, Pan C. Regulation of cellular glutathione peroxidase by different forms and concentrations of selenium in primary cultured bovine hepatocytes. J Nutr Biochem. 2010;21:153–61.
31. Wang D, Wang LJ, Zhu FX, Zhu JY, Chen XD, Zou L, et al. In vitro and in vivo studies on the antioxidant activities of the aqueous extracts of Douchi (a traditional Chinese salt-fermented soybean food). Food Chem. 2008;107:1421–8.
32. Chen X, Ren F, Hesketh J, Shi X, Li J, Gan F, et al. Interaction of porcine circovirus type 2 replication with intracellular redox status in vitro. Redox Rep. 2013;18:186–92.
33. Kleen J, Hooijer G L, Rehage J A, Noordhuizen JP. Subacute ruminal acidosis (SARA): a review. J Vet Med A Physiol Pathol Clin Med. 2003;50:406–14.
34. Steele MA, Croom J, Kahler M, AlZahal O, Hook SE, Plaizier K, et al. Bovine rumen epithelium undergoes rapid structural adaptations during grain-induced subacute ruminal acidosis. Am J Physiol Regul Integr Comp Physiol. 2011;300:R1515–23.
35. Dong G, Liu S, Wu Y, Lei C, Zhou J, Zhang S. Diet-induced bacterial immunogens in the gastrointestinal tract of dairy cows: impacts on immunity and metabolism. Acta Vet Scand. 2011;53:48.
36. Bilal MS, Abaker JA, Ul Aabdin Z, Xu T, Dai H, Zhang K, et al. Lipopolysaccharide derived from the digestive tract triggers an inflammatory response in the uterus of mid-lactating dairy cows during SARA. BMC Vet Res. 2016;12:284.
37. Schlau N, Guan LL, Oba M. The relationship between rumen acidosis resistance and expression of genes involved in regulation of intracellular pH and butyrate metabolism of ruminal epithelial cells in steers. J Dairy Sci. 2012;95:5866–75.
38. Huhtanen P, Miettinen H, Ylinen M. Effect of increasing ruminal butyrate on milk yield and blood constituents in dairy cows fed a grass silage-based diet. J Dairy Sci. 1993;76:1114–24.
39. Norton B, Janes A, Armstrong D. The effects of intraruminal infusions of sodium bicarbonate, ammonium chloride and sodium butyrate on urea metabolism in sheep. Br J Nutr. 1982;48:265–74.
40. Dionissopoulos L, Laarman AH, Alzahal O, Greenwood SL, Steele MA, Plaizier JC, et al. Butyrate-mediated genomic changes involved in non-specific host defenses, matrix remodeling and the immune response in the rumen epithelium of cows afflicted with subacute ruminal acidosis. Am J Anim Vet Sci. 2013;8:8.
41. Laarman AH, Alzahal O, Mcbride BW, Dionissopoulos L, Steele MA, Greenwood SL, et al. Butyrate supplementation affects mRNA abundance of genes involved in glycolysis, oxidative phosphorylation and lipogenesis in the rumen epithelium of Holstein dairy cows. Am J Anim Vet Sci. 2013;8:239.
42. Herrick K, Hippen A, Kalscheur K. Schingoethe D. Metabolism of ruminally-dosed butyrate and lactose in lactating dairy cows. J dairy Sci. E-Suppl. 2011;1:64.
43. Mimura Y, Sakisaka S, Harada M, Sata M, Tanikawa K. Role of hepatocytes in direct clearance of lipopolysaccharide in rats. Gastroenterology. 1995;109:1969–76.
44. Bieghs V, Trautwein C. The innate immune response during liver inflammation and metabolic disease. Trends Immunol. 2013;34:446–52.
45. Ma X, Fan PX, Li LS, Qiao SY, Zhang GL, Li DF. Butyrate promotes the recovering of intestinal wound healing through its positive effect on the tight junctions. J Anim Sci. 2012;90:266–8.
46. Vanhoutvin SA, Troost FJ, Hamer HM, Lindsey PJ, Koek GH, Jonkers DM, et al. Butyrate-induced transcriptional changes in human colonic mucosa. PLoS One. 2009;4:e6759.
47. Sauer J, Richter KK, Pool-Zobel BL. Physiological concentrations of butyrate favorably modulate genes of oxidative and metabolic stress in primary human colon cells. J Nutr Biochem. 2007;18:736–45.
48. Watson AM, Warren G, Howard G, Shedlofsky SI, Blouin RA. Activities of conjugating and antioxidant enzymes following endotoxin exposure. J Biochem Mol Toxicol. 1999;13:63–9.
49. Guo Y, Xu X, Zou Y, Yang Z, Li S, Cao Z. Changes in feed intake, nutrient digestion, plasma metabolites, and oxidative stress parameters in dairy cows with subacute ruminal acidosis and its regulation with pelleted beet pulp. J Anim Sci Biotechnol. 2013;4:1.
50. Zhang WH, Gao F, Zhu QF, Li C, Jiang Y, Dai SF, et al. Dietary sodium butyrate alleviates the oxidative stress induced by corticosterone exposure and improves meat quality in broiler chickens. Poult Sci. 2011;90:2592–9.

51. Peavy DL, Fairchild EJ. Evidence for lipid peroxidation in endotoxin-poisoned mice. Infect Immun. 1986;52:613–6.
52. Slatter DA, Bolton CH, Bailey AJ. The importance of lipid-derived malondialdehyde in diabetes mellitus. Diabetologia. 2000;43:550–7.
53. Niture SK, Khatri R, Jaiswal AK. Regulation of Nrf2-an update. Radic Biol Med. 2014;66:36–44.
54. Pool-Zobel BL, Selvaraiu V, Sauer J, Kautenburger T, Kiefer J, Kiefer J, Richter KK, et al. Butyrate may enhance toxicological defence in primary, adenoma and tumor human colon cells by favourably modulating expression of glutathione S-transferases genes, an approach in nutrigenomics. Carcinogenesis. 2005;26:1064–76.
55. Millard AL, Mertes PM, Ittelet D, Villard F, Jeannesson P, Bernard J. Butyrate affects differentiation, maturation and function of human monocyte-derived dendritic cells and macrophages. Clin Exp Immunol. 2002;130:245–55.
56. Ahmad MS, Krishnan S, Ramakrishna BS, Mathan M, Pulimood AB, Murthy SN. Butyrate and glucose metabolism by colonocytes in experimental colitis in mice. Gut. 2000;46:493–9.
57. Russo I, Luciani A, De Cicco P, Troncone E, Ciacci C. Butyrate attenuates lipopolysaccharide-induced inflammation in intestinal cells and Crohn's mucosa through modulation of antioxidant defense machinery. PLoS One. 2012;7:e32841.
58. Yaku K, Enami Y, Kurajyo C, Matsui-Yuasa I, Konishi Y, Kojima-Yuasa A. The enhancement of phase 2 enzyme activities by sodium butyrate in normal intestinal epithelial cells is associated with Nrf2 and p53. Mol Cell Biochem. 2012;370:7–14.

23

Why do farmers and veterinarians not report all bovine abortions, as requested by the clinical brucellosis surveillance system in France?

Anne Bronner[1*], Viviane Hénaux[1], Nicolas Fortané[2], Pascal Hendrikx[3] and Didier Calavas[1]

Abstract

Background: Since 2005, France has been officially free of brucellosis, an infectious disease that causes abortion in cattle and can be transmitted from cattle to humans. Recent animal and human cases have drawn attention to the need to prevent infection of humans and animals from any primary outbreaks. In order to detect any new outbreaks as soon as possible, a clinical surveillance system requires farmers and veterinarians to report each abortion and to test the aborting cow for brucellosis. However, under-reporting limits the sensitivity of this system. Our objective was to identify the barriers and motivations influencing field actors in their decision to report or not to report bovine abortions. We used a qualitative approach with semi-structured interviews of 12 cattle farmers and their eight veterinarians.

Results: Our analysis showed that four main themes influence the decision-making process of farmers and veterinarians: 1) the perceived risk of brucellosis and other abortive diseases; 2) the definition of a suspected case of brucellosis and other abortive diseases adopted by field actors, which is less sensitive than the mandatory definition; 3) the cost-benefit analysis conducted by actors, taking into account regulatory and health aspects, economic and financial losses, technical and practical factors; 4) the level of cooperation within the socio-technical network. We discussed how early detection may be improved by revising the definition of abortion, extending the time frame for notification and generalising the differential diagnosis of the causes of abortion.

Conclusions: In contrast to quantitative approaches, qualitative studies can identify the factors (including unknown factors) influencing the decision-making process of field actors and reveal why they take those factors into consideration. Our qualitative study sheds light on the factors underlying the poor sensitivity of clinical brucellosis surveillance system for cattle in France, and suggests that early detection may be improved by considering actors' perceptions. We believe our findings may provide further insight into ways of improving other clinical surveillance systems and thus reduce the risk of disease.

Background

In the context of increasing cross-border movements of people and growing international trade of animals and animal products, the identification of health hazards before they emerge and spread is of utmost importance for both human and animal communities. From the late 1990s, the number of emerging and re-emerging infectious diseases has dramatically increased [1]. About 75% of the new diseases that have affected humans over the past ten years have been caused by zoonotic pathogens, i.e. pathogens that can be transmitted from animals (the main reservoir of the disease) to humans [2]. It is thus essential to detect any outbreak of a disease in animals as early as possible to prevent primary sources spreading the disease to other animals or humans. Clinical surveillance systems have therefore been implemented for several animal diseases, all of them relying on the mandatory notification of suspected clinical cases by farmers and veterinarians. However, under-reporting is regularly cited as one of the main limitations of these networks [3,4]. In order to improve the sensitivity of these surveillance systems, it is essential to understand

* Correspondence: anne.bronner@anses.fr
[1]Anses-Lyon, Unité Epidémiologie, 31 avenue Tony Garnier, 69364 Lyon Cedex 07, France
Full list of author information is available at the end of the article

the decision-making process of field actors, and identify the factors they perceive as incentives or barriers to reporting suspected cases.

Brucellosis in cattle is an infectious disease caused by *Brucella abortus* (and less frequently to *B. melitensis* and *B. suis*) which primarily affects the reproductive organs of infected animals. It can be transmitted to humans, causing a febrile syndrome and complications such as orchitis, endocarditis or arthritis. In France, the disease was eradicated in the bovine population in 2003 and the country has been declared officially free of the disease since 2005. However, there remains a chance that the disease will reoccur: in 2012, two bovine brucellosis outbreaks were detected [5]. The first, in a beef cattle herd, was due to the introduction of an infected animal from Belgium, another officially disease-free country. The second, in a dairy cattle herd, may have been infected by wildlife (unpublished results). This second outbreak led to a human case in a child who ate raw milk cheese produced from the infected herd. This human case was diagnosed three months before the disease was detected on the cattle farm [5,6].

These recent cases remind us that the early detection of primary cases and prevention of any bovine brucellosis outbreaks remain crucial for both public and animal health considerations. Once introduced into a cattle herd, abortion is not only the main clinical sign of the disease, but also its main source of dissemination, large quantities of bacteria being excreted in the foetus and uterine fluids [7]. The clinical surveillance system therefore relies on the mandatory notification of any abortions. According to national regulations, farmers have to call their sanitary veterinarian (mandated by veterinary services to carry out regulatory interventions such as vaccination or sample collection) in the event of a bovine abortion, defined by the French "Code Rural" as the expulsion of the foetus or calf, stillborn or dying within 48 hours of birth [8]. The sanitary veterinarian then has to report the abortion and take a blood sample from the aborting cow to test for *Brucella* spp. In practice, an abortion is reported to the veterinary services when the blood sample is sent to the departmental laboratory (a French department being an administrative and territorial unit with a mean surface area of 5,800 km^2). The farm where a suspected case has been reported is not isolated and both the veterinarian's visit and brucellosis analysis are financed by public funds. The failure to report a detected abortion is punishable by a fine of 1,500 euros [9]. However, in practice, it is extremely difficult to identify farmers and veterinarians not complying with regulations, so this sanction is never actually applied. Clinical surveillance of brucellosis is complemented by active surveillance, which consists of annual serological tests of each herd based on either bulk milk samples from dairy cattle or serum samples from 20% of beef cattle over 24 months old. However, its objective is more to prove the official disease-free status of France for bovine brucellosis than to ensure early detection [10].

Abortion is not a clinical sign specific to brucellosis. Several abortive endemic diseases such as Q fever, neosporosis and bovine viral diarrhoea may also cause abortions and thus direct economic losses for farmers. To help identify the cause of abortion, the GDS animal health groupings (*Groupements de Défense Sanitaire*, a departmental association of stock farmers addressing health issues, recognised in an official capacity under French law) for some departments have developed a differential abortion diagnosis protocol including alternative abortive diseases endemic to France. The GDS funds part of the analyses should the differential diagnosis protocol be followed.

However, despite national regulations and the importance to farmers of preventing health and animal risks related to brucellosis and other abortive diseases, the under-reporting of bovine abortion remains of major concern. In a previous study using capture-recapture methods, we found that the overall surveillance sensitivity, i.e. the proportion of farmers who reported abortion(s) out of all the farmers who had detected abortion(s), was about 20% for beef cattle herds and 39% for dairy cattle herds [3].

It is crucial to assess the willingness of farmers and veterinarians to participate in the system and the constraints that may influence their decision in order to identify the best ways of improving brucellosis risk management [11]. Yet, although effective and reliable surveillance requires motivated participants, there is still much research to be done on the social aspects of participation in animal disease surveillance systems [12]. Previous studies on the participation of these actors in clinical surveillance systems for avian influenza or scrapie in small ruminants have highlighted several potential barriers to reporting. These include a lack of knowledge and awareness of the disease; guilt, shame and prejudice; a negative opinion of control measures; dissatisfaction with post-reporting procedures; a lack of trust in veterinarians and government; a lack of transparency in reporting procedures, and finally, uncertainty about the notification process [13-15]. However, in our case, we hypothesise that these reasons do not fully explain the low abortion reporting rate because of the existence of precise criteria to define a suspected case and the absence of farm isolation after notification.

The objectives of this study were to understand farmers' and veterinarians' decision-making process when choosing whether or not to report abortions, and to analyse the role of differential diagnosis in this process. We used a qualitative approach with semi-structured interviews of field actors. Open questions were designed to encourage them to talk about their attitudes and perceptions [16]. We sought to identify the factors influencing their decision, potential interactions among those factors, and to contextualise the reporting or non-reporting decision in terms of multiple factors, such as institutional, social, psychological, technical or economic factors. Ultimately, these consultations with field actors were designed to reveal the barriers and motivations influencing the decision of farmers and veterinarians to participate in the clinical brucellosis surveillance system.

Methods

Study design

The study was conducted in two French departments named, for the purpose of this study, "A" and "B" and located in north-eastern and south-eastern France respectively. Department A hosted 1,674 cattle herds at the beginning of 2010 and the last bovine brucellosis outbreak was recorded in 1993. Department B hosted 4,177 cattle herds at the beginning of 2010 and the last bovine brucellosis outbreak was recorded in 2001. Taking into account production type and herd size, the proportion of farmers notifying abortions in department A did not differ significantly from the national level, whereas in department B, it was about 1.5 times higher than the national level. In both departments, the GDS animal health grouping provides financial support for differential diagnosis. In department B, a differential diagnosis protocol has to be followed for the farmer to benefit from this financial support.

Pre-selection of participants

In each department, representatives of the veterinary services, GDS and GTV technical veterinarian association were interviewed separately to gather information about certain aspects of the implementation of the mandatory abortion notification system (including the roles of departmental veterinary services and the GDS, relationships between actors involved in surveillance and information provided by the GDS to farmers) and the differential diagnosis following abortion (for instance the existence of a protocol and extent of financial support). According to the "grounded theory" approach [17], participants were chosen purposively to include a variety of herd characteristics and different attitudes to having to report abortions: farmers were pre-selected taking into account the herd production type (dairy, beef or a mixture), the number of abortion(s) reported over the last two years, and participation in cattle performance recording programmes. In these programmes, farmers are subject to frequent checks on cattle performance, including milk production for dairy cattle, and maternal qualities or morphology for beef cattle. Committed to the long-term improvement of their herd, these farmers are assumed to have better farming practices and thus be more prone to reporting abortions than other farmers.

The GDS contacted potential participants by phone to provide information on the purpose and nature of the study. One farmer wished not to be interviewed due to lack of time and one veterinarian refused categorically to participate in the study (and did not explain why). Participants who agreed to be interviewed were then contacted by phone by the person in charge of the study. The aim, nature and background of the study were explained in detail, and potential participants were informed that their data would remain anonymous and that any material potentially leading to individual identification would be removed. Once verbal consent was obtained, a time was arranged for the survey to be conducted. Farmers and veterinarians had no financial incentive for participating in the study.

Data collection

In-depth interviews were conducted by the same person at the participants' location of choice (office or home) from October to December 2012. At the beginning of the interview, the aim and background of the study were recalled, as was the confidentiality of the interview. These issues were presented in a document that was given to each participant. In accordance with [18], participants were again asked to provide verbal informed consent prior to the interview. It was made clear that by agreeing to be interviewed, they were agreeing to be part of the study. All participants agreed to the interview being recorded.

Interviews lasted between 50 and 105 minutes. Through open-ended questions, the interviewer asked the farmer or veterinarian to talk about their knowledge of abortive diseases, their perception of this issue, and the difficulties and barriers to participating in the surveillance system (Table 1). In all, 12 farmers and their eight sanitary veterinarians were involved in the study; a sample size justified by interviewing participants until "theoretical saturation" [17] was achieved (i.e. no novel idea was raised during the most recent interviews). For all but one, the farmers' sanitary veterinarian was the same as their private veterinarian practitioner who treated their animals. The characteristics of the interviewed farmers are provided in Table 2.

Data analysis

The interview notes were transcribed from the audio recording and analysed using thematic analysis [19]. They were reviewed and a code given to each key word or sentence. Similar codes were grouped into categories, and categories were gathered into sub-themes and themes [19]. The credibility and rigor of the analysis

Table 1 Topics of discussion during farmer and veterinarian in-depth interviews

Participants	Topics of discussion
Farmers	Number of abortions detected in the last two years, circumstances of detection
Veterinarians	Circumstances of a farmer's call in the event of abortion(s)
Farmers and veterinarians	Definition of abortion
	Measures taken in the event of abortion and reasons
	Knowledge about the mandatory abortion notification system, and the differential diagnosis protocol and/or financial and technical support by the GDS
	Type of information obtained about bovine abortions, the mandatory bovine abortion notification system and differential diagnosis actions
	Expectations about abortion surveillance actions or information

were aided by co-analysis of transcripts by two of the researchers and continual re-examination of the emergent data throughout the research process. Discrepancies in interpretation were discussed and resolved during consensus meetings. All the data presented in the results section reflect the observations, insights and opinions expressed by participants. In addition, a typology of the interviewed farmers was constructed to study the main factors driving their decision-making process based on two considerations: their perception of risk and their attitude towards abortion notification. They were scored on these two considerations as negative or positive, i.e. low versus high perceived risk of introducing an animal disease into their herd for the first consideration, and low versus high rate of notification of detected abortions for the second consideration.

Data collection and analysis were conducted in accordance with the research ethics requirements of the American Sociological Association [18] and qualitative research review guidelines (Additional file 1).

Results

Factors influencing farmers' and veterinarians' decision-making process

We defined health risk as the likelihood of the occurrence of a negative health event for people or animals. Four main themes emerged from the analysis of farmers' and veterinarians' interviews: their perception of the risk of brucellosis and other abortive diseases; the definition of a suspected case related to the risk of brucellosis and other abortive diseases; cost-benefit analysis and socio-technical factors.

Theme 1: Risk perception of brucellosis and other causes of abortion

Brucellosis was perceived by most farmers and veterinarians as a serious disease, particularly when they had experienced brucellosis in the past. However, the probability of an outbreak occurring was usually perceived as negligible, i.e. most farmers did not fear a potential introduction of brucellosis when an abortion occurred in their herd. The detection of two brucellosis cattle outbreaks and one human case in France in 2012 did not change their perception of this risk because these cases were distant (i.e. not in a neighbouring department). In contrast, some farmers were conscious of the potential role of wildlife in the transmission of diseases

Table 2 Characteristics of farmers

Production type[1]	Start of the farmer's activity	Number of breeding cattle	Number of reported abortions (detected abortions[2])		Department
			in 2011	in 2012	
Dairy	2005	70-80	0 (3)	1 (1)	A
Dairy*	1990	130	5 (5)	4 (4)	A
Dairy*	1990	130-150	1 (9)	2 (2)	A
Dairy*	2008	65	0 (1)	0 (0)	A
Mixed	1992	65	1 (1)	0 (4)	A
Mixed*	1997	270	0 (about 14)	0 (about 14)	A
Mixed	1992	30-40	0 (5)	0 (2)	A
Mixed*	1991	110	0 (2)	0 (5 to 8)	A
Mixed*	2003	160	2 (3–4)	0 (1)	B
Beef*	1993	65-70	0 (1)	0 (0)	B
Beef*	2006	50	0 (0)	7 (7)	B
Beef	2001	85-90	0 (1)	1 (1)	B

[1]Herds included beef cattle, dairy cattle, or a mixture of both types of production.
[2]The number of detected abortions was collected during the interviews with farmers.
*Farmers participating in the cattle performance recording programme.

(e.g. tuberculosis) to cattle and feared a brucellosis infection of their cattle herd by wild animals, or the introduction of an endemic disease such as bovine viral diarrhoea by contact with an infected neighbouring cattle herd. In these cases, farmers usually requested a test for brucellosis or another disease.

Theme 2: Definition of a suspected case related to the risk of brucellosis and other abortive diseases

The interviews revealed that the word "abortion" had a negative connotation. Farmers and veterinarians defined an abortion as the direct observation of a foetus or placental tissues that had just been expelled from the cow, which is a more restrictive definition than the official one (Figure 1). Although a cow returning to heat more than 42 days after an attempt at fertilisation may be a sign of an interruption of pregnancy, farmers and veterinarians did not usually consider this event as the consequence of an abortion, arguing that unsuccessful mating causes returns to heat that may be detected late (more than 42 days after an attempt at fertilisation). Thus, as most of the abortions occurring before 5–6 months of pregnancy or those in beef cows at pasture are usually detected by a return to heat, these events were not reported. Furthermore, farmers considered that abortion was the fault of the cow, whether due to an abortive disease or other cause. Stillbirths or death of the calves soon after birth, which are also officially considered abortions, were not reported because they were believed to result from calving difficulties or calf illness rather than from an abortive disease.

Abortion (as defined by farmers and veterinarians) was considered a normal event as long as it remained sporadic and under a "threshold" proportion in the herd, which ranged among interviewed farmers from 1.5 to 5% a year. Most farmers were confident in diagnosing abortions themselves, with common causes including accidents or feed-, medication- or health-related issues. Therefore, they did not feel the need to consult their veterinarian in the event of a sporadic abortion or if a non-infectious cause was suspected. All the farmers and even some veterinarians were more prone to carry out biological analyses in the event of recurrent abortions: one veterinarian was not in the habit of reporting abortion in the case of a return to heat, but mentioned reporting abortions on a farm where more than half of a group of heifers had returned to heat.

Theme 3: Cost-benefit analysis

Farmers and veterinarians weighed the benefits and costs of both reporting abortions and conducting a differential

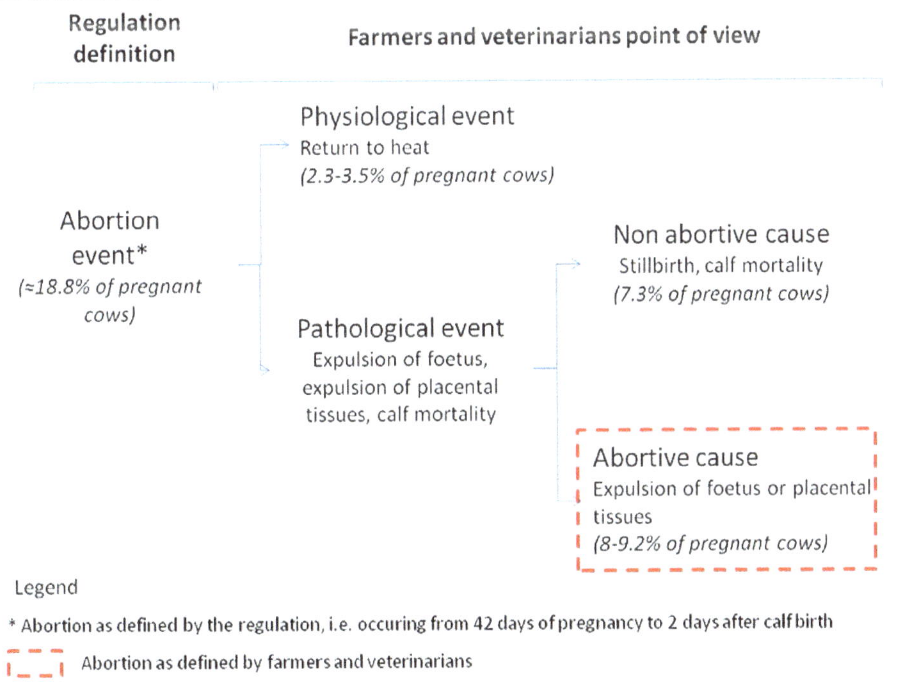

Figure 1 Difference between the official definition of abortion and the definition of farmers and veterinarians. Abortion is defined by French regulations as an interruption of pregnancy occurring from 42 days of pregnancy to term, or as the death of a calf within 48 hours of its birth. A recent study of the time between artificial insemination and calving in dairy cattle estimated that the rates of abortion occurring in mid-pregnancy and late pregnancy were about 6.4% and 5.1% respectively [20]. As only 20 to 30% of abortions are detected visually [21], only 70% to 80% of these aborting cows are detected by field actors, i.e. 8 to 9.2% of pregnant cows. Furthermore, the 7.3% or so of calves that die within 48 hours after birth are supposed to be systematically detected [22].

diagnosis. Multiple factors, including regulations, health, economic, financial, technical, and practical considerations, were taken into account in that process (Tables 3 and 4). The role of regulations appears to be mitigated by the absence of sanctions and the perceived lack of relevant measures to ensure early detection of brucellosis. Farmers and veterinarians did not report abortions in order to meet the objective of the brucellosis surveillance system—detecting an outbreak as soon as possible—but to comply with their direct preoccupations. They were more concerned about their animals' health (for economic and professional reasons) than about public health issues.

The possibility of identifying the cause of abortion through a differential diagnosis motivated their decision to report. However, technical difficulties such as a lack of training, no standard differential diagnosis protocol, and absence of the placenta or foetal material make identifying the cause harder and thus discouraged some farmers and veterinarians from reporting abortions. Therefore, some veterinarians conduct a differential diagnosis test only if the placenta or foetal material is available. Furthermore, veterinarians pointed out the cost to farmers of additional analyses as a barrier to evaluating other abortive diseases, though most farmers mentioned that financial factors did not greatly influence their decision when they were worried about the occurrence of abortions. Nevertheless, their cost-benefit analysis varied with the herd production type: beef cattle farmers mentioned the practical difficulties in catching a cow at pasture for a serological analysis.

Theme 4: Socio-technical network

As sanitary veterinarians are mandated by French authorities, they were more prone to apply the mandatory measures than farmers. However, they underlined the difficulty of convincing reluctant farmers to report abortions. Furthermore, some veterinarians were positive about their relationships with veterinary services, others criticised the absence of technical support should they have difficulties in identifying the cause of abortion, and the lack of information about surveillance results (Table 5). The decision to report an abortion and carry out the differential diagnosis protocol was either the result of a consensus between the farmer and his veterinarian, or the initiative of the veterinarian. However, in the case of abortions, most farmers did not know if a differential diagnosis test was conducted or which diseases were included. Because of the possible negative consequences of a positive result for the farmer and to preserve their relationship with their clients, veterinarians took farmers' expectations and difficulties into consideration. This was to the detriment of their obligation to report abortion, their financial interest in participating in the surveillance system, and their technical interest in conducting differential diagnosis.

Typology of farmers

The classification of farmers according to risk perception and decision on abortion notification underlined the influence of regulation incentives, health incentives, practical difficulties, and poor integration in socio-technical networks (due to a lack of communication with their sanitary veterinarian and the GDS (Table 6). The perceived risk of brucellosis occurrence did not influence farmers' decision on abortion notification in a deterministic manner. Farmers who perceived the risk of introducing brucellosis or another enzootic abortive disease into their herd as low reported abortions to comply with the law except when practical issues predominated. In the latter case, one farmer suggested the possibility of reporting abortions only if a disease was detected in his

Table 3 Cost-benefit analysis by farmers deciding whether to call their veterinarian for an abortion

Factors	Benefits	Costs
Regulations	Professional conscientiousness	No added-value for the farmer as there is no enforcement
	Avoid sanctions	
		Lack of technical justification: they believed a brucellosis outbreak would be detected by a significant abortion episode or by active surveillance
Health	Identify cause of abortion	Difficulties in identifying the cause: one farmer stopped reporting abortions after an unsuccessful differential diagnosis to identify the cause
	Ensure the absence of a specific disease or diseases in general (some farmers were not aware that brucellosis is the only disease tested)	
	Care for the aborting cow	
Financial	Free visit	Financial costs of additional analyses and sanitary/medical measures to prevent further abortions
Economic	Prevent further abortions	Lower sales of animals from a herd with seropositive animals
Practical		Animal has to be caught
		Time-consuming

Table 4 Cost-benefit analysis by veterinarians deciding whether to report abortions and make a differential diagnosis

Factors	Benefits	Costs
Regulations	Professional conscientiousness Technical justification: brucellosis may cause late abortion (after six months of pregnancy	Lack of technical justification: a brucellosis outbreak would be detected by a significant abortion episode or by active surveillance
Technical	Technical interest in identifying the cause of the abortion	Difficulties in identifying the cause of abortion Lack of knowledge: when veterinarians carried out a differential diagnosis, they included known abortive diseases with a known diagnosis protocol and effective measures to reduce the occurrence of abortions; their diagnosis protocol sometimes differed from scientific requirements. One veterinarian did not report an abortion because he did not know which diseases other than brucellosis to include Low impact of sanitary and medical measures to prevent further abortions due to enzootic diseases
Financial		The farmer refused to pay additional costs for analyses and sanitary/medical measures to prevent further abortions
Practical		Time schedule with farmer Lack of time to seek advice about differential diagnosis

neighbourhood. In contrast, farmers who perceived the risk of introducing a disease into their herd as high did not consider abortion as a banal event and were better informed about diseases than the others (from GDS leaflets, information on the Internet or meetings organised by their veterinarian or GDS). However, some of them did not notify all abortions due a lack of trust in the sanitary veterinarian due to previous inconclusive diagnosis and expensive veterinary visits.

Discussion

The low sensitivity of the mandatory notification system for bovine abortion limits the early detection of any potential introduction of brucellosis or other abortive diseases into France [3], with potentially important animal and health risks. In that context, our qualitative study investigated the barriers and difficulties that hinder the notification of a suspected case, and possible solutions or incentives to encourage farmers' and

Table 5 Socio-technical factors taken into account by farmers and veterinarians in their decision-making process

Factor		Reasons for reporting abortions[1]	Reasons for not reporting abortions[1]
Farmer	Relationships with veterinary services and the GDS		Limited interactions Farmers did not feel responsible for early detection of brucellosis
	Relationships with veterinarians	Trust in the veterinarian's expertise (even though one farmer required further advice from the GDS) Explanations by the veterinarian of the advantages and limits of differential diagnosis	No trust in the expertise of the sanitary veterinarian and consultation of another practitioner in the event of health problems Difficulties due to the absence of consideration by veterinarians: for example, animal sales had been stopped for several weeks after a seropositive result obtained from a differential diagnosis about which the farmer had not been informed
Veterinarians	Relationships with veterinary services and the GDS	Role of the sanitary veterinarian	Absence of a technical added-value Feeling of being under the supervision of veterinary services and the GDS Dissatisfaction with veterinary services including lack of information on surveillance results, lack of technical training, lack of discussion about their difficulties
	Technical network	Some veterinarians have their own expert network	Lack of technical support should they have difficulties in identifying the cause of abortion
	Relationships with farmers		Blame farmers for not systematically consulting them in the event of abortion despite their messages to increase farmers' awareness Farmers' expectations and difficulties taken into account: no differential diagnoses were performed on farms where animals were sold abroad or if technical difficulties in determining the cause of abortion were feared

[1]For farmers, reasons for calling their veterinarian in the event of abortions; for veterinarians, reasons for reporting abortion and performing a differential diagnosis.

Table 6 Main factors driving the decision process of farmers according to risk perception and abortion notifications

		Rate of abortion notifications	
		Low	High
Perceived risk of introduction of an animal disease in their herd	Low	Practical difficulties	Respect for the law
	High	Lack of relationship	Sanitary and economical factors

The typology of interviewed farmers was based on two considerations: their perception of risk and their attitude towards abortion notification. They were scored on these two considerations as negative or positive, i.e. low versus high perceived risk of introducing an animal disease into their herd, and low versus high rate of abortion notification. On the basis of these two considerations, four groups of farmers were identified, being especially influenced by incentives, health and economic factors, practical difficulties, or poor integration in socio-technical networks respectively.

veterinarians' participation in brucellosis risk prevention. Although our study focused on the sensitive topic of fulfilling mandatory requirements, the confidentiality of the interviews warranted the trustworthiness of participants' answers [16], as suggested by farmers admitting that they report none or only some of the detected abortions (Table 2). A review of data on abortion notifications enabled us to validate farmers' assertions about their participation in the surveillance system.

Specificity of the qualitative approach

In contrast to a quantitative approach, commonly used in epidemiology to estimate the proportion of actors influenced by specific predetermined factors, the goals of our study were to identify which factors (including unknown factors) influence their decision and to understand why those factors are taken into consideration by field actors [23]. The qualitative approach relies on "purposive sampling" to maximise diversity [17] by covering a broad spectrum of experiences and positions relative to a given phenomenon [24]. In our study, we selected a wide range of participants with different herd characteristics and a variety of attitudes towards their duty to report abortions. We found that "theoretical saturation" was reached after ten interviews for farmers and six interviews for veterinarians, which means that no new information was raised in the last interviews [17]. This finding is in agreement with other studies, for which saturation occurred within the first 12 interviews [25].

Purposive sampling and theoretical saturation ensured the robustness and trustworthiness of our study and enabled us to generalise the information provided by study participants [24] into four themes that may be used to interpret the decision-making process of other farmers and veterinarians or decision-making process concerning other clinical surveillance systems such as classical swine fever or foot and mouth disease [26]. In addition, within the four themes, factors influencing the decision-making process may be transferable to other related topics [26], as suggested by the similarity between our findings and concepts developed in other studies on notification decisions or behaviour regarding vaccination [27,28].

Farmers' and veterinarians' decision-making process for reporting abortions

The typology of farmers according to their risk perception and attitudes towards the abortion notification process underlines little diversity in the driving factors once these two considerations are taken into account [14,29,30]. Furthermore, the same themes appear for both farmers and veterinarians. This result suggests that despite profession-related differences (interests, knowledge, position in the surveillance network), these actors share common interests in participating in the surveillance system. This contrasts with their divergent views on other issues such as the implementation of vaccination and culling strategies to control animal epidemics, where farmers have more to lose than veterinarians [31].

The decision of farmers and veterinarians to report or not to report abortions is driven by three main considerations: the perceived risk of brucellosis (or other abortive diseases); an evaluation of the individual benefits and costs linked to the abortion notification and differential diagnosis; the socio-technical network to which the farmers and veterinarians belong. The first two concepts—risk perception and cost-benefit analysis—are the backbone of the Health Belief Model (HBM), which is a framework widely used to describe health-related actions [32].

Risk perception is essentially based on lay knowledge and "value and belief-oriented rationality" [33], especially for farmers who rely on their experience and regular observations of their cattle herd. We also noticed the role of experience for veterinarians, although their judgement depends primarily on their technical and scientific knowledge. In contrast to risk perception, cost-benefit analysis is essentially based on "purposive and instrumental rationality" [33]. Farmers and veterinarians considered calculated reasons (including regulations and health aspects, economic and financial losses, technical and practical factors) and their own self-interest to guide their decision. The importance of some factors and thus the outcome of the cost-benefit analysis varies with the type of production. Indeed, the technical features of beef production compared to dairy cattle [34] reduce beef cattle farmers' *margin for manoeuvre,* as they are bound by both practical and financial constraints. The grazing season increases their difficulties in detecting abortions and carrying out a serological analysis on cows, as they do not have the daily contact that dairy cattle farmers do with their cows. Moreover, in the event of an abortion, the cow is usually immediately culled and sent to the slaughterhouse, and farmers do not feel the need to notify the abortion.

Besides risk perception and cost-benefit analysis, there is a third consideration. The French veterinary services, GDS, GTV, sanitary veterinarians and farmers are all linked through a social and technical network [35]. The mandatory abortion notification surveillance system defines what should be done by each actor to facilitate abortion notification and ensure early detection of a brucellosis outbreak. However, even if farmers are theoretically best placed to report abortions and benefit the most from the surveillance system, it does not make sense for most of them to be required to report abortion as a public duty. This reluctance to report abortions may be compared to the reluctance of humans to vaccinate themselves [28]: in both cases, the decision-making process of most people is not driven by the need to fulfil a public duty but by self-interest. Moreover, some farmers decide not to report an abortion because other farmers do not (peer influence), considering they have performed the necessary cost-benefit analysis to make a wise choice [27,36].

Perception of brucellosis risk and preventive measures

Like most clinical surveillance systems, the mandatory abortion notification system gives priority to sensitivity rather than specificity, given the sensitive definition of a suspected case. However, farmers and veterinarians do not feel the need to report every abortion to detect a brucellosis outbreak early, as most of them perceive the risk of brucellosis as low. The epidemiological situation regarding brucellosis in France has dramatically changed since the implementation of the surveillance system in 1965, when brucellosis was enzootic. Today, in a brucellosis-free context, farmers are more concerned about abortive enzootic diseases such as bovine viral diarrhoea or Q fever than brucellosis. Furthermore, as it has been cited elsewhere [37], high levels of confidence reduce perceived risks. In our case, farmers feel confident in their ability to detect a brucellosis case in their herd or in the effectiveness of the annual serological surveillance system for dairy and beef cattle herds. Abortion is a non-specific clinical sign that may or may not be caused by an infectious disease. Thus, reporting the abortion, illness, or death of a single animal does not make sense to them if the objective is to detect an infectious disease [13].

These pragmatic feelings are supported to some extent by the results of a simulation study which showed that abortion notification (with a reporting rate varying from 20% to 80%) tends to be a less effective method for the early detection of brucellosis than annual bulk milk surveillance [38]. However, some infected animals may not be identified by serological tests as infected cows may be serologically negative for several months until they give birth to a calf or abort [7,39]. Moreover, brucellosis infection may spread slowly, depending on how the disease is introduced into the herd (from an infected animal or by indirect transmission) and herd management factors, which may influence the amount of contact the animals have with each other, such that only a few animals in the herd may become seropositive or abort, even though brucellosis is a contagious disease [5,40].

Prospects for improving brucellosis risk prevention

The mandatory notification of each abortion is viewed by most farmers as an externally imposed tool for an externally imposed issue (the need to detect a brucellosis outbreak early) that they are not actually worried about [41]. However, the multiple patterns of brucellosis transmission and dissemination underline the importance of monitoring abortions. In addition, this system is useful for the surveillance of other diseases causing bovine abortion such as Q fever or Rift Valley Fever. Efforts by veterinary services and the GDS to increase farmers' and veterinarians' awareness about the need to report abortions has not reduced the probability of under-reporting, which has remained stable since 2009 [3]. As underlined by psychologists, messages to change people's attitude require three factors: a credible communicator, a high level of "similarity" between the audience and communicator, and finally both the message and communicator must be perceived as trustworthy [42,43]. The "traditional" communication strategy is based on rational arguments, but requires the actors' basic motivation and an interest in the topic [44]. Our findings suggest that the difficulties in increasing abortion notification rates may be related to the low level of trust in the communicators (veterinary services and GDS) and the low level of concern about abortions as long as they remain sporadic. Moreover, we found that both farmers and veterinarians do not agree with the decision to report every single abortion, which they consider irrelevant and of no real use [45]. Thus, enhancing risk communication requires not only relying on the "rational choice" model of decision but also taking into account the actors' social context as well as their values, beliefs, and how much they trust the different sources of information [46].

Our analysis suggests that putting a great deal of effort into increasing the number of reported abortions without updating the surveillance procedure would be inefficient. First, brucellosis infection causes mid-term or late abortions, which occur between five months and the end of pregnancy [7]. Second, the current situation regarding brucellosis in France suggests that the human and financial resources required to meet national and international regulations, i.e. testing about 18.8% of cows (which suffered mid-term or late abortions, or had a calf that died within 48 hours of birth, [20,47]), seem disproportionate in comparison with the number of brucellosis outbreaks that would be detected promptly (one outbreak in 2003 and two cases in 2012). Third, the performance of such a system will always rely on farmers' and veterinarians' willingness to participate in it, and these field actors are not prone to report cows returning to heat or the death of newborn calves. Therefore, there is a real need to revise the surveillance procedure so that the system is more efficient. Considering OIE requirements [48], these changes need to be considered on an international scale. An initial suggestion would be to revise the definition of abortion by excluding stillborn and newborn calves. Furthermore, a brucellosis analysis could be requested only beyond a certain frequency threshold of abortions within the herd. Farmers could be requested to register each abortion themselves (in their own record or in a national record system) but call their veterinarian only in the event of recurrent abortions (for example, two abortions or more within a month). Besides, cows suspected to have aborted could be sampled after a certain delay should the abortion not be detected early enough, or if practical issues hinder the immediate sampling of the cow.

By helping farmers and veterinarians identify the cause of abortion, the differential diagnosis with respect to other abortive enzootic diseases was supposed to improve their willingness to contribute to the surveillance system and participate in brucellosis risk prevention. However, it has been estimated that the cause was only diagnosed for about one third of the biological submissions [49]. The difficulties in identifying the abortive pathogen arise from the wide range of potential pathogens, the ubiquity of pathogens such as Q fever or salmonellosis, and veterinarians' lack of knowledge about the type of samples to collect for the analyses. Therefore, a differential diagnosis protocol has been recently drawn up nationally in order to help veterinarians with the sampling process and identification of the cause of the abortion. Improving the coordination of veterinarians by official bodies providing technical support, training and information on the results of the differential diagnosis protocol (in addition to the results from the mandatory surveillance system) is also expected to increase participation in the surveillance system [12].

These prospects for improving brucellosis risk prevention could be useful for improving other clinical surveillance systems. Indeed, while some factors related to the reporting bias are specific to brucellosis or cattle production, most of them may influence the decision-making process for other diseases. Many exotic diseases with clinical surveillance are seen as low-risk, and the decision-making process is likely to be driven by regulations, health, economic, financial and technical factors. Thus, we expect that farmers and veterinarians would be more likely to report avian influenza or classical swine fever suspicions in the event of high mortality or morbidity than if only a few animals fall ill or die (which is the case if the outbreak is due to low pathogenic AI virus or pestivirus) [14]. Enhancing risk communication, developing differential diagnosis and revising the surveillance procedure to find the best compromise between sensitivity, expenditure and acceptability are ways of improving clinical surveillance systems that should be taken into consideration.

Nonetheless, performance will always rely on farmers' and veterinarians' participation, and consequently under-reporting will remain the major limitation of such systems. In this context, it might be useful to develop other surveillance procedures, such as syndromic surveillance. As regards brucellosis, it might be useful to develop an indicator to identify the occurrence of abortions in cattle by using information from additional sources, such as the dates of artificial insemination or calving intervals.

Conclusions

According to the U.S. Centres for Disease Control and Prevention, the acceptability of public health surveillance systems by participants is one of the attributes to be assessed when evaluating a surveillance system [11]. To our knowledge, our study is the first to investigate the factors underlying the participation of farmers and veterinarians in a mandatory clinical surveillance system where there is no farm isolation in the event of notification of a suspected case; this issue is one of the main barriers to reporting in other passive surveillance systems [14]. Our qualitative study sheds light on the factors underlying the high proportion of under-reporting farmers, and differences in the reporting rates between dairy and beef cattle farmers [3]. Several recommendations, including revising the definition of a suspected case, extending the time frame for notification, and providing adequate diagnostic tools, support and information to field actors, may improve their participation in the surveillance system. We believe these incentives and measures should also be considered in other clinical surveillance systems to improve the rate of notification of suspected cases, facilitate the detection of emerging pathogens, and improve animal and public health risk prevention.

Competing interests

The authors declare that they have no competing interests.

Authors' contributions

AB performed the study and drafted the manuscript. NF, DC and PH participated in the conception, design and analysis of the study. VH contributed to the analysis and helped draft the manuscript. All the authors read and approved the final manuscript.

Acknowledgements

The authors wish to thank the representatives of the French departmental veterinary services, GDS, GTV and all the farmers and veterinarians involved in this study.

Author details

[1]Anses-Lyon, Unité Epidémiologie, 31 avenue Tony Garnier, 69364 Lyon Cedex 07, France. [2]Unité RiTME, Inra, 65 Boulevard de Brandebourg, 94 200 Ivry-sur-Seine, France. [3]Anses, Unité de surveillance épidémiologique SURVEPI, Direction scientifique des laboratoires, 22 Rue Pierre Curie, 94700 Maisons-Alfort, France.

References

1. Figuié M: **Towards a global governance of risks: international health organisations and the surveillance of emerging infectious diseases.** *J of Risk Res* 2013, **17**:469–483.
2. **Veterinary pubic health.** [http://www.who.int/zoonoses/vph/en/]
3. Bronner A, Hénaux V, Vergne T, Vinard J-L, Morignat E, Hendrikx P, Calavas D, Gay E: **Assessing the mandatory bovine abortion notification system in France using unilist capture-recapture approach.** *Plos one* 2013, **8**:e63246.
4. Doyle TJ, Glynn MK, Groseclose SL: **Completeness of notifiable infectious disease reporting in the United States: an analytical literature review.** *Am J Epidemiol* 2002, **155**:866–874.
5. **Country reports.** [http://www.oie.int/wahis_2/public/wahid.php/Countryinformation/Countryreports]
6. Mailles A, Rautureau S, Le Horgne JM, Poignet-Leroux B, D'Arnoux C, Dennetiere G, Faure M, Lavigne JP, Bru JP, Garin-Bastuji B: **Re-emergence of brucellosis in cattle in France and risk for human health.** *Euro Surveillance* 2012, **17**:2–4.
7. Nicoletti P: **The epidemiology of bovine brucellosis.** *Adv Vet Sci Comp Med* 1980, **24**:69–98.
8. Anonymous: **Article R. 223–79 du Code rural et de la pêche maritime.** 1965. JORF 15.03.2013.
9. Anonymous: **Article *L. 223-5 du* Code rural et de la pêche maritime.** 2003. JORF 07.08.2003.
10. Anonymous: **Arrêté du 22 avril 2008 fixant les mesures techniques et administratives relatives à la prophylaxie collective et à la police sanitaire de la brucellose des bovinés.** JORF 02.05.2008.
11. **Updated guidelines for evaluating public health surveillance systems - Recommendations from the Guidelines Working Group.** [http://www.cdc.gov/mmwr/preview/mmwrhtml/rr5013a1.htm]
12. Sawford K: *Animal health surveillance for early detection of emerging infectious disease risks.* Calgary: Department of Medical Science; 2011.
13. Palmer S, Fozdar F, Sully M: **The effect of trust on West Autralian farmers' responses to infectious livestock diseases.** *ESRS* 2009, **49**:4.
14. Elbers A, Gorgievski-Duijvesteijn M, Zarafshani K, Koch G: **To report or not to report: a psychosocial investigation aimed at improving early detection of avian influenza outbreaks.** *Rev Sci Tech Off Int Epiz* 2010, **29**:435–449.
15. Hopp P, Vatn S, Jarp J: **Norwegian farmers' vigilance in reporting sheep showing scrapie-associated signs.** *BMC Vet Res* 2007, **3**:34.
16. Quivy R, Campenhoudt L: *Manuel de recherche en sciences sociales.* Paris: Dunod; 2006.
17. Glaser B, Strauss A: *The discovery of grounded theory. Strategies for qualitative research.* New York: Aldine de Gruyter; 1967.
18. **American Sociological Association, Code of Ethics and policies and procedures.** 2008. http://www.asanet.org/. 30 pp.
19. Graneheim U, Lundman B: **Qualitative content analysis in nursing research: concepts, procedures and measures to achieve trustworthiness.** *Nurs Educ Today* 2004, **24**:105–112.
20. Nusinovici S, Seegers H, Joly A, Beaudeau F, Fourichon C: **Increase in the occurence of abortions associated with exposure to the Bluetongue virus serotype 8 in naïve dairy herds.** *Theriogenology* 2012, **78**:1140–51.
21. Forar A, Gay J, Hancok D: **The frequency of endemic fetal loss in dairy cattle: a review.** *Theriogenology* 1995, **43**:989–1000.
22. Raboisson D, Delor F, Cahuzac E, Gendre C, Sans P, Allaire G: **Perinatal, neonatal, and rearing period mortality of dairy calves and replacement heifers in France.** *J Dairy Sci* 2013, **96**:2913–2924.
23. Vaarst M, Paarup-Laursen P, Houe H, Fossing C, Andersen H: **Farmers' choice of medical traitment of mastitis in danish dairy herds based on qualitative research interviews.** *J Dairy Sci* 2002, **85**:10.
24. Côté L, Turgeon J: **Comment lire de façon critique les articles de recherche qualitative en médecine.** *Pédagogie Médicale* 2002, **3**:81–90.
25. Guest G, Bunce A, Johnson L: **How many interviews are enough? An experiment with data saturation and variability.** *Field Methods* 2006, **18**:59–82.
26. Polit D, Beck C: **Generalization in quantitative and qualitative research: myths and strategies.** *Int J Nurs Stud* 2010, **47**:1451–1458.
27. Lam T, Jansen J, Borne B, Veersen J: **A structural approach of udder health improvement via private practitioners: ups and downs.** In *46th NMC Annual Meeting Proceedings.* San Antonio, Texas, USA; 2007:142–151.
28. Hobson-West P: **Understanding vaccination resistance: moving beyond risk.** *Health, Risk Soc* 2010, **5**:273–283.
29. Dockès A, Kling-Eveillard F: **Les représentations de l'animal et du bien-être animal par les éleveurs français.** *Inra Prod Anim* 2007, **20**:23–28.
30. Jansen J, Steuten C, Renes R, Aarts N, Lam T: **Debunking the myth of the hard-to-reach farmer: effective communication on udder health.** *J Dairy Sci* 2010, **93**:1296–1306.
31. Zinng A, Siegrist M: **Lay people's and experts' risk perception and acceptance of vaccination and culling strategies to figght animal epidemics.** *J Risk Res* 2012, **15**:53–66.
32. Rosenstock IM, Strecher VJ, Becker MH: **Social learning theory and the Health Belief Model.** *Health Educ Q* 1988, **15**:175–183.
33. Weber M: *Economy and society: an outline of interpretive sociology.* Berkley: University of California press; 1979.
34. Commandeur M: **Diversity of pig farming styles: understanding how it is structured.** *NJAS* 2006, **54**:111–127.
35. Strauss A, Schatzman L, Ehrlich D, Bucher R, Sabshin M: **The hospital and its negotiated order.** In *The hospital in modern society.* Edited by Freidson E. New York: Free Press; 1994:147–169.
36. Hershey J, Ash D, Thumasathit T, Meszaros J, Walters V: **The roles of altruism, free riding and bandwagoning in vaccination decisions.** *Organ Behav Hum Dec* 1994, **59**:177–187.
37. Siegrist M, Gutscher H, Earle T: **Perception of risk: the influence of general trust and general confidence.** *J Risk Res* 2005, **8**:145–156.
38. Yamamoto T, Tsutsui T, Nishiguchi A, Kobayashi S: **Evaluation of surveillance strategies for bovine brucellosis in Japan using a simulation model.** *Prev Vet Med* 2008, **86**:57–74.
39. Cunningham B: **A difficult disease called brucellosis.** In *Bovine brucellosis, an international symposium.* Edited by Crawford RP, Hildalgo RJ. Texas USA: A & M University Press, College Station; 1977:11–20.
40. Bronner A, Rautureau S, Jay M, Garin-Bastuji B: **Un nouveau foyer de brucellose bovine identifié en 2013 en Belgique.** *Bulletin Epidémiologique Santé Animale- Alimentation* 2013, **57**:9.
41. Heffernan C, Nielsen L, Thomson K, Gunn G: **An exploration of the drivers to bio-security collective action among a sample of UK cattle and sheep farmers.** *Prev Vet Med* 2008, **87**:358–372.
42. Petty R, Cacioppo J: *Attitudes and persuasion: classic and contemporary approaches.* Dubuque: Westview Press; 1981.
43. Poortinga W, Staff KB, Langford I, Newohner J, Pidgeon N: **The british, Foot and mouth crisis: a comparative study of public risk perceptions, trust and beliefs about government policy in two communities.** *J Risk Res* 2001, **2010**(7):73–90.
44. Jansen J, Renes RJ, Lam TJ: **Evaluation of two communication strategies to improve udder health management.** *J Dairy Sci* 2010, **93**:604–612.

45. Woods A: **Understanding BTV vaccination behaviours.** *Govern Vet J* 2011, **21**:30–44.
46. Alaszewski A: **Risk communication: identifying the importance of social context.** *Health, Risk and Society* 2010, **7**:101–105.
47. Abortion in dairy cattle.I: **Diagnosing and preventing abortion problems.** http://pubs.ext.vt.edu/404/404-289/404-289.html.
48. OIE, **Terrestrial animal health code. Chapter 11.3. Bovine brucellosis.** http://www.oie.int/index.php?id=169&L=0&htmfile=chapitre_1.11.3.htm; 2012.
49. Carpenter TE, Chriel M, Andersen MM, Wulfson L, Jensen AM, Houe H, Greiner M: **An epidemiologic study of late-term abortions in dairy cattle in Denmark, July 2000-August 2003.** *Prev Vet Med* 2006, **77**:215–229.

24

Non-invasive indicators associated with the milk yield response after anthelmintic treatment at calving in dairy cows

Sien H Verschave[1*], Jozef Vercruysse[1], Andrew Forbes[2], Geert Opsomer[3], Miel Hostens[3], Luc Duchateau[4] and Johannes Charlier[1]

Abstract

Background: Gastrointestinal nematodes are an important cause of reduced performance in cattle. Previous studies in Europe showed that after anthelmintic treatment an average gain in milk production of around 1 kg per day/cow can be expected. However, (1) these studies have mainly evaluated group-based anthelmintic treatments during the grazing season or at housing and (2) little is known about parameters affecting variations in the treatment response amongst cows. A better knowledge of such parameters could help to select animals that benefit most from treatment and thus lead to a more rational use of anthelmintics. Therefore, a randomized, non-blinded, controlled clinical trial was performed on 11 commercial dairy farms (477 animals) in Belgium, aiming (1) to study the effect of eprinomectin treatment at calving on milk production and (2) to investigate whether the milk yield response was related to non-invasive animal parameters such that these could be used to inform targeted selective treatment decisions.

Results: Analyses show that eprinomectin treatment around calving resulted in an average (± standard error) increase of 0.97 (±0.41) kg in daily milk yield that was followed up over 274 days on average. Milk yield responses were higher in multiparous compared to primiparous cows and in cows with a high (4th quartile) anti-*O. ostertagi* antibody level in a milk sample from the previous lactation. Nonetheless, high responses were also seen in animals with a low (1st quartile) anti-*O. ostertagi* antibody level. In addition, positive treatment responses were associated with higher faecal egg counts and a moderate body condition score at calving (2nd quartile).

Conclusions: In conclusion, this study provides novel insights into the production response after anthelmintic treatment at calving and factors which influence this. The data could be used to support the development of evidence-based targeted selective anthelmintic treatment strategies in dairy cattle.

Keywords: Dairy cattle, Gastrointestinal nematodes, Targeted selective treatment, Anti-*O. ostertagi* antibody level, Faecal egg counts, Eprinomectin

Background

Gastrointestinal nematodes (GI) are an important cause of reduced performance in grazing cattle. A review of studies on the effect of anthelmintic treatment on milk production from 2000 onwards reported an average milk yield response of 1 kg/cow per day [1]. At present, two innovative concepts are described to prevent nematode-associated production losses while preserving anthelmintic drug efficacy, namely targeted treatment (TT) and target selective treatment (TST) [2]. Using TT, the whole group or herd is treated at an optimal time based on parameters that quantify the risk of infection and/or production losses. A useful parameter currently available for TT in adult dairy cattle is the anti-*O. ostertagi* antibody level in milk [3]. When applying TST, only selected individuals are treated, the aim being to lower the risk for the development of anthelmintic resistance by increasing the size of the parasite population in *refugia*. However, practical implementation of TST at the farm level is still limited by the lack of data on which indicators are useful to identify animals that would benefit most

* Correspondence: sien.verschave@ugent.be
[1]Department of Virology, Parasitology and Immunology, Faculty of Veterinary Medicine, Ghent University, Salisburylaan 133, 9820 Merelbeke, Belgium
Full list of author information is available at the end of the article

from treatment. In Europe, all recent clinical studies, evaluating the effect of anthelmintic treatment on milk production can be considered as TT, as they target whole herds, and have been restricted to mid-season or housing treatments [4-8]. However, a cost-benefit analysis suggested that the economic benefit of anthelmintic treatment of dairy cows is considerably larger when cows are treated around calving compared to treatment at housing [9]. Studies assessing the effect of anthelmintic treatment around calving have been carried out in Canada [10-12] and New Zealand [13], but significant differences in climate and farm management practices may result in different parasite epidemiology and different production effects when compared to European circumstances.

Most clinical trials evaluating the effect of anthelmintic treatment, report the average effect in the study population. It is well recognized that treatment responses can vary largely within and amongst herds [14] and depend on factors such as infection level and management [15]. Less is known about easy-to-use animal parameters that are potentially related to the within-herd variation of production responses, or that could act as a proxy for such responses. A better understanding of these relationships is key to identifying the animals that would benefit most from anthelmintic drug administration, consistent with the TST approach. Previously some indicators have been associated with the milk yield response after anthelmintic treatment. The parameter yielding most consistent results is the anti-*Ostertagia ostertagi* antibody level in individual milk samples, with studies showing higher treatment responses in cows with a high antibody level [3,11,12,16]. As for the age of the cow, some studies found no relationship with treatment effect [10,14,17], whereas others reported a higher milk yield gain in older cows [3,13] or conversely, in younger cows [5]. Finally, it was suggested that treatment responses were higher in high producing animals [14,18], but this was not observed by others [10].

A problem associated with previous studies is that mostly the study design was aimed at investigating the overall treatment effect, leading to low statistical power or unbalanced data for evaluating the effect of indicators associated with the treatment response. Therefore, we performed a randomized, non-blinded, controlled clinical trial with herd (< vs. > average herd production level), *O. ostertagi*-antibody levels in individual milk samples (optical density ratio (ODR) < vs. ≥0.5) and age (2nd vs. ≥3rd lactation) as blocking factors for treatment assignment. The aim were (1) to investigate the effect of anthelmintic treatment at calving on milk production in dairy cattle and (2) to evaluate if some easy-to-use animal parameters (i.e. parity, body condition score, pre-treatment anti-*O. ostertagi* antibody levels) are associated with the milk production response following anthelmintic treatment.

Methods

Following the recommendations of Belgian and European legislation (KB 11/5/2007; 86/609/EEC), this field study conducted on commercial dairy herds did not require ethical approval.

Selection of farms

The study was conducted on 11 dairy herds located in Flanders, Belgium. On average 73 lactating animals, mainly Holstein Friesians, were present on these farms, while the rolling herd average ranged between 7262 and 10920 kg milk. Cows calved all year round. The following criteria were applied for the selection of farms: (1) cows had previous access to pasture and were naturally infected with gastrointestinal nematodes (i.e. anti-*O. ostertagi* ELISA results on bulk milk samples of April 2011 ≥ 0.6 ODR), (2) the last treatment of the cows against GI nematodes had been performed ≥6 months before the experimental treatment and (3) participation in the milk production registration program of CRV (Arnhem, The Netherlands) to enable standardized data collection.

Study design

A non-blinded, randomized, controlled clinical trial was performed to evaluate the effect of anthelmintic treatment at calving on milk production and to evaluate the association with easy-to-use indicators. Both first-calving heifers and older animals were included in the trial. Within 15 days after calving the animals either received treatment with eprinomectin (Eprinex® pour-on, Merial) at a dosage of 0.1 ml per kg bodyweight (500 μg/kg) or received no treatment. Treatments were performed by the farmer, based on the estimated bodyweight. The capability of the participating farmers to estimate the body weight of their animals was tested during farm visits on which the body weight of several random animals was measured based on the heart girth. These measurements revealed that when body weights were underestimated the underestimations were generally low (<50 kg).

The multiparous cows were randomly assigned within herd using *O. ostertagi* antibody level in individual milk sample measured in April 2011 (< vs. ≥0.5 ODR), cow parity (2nd vs. ≥3rd lactation) and previous production level (< vs. > herd-average) as blocking factors. The treatment assignments were communicated to the farmer through a hard-copy attention list, generated with the aid of the herd management software (Veemanager, CRV, Arnhem, The Netherlands) and specifying cow identification, expected calving date and the randomly assigned group (treatment or no treatment).

Due to the lack of an ODR measurement from the previous lactation, heifers could not be included in the randomized list. Therefore, the farmer was asked to alternate treatment assignments (yes/no) for this group of animals following the order in which they calved.

The farmer was requested to complete the hard-copy attention list at the time of treatment by noting the body condition score (BCS; according to [19]), the estimated body weight, dosage applied and the date of treatment. A scale from 1 to 5 was used for the body condition scoring, where 1 stands for severe under-conditioned and 5 for severe over-conditioned. At the start of the trial all farmers received a body condition scoring chart based on Edmonson et al. [19], as a visual aid.

It is acknowledged that the use of anthelmintic pour-on treatments in individual cattle potentially compromises the precision of delivery of the recommended dosage to the intended animal and may result in sub-optimal dosing to in-contact animals too [20]. Therefore, farmers were advised to prevent close contact between treated and untreated animals in the first hours following treatment. Most did this by treating animals in a separate calving box several hours before allowing them to rejoin the lactating animals. This is an important logistical consideration if TST is to be practised on commercial farms, because this is an inevitable compromise as it is generally impractical to keep recently treated and untreated cattle apart for adequate periods.

All concomitant anthelmintic and ectoparasiticide treatments, other than those defined in the study protocol were documented, so that these animals could be excluded from the analysis. Cattle were enrolled from July 2011 until September 2012. The progress and compliance with the study protocol was monitored through monthly telephone contacts and three farm visits during the study period.

Collection of faecal samples and faecal egg counts

On two of the 11 dairy herds, the farmers took faecal samples at the moment of treatment and these were stored in a refrigerator until collection on a weekly basis. Samples were processed immediately after arrival at the lab and examined for nematode eggs by the FLOTAC® method based on the manufacturer's instructions. A saturated sucrose-salt solution was used as flotation solution (density 1.27) and the suspension was distributed over one chamber (1 × 5 ml), resulting in an analytic sensitivity of 2 eggs per gram faeces (EPG; [21]).

Collection of milk samples and *O. ostertagi* milk ELISA

Bulk tank milk samples were collected at monthly intervals from July 2011 until July 2012. For logistical reasons, bulk milk data for the month of April 2012 were not available. Individual milk samples from all lactating animals were collected at three-monthly intervals starting from July 2011 until July 2012 as part of routine sampling for quality control and milk production registration, in cooperation with the Milk Control Centre, Flanders (MCC, Lier, Belgium) and CRV (Arnhem, The Netherlands).

The collected milk samples were subjected to a commercially available antibody-detection *O. ostertagi* ELISA (SVANOVIR®*O. ostertagi*-Ab, Boehringer Ingelheim Svanova, Uppsala) according to the manufacturer's instructions, at the laboratories of the MCC. The results were expressed as an optical density ratio (ODR) that is calculated following the formula ODR = (OD - NC)/(PC - NC), where OD is the result of the optical density reading of the sample at 405 nm, and NC and PC are the OD of the negative and positive control samples, respectively.

Collection of milk production data and processing

Individual milk production records were collected from the CRV milk recording program, with a 4 to 6 week sampling interval; the parameters used were: kg milk, somatic cell count (SCC)/1000, breed, days in milk and lactation number. The milk production records were also subjected to the MilkBot® lactation model (DairySight LLC, Argyle, New York) to create a lactation curve [22]. Both the shape and magnitude of the lactation curve are quantified by the model as a set of parameter values, each of which is associated with a single aspect of lactation curve shape. Analysis of the Milkbot® parameters allows the detection of changes in the distribution of milk production that are not apparent when only daily milk weights or totals are analyzed. The parameter "scale" is a measure of magnitude, without changing the shape of the curve. The parameter "ramp" measures the steepness of the post-parturient rise in production. The parameter "decay" is used to measure the rate of decline in production after the peak in milk production.

Statistical data-analysis

The effect of eprinomectin treatment on the anti-*O. ostertagi* antibody levels was analysed through a linear mixed model with herd and cow nested in herd as random effects. Treatment (yes/no), days after treatment, the month at which the milk was tested and an interaction term between treatment and days after treatment were used as fixed effects in the model. Because the herds were sampled at 3-monthly intervals, the variable 'days after treatment' was categorized in 3 intervals: '0–3 months', '3–6 months', '>6 months'. To investigate the correlation between the faecal egg counts, measured around the moment of calving, and the anti-*O. ostertagi* antibody level, measured in individual milk samples within 30 days post calving, the Spearman rank correlation coefficient was used.

The effect of eprinomectin treatment on milk production parameters was first analysed through a linear mixed model with the test day milk production (kg milk) as outcome variable and herd and cow nested in herd as random effects. Treatment (yes/no), lactation number ('1st', '2nd', '3rd or higher'), the natural logarithm of SCC/1000, number of days in milk (dim), wilmink's function ($dim^{-0.05}$) and year quarter in which calving occurred, were introduced as fixed effects in the model.

In order to investigate between-herd variation in treatment responses, the analysis was repeated for each herd separately. Results were presented by a forest plot and an overall treatment effect was obtained by the precision weighted average, using a random effects model with herd as the random effect.

The relationship between the indicators and the milk production response after treatment was evaluated using the model mentioned above. Linear interaction terms were evaluated between the indicators and treatment effect. These interaction terms were not significant (α =0.05), except for faecal egg counts. Non-linear effects were explored by categorizing the data. Continuous parameters (anti-*O. ostertagi* antibody level and BCS centered to the herd mean) were categorized according to their quartiles. The moment of treatment was categorized in housing (November until March) and pasture period (April until October) based on Bennema et al. [23]. When several pre-treatment *O. ostertagi* ELISA results for a cow were available, the result of the sample closest to calving was used. The treatment effect was estimated within each category of the pre-treatment anti-*O. ostertagi* antibody level, lactation number, the BCS centered to the herd mean and season (pasture vs. housed) in which treatment occurred.

To evaluate heteroscedasticity and the normality of the residuals, plots of the residuals and predicted values were performed. All analyses were carried out with the PROC MIXED or PROC CORR command in the software package SAS version 9.3 (SAS institute Inc., Cary, NC, USA), except for the forest plot and precision weighted treatment effect, which was computed by the 'metafor' package in R (Cran).

Results

Herd characteristics and treatment allocation

Analysis of the effect of treatment on the anti-*O. ostertagi* antibody levels was based on data obtained from 498 animals present in the participating herds. The data used to analyse the effect of treatment on milk production, were based on 477 cows from all participating herds. Of these, 234 belonged to the treatment group and 243 to the control group. The treatment response on the test day milk production records was followed up during 274 days of lactation on average. Table 1 shows that the average anti-*O. ostertagi* antibody level of individual milk samples before treatment of both groups, lactation number, breed and the year quarter in which calving occurred were evenly distributed between the treated and untreated animals indicating a successful treatment allocation. Concomitant treatments were recorded on 3 of the participating herds, for a total of 39 animals. All these treatments were related to liver fluke infections and the anthelmintics used were either closantel or oxyclozanide, neither of which has any activity against *O. ostertagi*.

Parasitological parameters

Figure 1a shows the course of the anti-*O. ostertagi* antibody levels recorded in bulk-tank and individual milk samples during the study period. Figure 1b shows the course of the anti-*O. ostertagi* antibody levels in individual milk samples relative to the month of treatment. In both the treated and untreated group, antibody levels decreased after calving. However, there was a significant interaction ($P = 0.02$) between treatment and the time variable "month after treatment" indicating that antibody levels dropped quicker and remained low for a longer period in treated compared to untreated animals (Figure 1b). The proportion of the total variation in anti-*O. ostertagi* antibody levels that resided at the herd, cow and residual level was 12, 52% and 36%, respectively.

Table 2 lists the faecal egg count results of the two studied herds. The EPG of trichostrongyle-type eggs was, on average (± standard error) 26 (±42) and 13 (±29), for the two herds, respectively. The Spearman rank correlation coefficient between FEC and anti-*O. ostertagi* antibody level of individual milk samples taken within 30 days post calving was $R = 0.42$ ($P = 0.03$).

Table 1 Number of cows, average ± standard deviation of anti-*O. ostertagi* antibody levels in individual milk before treatment registration and distribution of breed, lactation number and year quarter of calving in the treated and untreated group

Parameter	Eprinomectin	Control
Number of cows	234	243
Anti-*O. ostertagi* antibody level (ODR)	0.49 ± 0.3	0.52 ± 0.3
Breed (%):		
Holstein Friesian	43	46
Other	6	5
Lactation number (%):		
First	14	12
Second	13	13
Third or higher	26	22
Year quarter:		
Jan - Mar	10	14
Apr - Jun	13	8
Jul - Sep	13	12
Oct - Dec	15	15

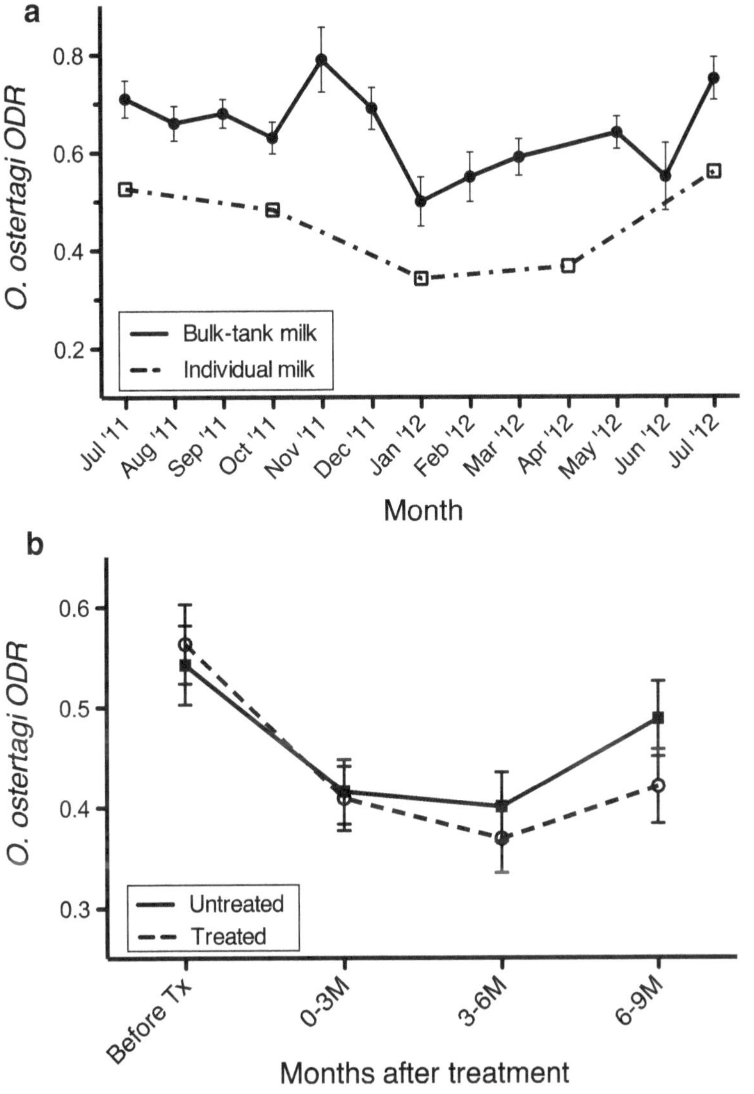

Figure 1 Anti-O. ostertagi antibody levels (ODR) in bulk-tank and individual milk samples during the study period. (a) Course of the anti-*O. ostertagi* antibody levels (ODR) during the study period in bulk-tank and individual milk samples on the 11 herds. **(b)** The course of the anti-*O. ostertagi* antibody levels (ODR) relative to the month of calving in individual milk samples of 1274 cows coming from all the 11 herds. Bars represent standard error of the mean.

Overall effect of eprinomectin treatment on milk production

Analysis of the treatment response on the test day milk production records by a linear mixed model showed a significant increase of 0.97 (95% confidence interval: 0.17 to 1.77) kg milk/day per cow in eprinomectin treated animals ($P < 0.02$; Table 3). The model controlled for the factors lactation number, somatic cell count, number of days in milk and the year quarter in which calving occurred. The herd, cow and residual level explained 27, 35 and 38% of the total variation, respectively.

The forest plot in Figure 2 shows the variation in treatment effects between herds. The precision-weighted

Table 2 Results of the faecal egg counts measured around calving on two dairy herds (N: the number of samples; P25 and P75: the first and third quartile, respectively)

	Fecal egg count results (EPG)				
	N	Median	P25	P75	Range
Herd 1	34	14	2	28	0 - 228
Herd 2	49	2	0	8	0 - 128

Table 3 The results of a linear mixed model to estimate the effect of eprinomectin treatment around calving on daily milk production in 11 herds (based on 477 cows)

Variable	Estimate	Lower limit 95% CI	Upper limit 95% CI	t Value	*P*
Intercept	95.80	88.66	102.93	26.36	<0.001
Eprinomectin (vs. control)	0.97	0.16	1.77	2.36	0.019
Lactation number (baseline is third lactation or higher)					<0.001
First	−7.15	−8.17	−6.11	−13.66	<0.001
Second	−3.05	−4.03	−2.07	−6.13	<0.001
Ln (SCC/1000)	−1.21	−1.36	−1.06	−16.12	<0.001
DIM	−0.07	−0.07	−0.06	−43.02	<0.001
Wilmink	−58.02	−66.05	−49.99	−14.16	<0.001
Year quarter (baseline is 4th year quarter)					0.0064
First	−1.23	−2.31	−0.15	−2.23	0.026
Second	−1.48	−2.68	−0.27	−2.41	0.016
Third	−1.84	−2.94	−0.74	−3.29	0.001
Random Effects	**Variance**	**S.E.**	**Proportion of total variance (%)**	**Z value**	***P***
Herd	12.95	6.00	27	2.16	0.015
Animal	16.50	1.28	35	12.90	<0.001
Residual	17.92	0.44	38	40.48	<0.001

average treatment effect was 0.83 (95% confidence interval: 0.02 to 1.64) kg milk/cow per day.

Although the Milkbot® parameters (scale, ramp, decay) did not differ significantly between groups, the lactation curve (Figure 3) suggested that eprinomectin treatment at calving resulted in a higher peak production and that this effect was maintained throughout the entire lactation.

The association between non-invasive indicators and the milk yield response after anthelmintic treatment

An overview of the effect of eprinomectin treatment on daily milk yield per animal in relation to the evaluated parameters is provided in Figure 4. Animals in third lactation or higher had a significant increase in daily milk production after treatment of 1.24 (95% confidence interval: 0.08 to 2.40) kg milk/cow per day. A large and almost significant effect of 2.05 (95% confidence interval: −0.03 to 4.12) kg milk/cow per day, was also found in the cows in the 4th quartile of anti-*O. ostertagi* antibody level pre-treatment (ODR ≥0.72). A significant treatment effect was observed in animals with a BCS centered to the herd mean in the second quartile. The treatment effects were similar in both seasons. Finally, on the 2 herds where FECs were performed, a significant linear association was found between treatment effect and FEC ($P = 0.01$). This relationship is illustrated in Figure 5.

Discussion

On commercial dairy farms under European conditions, anthelmintic treatment administered shortly after calving, resulted in an average increase of 0.97 kg milk/day per cow. A recent study on anthelmintic treatment at calving in Canada found a maximal treatment effect of 0.67 kg milk/day per cow, which is considerably lower than some of the observations reported here [12]. A study performed in New Zealand, found a positive effect after treatment in only one of the three participating herds of 0.78 kg energy corrected milk/day per cow [24]. Such differences from our observations underline the need to evaluate treatment effects on milk production in different geographical locations and under different management conditions.

Although the treatment was randomly assigned, it was not blinded to the farmer which could have led to a biased treatment effect estimate. Non-blinding may result in occasional use of the anthelmintic product for animals assigned to the control group, however this was rarely recorded. At the end of the trial, the prescribed treatment assignment ('treat' or 'not treat') had not been followed for a total of 45 animals. The farmer also had to assess the appropriate dosing based on the estimated body weight, which is likely to lead to underdosing as compared to dosing under controlled conditions by the researcher. Furthermore, there is a higher risk of incorrect storage of the anthelmintic product on the farm, which would also lead to a reduced treatment effect.

Previously, we observed a significant effect of anthelmintic treatment at housing on the anti-*O. ostertagi* antibody level in bulk milk samples [7]. In that study, no re-exposure to GI nematodes in the months following treatment took place and the effect was largest 3–4

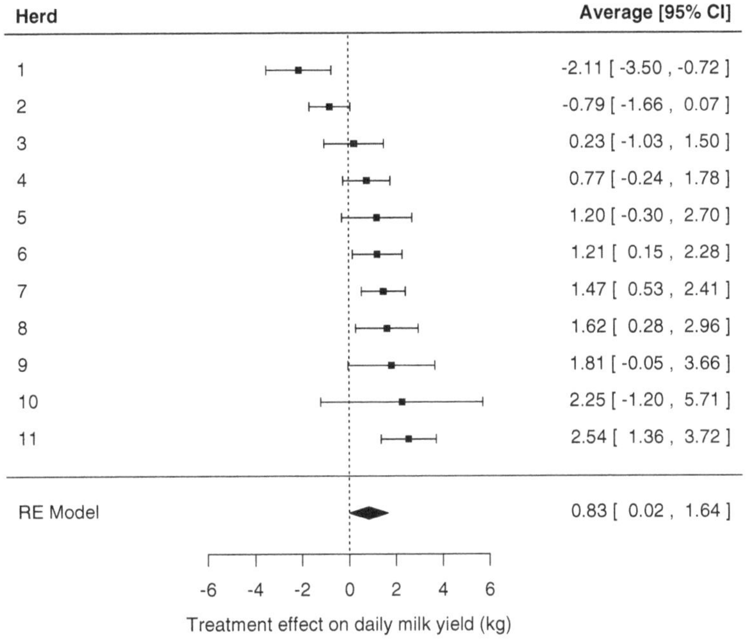

Figure 2 Average response in daily milk yield after eprinomectin treatment at calving for each herd separately and the overall inverse variance weighted average computed with a random effects model. Bars represent the 95% confidence interval.

months after treatment. In the present study, we observed a significant treatment effect on the anti-*O. ostertagi* antibody level in individual milk samples. However, the anti-*O. ostertagi* antibody levels dropped in both the treated and untreated group and the differences only became clear after >6 months, when the anti-*O. ostertagi* antibody levels started to rise again. The concurrent drop in both the treated and untreated group may be explained by a dilution effect since milk production is highest 2–3 months following treatment. However, this effect has previously been estimated to be small [25]. A second possible reason is a generally lower exposure to GI nematodes in the herd caused by treatment approximate 50% of the animals.

There was a considerable variation in treatment response between the different herds. In one herd, there was a significant negative effect of treatment on milk production. Interestingly, although the animals were grazed up to the start of the trial, further investigation revealed that the animals in this herd had not been at pasture during the entire study period. This, together with the low average (± standard deviation) bulk tank milk *O. ostertagi* ELISA results (0.52 ODR (±0.09)) suggests a very low worm exposure in this herd. When this herd was excluded from the analysis, the overall treatment effect (95% confidence interval) on milk yield was 1.09 (0.42 – 1.76) kg milk per day per cow.

Currently, targeted selective treatments (TST), where anthelmintic treatments are given to selected individuals

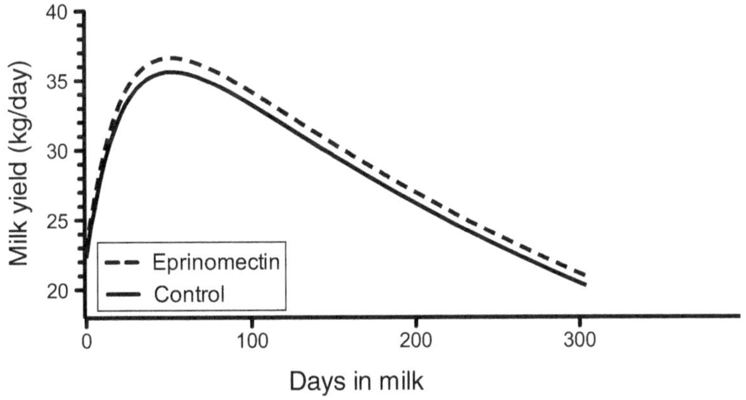

Figure 3 Average lactation curve of cows treated with eprinomectin at calving versus untreated cows.

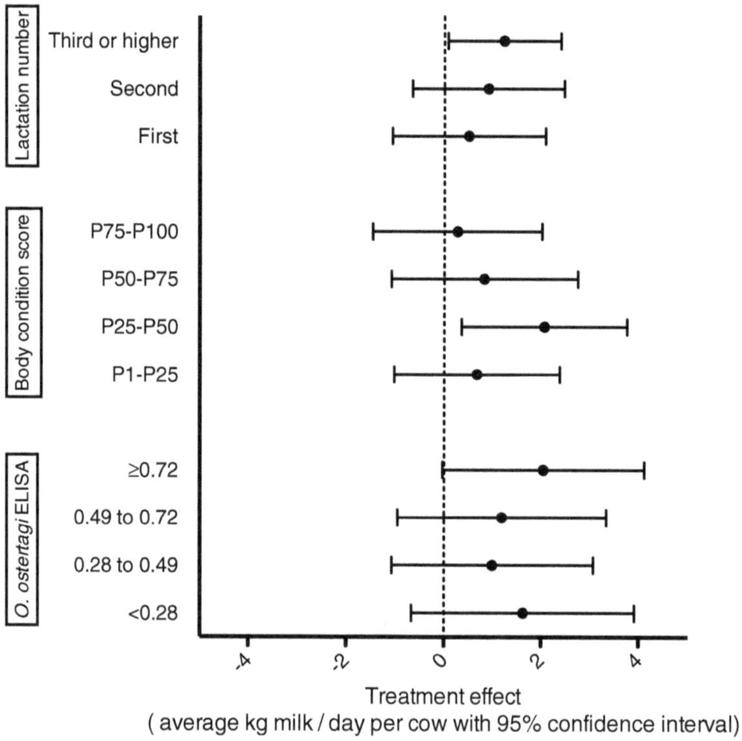

Figure 4 Estimated effect of eprinomectin treatment at calving on daily milk yield (kg) per cow according to different potential selection parameters for anthelmintic treatment. Error bars represent the 95%-confidence interval. Categories for body condition score and *O. ostertagi* ELISA are based on quartiles.

within a herd, are promoted as a method of maintaining animal performance whilst reducing the selection of anthelmintic resistant nematodes [2]. However, until now, TST have been mostly evaluated in sheep [26-31] and few data are available to determine which parameters are useful in identifying cattle that would benefit most from anthelmintic treatment. Previously the anti-*O. ostertagi* antibody level in individual milk samples has been proposed as a useful indicator of which animals would benefit most from treatment [3,11,12,16]. Here, we confirm that the highest treatment responses were observed in the cows with the highest anti-*O. ostertagi* antibody level. However, considerable responses were also observed in animals with low anti-*O. ostertagi* antibody levels, which corresponds to a recent study of Ravinet et al. [8]. This indicates that significant economic potential may be lost if only animals with a high anti-*O. ostertagi* antibody level are treated and highlights the need to assess the broader economic implications on-farm when this indicator would be used as the principal treatment criterion.

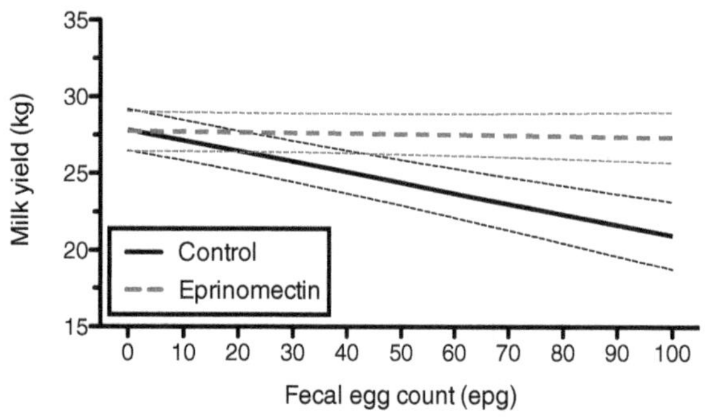

Figure 5 Relationship between faecal egg count results and daily milk yield (kg) per cow measured at two herds. The dotted lines represent the standard error of the mean.

In addition to the anti-*O. ostertagi* antibody level, we identified 2 other promising candidates for targeting individual treatments: age and FEC. Although higher treatment responses in older cattle may appear counterintuitive, as these animals are considered to be functionally immune to gastrointestinal nematodes, several previous studies reported higher treatment responses with increasing cow age [3,13,25]. As discussed before [3], the latter may be attributed to a different priority of nutrient allocation between primi- and multiparous cows [13] or the higher energy requirements of the immune response to GI nematodes in older animals, as observed in sheep [32]. Finally, it may also be considered that the level of GI nematode infection in primiparous cows is generally lower than in older cows. A recent survey in Europe showed that anthelmintic control measures in young stock are relatively intensive resulting in low infection levels of first-season grazing calves and a compromised build-up of immunity against these parasites [33].

The milk production response increased in those animals with higher FECs at calving. This is surprising because FECs are generally considered as a poor indicator of the GI nematode infection level in cattle [34,35]. O'Farrell et al. [17] and Walsh et al. [36] found no relationship between milk yield response after anthelmintic treatment and FEC. However, in a recent study, a significantly lower milk production was observed in animals with a FEC >10 EPG, although whether these animals would also have a higher milk yield after anthelmintic treatment was not investigated [37]. Our results require to be confirmed because they are only based on observations in 2 herds. A possible reason why we observe this relationship is because a much more sensitive technique (FLOTAC with analytical sensitivity of 2 EPG) was used than in the previous studies.

A significant positive treatment effect on milk production was found for cows with a BCS between the 25th and 50th percentile. Despite our attempt to standardize this variable by centering it to the herd mean, the results for the BCS should still be interpreted with caution since important between-farmer variability may remain present and the study design did not take BCS into account as blocking parameter. Based on the described relationship between BCS around calving and subsequent milk yield [38], it could be expected that production responses are lower in cows with either a very low or very high BCS. Such a relationship was previously observed in a study investigating milk yield responses after flukicide treatments [9].

Conclusions

In conclusion, this randomized controlled field trial demonstrates that eprinomectin treatment at calving reduces the infection levels with gastro-intestinal nematodes and increases the daily milk yield in the following lactation. Treatment responses were highest in animals in their third or higher lactation, with a high pre-treatment anti-*O. ostertagi* antibody level in a milk sample of the previous lactation and in animals with a moderate body condition score. Further research needs to be done to assess the economic impact of selective treatment approaches, which greatly influences the eventual uptake of these indicators in practice.

Competing interests

The authors declare that the SVANOVIR® *O. ostertagi*-Ab ELISA kit is commercialized by Boehringer Ingelheim Svanova, Uppsala under a license agreement with Ghent University.

Authors' contributions

Design of the study: JC, JV, AF. Coordination of the study: SV, JC. Collection of fecal samples and fecal egg counts: SV. Statistical analysis: JC, SV. MilkBot® lactation analysis: MH, GO. Supervision statistical analysis: LD. Manuscript draft: SV, JC. Revision of the manuscript: JV, AF, GO, MH, LD. All authors read and approved the final manuscript.

Acknowledgments

We are very grateful to all participating farmers and their veterinarians; Tom Meyns (Merial SAS); Hiemke Knijn (CRV); Luc De Meulemeester and Jean- Marie Van Crombrugge (Milk Control Centre Flanders) without whose support this study would not have been possible. This study received funding from the E.U. FP7 GLOWORM project (Grant Agreement No. 288975CP-TP-KBBE.2011.1.3-04) and Merial.

Author details

[1]Department of Virology, Parasitology and Immunology, Faculty of Veterinary Medicine, Ghent University, Salisburylaan 133, 9820 Merelbeke, Belgium. [2]Merial SAS, 29 Avenue Tony Garnier, Lyon 69007, France. [3]Department of Reproduction, Obstetrics and Herd Health, Faculty of Veterinary Medicine, Ghent University, Salisburylaan 133, 9820 Merelbeke, Belgium. [4]Department of Physiology and Biometrics, Faculty of Veterinary Medicine, Ghent University, Salisburylaan 133, 9820 Merelbeke, Belgium.

References

1. Charlier J, Höglund J, von Samson-Himmelstjerna G, Dorny P, Vercruysse J: **Gastrointestinal nematode infections in adult dairy cattle: impact on production, diagnosis and control.** *Vet Parasitol* 2009, **164**:70–79.
2. Kenyon F, Jackson F: **Targeted flock/herd and individual ruminant treatment approaches.** *Vet Parasitol* 2012, **186**:10–17.
3. Charlier J, Vercruysse J, Smith J, Vanderstichel R, Stryhn H, Claerebout E, Dohoo I: **Evaluation of anti-*Ostertagia ostertagi* antibodies in individual milk samples as decision parameter for selective anthelmintic treatment in dairy cows.** *Prev Vet Med* 2010, **93**:147–152.
4. Reist M, Medjitna TD, Braun U, Pfister K: **Effect of a treatment with eprinomectin or trichlorfon on the yield and quality of milk produced by multiparous dairy cows.** *Vet Rec* 2002, **151**:377–380.
5. Forbes AB, Huckle CA, Gibb MJ: **Impact of eprinomectin on grazing behavior and performance in dairy cattle with sub-clinical gastrointestinal nematode infections under continuous stocking management.** *Vet Parasitol* 2004, **125**:353–364.
6. Gibb MJ, Huckle CA, Forbes AB: **Effects of sequential treatments with eprinomectin on performance and grazing behaviour in dairy cattle under daily-paddock stocking management.** *Vet Parasitol* 2005, **133**:79–90.
7. Charlier J, Duchateau L, Claerebout E, Vercruysse J: **Predicting milk-production responses after an autumn treatment of pastured dairy herds with eprinomectin.** *Vet Parasitol* 2007, **143**:197–390.
8. Ravinet N, Bareille N, Lehebel A, Ponnau A, Chartier C, Chauvin A: **Change in milk production after treatment against gastrointestinal nematodes according to grazing history, parasitological and production-based indicators in adult dairy cows.** *Vet Parasitol* 2014, **201**(1–2):95–109.
9. Charlier J, Levecke B, Devleesschauwer B, Vercruysse J, Hogeveen H: **The economic effects of whole-herd versus selective anthelmintic treatment strategies in dairy cows.** *J Dairy Sci* 2012, **95**:2977–2987.
10. Nødtvedt A, Dohoo I, Sanchez J, Conboy G, DesCôteaux L, Keefe G: **Increase in milk yield following eprinomectin treatment at calving in pastured dairy cattle.** *Vet Parasitol* 2002, **105**:191–206.
11. Sithole F, Dohoo I, Leslie K, DesCôteaux L, Godden S, Campbell J, Stryhn H, Sanchez J: **Effect of eprinomectin treatment at calving on milk production in dairy herds with limited outdoor exposure.** *J Dairy Sci* 2005, **88**:929–937.
12. Vanderstichel R, Dohoo I, Sanchez J, Sithole F, Keefe G, Stryhn H: **Predicting the effect of anthelmintic treatment on milk production of dairy cattle in Canada using an *Ostertagia ostertagi* ELISA from individual milk samples.** *Prev Vet Med* 2013, **111**:63–75.
13. McPherson WB, Gogolewski RP, Slacek B, Familton AS, Gross SJ, Maciel AE, Ryan WG: **Effect of a peri-parturient eprinomectin treatment of dairy cows on milk production.** *New Zeal Vet J* 2001, **49**:106–110.
14. Ploeger HW, Schoenmaker GJ, Kloosterman A, Borgsteede FH: **Effect of anthelmintic treatment of dairy cattle on milk production related to some parameters estimating nematode infection.** *Vet Parasitol* 1989, **34**:239–253.

15. Bisset SA, Marshall ED, Morrison L: **Economics of a dry-cow anthelmintic drenching programme for dairy cows in New Zealand. Part 1. Overall response in 47 herds.** *Vet Parasitol* 1987, **26:**107–118.
16. Sanchez J, Dohoo I, Leslie K, Keefe G, Markham F, Sithole F: **The use of an indirect *Ostertagia ostertagi* ELISA to predict milk production response after anthelmintic treatment in confined and semi-confined dairy herds.** *Vet Parasitol* 2005, **130:**115–124.
17. O'Farrell JK, Downey NE, Sherington J: **The effect of anthelmintic treatment at calving on the subsequent milk production characteristics of dairy cows.** *Irish Vet J* 1986, **40:**116–123.
18. Ploeger HW, Kloosterman A, Bargeman G, Vonwuijckhuise L, Vandenbrink R: **Milk yield increase after anthelmintic treatment of dairy cattle related to some parameters estimating helminth infection.** *Vet Parasitol* 1990, **35:**103–116.
19. Edmonson AJ, Lean IJ, Weaver LD, Farver T, Webster G: **A body condition scoring chart for Holstein dairy cows.** *J Dairy Sci* 1989, **72:**68–78.
20. Bousquet-Mélou A, Jacquiet P, Hoste H, Clément J, Bergeaud J, Alvinerie M, Toutain P: **Licking behaviour induces partial anthelmintic efficacy of ivermectin pour-on formulation in untreated cattle.** *Int J Parasitol* 2011, **41:**563–569.
21. Cringoli G, Rinaldi L, Maurelli MP, Utzinger J: **FLOTAC: new multivalent techniques for qualitative and quantitative copromicroscopic diagnosis of parasites in animals and humans.** *Nat Protoc* 2010, **5:**503–515.
22. Ehrlich J: **Quantifying shape of lactation curves, and benchmark curves for common dairy breeds and parities.** *Bov Pr* 2011, **45:**88–95.
23. Bennema S, Vercruysse J, Morgan E, Stafford K, Hoglund J, Demeler J, von Samson-Himmelstjerna G, Charlier J: **Epidemiology and risk factors for exposure to gastrointestinal nematodes in dairy herds in northwestern Europe.** *Vet Parasitol* 2010, **173:**247–254.
24. Mason WA, Pomroy WE, Lawrence KE, Scott I: **The effect of repeated, four-weekly eprinomectin treatment on milk production in pasture-based, seasonally-calving dairy cattle.** *Vet Parasitol* 2012, **189:**250–259.
25. Sanchez J, Markham F, Dohoo I, Sheppard J, Keefe G, Leslie K: **Milk antibodies against *Ostertagia ostertagi*: relationships with milk IgG and production parameters in lactating dairy cattle.** *Vet Parasitol* 2004, **120:**319–330.
26. Stafford KA, Morgan ER, Coles GC: **Weight-based targeted selective treatment of gastrointestinal nematodes in a commercial sheep flock.** *Vet Parasitol* 2009, **164:**59–65.
27. Besier RB, Love RA, Lyon J, van Burgel AJ: **A targeted selective treatment approach for effective and sustainable sheep worm management: investigations in Western Australia.** *Anim Prod Sci* 2010, **50:**1034–1042.
28. Ouzir M, Berrag B, Benjouad A, Cabaret J: **Use of pathophysiological indicators for individual decision of anthelmintic treatment of ewes against gastro-intestinal nematodes in Morocco.** *Vet Parasitol* 2011, **180:**372–377.
29. Busin V, Kenyon F, Laing N, Denwood MJ, McBean D, Sargison ND, Ellis K: **Addressing sustainable sheep farming: application of a targeted selective treatment approach for anthelmintic use on a commercial farm.** *Small Ruminant Res* 2013, **110:**100–103.
30. Kenyon F, McBean D, Greer AW, Burgess CG, Morrison AA, Bartley DJ, Bartley Y, Devin L, Nath M, Jackson F: **A comparative study of the effects of four treatment regimes on ivermectin efficacy, body weight and pasture contamination in lambs naturally infected with gastrointestinal nematodes in Scotland. International journal for parasitology.** *Drugs Drug Resist* 2013, **3:**77–84.
31. Busin V, Kenyon F, Parkin T, McBean D, Laing N, Sargison ND, Ellis K: **Production impact of a targeted selective treatment system based on liveweight gain in a commercial flock.** *Vet J* 2014, **200:**248–252.
32. Greer AW: **Trade-offs and benefits: implications of promoting a strong immunity to gastrointestinal parasites in sheep.** *Parasite Immunol* 2008, **30:**123–132.
33. Charlier J, Demeler J, Höglund J, von Samson-Himmelstjerna G, Dorny P, Vercruysse J: ***Ostertagia ostertagi* in first-season grazing cattle in Belgium, Germany and Sweden: general levels of infection and related management practices.** *Vet Parasitol* 2010, **171:**91–98.
34. Eysker M, Ploeger HW: **Value of present diagnostic methods for gastrointestinal nematode infections in ruminants.** *Parasitology* 2000, **120:**109–119.
35. Forbes AB: **Parasite myth busting.** *Cattle Pract* 2012, **20:**128–136.
36. Walsh A, Younis PJ, Morton JM: **The effect of ivermectin treatment of late pregnant dairy cows in south-west Victoria on subsequent milk production and reproductive performance.** *Aust Vet J* 1995, **72:**201–207.
37. Perri AF, Mejía ME, Licoff N, Lazaro L, Miglierina M, Ornstein A, Becu-Villalobos D, Lacau-Mengido IM: **Gastrointestinal parasites presence during the peripartum decreases total milk production in grazing dairy Holstein cows.** *Vet Parasitol* 2011, **178:**311–318.
38. Roche JR, Friggens NC, Kay JK, Fisher MW, Stafford KJ, Berry DP: **Invited review: Body condition score and its association with dairy cow productivity, health, and welfare.** *J Dairy Sci* 2009, **92:**5769–5801.

Permissions

All chapters in this book were first published in VR, by BioMed Central; hereby published with permission under the Creative Commons Attribution License or equivalent. Every chapter published in this book has been scrutinized by our experts. Their significance has been extensively debated. The topics covered herein carry significant findings which will fuel the growth of the discipline. They may even be implemented as practical applications or may be referred to as a beginning point for another development.

The contributors of this book come from diverse backgrounds, making this book a truly international effort. This book will bring forth new frontiers with its revolutionizing research information and detailed analysis of the nascent developments around the world.

We would like to thank all the contributing authors for lending their expertise to make the book truly unique. They have played a crucial role in the development of this book. Without their invaluable contributions this book wouldn't have been possible. They have made vital efforts to compile up to date information on the varied aspects of this subject to make this book a valuable addition to the collection of many professionals and students.

This book was conceptualized with the vision of imparting up-to-date information and advanced data in this field. To ensure the same, a matchless editorial board was set up. Every individual on the board went through rigorous rounds of assessment to prove their worth. After which they invested a large part of their time researching and compiling the most relevant data for our readers.

The editorial board has been involved in producing this book since its inception. They have spent rigorous hours researching and exploring the diverse topics which have resulted in the successful publishing of this book. They have passed on their knowledge of decades through this book. To expedite this challenging task, the publisher supported the team at every step. A small team of assistant editors was also appointed to further simplify the editing procedure and attain best results for the readers.

Apart from the editorial board, the designing team has also invested a significant amount of their time in understanding the subject and creating the most relevant covers. They scrutinized every image to scout for the most suitable representation of the subject and create an appropriate cover for the book.

The publishing team has been an ardent support to the editorial, designing and production team. Their endless efforts to recruit the best for this project, has resulted in the accomplishment of this book. They are a veteran in the field of academics and their pool of knowledge is as vast as their experience in printing. Their expertise and guidance has proved useful at every step. Their uncompromising quality standards have made this book an exceptional effort. Their encouragement from time to time has been an inspiration for everyone.

The publisher and the editorial board hope that this book will prove to be a valuable piece of knowledge for researchers, students, practitioners and scholars across the globe.

List of Contributors

N. Byrne, A. B. Garcia and R. G. Sayers
Animal & Bioscience Research Department, Animal & Grassland Research and Innovation Centre, Teagasc, Moorepark, Fermoy, Co. Cork, Ireland

A. E. Kennedy
Animal & Bioscience Research Department, Animal & Grassland Research and Innovation Centre, Teagasc, Moorepark, Fermoy, Co. Cork, Ireland
Department of Biological Sciences, Cork Institute of Technology, Bishopstown, Co. Cork, Ireland

J. O'Mahony
Department of Biological Sciences, Cork Institute of Technology, Bishopstown, Co. Cork, Ireland

Marie-Eve Cousin
Epidemiology, Biostatistics and Prevention Institute, University of Zürich, Hirschengraben 84, 8001 Zürich, Switzerland

Maria Christina Härdi-Landerer
Institute of agricultural science, Animal Physiology, Swiss Federal Institute of Technology (ETH), Universitätsstrasse 2, 8092 Zürich, Switzerland

Verena Völk
Tierarztpraxis Dr. med. vet. Markus Nydegger, Scherlihalde 5, 3145 Niederscherli, Switzerland

Michèle Bodmer
Clinic for Ruminants, Vetsuisse-Faculty, University of Bern, Bremgartenstrasse 109a, 3001 Bern, Switzerland

Richard E. Booth and Joe Brownlie
Royal Veterinary College, Hawkshead Lane, North Mymms, Hertfordshire AL9 7TA, UK

Muhammad Shahid Bilal, Juma Ahamed Abaker, Zain ul Aabdin, Tianle Xu, Hongyu Dai, Kai Zhang, Xinxin Liu and Xiangzhen Shen
College of Veterinary Medicine, Nanjing Agricultural University, Nanjing 210095, People's Republic of China

S. Yavari, G. Szura, M. Schmicke, W. Kehler, M. Heppelmann and J. Rehage
Clinic for Cattle, University of Veterinary Medicine Hannover, Foundation, Hannover, Germany

N. Khraim
Department for Veterinary Surgery, College of Veterinary Medicine, An-Najah National University, Nablus, Israel

A. Starke
Clinic for Ruminants, University of Leipzig, Leipzig, Germany

E. Engelke and C. Pfarrer
Institute for Anatomy, University of Veterinary Medicine Hannover, Foundation, Hannover, Germany

K. Hopster
Clinic for Horses, University of Veterinary Medicine Hannover, Foundation, Hannover, Germany

S. B. R. Kästner
Clinic for Small Animals, University of Veterinary Medicine Hannover, Foundation, Hannover, Germany

Ye Lin, Xiaoxu Sun, Xiaoming Hou, Bo Qu, Xuejun Gao and Qingzhang Li
Key Laboratory of Dairy Science of Education Ministry, Northeast Agricultural University, Harbin 150030, China

Sławomir J. Kowalczyk, Michał Czopowicz, Tomasz Nalbert, Andrzej Bereznowski and Jarosław Kaba
Laboratory of Veterinary Epidemiology and Economics, Faculty of Veterinary Medicine, Warsaw University of Life Sciences, Nowoursynowska 159c, 02-776 Warsaw, Poland

Corinna N. Weber and Elisabeth Müller
Laboklin GmbH & Co. KG, Steubenstrasse 4, 97688 Bad Kissingen, Germany

H. Müller, M. Heinrich and A. Starke
Clinic for Ruminants and Swine, Faculty of Veterinary Medicine, University of Leipzig, An den Tierkliniken 11, 04103 Leipzig, Germany

N. Mielenz
Institute for Agriculture and Nutrition, Biometrics and Informatics in Agriculture Group, Martin-Luther-University, Karl-Freiherr-von-Fritsch-Straße 4, 06120 Halle (Saale), Germany

S. Reese
Institute of Anatomy, Histology and Embryology, Department of Veterinary Sciences, Faculty of Veterinary Medicine, Ludwig-Maximilians-University, Veterinaerstraße 13, 80539 Munich, Germany

A. Steiner
Clinic for Ruminants, Vetsuisse-Faculty, University of Bern, Bremgartenstrasse 109a, 3012 Bern, Switzerland

Cathy A. Bauman, Andria Jones-Bitton, David Kelton and Paula Menzies
Department of Population Medicine, University of Guelph, Guelph, Ontario N1G 2W1, Canada

Jocelyn Jansen
Veterinary Science & Policy Unit, Animal Health and Welfare Branch, Ontario Ministry of Agriculture, Food & Rural Affairs, Elora, Ontario N0B 1S0, Canada

Pedro Melendez
College of Veterinary Medicine, University of Missouri, Columbia, MO 65211, USA

Pablo Pinedo
Texas A&M AgriLife Research & Extension Center, College of Veterinary Medicine and Biomedical Sciences, Texas A&M University System, Amarillo, TX 79106-1769, USA
Department of Animal Sciences, Colorado State University, Fort Collins, CO 80523, USA

Consuelo Bustamante, Natalia Adaro, José Bastias, Maria Paz Marin and Carolina Rios
College of Veterinary Medicine, University Santo Tomas, Viña del Mar, Chile

Mario Duchens
College of Veterinary Medicine, University of Chile, Santiago, Chile

S. A. Mekonnen, A. E. Riesebos and G. Koop
Department of Farm Animal Health, Faculty of Veterinary Medicine, Utrecht University, Yalelaan 7, 3584, CL, Utrecht, The Netherlands

T. J. G. M. Lam
Department of Farm Animal Health, Faculty of Veterinary Medicine, Utrecht University, Yalelaan 7, 3584, CL, Utrecht, The Netherlands
GD Animal Health, 7400 AA Deventer, The Netherlands

J. Hoekstra
Department of Farm Animal Health, Faculty of Veterinary Medicine, Utrecht University, Yalelaan 7, 3584, CL, Utrecht, The Netherlands
Department of Infectious Diseases and Immunology, Faculty of Veterinary Medicine, Utrecht University, Yalelaan 1, 3584 CL Utrecht, The Netherlands

M. P. Spaninks and E. M. Broens
Department of Infectious Diseases and Immunology, Faculty of Veterinary Medicine, Utrecht University, Yalelaan 1, 3584 CL Utrecht, The Netherlands

V. P. M. G. Rutten
Department of Veterinary Tropical Diseases, Faculty of Veterinary Science, University of Pretoria, Onderstepoort, Pretoria 0110, South Africa T. S. Tessema
Institute of Biotechnology, Addis Ababa University, Addis Ababa, Ethiopia

Zhoumeng Lin, Jim E. Riviere and Ronette Gehring
Anatomy and Physiology, College of Veterinary Medicine, Kansas State University, 1800 Denison Avenue, P200 Mosier Hall, Manhattan, KS 66506-5802, USA

Mengjie Li
Anatomy and Physiology, College of Veterinary Medicine, Kansas State University, 1800 Denison Avenue, P200 Mosier Hall, Manhattan, KS 66506-5802, USA
Department of Pharmaceutical Science, College of Pharmacy, University of Oklahoma Health Sciences Center, Oklahoma City, OK 73117, USA

Matthew Cuneo and Joan D. Rowe
Department of Population, Health and Reproduction, College of Agricultural and Environmental Sciences, University of California, Davis, CA 95616, USA

Lisa A. Tell
Department of Medicine and Epidemiology, College of Agricultural and Environmental Sciences, University of California, Davis, CA 95616, USA

Shayna Allison and Jan Carlson
School of Veterinary Medicine and Department of Animal Science, College of Agricultural and Environmental Sciences, University of California, Davis, CA 95616, USA

Michael K. Ghebremariam
Department of Infectious Diseases and Immunology, Faculty of Veterinary Medicine, Utrecht University, Utrecht, The Netherlands
Department of Veterinary Sciences, Hamelmalo Agricultural College, Keren, Eritrea

V. P. M. G Rutten
Department of Infectious Diseases and Immunology, Faculty of Veterinary Medicine, Utrecht University, Utrecht, The Netherlands
Department of Veterinary Tropical Diseases, Faculty of Veterinary Science, University of Pretoria, Pretoria, South Africa

A. L. Michel
Department of Veterinary Tropical Diseases, Faculty of Veterinary Science, University of Pretoria, Pretoria, South Africa

J. C. M. Vernooij and M. Nielen
Department of Farm Animal Health, Faculty of Veterinary Medicine, Utrecht University, Utrecht, The Netherlands

K. Uqbazghi
Veterinary Services, Ministry of Agriculture (MOA), Mendefera, Debub Region, Eritrea

T. Tesfaalem
National Animal and Plant Health Laboratory, MOA, Asmara, Eritrea

T. Butsuamlak
Veterinary Services, MOA, Keren, Anseba Region, Eritrea

A. M. Idris
Veterinary Services, MOA, Asmara, Maekel Region, Eritrea

Abd Al-Bar Al-Farha
School of Animal and Veterinary Sciences, The University of Adelaide, Roseworthy, South Australia 5371, Australia
Northern Technical University, Technical Foundation, Mosul, Iraq

Farhid Hemmatzadeh and Manouchehr Khazandi
School of Animal and Veterinary Sciences, The University of Adelaide, Roseworthy, South Australia 5371, Australia
Australian Centre for Antimicrobial Resistance Ecology, The University of Adelaide, Adelaide, South Australia 5000, Australia

Kiro Petrovski
School of Animal and Veterinary Sciences, The University of Adelaide, Roseworthy, South Australia 5371, Australia
Australian Centre for Antimicrobial Resistance Ecology, The University of Adelaide, Adelaide, South Australia 5000, Australia
Davies Centre, School of Animal and Veterinary Sciences, The University of Adelaide, Roseworthy, South Australia 5371, Australia

Andrew Hoare
South East Vets, Mt Gambier, South Australia 5290, Australia

Jacopo Guccione, Luigi D'Andrea, Angela De Rosa and Paolo Ciaramella
Department of Veterinary Medicine and Animal Productions, University of Napoli "Federico II", Via F. Delpino 1, 80137 Naples, Italy

Christian Carcasole
Veterinary practitioner, Latina District 04100, Italy

Adrian Steiner and Maher Alsaaod
Clinic for Ruminants, Department of Clinical Veterinary Medicine, Vetsuisse-Faculty, University of Berne, 3012 Bern, Switzerland

Antonio Di Loria
Department of Health Science, University of Magna Graecia of Catanzaro, 88100 Catanzaro, Italy

Jianguo Wang, Guanghui She, Yezi Kong, Yazhou Guo and Baoyu Zhao
College of Veterinary Medicine, Northwest A&F University, Yangling 712100, Shaanxi, China

Xiaoyan Zhu
UMR 1195 Inserm and University Paris-Saclay, Kremlin-Bicêtre, France

Zhe Wang and Guowen Liu
College of Veterinary Medicine, Jilin University, Changchun 130062, Jilin, China

Barbara Colitti, Elvira Muratore, Luigi Bertolotti, Bryan Iotti, Mario Giacobini, Margherita Profiti, Chiara Nogarol and Sergio Rosati
Department of Veterinary Science, University of Turin, Largo Paolo Braccini, 2 10095 Grugliasco, Turin, Italy

Maria Elena Careddu
Istituto Zooprofilattico Sperimentale del Piemonte della Liguria e della Valle D'Aosta (IZSPLV), 12100 Cuneo, Italy

Jens Böttcher
Tiergesundheitsdienst bayern e.V, 85586 Poing, Germany

Andreino Ponzo
Azienda sanitaria locale Cuneo (ASL CN1), 12100 Cuneo, Italy

Roberto Facelli
Associazione Regionale Allevatori Piemonte (ARAP), 12020 Madonna dell'Olmo, Cuneo, Italy

Petra Bandelj, Marko Cotman, Matjaz Ocepek and Modest Vengust
Veterinary faculty, University of Ljubljana, Cesta v Mestni log 47, SI-1115 Ljubljana, Slovenia

Céline Harmanus and Ed J. Kuijper
Department of Medical Microbiology, Center of Infectious Diseases, Leiden University Medical Center, Leiden, Netherlands

Rok Blagus
Institute for biostatistics and Medical informatics, University of Ljubljana, Vrazov trg 2, SI-1104 Ljubljana, Slovenia

Amilcar Arenal, Yipsi García, Lídice Quesada, Dayamis Velázquez, Diamela Sánchez, Mayelin Peña, Asnaldo Suárez, Arnielis Díaz and Yuliet Sánchez
Department of Morpho-Physiology, University of Camagüey Ignacio Agramonte Loynaz, Km 5 ½, 74650 Camagüey, Cuba

Stijn Casaert and Jozef Vercruysse
Department of Virology, Parasitology and Immunology, Faculty of Veterinary Medicine, Ghent University, Salisburylaan 133, 9820 Merelbeke, Belgium

Jan van Dijk
Centre for Preventive Medicine, Animal Health Trust, Lanwades Park, Kentford, Newmarket, Suffolk CB8 7UU, UK

Johannes Charlier
Kreavet, Hendrik Mertensstraat 17, 9150, Kruibeke, Belgium

Kaiwen Chen, Yuxuan Song, Xiaochuan Zhang, Gonghai Zhang, Kai Wen, Haidong Ma, Guang Li, Binyun Cao and Xiaopeng An
College of Animal Science and Technology, Northwest A&F University, No. 22 Xinong Road, Yangling, Shaanxi 712100, People's Republic of China

Yuhan Liu
College of Animal Science and Technology, Northwest A&F University, No. 22 Xinong Road, Yangling, Shaanxi 712100, People's Republic of China
Animal Engineering Branch, Yangling Vocational & Technical College, No. 10 Xinong Road, Yangling, Shaanxi 712100, People's Republic of China

Jinxing Hou
Animal Engineering Branch, Yangling Vocational & Technical College, No. 10 Xinong Road, Yangling, Shaanxi 712100, People's Republic of China

Nana Ma, Juma Ahamed Abaker, Muhammad Shahid Bilal, Hongyu Dai and Xiangzhen Shen
College of Veterinary Medicine, Nanjing Agricultural University, Nanjing 210095, People's Republic of China

Index

A

Accessory Gene Regulators, 14

Antimicrobial Susceptibility Testing, 110, 116, 175

B

Bayesian Hierarchical Model, 95, 98

Bayesian Information Criterion, 4

Body Condition Score, 142-144, 146, 148, 158

Bovine Herpesvirus 1, 41, 159-161, 164-165

Bovine Tuberculosis, 127, 132-133

Bovine Viral Diarrhoea Virus, 11, 26-27, 40-41

Bubalus Bubalis, 69, 142, 149

Bulk Milk Antibody, 26, 31-33, 36, 41, 77

Bulk-tank Milk, 70-71, 73, 76-77, 140, 165, 176-177, 183

C

Claw Diseases, 50

Claw Horn Lesions, 50, 58, 78

Cleanliness Score, 142-144, 146-148

Clostridioides Difficile, 166-167, 170

Coagulase-positive Staphylococci, 136

Comparative Tuberculin Test, 127, 130, 132

Conjugated Linoleic Acid, 104, 107-108

Contagious Mastitis, 13

Corkscrew Claw, 142, 144-145

D

Dairy Sheep Shedding, 95

Distal Hind Limb, 49-50, 54

Doppler Ultrasonography, 78-79, 82, 85-86

Dry Matter Intake (dmi), 42, 153

E

Enzyme Linked Immunosorbant Assay (elisa), 1

F

Fasciola Hepatica, 70-71, 73-77, 176-177, 183

Fatty Acid, 48, 104-105, 107, 157-158, 184-185, 191, 195, 201, 205

Fecal Culture, 95-103

Fecal Pcr, 95-101

Fibroblast Growth Factor 21, 150, 157-158

Foot Disorders, 78, 142-143, 145-146

G

Glucose, 49, 52-55, 60-69, 78, 88, 90, 153-154, 158, 206

H

Haylage, 70-71, 73-76

Herd Map Status, 1

Herd-level Seroprevalence, 70-73

Hindfoot Nerve Block Anaesthesia, 49-50, 52, 54-55

I

Id Screen Paratuberculosis Indirect Screening Test, 1

Immune Gene Expression, 42, 48

Indirect Immunoenzymatic Tests, 70, 72

Infectious Bovine Rhinotracheitis, 159, 164-165

Interdigital Hyperplasia, 142, 146, 148

J

Johne's Disease, 1, 11-12, 103

K

Ketogenesis, 104-105, 151, 154, 158

L

Lactating Mammary Gland, 60-62, 65, 69

Lactose, 60-67, 69, 188, 205

Least Squared Mean Concentration, 105

Lipopolysaccharide, 42, 47-48, 79, 87, 93-94, 194-195, 199, 205-206

Local Anaesthesia, 49-50, 52-54, 57

M

Mammary Epithelial Cells, 60-66, 68-69, 158, 184-186, 191-192

Map Elisa, 1-2, 4-5, 7, 10, 102

Maximum Systolic Velocity, 78, 80-81, 83-85

Milk Composite Sample, 107

Multilocus Sequence Typing, 116, 175

Mycobacterium Avium Subspecies Paratuberculosis, 1, 11-12, 103

Mycobacterium Bovis, 127, 132-133

Mycoplasma Mastitis, 134-135, 137-140

N

Non-forage Carbohydrates, 42-43

Nonlinear Mixed-effects Pharmacokinetic, 117

O

Optical Density Ratio, 70, 73, 75-76, 178

Ostertagia Ostertagi, 70-71, 74, 76-77, 183

P

Paratuberculosis, 1-2, 11-12, 95-103

Physiologically Based Pharmacokinetic, 118, 125-126

Plasma Tulathromycin, 119-120, 122-125

R

Receiver Operating Characteristic, 26

Rumenitis, 87, 90, 93

S

Sem, 45-46, 52, 54-55, 61-65, 68, 104-106, 199-200, 202-203

Serum Amyloid-a, 87, 93

Serum Hepatokines, 150-151

Single Intradermal Comparative Tuberculin Test, 127, 132

Sodium Butyrate, 194-196, 198-206

Somatic Cell Count, 1-2, 11, 15, 17, 20, 25, 107, 134-135, 140-141

Staphylococcus Aureus, 13, 24-25, 109, 111-112, 115-116, 149, 174

Subacute Ruminal Acidosis, 42, 47-48, 87, 93-94, 194, 205

T

Tnf Receptor-associated Factor 6, 43, 47

Tulathromycin, 117-126

V

Virulence Genes Detection, 110

W

White Line Disease, 50, 78, 142-143, 148-149

Whole Herd Testing, 35, 39

Y

Youngstock Check Test, 26

www.ingramcontent.com/pod-product-compliance
Lightning Source LLC
LaVergne TN
LVHW080737210926
848485LV00003B/573